Fortune

W9-CEQ-284

SECOND EDITION

THE PHYSICS
AROUND YOU

Dale D. Long

*Virginia Polytechnic Institute
and State University*

*Wadsworth Publishing Company
Belmont, California
A Division of Wadsworth, Inc.*

Picture Credits

Part-opening pictures: I, II, IV, V, Historical Picture Service, Chicago; III, Culver Pictures, Inc.; VI, UPI/Bettmann Newsphotos.

Figure 2.1 Dale D. Long; 2.7 Dr. Harold Edgerton, MIT, Cambridge, Ma.; 3.8 Matt Berg; 3.9 Reprinted from Popular Science with permission © 1983, Times Mirror Magazine, Inc.; 3.15 Busch Gardens, Williamsburg, Va.; 4.6 Courtesy Pulaski Furniture Co.; 4.10 Courtesy Thomas Jefferson Memorial Foundation; 4.12 Tennessee Valley Authority; 4.14 Dale D. Long; 4.20 Wide World Photos; 4.25 Wide World Photos; 4.26 Photo by Wm. A. Bernhill/New York Public Library; 6.16(a) Photo by Marta Kongsle; 6.16(b) Willa Percival Photo, Photo Researchers, Inc.; 6.18 Mano Photo; 6.22 Wide World Photos; 6.23 Courtesy Sikorsky Aircraft; 8.8 Wide World Photos; 8.9 Port Authority of New York and New Jersey; 10.12(a) W. J. Wallace, from *Oceanography: An Introduction*, Third Edition, by Dale E. Ingmanson and William J. Wallace, © 1985 Wadsworth Publishing Company, reprinted by permission; 10.12(b) Va. Tech. Photo; 10.15 American Cancer Society; 12.11 Education Development Center; 12.14 Wide World Photos; 12.16 Virginia Department of Conservation and Economic Development; 12.17 © 1976 Memorex Corp.; 13.6 Dale D. Long; 13.9 Donald E. Hall; 13.11 C.G. Conn., Ltd., Oakbrook, Il.; 13.12 Grayson Wright; 13.13 Joan K. Simons, R.T. (R) RDMS; 13.16 Education Development Center; 14.3 JOEL, Ltd., Tokyo, courtesy of Prof. H. Hashimoto; 14.9 Photograph by Bob Veltri; 15.1 Battery Products Division, Union Carbide, Inc.; 16.14 Dale D. Long; 16.19 Wide World Photos; 16.22 UPI/Bettmann Newsphotos; 19.5 Sekai Bunka Photo; 20.7 General Electric Company; 21.1 Va. Tech. Photo; 21.5 Loren Long; 21.7 Rosemary Eakins, Research Reports; 21.9 Loren Long; 21.11(b) Fundamental Photographs; 21.18 © 1979 Ian Lucas/Photo Researchers, Inc.; 21.35 © Joel Gordon, 1979; 22.13 Polaroid Sunglasses, Inc.; 22.14 Polaroid Sunglasses, Inc.; 23.1 The Family Circus, reprinted courtesy of The Register and Tribune Syndicate, Inc.; 23.18 Don Leonard; 24.5(a) Amray, Inc.; 24.5(b) Dr. R. L. Grayson; 25.4(a) © Joel Gordon; 25.4(b), (c) St. Vincent's Hospital, New York; 26.7(a) Wide World Photos; 26.7(b) University of Chicago

Physics Editor: Jack C. Carey
Development Editor: Mary Arbogast
Editorial Assistant: Judi Walcom
Production Editor: Deborah McDaniel
Designer: Donna Davis
Print Buyer: Barbara Britton
Art Editor: Marta Kongsle
Copy Editor: Caroline Arakelian
Technical Illustrators: Ayxa Art, Darwen and Vally Hennings, Wayne Clark, Cyndie Clark-Hugel
Cover Designer: Donna Davis
Cover Photograph: Eruption of Kilauea Volcano, taken by Douglas Peebles in 1986

© 1988, 1980 by Wadsworth, Inc. All rights reserved. No part of this book may be reproduced, stored in a retrieval system, or transcribed, in any form or by any means, electronic, mechanical, photocopying, recording, or otherwise, without the prior written permission of the publisher, Wadsworth Publishing Company, Belmont, California 94002, a division of Wadsworth, Inc.

Printed in the United States of America   50

1 2 3 4 5 6 7 8 9 10—92 91 90 89 88

Library of Congress Cataloging-in-Publication Data

Long, Dale D.
The physics around you / Dale D. Long. -- 2nd ed.
    p.   cm.
Includes bibliographies and index.
ISBN 0-534-08670-5
1. Physics.   I. Title.
QC23.L85 1988
530--dc19   87-20146
                                                                   CIP

BRIEF CONTENTS

DETAILED CONTENTS

To my mother
MYRTLE I. LONG
In memory

TO THE INSTRUCTOR

The Physics Around You is designed to introduce the nonscientist to physics. By focusing on the physics we encounter every day, it encourages readers to wonder about the physical causes for everything that goes on and to look for an explanation in the principles of physics. This book can serve as a text for a purely conceptual physics course or for one that limits its mathematics to simple algebra. The following features are designed to aid students in gaining an introductory knowledge of physics and an appreciation of the generality and unity of its principles.

Flexible Format

Courses for nonscience students vary over a broad range in mathematical depth and course length. Many instructors feel that their courses should contain no equations or calculations, while others want to include quantitative features. Courses may also vary in length from one quarter to a full academic year. Adaptations of this book can allow for any of these situations.

Mathematics level. Even though this book contains mathematical examples, quantitative exercises within the chapters, and end-of-chapter problems, a course that is purely conceptual can skip over these features, and the student will benefit from reading the descriptive material alone. At an intermediate level, students can read the examples to acquire a more concrete understanding of the physics principles, or they can enhance their understanding by perusing the exercises and problems whose answers are given in the text, even if they are not expected to solve the exercises and are assigned only the simplest problems. (The course I teach is at this level.) For more quantitative courses, instructors can expect students to study the examples as models for their own work and solve the in-chapter exercises; they can also assign problems of varying difficulty.

Course structure. Instructors can choose the topics of most interest or benefit to their students. Only Newton's laws (Chapter 2) and energy concepts (Chapter 4) are essential for all readers. The general study of waves (Chapter 12) should precede the study of light waves (Chapter 22), and the particle nature of light (Chapter 23) logically follows it. Also, knowledge of qualitative aspects of electrostatics (Chapter 14) facilitates study of the atom (Part Six). Otherwise, Parts One through Six are essentially independent and can be chosen to best suit the course.

Furthermore, parts of the text can be rearranged to suit an instructor's preference. For example, waves and sound (Part Three) can be placed just before light waves (Part Five) to present a unified treatment of waves. (Other areas that include references to waves—Chapters 10 and 19—can stand on their own.)

Within parts, some chapters can be skipped without concern that the material will be needed later. For example, any or all chapters from mechanics except chapters 2 and 4 could be omitted. Instructors interested in the second law of thermodynamics (Chapter 11) might cover Chapter 8 but skip chapters 9 and 10. All or parts of magnetism (Chapters 17–19) could be omitted. Chapter 20 ties together various aspects of electrical and thermal energy and is not needed for any other chapter. Relativity (Chapter 27) could be left off.

Within chapters, examples of sections that could be skipped include: Sections 2.7–2.8, 4.8, 6.3–6.4, 7.6–7.7, 8.4, 9.6, 11.5, 13.6, 14.5, 14.6 (for qualitative courses), 15.7–15.8, 16.5, 23.3, 25.4–25.5, 27.6, and 27.8. An occasional reference may be made to material that is skipped, but such references are not fundamental to the discussion at hand.

Clarity of Writing and Illustrations

The writing style is aimed at the nonscientist. Emphasis is on clarity of expression without an excess of words. Numerous illustrations, examples, and analogies from the everyday experience of students assist readers in identifying with the principles and understanding their significance. The tone is conversational. Illustrations are interesting and visually appealing, without being cartoons; the goal of the illustrations is to instruct and elucidate, not to entertain.

Study Aids

A number of features supplement the basic text and make the book a valuable teaching resource. These features are discussed and explained in the section *To the Student*. However, I would like to make a few comments about them to instructors.

Many more *Questions* and *Problems* are included than you are likely to assign. Undoubtedly you will assign those questions that are most appropriate to your course, but I suggest that you encourage students to read all the questions in an assigned chapter and answer those that interest them. Problems span a range of difficulty. I suggest that you assign only those appropriate to the mathematical level of your course—for purely conceptual courses, perhaps no problems at all.

The sections called *Application* contain little or no new physics but represent interesting or practical applications of the principles being discussed. These sections are presented in context, allowing students to get a better understanding of which principles are being applied, rather than placed as boxed essays that might disrupt their train of thought.

Supplements

Supplements available to the instructor include an *Instructor's Manual* and a computerized test bank. The instructor's manual includes for each chapter: a chapter overview, suggested demonstrations, answers to selected questions, solutions to the more difficult problems, sample test questions, and suggested term-paper topics. The computerized test bank comes with a set of questions that the instructor can add to or delete from and a program that facilitates the effective generation of tests with this bank.

The Second Edition

Several significant changes make the second edition of *The Physics Around You* even more effective, while maintaining its flavor as a thorough, conceptual-level book that connects the principles of physics with the real world. Many of these revisions stem from the suggestions of reviewers who have carefully considered all aspects of the book. The principal changes are the following:

Organization.　Energy is now introduced in Chapter 4, a more traditional position that most instructors prefer. Gravitation and curvilinear motion appear in a separate chapter to emphasize their importance and the connection between gravity and many forms of such motion. Momentum and impulse are separated into a chapter of their own. The discussion of waves and sound is placed between thermodynamics and electricity, in keeping with what has become the traditional order in physics books of this level. The entire discussion of electrostatics appears at the beginning of the section on electricity. The material has been divided into shorter chapters to increase scheduling flexibility for instructors.

New and updated material.　A new section on projectiles has been added. New applications feature recent developments in superconductors and the physics of how a computer functions. The discussion of quantum mechanics and how it relates to atomic and subatomic phenomena has been expanded considerably. A section on the search for the most elementary of particles has been added, including a discussion of the quark model. Finally, a chapter on the theory of relativity has been added.

Writing. A substantial amount of rewording and polishing has been done to improve the clarity of the writing. Almost every paragraph originally in the first edition has incorporated some changes.

Acknowledgments

I first want to thank the many reviewers whose comments and thoughtful suggestions have been responsible for major improvements in the text. These reviewers include: Royal G. Albridge, Vanderbilt University; Joseph Aschner, City College of the C.U.N.Y.; Robert Boughton, Bowling Green State University; Nicholas Brown, California Polytechnic State University, San Luis Obispo; Ronald A. Brown, S.U.N.Y. College, Oswego; Anthony J. Buffa, California Polytechnic State University, San Luis Obispo; Roger Creel, University of Akron; H. H. Forster, University of Southern California; Patrick C. Gibbons, Washington University; Charles H. Hunt, American River College; Andrew Kowalik, Nassau Community College; Robert Marchini, Memphis State University; Victor E. Michalk, Southwest Texas State University; Allen Miller, Syracuse University; Thomas M. Miller, University of Oklahoma; Thor F. Stromberg, New Mexico State University; Leo Takahashi, Pennsylvania State University, Beaver Campus; John K. Wood, Utah State University; Stanley J. Yarosewick, West Chester University.

I appreciate the contributions of several of my colleagues at Virginia Tech—John Ficenec, Dale Schutt, Ray Tipsword, and Royce Zia, in particular—who have read and commented on new material in this edition. Linda Mattox has always been supportive when I needed help with typing or other clerical matters. My family has encouraged and endured as I struggled through the much-longer-than-expected revision process. I especially thank my son Doug who always came through when I couldn't quite make a paragraph say what it needed to say; I could request, "Please fix it," and he would. Finally, I express my appreciation to the following editors at Wadsworth Publishing Company: Mary Arbogast who, as development editor, provided much-needed feedback and encouragement, as well as responses from the reviewers; Jack Carey for his suggestions for making the book more useful to instructors and students; Marta Kongsle for her work in coordinating the illustrations; and Debbie McDaniel for overseeing the process of turning the manuscript into a textbook.

T O   T H E   S T U D E N T

This book introduces you to physics by focusing on *the physics around you*. It aims to broaden your understanding and increase your insight into the interaction between physics and your life. On these pages you'll find not only a discussion of the principles of physics but also a vast array of analogies and illustrations from everyday experiences that demonstrate the meaning and significance of those principles.

Study Supplements

The various elements of this book are designed to enhance your appreciation of the material and to make the book a valuable and interesting learning tool. I will list these elements here, along with some suggestions for maximizing their benefit to you. The kind of course in which you are enrolled—whether purely descriptive or oriented toward problem solving—will determine which elements will be useful to you.

Distributed throughout the chapters you will find the following items:

1. *Questions* give you a chance to think about the meaning and some applications of the principles being discussed. They also provide some dialogue between you and me. As you come to them, please take the time to propose an answer.

2. The *Application* sections highlight important applications or interesting sidelights of the text material that precedes them.

3. An occasional *News Release* brings out historical developments related to the surrounding text. Though not actual news articles, they accurately reflect historical aspects of the physics being discussed.

4. Many of the sections that express mathematical relationships between physical quantities have *Examples*, which illustrate how to solve problems on that particular topic. If you will be solving problems in your course, they provide models. These examples also demonstrate the kinds of useful information that can result from a problem solution.

5. The *Exercises* give you a chance to practice a calculation using the principles being discussed. Even if you are not required to solve these exercises, I encourage you to read them and their answers in order to increase your insight into the practical applications of the principles under investigation.

At the end of each chapter you will find the following items:

1. The *Key Concepts* sections summarize, in paragraph form, the major new physics principles in the chapter.

2. Numerous *Questions* give you an opportunity to apply physics to interesting everyday situations. Answering these questions will help to clarify the meaning of the principles of the chapter, as well as give you a better understanding of the things going on around us. Your instructor will probably not assign every question. I encourage you to glance through all of them, though, and look for answers to those that interest you.

3. A broad range of *Problems* with various levels of difficulty are available. If your course includes the solving of problems, notice that Appendix I reviews the algebra you will need, and Appendix IV gives some suggestions for problem solving. Answers are given for all problems so that you get feedback on the correctness of your solution.

4. Suggested *Home Experiments* for each chapter will assist you in demonstrating the validity of the physics principles. Even if your instructor does not assign them and you don't have time to carry them out, I suggest you try to predict the outcome of these investigations.

5. *References for Further Reading* point out some other sources of information related to each chapter and describe what you expect to find in each one.

Answers to Questions

There is one very important idea that I wish to leave you with, even if you get nothing else from this preface: **The process of searching seriously for the answer to a question is far more important than getting the right answer**. Many students feel compelled to get the "right" answer. However, many of the questions posed in this book have very complex answers for which *nobody* is likely to know all the details. The purpose of posing a question is for you to search for the ways in which physics principles explain the main, or at least some major, features of the circumstance the question addresses. Your answer may differ from that of another student or your instructor, yet it may still contain substantial correct insights into the physics of the question.

Your Comments

I invite you to write to me with your comments about the book. Tell me what you liked and didn't like. If you find a section that is not as clear as you would like, tell me what you think should be done about it. Your comments will tell me how well I have succeeded in presenting physics in a clear and interesting a manner. They will also be valuable to students who follow after you, for whom I may write in the future.

P H Y S I C S   A N D   Y O U

"I hate physics; it's got to be the most boring subject in the world!" "I'm terrified of physics! You have to know all that math." "Physics is just memorizing formulas, and then guessing the right one to plug numbers into." "Physics has nothing to do with my major. I don't know why they make me take it."

How many times have you heard statements like these, maybe with a few expletives thrown in for emphasis? You may even have had such thoughts yourself. If so, cherish those thoughts while you have them. Because the goal of this book is to banish any such ideas from your consciousness, and to replace them with the conviction that physics describes the world in which *you* live.

I want to change the way you look at your world. By the time you've completed this course, I hope you'll no longer be able to marvel at the beauty of a sunset . . . without wondering what causes the spectacular colors. It's my wish that never again will you be exhilarated by your favorite music . . . and not ponder the physical source of its enchantment. I want you unable to watch a football game, excited by the absolute perfection of your team's players, . . . without seeing the laws of physics in action. You may not use a bottle-opener to open a drink, or even turn on a light switch . . . unless you pause to consider the natural principles that form the foundation of these practical devices. And finally, you may never again discuss the energy shortage, nuclear power, or any other technological issue relevant to our society without first reaching for an understanding of the physics pertinent to the situation. In short, besides responding to the beauty you experience or getting a job done correctly or making your point in a discussion, I hope you'll also be fascinated by the orderliness and simplicity of our natural world, even in the midst of its indescribable complexity. And I am confident you'll be impressed by the ability of your mind to glimpse the harmony of its laws.

I view this book as a conversation between you and me. I try to imagine that we're sitting in comfortable chairs discussing the rules that govern our world and how those rules influence our lives. You may be surprised by a physics book in which the author speaks directly to the reader in a conversational tone. But that's what I try to do in this book. And along the way I ask you a number of questions, the answering of which will help to fix the laws of physics firmly in your mind.

1.1 WHAT IS PHYSICS?

Ask a physicist what physics is, and you're likely to get the quick answer: "It's what physicists do." Since different physicists do lots of different things, that answer won't help you very much. But when we lump together all of those diverse projects, we can get a general description of what physics is all about.

The **goal of physics** is to set down in *simple* and *precise* form a *few* rules that govern the *entire* behavior of our universe. That's a tall order! Let's pick that goal apart by looking at the italicized words.

First of all, physics should describe the *entire* behavior of everything that goes on in our universe—from the interactions between the galaxies to an explanation for the colors in a butterfly's wings to the characteristics of the particles making up the atoms of our bodies.

The catch is that physicists want to provide this description using as *few* rules as possible. Those who see physics as a massive collection of formulas miss this point. In fact, a relatively small number of fundamental rules describe what we know about the universe. It's certainly true that you need to understand a number of details to know physics. But the secret to learning physics is acquiring the ability to pick out the few basic ideas and apply them to a wide variety of situations.

Ideally, these rules need to be *simple*. A description must be complex enough to correctly explain the situation it applies to, but no more complex than absolutely necessary. For example, it is possible in principle to provide a mathematical description of our solar system in which the earth is assumed to be at the center. Such a representation could describe the orbit of every object in the solar system, but it would be extremely complicated. However, the usual explanation using a sun-centered system is so much simpler that most of us accept it as the "correct" one.

Finally, the rules provided by physics need to be *precise* so that everyone can agree on their meanings. For that reason, physicists usually formulate them in mathematical language. But the content of that mathematical language can be expressed in words that provide all of us with a qualitative understanding of the behavior of our world. When you finish this book I want you to have that qualitative understanding.

Laws, Principles, and Theories

When a description of some aspect of the physical world has been around long enough to be tried, tested, and proven valid under a broad range of circumstances, we usually call it a *law*. Sometimes the word *principle* is used instead. These two words mean basically the same thing and can usually be interchanged.

The fact that we call something a law does not mean it is known to be true under *all* circumstances. Some laws are valid under certain conditions but are not valid under others. For example, Newton's laws of motion (which we'll study in Chapter 2) are very precise for ordinary objects at ordinary speeds. But if the object is moving at a speed near the speed of light, the laws break down. Often laws must be modified as new facts are learned, or they must be expanded to include situations not considered when the laws were formulated.

The word *theory* usually implies a proposed description whose validity is less certain. People use it for everything from proposals that have little experimental verification to those that have passed a vast array of practical tests. An example in the last category is Einstein's theory of relativity. Since its proposal in 1905 this theory has been put to some of the most rigorous tests the human mind could formulate, and it has held up in every respect. Yet, because of long-standing tradition, we continue to refer to it as "theory."

Are Laws Discovered or Invented?

The laws we're talking about are not the type passed by some legislative body. (Congress cannot change π from 3.14. . . to 3, even if it would seem more convenient.) Instead they are rules describing the way physical entities behave or interact under certain circumstances. A theory is accepted as law only when we *discover* that nature really does work in the way described by the theory. Therefore, as statements describing the behavior of the physical world, laws have to be discovered.

But what we discover are relationships between properties that we humans have *defined*. Let me illustrate with a law from Chapter 2. (We'll worry later about what the law actually means.) Newton's second law of motion says that force = mass × acceleration. That law describes how nature behaves. But the terms *mass* and *acceleration* have been defined by humans. True, they describe a certain physical reality, but they nevertheless have been defined by us in a certain way. Furthermore, as we'll see when we study this law, the law itself defines what we mean by force. Thus, since we define the terms in the law, we have actually *invented* it to describe a *discovery* about the way nature works.

Physical laws say nothing about *why* parts of our universe interact as they do. Rather, the laws describe the nature of these interactions. For example, in Chapter 3 we will discuss the law of gravitation discovered-invented by Sir Isaac Newton. This law tells us that a force of attraction exists between any two pieces of matter, and it tells us how that force depends on both the amount of matter present and its relative separation. But it gives no explanation for *why* the objects attract each other. All we can say is that that's the way nature is.

You may find this state of affairs a bit disappointing if you were looking forward to having a lot of your whys answered by reading this book. However, I think you will find it rather fascinating that you can understand so many of the complex processes that go on around you in terms of a few physical laws. Most of these laws should seen reasonable in terms of your own intuition and experiences.

Physics, Engineering, and Technology

This book focuses on *the physics around you*. By that, I mean that it not only discusses the basic principles of physics but also indicates how those principles influence *your* life. It shows how they form the foundation for natural phenomena—orbiting planets, weather conditions, lightning, and sunsets, for example—as well as how they form the foundation for the practical devices that are the bases of our technological way of life.

The term *technology* encompasses the methods and materials used to achieve practical objectives. A claw hammer or a hair drier are parts of our technology. These devices operate according to certain physical laws discussed in this book. The engineer's job is to understand and use known physical laws to develop devices and systems for practical purposes. This book uses many everyday practical devices to illustrate the principles of physics.

1.2 CONSERVATION LAWS

Suppose a friend of yours suspects his roommate of pilfering money from his billfold at night. To check out his suspicions, your friend begins keeping an accurate record of his funds. Let's say that in a given month he has $510 available. He spends $194 for rent, $122 for food, $39 for utilities, $41 for books, and $26 for other miscellaneous expenses. But he has only $2 left at the end of the month. Since the total amount of money was known, your friend could conclude that $86 had disappeared under mysterious circumstances.

The principle used in reaching this conclusion might be called the "principle of conservation of money": If no money is added to or leaves the fund, the total amount of money remains constant. That knowledge lets you determine the amount of missing money, even though you do not know the details of its status. This principle has nothing to do with conserving money by being thrifty; it says that the *total* money *is conserved*—that is, it stays constant.

Conservation principles or laws are among the most useful tools of physics. We can understand a great deal about what is going on in a system merely by knowing that the total amount of some entity remains constant. Among the more important conservation laws are those for energy, momentum, and electric charge. In particular, the energy conservation principle pervades all areas of physics, and it will be extremely useful in providing insights into every area of our study.

1.3 MATHEMATICS IN THIS BOOK

I mentioned earlier that, in order to ensure precision, physicists usually write laws in mathematical language. Let me illustrate the advantage of doing so: Suppose you're out for a Sunday afternoon drive and you travel continuously for several hours at about the same speed. If you get curious about how far you've driven, it's easy enough to figure out the answer if you know your speed and how long you've been driving. It isn't good enough just to say that the faster and longer you drive the farther you go. That gives you a qualitative hint at the distance. But to get the answer you want, you need to multiply the speed by the time traveled. This quantitative mathematical relationship gives you an answer that's as precise as the values you have for speed and time. In expressing *all* the laws of physics mathematically, we gain the same advantage as we did in this distance example.

If you don't consider mathematics your strong point, you may feel somewhat apprehensive about the mathematics in this book. You will find here no math more complicated than simple algebra. And if your algebra is a bit rusty, Appendix I reviews the rules of algebra needed in this book. You can benefit from reading the book without working the numerical problems, but if you expect to be solving problems (and that's how you'll benefit most from this study), you will find suggestions for problem solving in Appendix IV. As you begin working on problems, I urge you to read those suggestions and to follow them or a similar systematic approach. Doing so will make you a much more successful problem-solver.

In summary, to get the maximum grasp of the laws of physics, we need to express them quantitatively (mathematically) and to express physical properties in quantitative terms. Let me illustrate two important aspects of quantitativeness. If you were applying for a job and the interviewer said, "We intend to pay you *every* month," you probably would feel comforted. But unless you heard exactly *how much* they intended to pay every month, you might not leave with a very secure feeling. Or suppose you are visiting the Washington Monument and are considering walking to the top rather than riding the elevator. If you ask a woman just returning from the walk how long it takes, and she says, "Twenty," you still won't know very much. She might mean 20 minutes, 20 hours, or 20 days, depending on her physical condition. Thus, to answer the question "How much?" you need both a number and some particular units: dollars, minutes, or whatever is appropriate.

Numbers and Units

Numbers.   When we specify the value of some quantity, we will usually round it off to two significant digits: for example, 7800 rather than 7764; or 0.041 rather than 0.04129. Also, when we do a calculation using two-digit numbers, we will usually round off the answer to two digits. When one answer will be used in another calculation, we keep an extra digit in the first answer to avoid confusion in the later one. If rounding off seems to be throwing away all those good numbers your calculator gives you, I suggest you read Part A of Appendix II. It describes the use of significant digits and how they give you information about how accurate a number is.

For some topics, we will need to use very large or very small numbers. It will save you and me both a lot of trouble if we use the exponential (sometimes called "scientific") notation for such numbers. That means expressing 2,900,000 as 2.9×10^6 and 0.000071 as 7.1×10^{-5}. If you're rusty on or unfamiliar with this notation, spend a few minutes with Part B of Appendix II and brush up on it.

Units.   As we illustrated, a number for a physical quantity means nothing without specifying the units used. We will stick with metric units, with an occasional comparison in the old "British" system to aid your intuition when the British unit may be more familiar. Our basic units will be a particular subgroup of metric units known as the SI system, where SI stands for *Système International*. This system has been internationally adopted for scientific purposes. In it, we express distance in meters (m), time in seconds (s), and weight in newtons (N), for example, where the accepted symbols for these units are in parentheses. As we define new quantities, we will give their appropriate SI units.

Appendix III gives you more information on metric units and on conversion factors between various units. It also includes a definition of the metric prefixes that express various multiples of 10 of the basic units. These prefixes are very handy, and we'll use them where appropriate. It's much easier, for example, to say that the radius of a uranium nucleus is 7.4 fm than to say it is 7.4×10^{-15} m.

Part D of Appendix III reviews how to convert a quantity from one unit to another, and it also shows how to use the units in a problem to check your solution to the problem.

The Physics around You

Now you have a hint of what physics is all about and of the kind of language we use to discuss it. Before you get very far into this book, I hope it will be obvious that physics *is where you are*, that it deals with the things going on around you and within you, as well as with the things physicists have tucked away in secluded laboratories. As you read you'll get a peep at the work done by physicists in their labs, but our primary goal is to focus on the physics *around you*.

PART ONE

MECHANICS

Every body perseveres in its state of rest, or of uniform motion in a right line, unless it is compelled to change that state by forces impressed thereon.   ISAAC NEWTON

Isaac Newton, born at Woolsthorp, Lincolnshire, England in 1642, received a BA degree from Cambridge University in 1665, an MA degree in 1668, and became professor of mathematics there in 1669. Because of the Plague, he spent the years 1665 and 1666 at his home in Woolsthorp, where he laid the foundation for his later publications in optics, mechanics, and mathematics. His *Principia*, published in 1687, contained his renowned laws of motion and the law of gravitation, which he could verify only after inventing calculus. His many nonacademic accomplishments included serving briefly as a Member of Parliament and as Master of the Mint in London for the last 27 years of his life.

CHAPTER 2

FORCES AND MOTION

Motion is an integral part of the universe in which we live. To study physics—the laws that govern our universe—we have to study motion and its changes. That means defining and studying *forces*, the root cause of all changes in motion. A **force** is what we mean, in everyday terms, by a push or a pull. You know that if you want something to move, you push or pull on it. But just how much does it move, or how fast, as related to how hard you push or pull? Or does it even move at all? (If you push on a wall, it usually doesn't move.) As we find the answer to these questions we will arrive at a more precise meaning for force— a meaning based on the relationship between force and motion.

We call the laws that relate force and motion *Newton's laws of motion*. To understand these laws, we will need to define the terms used to describe motion. We will also encounter the concept of *mass*. Newton's laws of motion are basic to all of physics because they tell us how matter reacts to *any* force, whatever the origin of that force.

When you exert a force on something—when you push or pull on it—you usually have to make contact with the "something." Figure 2.1 illustrates this point. Most of the ordinary forces in our everyday experience result from contact between two bodies. The most prominent exception is the gravitational force that holds you and your surroundings safely on earth. That force acts on you whether or not you are in contact with the earth. One question you might want to ponder is: Just what *is* "contact"? What does it mean to touch something? That's a question whose answer we can hint at when we're about half through this book.

Figure 2.1 The soccer player's foot exerts a force on the ball by *contact* with the ball.

2.1 THE LAW OF INERTIA: NEWTON'S FIRST LAW

A golf ball at rest on a tee does not start to move until it is struck by the club. Neither does this book lying untouched on your desk suddenly begin to move. In general, any object that is not moving does not start to move unless some force acts on it.

But what if an object is already moving? A baseball leaving a pitcher's hand continues to move until it is hit by the batter or caught by the catcher. If you apply your brakes on an icy road, your car may continue to move at almost the same speed. The ice offers very little frictional force to slow your motion. In these two examples (and in any other situation) a force is needed to change either the speed or the direction of a moving object.

Sir Isaac Newton's *Mathematical Principles of Natural Philosophy*, published in 1687, included three laws that describe the relationship between force and motion. These are now called **Newton's laws of motion**. The first law covers the situations we have just described, and may be stated as follows: *Every body at rest stays at rest unless acted on by an outside force; every body in motion stays in motion in a straight line with the same speed unless acted on by an outside force.* (Outside force means a force that originates outside the body.)

The first part of this law should make sense to you at once. The second part may not seem quite right. You know, for example, that the pitched baseball will eventually curve down and hit the ground even if it is neither hit nor caught. The car sliding on ice will eventually stop, even if it never hits clear pavement. A moving car taken out of gear on a smooth level highway will slowly come to a complete stop without the brakes being applied. Satellites in orbit around the earth or planets in orbit around the sun move in curved rather than straight paths.

Each of these seeming conflicts with Newton's first law involves one or more outside forces. The pitched baseball curves down and eventually hits the ground only because the earth pulls on it with a force—the baseball's weight. Even slick ice offers some frictional resistance that slows the car. The car moving in neutral on the level highway experiences friction from the air, from the interaction between the tires and the road, and from around the wheel axles as they rotate.

· ·

QUESTION   When you're riding in a car and the driver slams on brakes there seems to be a force trying to slam you against the windshield. Is there such a force? If not, why does this happen?*

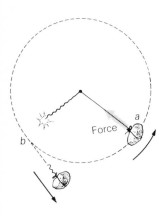

Figure 2.2 At point *a*, the string exerts a force on the rock that keeps it in circular motion. If the string breaks when the rock is at *b*, the rock moves off in a straight line because there is no longer a force acting on it.

An object traveling in a circular path can continue to do so only if there is a force acting on it that continuously changes its direction. For example, if you twirl a rock on a string (Fig. 2.2), the string exerts a force on the rock that keeps it going in a circle. If the string breaks, the rock goes off in a straight line in the direction of its motion at the instant the string broke. Satellites move in curved paths around the earth because of the gravitational force exerted on them by the earth.

The property that makes matter tend to continue in a straight line without changing speed or direction is called *inertia*. This term means essentially the same here as it does in everyday use—resistance to change. Because Newton's first law of motion deals with the inertia of matter, it is sometimes called the *law of inertia*.

If an object is to *change* its speed or direction, or both, there must be an outside force acting on it. Our next job, then, is to describe how such a force changes the motion. And in order to do that, we must first find out how to describe *motion*.

2.2 THE DESCRIPTION OF MOTION

In order to say precisely how forces change the motion of an object, we need to define the quantities that describe motion: *speed, velocity,* and *acceleration.*

__To the student:__ Throughout this book you will find small sections called "Question" or "Exercise." These items provide dialogue between you and me, the author. They are to encourage you to think about the meaning of the principle being discussed and how it applies in a particular situation. The questions challenge you to use physics principles to explain that situation. The answers to the questions may not be immediately obvious; in fact the precise answer may be much too complicated to describe. But that's all right; the idea is to get you to think about the physics of the situation. The exercises give you a chance to calculate a quantitative answer for a problem.

Speed

Speed means how fast an object is moving. More precisely, speed is the rate at which an object changes position:

$$\text{speed} = \frac{\text{distance traveled}}{\text{time required}}$$

or

$$\text{speed} = \frac{d}{t}$$

where the symbols d and t are shorthand for distance and time, respectively.

If you are traveling at a steady speed of 80 kilometers/hour (km/h), you travel a distance of 80 kilometers in one hour. Knowing any two of the three quantities in this equation, you can calculate the third.

• •

EXAMPLE 2.1

(a) In the 1936 Olympics Jesse Owens ran the 200-meter dash in 20.7 seconds. What was his speed in meters/second (m/s)?

From the definition of speed,

$$\text{speed} = \frac{d}{t}$$

Using the distance in meters and the time in seconds in this equation gives

$$\text{speed} = \frac{200 \text{ m}}{20.7 \text{ s}} = 9.66 \text{ m/s}$$

(b) If Owens could have maintained this speed, how long would it have taken him to run the 375 km from New York City to Washington, D.C.?

Rearranging our speed equation (see the review section of Appendix I called Multiplication and Division), we get

$$t = \frac{d}{\text{speed}}$$

$$= \frac{375,000 \text{ m}}{9.66 \text{ m/s}} = 38,800 \text{ s, or } 10.8 \text{ h}$$

Velocity

If you are moving, you have to be moving in some direction. To describe your motion completely, you have to say which way you are moving. The term **velocity** includes both speed and direction. Velocity is the *rate of change of position, with the direction indicated*. If you say you are traveling at 80 km/h, you have given your speed. If you say you are traveling at 80 km/h due north, you have given your velocity. If you change either your speed *or* direction, you change your velocity. Speed, then, is just the "size" of the velocity, without any direction given. We will use the symbol *v* for velocity (and also for speed if the direction is understood).

Acceleration

Velocity can change, of course, and we describe a change in velocity by the term **acceleration**. An acceleration can involve a change in speed, a change in direction, or both. When we talk about motion along a curved path, we will consider accelerations that are changes in direction. For now, we will concentrate on accelerations that are changes in speed only. That is, for the time being, we will consider only motion in a straight line. For this situation only, since the direction is then either unimportant or implicitly understood, we can use the terms speed and velocity interchangeably.

Acceleration is, by definition, *the rate of change of velocity:*

$$\text{acceleration} = \frac{\text{change in velocity}}{\text{time required}}$$

or, in symbols,

$$a = \frac{\text{change in } v}{t}$$

This equation means that if you start to move your car from a complete stop and reach a speed of 80 km/h in 10 s, your acceleration is given by

$$a = \frac{\text{change in } v}{t}$$

$$= \frac{80 \text{ km/h}}{10 \text{ s}}$$

$$= 8 \text{ (km/h)/s}$$

or 8 kilometers per hour per second. This measurement means that *in each second the speed changes by an average of 8 km/h:* After 1 s your speed is 8 km/h; after 2 s, it is 16 km/h; after 3 s, it is 24 km/h; and so on.

Chapter 2   Forces and Motion

EXAMPLE 2.2   (a) A manufacturer claims his car is capable of acceleration of 10 (km/h)/s. How long would it take this car to reach a speed of 130 km/h, starting from rest?

Since the car's speed increases by 10 km/h each second, it would be 10 km/h after 1 s, 20 km/h after 2, on up to 130 km/h after 13 s. Alternatively, we could find the answer by using the definition of acceleration,

$$a = \frac{\text{change in velocity}}{t}$$

by rearranging it to get

$$t = \frac{\text{change in } v}{a}$$

$$= \frac{130 \text{ km/h} - 0 \text{ km/h}}{10 \text{ (km/h)/s}}$$

$$= 13 \text{ s}$$

(b) What speed would this car reach in 4 s?

We can rearrange the equation that defines acceleration to give

$$\text{change in } v = a \times t$$

$$= 10 \text{ (km/h)/s} \times 4 \text{ s}$$

$$= 40 \text{ km/h}$$

EXAMPLE 2.3   When a flea jumps, it takes about 0.0008 s to push off. In this time it reaches a speed of about 1 m/s. What is this flea's acceleration? We use the equation

$$a = \frac{\text{change in } v}{t}$$

$$= \frac{1 \text{ m/s} - 0 \text{ m/s}}{0.0008 \text{ s}}$$

$$= 1250 \text{ (m/s)/s}$$

This acceleration of 1250 meters per second per second is about 30 times that felt by Neil Armstrong during the launch of the *Saturn 5* moon rocket.

2.3 A COMMENT ABOUT UNITS

From our experience with cars, most of us are probably used to measuring speeds in units such as kilometers/hour or miles/hour. For the same reason, it's probably easiest to think of acceleration in units such as kilometers/hour per second. However, speed can be expressed in any units of length per any unit of time—for example, miles/hour, meters/second, centimeters/second, or furlongs/fortnight. Acceleration can be expressed in any units of length per unit of time per unit of time.

Any one physical quantity alone can be written in whatever units are most convenient. But when that quantity is combined with other quantities in an equation, we have to be very careful about the units we use. All the quantities must be expressed in the same **system of units**. For example, to find the distance traveled in 30 minutes at the speed of 80 km/h, you know that you do *not* multiply 80 km/h by 30 min. You must either change the time to hours or the speed to km/min before multiplying.

To avoid errors caused by using mixed units, we will usually express and calculate quantities in what are called *SI units* (The International System of units. See Appendix III, Section B). When we use metric prefixes (see Appendix III, Section A) to specify the decimal point—for example, 3 km rather than 3000 m—we need to convert to the actual SI unit before using the quantity in a calculation. Appendix III, Section D shows how to use the units algebraically to convert from one unit to another, and it also shows how to use them to check the correctness of your calculations.

In Example 2.3 we found the flea's acceleration to be 1250 (m/s)/s. We can express these units more compactly. Recall from algebra that

$$(x/y)/y = x/y^2$$

Units can be treated in the same way. Thus,

$$(m/s)/s = m/s^2$$

Then the flea's acceleration can be expressed as $a = 1250$ m/s^2. You do not have to worry about what a "square second" is. The m/s^2 is merely a shorthand way of saying meters per second per second. This possibility didn't arise in using the "mixed" units (km/h)/s.

· ·

EXERCISE   Many major league pitchers can throw a baseball with a speed up to 94 mi/h. What is this speed in m/s?   *Answer*: 42 m/s

• •

EXERCISE   Assume it takes 0.1 s to throw this ball (that is, the time from backmost position of the ball in the windup to its release). What was the ball's acceleration during the throwing process? Give the answer in (mi/h)/s and m/s². *Answer*: 940 (mi/h)/s; 420 m/s²

2.4 CHANGES IN THE STATE OF MOTION: NEWTON'S SECOND LAW

Newton's first law describes the behavior of a body that has no outside forces acting on it. According to this law, such a body either stays at rest or continues moving with constant velocity—that is, constant speed in a straight line. That means its acceleration is zero. The law implies that if there *is* an outside force, the body will accelerate. We can now consider how the force and acceleration are related.

$F = 50$ units

$a = 7$ m/s²

$F = 100$ units

$a = 14$ m/s²

Acceleration $\propto$ force

Figure 2.3 The acceleration a of the ball is directly proportional to the force F applied to it. (Here we show a rather unorthodox way of throwing the ball, so that the force acts horizontally.)

Suppose you want to throw a bowling ball; that is, you want to accelerate it from rest to some velocity. You do this by applying a force with your hand. The harder you throw (or, to say it another way, the more force you apply), the higher the speed of the ball when you release it; that is, the greater the acceleration during the throwing process. This is an example of a general law of nature: *The acceleration of an object is directly proportional to the force applied to it*. In symbols,

$$a \propto F$$

where "$\propto$" means "is proportional to." In other words, if a force of 50 units gives the ball an acceleration of 7 m/s², a force of 100 units gives it an acceleration of 14 m/s². Figure 2.3 illustrates this point.

You probably know from experience that the acceleration of an object depends on more than just the force applied to it. Suppose you're throwing a softball, and you apply the same 50-unit force. The acceleration of the softball will be much higher than that of the bowling ball. The property of the two objects that accounts for this difference is their *mass*.

Mass *is a quantitative measure of the inertia of a body*. The greater a body's mass, the greater its inertia—that is, the greater its resistance to acceleration. The SI unit of mass is the kilogram (kg).

For a given force, the *acceleration is inversely proportional to the mass* of the object. We write this relationship mathematically in symbols as

$$a \propto \frac{1}{m}$$

$F = 70$ units

$a = 10$ m/s^2

$F = 70$ units

$a = 350$ m/s^2

Acceleration $\propto \dfrac{1}{\text{mass}}$

Figure 2.4 The acceleration a is inversely proportional to the mass for a given force F.

where *m* stands for the mass. Figure 2.4 illustrates the way acceleration depends on mass. Suppose enough force is applied to give the 7.0-kg bowling ball an acceleration of 10 m/s^2. That same force would give a 0.20-kg softball an acceleration that is higher in *inverse* proportion to the masses—(7.0 kg/0.20 kg) × 10 m/s^2—or 350 m/s^2.

We can combine the effects of force and mass together, and write

$$\text{acceleration} \propto \frac{\text{force}}{\text{mass}}$$

or

$$a \propto \frac{F}{m}$$

When one quantity is proportional to another, we can write that the first is *equal to* the product of a "proportionality constant" and the second. Therefore,

$$a = (\text{proportionality constant}) \times \frac{F}{m}$$

But the SI unit of force is defined so that the proportionality constant is 1 and can be omitted from the equation. Therefore,

$$a = \frac{F}{m}$$

We usually rearrange this equation to read

$$F = ma$$

that is, *the force equals the product: mass times acceleration*. This relationship is **Newton's second law of motion**. This second law describes the fundamental relationship between forces and the resulting accelerations. It actually includes Newton's first law, since zero force gives zero acceleration.

· ·

QUESTION   You could easily stop a softball moving at a speed of 50 m/h. Why would you have a little trouble stopping a freight train moving at the same speed? (Answer in terms of Newton's second law.)

Chapter 2   Forces and Motion

The second law of motion is the basis for the definition of the SI unit of force, the *newton* (N). Multiplying mass in kilograms by acceleration in meters per second per second gives force in newtons. In other words, the newton is the amount of force that gives a mass of 1 kg an acceleration of 1 m/s².

In the United States we've been accustomed to measuring forces in pounds (lb), whether to express our own weight, the weight of a sack of potatoes, or the thrust of a rocket on its way to the moon. One newton is equivalent to 0.225 lb. To get a feel for the newton, you can remember that it's about $\frac{1}{4}$ lb, or about the weight of a McDonald's "Big Mac." To convert in the other direction, notice that 1 lb equals 4.45 N. Thus a 10-lb bag of sugar weighs 44.5 N; a 125-lb math major weighs 556 N.

If the units of force in Figure 2.4 are newtons, the values of the quantities are in agreement with Newton's second law. For example, to give the 7.0-kg bowling ball an acceleration of 10 m/s² you need a force given by

$$F = ma$$
$$= 7.0 \text{ kg} \times 10 \text{ m/s}^2$$
$$= 70 \text{ N}$$

You can verify that the values also check for the softball.

The mass of an object has no direct relationship to its size. A soccer ball and a bowling ball are about the same size, but a bowling ball has much more mass. Thus, it needs much more force for a given acceleration. Have you tried kicking a bowling ball lately?

In Example 2.2 we considered a car capable of an acceleration of 10 (km/h)/s. In SI units, this acceleration is

$$\frac{10 \text{ (km/h)/s} \times 1000 \text{ m/km}}{3600 \text{ s/h}}$$

or 2.8 m/s². If the mass of this car is 1500 kg, we can, from Newton's second law, find the force needed to accelerate it:

$$F = ma$$
$$= 1500 \text{ kg} \times 2.8 \text{ m/s}^2$$
$$= 4200 \text{ N (or about 940 lb)}$$

· ·

EXAMPLE 2.4   Suppose the flea of Example 2.3 jumps from a sensitive device that can measure the excess force the flea exerts in jumping. If the flea exerts a force of 0.00056 N, what is its mass?

We can rearrange Newton's second law as

$$m = \frac{F}{a}$$

and substitute in the known quantities:

$$m = \frac{0.00056 \text{ N}}{1250 \text{ m/s}^2} = 0.45 \times 10^{-6} \text{ kg} = 0.45 \text{ mg}$$

· ·

EXERCISE   In the last exercise, you found that a baseball may be accelerated at 420 m/s^2. If the mass of the baseball is 0.15 kg, what average force must the pitcher exert on the ball?   *Answer*: 63 N (or 14 lb)

Net Force

You may have noticed that Newton's second law, in the way we have stated it, is not always true. For example, if you push on (exert an outside force on) the wall of your room, it probably doesn't accelerate in proportion to the force you apply. The reason it doesn't is that other objects—the floor, the other walls, the ceiling—exert forces in the opposite direction on the wall and thus prevent its acceleration. No net unbalanced force exists on the wall. The outside force that applies in Newton's second law is the *net*, or the *unbalanced*, force on the object. If Jim pulls on Suzie's right arm with a force of 100 N, and Joe pulls in the opposite direction on Suzie's left arm with a force of 300 N, Suzie accelerates to the left under a net force of 200 N (assuming she offers no resistance or assistance either way). Keep in mind, then, that in applying Newton's second law, *you must always use the **net** outside force on the object under consideration. The acceleration is always in the direction of the net force.*

You probably associate the concept of mass with that of weight. A relationship does exist between the two, but, as we'll see in the next section, mass and weight are very different quantities.

2.5 WEIGHT VERSUS MASS

Everything near the earth experiences a force directed toward the center of the earth. That "everything" includes you, me, this book, whatever you're sitting or lying on, an apple hanging on a tree, or an airplane flying overhead. We know that force exists because, if whatever holds the object in place is removed, it falls toward the center of the earth. This *gravitational force* that attracts every object toward the earth is defined as the **weight** of the object. We'll now contrast weight and mass, in order to make a clear distinction.

Suppose that you hold this book about waist high and then suddenly release it. It will, of course, fall to the floor because of the pull of the earth on it—its weight. But, because the book has mass, it also has inertia, and therefore resists that downward acceleration just as it would resist acceleration from any other force. We can apply Newton's second law to this situation. The force we are considering is the weight, which we'll give the symbol W. Then $F = ma$ becomes

$$W = ma$$

If you measured the acceleration of the book after you dropped it, you would find a value of 9.8 m/s^2. No matter where you dropped it near the surface of the earth, you would find the acceleration approximately the same. In fact, it doesn't matter what the mass of the book is. A book—or rock, or marble, or elephant—of any mass will have the same acceleration right after it is dropped. The acceleration doesn't change with mass because the weight is proportional to the mass. Doubling the mass, for example, doubles the weight so that the acceleration stays the same. This acceleration that a freely falling object experiences is called the *acceleration due to gravity*, and we give it the special symbol g. Substituting this symbol for a in the last equation, we get

$$W = mg$$

This equation represents the basic relationship between the weight and the mass of an object. Remember: *Weight is a force*, the particular force due to gravity; *mass is the measure of the body's inertia*. Since neither mass nor weight is influenced by whether or not the body is actually accelerating, this relationship is true regardless of the state of motion.

We express mass in units of kilograms. When we refer to a 10-kg rock, we are giving its **mass**. Since weight is a *force*, we express it in *force units* (newtons, in the SI system). The weight of the 10-kg rock is given by

$$W = mg$$
$$= 10 \text{ kg} \times 9.8 \text{ m/s}^2$$
$$= 98 \text{ N}$$

You may protest that you *know* the last paragraph is wrong—that only last week you bought a jar of peanut butter clearly marked "net weight 1.1 kilograms." We get by with saying that the "weight" is 1.1

kg or 1100 g in everyday use because the weight is directly related to the mass through $W = mg$. Let's consider an analogy. Suppose you're on the street in Macon, Georgia, and someone yells out a car window to you, "How far to Atlanta?" You might answer, "Hour and a half." You would not have answered the question you were asked, but you would have given the driver the required information based on the usual driving speed. In the same way we, in effect, know the weight of something if we know its mass. When your bathroom scale registers 52 kg, what it really measures is a force (your weight) proportional to your mass of 52 kg. Even though for simplicity we may say your "weight" is 52 kg, in reality your weight is (52×9.8)N or 510 N.

It's sometimes handy to notice that 9.8 is almost 10, a number much easier to work with. For most purposes we can then use

W (in newtons) $= 10 \times m$ (in kilograms), *approximately*

On the moon, the gravitational force is only about $\frac{1}{6}$ what it is on earth. That fact makes the acceleration due to the moon's gravity only 1.6 m/s². However, the mass of an object is the same anywhere: The force needed to give it a particular acceleration is the same anywhere in the universe. It would hurt just as much to kick a 10-kg rock on the moon as it would on earth! Figure 2.5 contrasts the weight and mass of such a rock on the earth and on the moon.

In connection with rocket travel, we sometimes hear that a traveler feels a "force" of several g's—say 4 g's. What this expression actually means is that the traveler experiences a force strong enough to give him an acceleration of four times what he would have if he were falling freely on earth. The force on his body would be four times his weight on earth.

(a)

On earth

(b)

On moon

Figure 2.5 The mass of an object is the same on the moon (or anywhere else) as on earth. The weight on the moon (b) is only $\frac{1}{6}$ that on earth (a).

2.6 THE ACTION-REACTION LAW: NEWTON'S THIRD LAW

We need to describe one other important characteristic of forces: You cannot have one single force; forces always come *in pairs*.

If you're sitting in a chair, your body pushes down on the chair. The chair in turn pushes up on your body. You can be sure of this because, if the chair were not there, you would fall to the floor. The force with which the chair pushes on you is exactly equal to the force with which you push on the chair. The force from the chair, however, is applied in the opposite direction of the force from your body (Fig. 2.6). When you hit a tennis ball with a racket, the racket exerts a force on the ball. The ball exerts a force of the same strength on the racket, but in the opposite direction. The bulging of the racket in Figure 2.7 results from the force of the ball on the racket.

What causes a bicycle to start forward when it is at rest? In order for

Chapter 2 Forces and Motion

Force that you exert
on the chair

Force that the chair
exerts on you

Figure 2.6 Whatever force you exert on the chair, the chair exerts an equal force back on you in the opposite direction.

the bicycle to begin moving, it must experience a force *in the direction of the acceleration*. When you push on the pedal, the back wheel tries to turn, causing the tire to exert a backward force on the road (Fig. 2.8). The road in turn pushes with an equal force in the forward direction on the tire. Since the tire is attached to the bike, this forward force causes it to accelerate forward. (We haven't shown the upward push of the road on the tire that results from the downward push of the tire on the road.) Why do *you* get tired when the road exerts the force to make you go?

These examples illustrate a general law of nature: *Whenever one object exerts a force on another, the second object simultaneously exerts an equal force in the opposite direction on the first*. This statement is **Newton's third law of motion**, even though it does not necessarily involve motion. Newton's wording was, "To every action there is always opposed an equal reaction." For this reason the two forces are sometimes called an action-reaction pair.

It may seem that action-reaction forces should cancel each other out, since they're equal but opposite in direction. If that were true, it would *never* be possible to have a net force on a body. However, notice that the action and reaction forces *always* act on different bodies: When your body pushes on the chair, the chair pushes on you; when the racket exerts a force on the ball, the ball exerts a force on the racket. These forces have nothing to do with whether or not there is a *net* force on any *one* body.

• •

QUESTION Why does it hurt when you stub your little toe on the bedpost in the night? After all, it's the post that gets kicked, not you.

Figure 2.7 When the racket pushes on the tennis ball, the ball pushes back on the racket with an equal force.

Force of road
on the tire

Force of tire
on the road

Figure 2.8 The bike accelerates forward because the road pushes forward on the tire as the tire pushes backward on the road.

2.7 FALLING BODIES

A body falling freely as it is acted on only by its own weight experiences a constant acceleration of g near the earth. Yet some bodies (feathers, for example) do not fall with that acceleration. If all did, it wouldn't be safe to be hit on the head by a raindrop, and it wouldn't help skydivers to carry a parachute. In this section we'll investigate falling bodies, and how their speeds and positions change with time.

Free Fall

The reason all bodies do not fall with the same acceleration is that they experience resistance from the air as they pass through it. For compact bodies that haven't yet reached high speeds, this resistance is small and can be neglected for most purposes. We say that such a body, acted on only by its weight, is in *free fall*.

Let's study free fall by means of a thought experiment. Imagine that you drop a rock from a window of the 87th floor of the Empire State Building. From windows below, observers precisely measure the time at which the rock passes, and its speed as it passes. (An observer might determine the speed by knowing the height of a window and measuring the time it takes for the rock to pass.)

Figure 2.9 shows the velocities and distances from the point of release as measured at several 1-s intervals from the time of release. These values are the same as those measured by other people in actual free-fall experiments. Time, velocity, and distance are all *zero* at the instant of release.

Since the acceleration is g (9.8 m/s^2), the velocity changes by 9.8 m/s each second. For example, after 4 s, the velocity is downward at 4×9.8 m/s, or 39.2 m/s.

The distance of fall is more complicated. Because the speed of the falling object is increasing, the distance traveled in each second gets progressively longer. The distance traveled in 2 s is four times that traveled in 1 s; the distance traveled in 3 s is nine times that traveled in 1 s; and so on. In other words, the distance fallen is proportional to the *square* of the time. Notice that you can get these distances if you multiply the square of each time by 4.9, or $\frac{1}{2}$ the value of g. All these distances d at each time t are consistent with the equation

$$d = \tfrac{1}{2}gt^2$$

If you measured the distance for other times and for other objects, you would find the results of any free-fall experiment where the object is dropped from rest to be described by this equation.*

---

*You can derive this equation algebraically by noticing that the average velocity from $t = 0$ to any time t is $\frac{1}{2}$ the velocity at time t. Then, $d = (average\ v) \times t = (v/2)t$. From the definition of acceleration, $v = gt$. Therefore, $d = (gt/2)t = \frac{1}{2}gt^2$.

Chapter 2   Forces and Motion

EXAMPLE 2.5 If you neglect air resistance, how long would it take you to hit the water after falling from a 5-m–high diving tower? Substituting $d = 5$ m, and approximating g as 10 m/s² in our distance equation, we get

$$5 \text{ m} = \tfrac{1}{2}(10 \text{ m/s}^2)t^2$$

We can solve this for t to get 1.0 s.

Distance	Downward velocity	Time
0	0	0
4.9 m	9.8 m/s	1 s
19.6 m	19.6 m/s	2 s
44.1 m	29.4 m/s	3 s
78.4 m	9.2 /s	4 s
122 m	49.0 m/s	5 s

Figure 2.9 The velocities and distances of fall are shown at various times after the rock is dropped.

2.7 Falling Bodies   **25**

EXERCISE   If there were no air resistance, how far would a sky diver fall in 10 s after jumping?

Answer: 500 m

Effects of Air Resistance

As we have said, if you drop this book, it accelerates with acceleration *g*. However, if you should (and please don't think about such an unconscionable act!) tear a page from the book, hold the page horizontally and drop it, it would drift down slowly. If you wad the page up, it would fall somewhat faster.

The three fall times differ because of air resistance. The acceleration depends on the *net* force, and these objects are not falling freely under the action of their weight *only*. The air resistance gives a force in the opposite direction to the motion, and this force retards the acceleration. For the book, the air resistance is small *compared with its weight*, and you notice no retardation for short heights. The force from air resistance on the open page almost equals its weight, so it falls slowly. The wadded paper experiences less air resistance because its shape is more compact.

If there were no air resistance, as in a vacuum, all objects would fall with the same acceleration. You may have seen the film that shows astronauts on the moon dropping a feather and a wrench at the same time. Since no atmosphere exists, these two objects fall at the same speed.

You may have noticed in riding a bicycle or motorcycle, or from sticking your arm out the window of a moving car, that air resistance increases with an increase in speed through the air. As an object picks up speed in falling, the air resistance increases. If it falls far enough, the force of the air pushing up equals the weight of the object pulling down. It then no longer accelerates, and we say it has reached its *terminal velocity*.

Spread-out objects, such as feathers and sheets of paper, reach their terminal velocity quickly. A skydiver (Fig. 2.10) falling through the air without a parachute, has a terminal velocity of about 200 km/h (125 mi/h). She reaches about 80% of this speed while falling 160 m, and gradually approaches the terminal velocity as she continues to fall. When she opens her parachute, the larger size greatly increases the air resistance. The net upward force slows her until the retarding force again exactly balances her weight. She then falls at a lower terminal velocity. The diver descends more slowly with an open parachute for the same reason that the open sheet of paper falls more slowly than the piece wadded up.

EXERCISE   Estimate the speed with which a raindrop, formed a few kilometers above, would hit your head if it experienced no air resistance.

(a) Parachute closed

Figure 2.10 (a) A skydiver reaches terminal velocity when the upward resisting force R exactly equals her weight W. (b) When her parachute opens, the higher resistance slows her until (c) R again equals W.

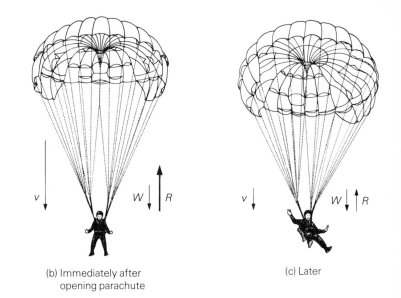

(b) Immediately after opening parachute

(c) Later

2.8 OTHER CONSTANT ACCELERATIONS

A freely falling body is just *one* example of constant acceleration. The acceleration of such a body is constant because the force on it—its weight—is constant. According to Newton's second law ($F = ma$), any body acted on by a constant net force will undergo a constant acceleration. This means that, for any body starting from rest and having constant acceleration, we can use our equation for distance of travel. However, we need to use the actual acceleration of the body, rather than g. In other words,

$$d = \tfrac{1}{2}at^2$$

where a is the constant acceleration of the body.

• •

EXERCISE How far does a car capable of accelerating at 2.2 m/s² travel in the 10 s needed to reach 80 km/h after starting from rest? *Answer*: 110 m

• •

Key Concepts

Speed is the rate of change of position,

$$\text{speed} = \frac{d}{t}$$

Velocity is the rate of change of position with the direction of motion specified or understood. **Acceleration** is the rate of change of velocity,

$$a = \frac{\text{change in } v}{t}$$

Acceleration may involve a change in speed, a change in direction, or both.

When you use any equation that states a definition or a physical law, all quantities in the equation must be expressed in the same **system of units**.

Newton's first law of motion: A body at rest stays at rest unless acted on by a net outside force; a body in motion stays in motion in a straight line with constant speed unless acted on by a net outside force.

Newton's second law of motion: The acceleration of a body is directly proportional to the net force acting on the body and inversely proportional to the mass of the body. In symbols,

$$F = ma$$

Newton's third law: For every force exerted by one object on another, the second object exerts an equal force in the opposite direction on the first.

The **mass** of a body is the quantitative measure of its inertia. The **weight** of a body is the force acting on it due to gravity. The relationship between the weight on earth and the mass is

$$W = mg$$

where g is the acceleration of an object in free fall near the earth. This acceleration is the same, whatever the mass of the body.

The distance d traveled in time t by a body that starts from rest and has constant acceleration a is

$$d = \tfrac{1}{2}at^2$$

Questions

Inertia

1. If you're riding a bicycle and hit a boulder in your path, why do you fly over the handlebars?

2. When you go around a curve in an automobile, why does the book on the seat beside you slide off the side of the seat?

3. Some people claim the talent of being able to yank the tablecloth from a set table without disturbing the place settings. If this is possible, why?

4. Why are football linemen usually heavier than backs? Is it their weight or their mass that is important?

5. If a pinball gets stuck, why do you hit the pinball machine in the opposite direction from the way you want the ball to go?

6. When you're shoveling snow, what removes the snow from the shovel?

Changes in the State of Motion

7. Is a runner's ability to accelerate usually a major factor in a 200-m race?

8. In what ways would bowling on the moon differ from bowling on earth? For example, could you throw the ball any faster?

9. Why is a person with a weak heart likely to feel dizzy while accelerating upward in an elevator?

10. Why can't you throw a shot-put with the same speed you can throw a softball?

11. Why is a basketball easier to throw than a bowling ball, even though it's bigger?

12. In terms of Newton's second law of motion, explain why inflatable airbags protect the passengers of an automobile during a collision.

13. When a woodpecker pecks, its beak experiences an acceleration of about 7000 m/s^2 when it stops in the wood. In terms of Newton's second law, explain how the spongy bone that makes up its brain case reduces the force on the woodpecker's brain during pecking.

14. Why is an expert athlete able to exert only 14 lb of force on a baseball? (See the Exercise in Section 2.4.) Hint: Compare the mass of his arm with the mass of the ball.

Weight versus Mass

15. When you stand on scales, you measure your weight. Which of the following statements is correct?

    (1) "The force exerted on the scales *is* your weight."

    (2) "The force exerted on the scales *is equal to* your weight."

16. What is your mass in kilograms and your weight in newtons?

17. In Figure 2.6 both forces shown are *equal* to the person's weight, but neither *is* the person's weight. Explain.

18. Why can't we express mass and weight in the same units?

19. Explain why, for everyday purposes, you might say that a package of hot dogs "weighs" 454 grams. Explain why your physics teacher would rather you not do so.

20. Why do a 10-kg rock and a 20-kg rock have the same acceleration when dropped?

The Action–Reaction Law

21. When you are sitting in a chair, the earth exerts a downward gravitational force on you. Is the corresponding reaction force the upward push of the chair on your body? Why not?

22. For a body to have an acceleration, there must be a net force on it *in the direction of the accel-eration.* What exerts that force on a car to accelerate it when you push down on the accelerator?

23. The action-reaction law says that if there is a force in one direction, there must be another equal force in the opposite direction. Why doesn't this mean that you can never have a *net* force on *any* body?

24. When a baseball batter ducks to avoid a "bean ball," why does he throw his arms up? Why do you also throw your arms up when you're ducking something coming rapidly at your head?

Falling Bodies

25. If you go over Niagara Falls in a barrel, do you fall as far during the third second of time as during the second second after you leave the top?

26. Could an astronaut bail out of a spacecraft above the moon and parachute safely to its surface?

27. Discuss the relative merits of skydiving on the earth and on the moon.

28. Why does snow fall more slowly than rain?

29. Legend says that Galileo dropped two stones, a large one and a small one, from the Leaning Tower of Pisa to demonstrate that they both fall with the same acceleration. Would the result of such an experiment be different if one stone were round and smooth, and the other large and flat?

Problems

1. In 1920, the thoroughbred racehorse Man O'War ran $1\frac{1}{8}$ miles (1.8 km) in the Preakness Stakes in 1 min:51 s. What was his speed in m/s and km/h? (See Appendix III for converting units of measure.)
   Answer: 16 m/s, 58 km/h

2. If the yellow light at the corner of Third and Elm stays on for 3.5 s, and you are traveling at 20 m/s (45 mi/h), how far can your tail light be from the far side of the intersection and make it through before the light turns red?
   Answer: 70 m

3. In women's championship softball, the pitcher might throw the ball the 12.2 m (40 feet) to

home plate with a speed of 40 m/s (90 mi/h). How much time does the batter have to decide if and where to swing?

Answer: less than 0.30 s

4. If the proverbial race between the tortoise and the hare took place over a 2.0-km course, how long was the hare's nap? Take the tortoise's and hare's speeds to be $\frac{1}{3}$ km/h and 60 km/h, respectively.

Answer: at least 5 h:58 min

5. Suppose you serve a tennis ball that travels at a speed of 34 m/s (76 mi/h) and hits the service line of your opponent's court 19 m away. How long after you hit the ball does it touch the court? (Neglect the slight downward slope and curvature of the ball's path.)

Answer: 0.56 s

6. A cheetah can run at a speed of 105 km/h. At that speed, how far could it run in 3 h:45 min?

Answer: 394 km

7. A swimmer can swim 1.5 km in 17 min. How far can she swim in 2 h averaging half this speed?

Answer: 5.3 km

8. A car can reach a speed of 70 km/h in 7 s. What is its acceleration in m/s²?

Answer: 2.8 m/s²

9. What speed could the car of Problem 8 reach in 10 s at the same acceleration?

Answer: 100 km/h

10. How long would it take the car of Problem 9 to reach 120 km/h?

Answer: 12 s

11. Suppose you hit a golf ball that leaves your club at a speed of 40 m/s. If the ball and club are in contact for 0.001 s, what is the average acceleration of the ball while it is being hit?

Answer: 40,000 m/s²

12. If the mass of the golf ball in Problem 9 is 45 g, what average force does the club exert on the ball?

Answer: 1800 N (400 lb)

13. A woodpecker's beak, while pecking, typically moves with a speed of 6.5 m/s (15 mi/h). If a peck takes about 1/1000 s, what is the acceleration of the beak as it stops in the wood?

Answer: 6500 m/s² (650*g*)

14. If the woodpecker's brain has a mass of 25 g, what is the force on its brain as its head stops during the peck? (See the previous problem.)

Answer: 160 N (37 lb)

15. A batter, disgusted after striking out, slams the 1.0-kg bat to the ground. What downward acceleration would the batter need to give the bat to lift himself off the ground in the process?

Answer: (for 80-kg batter): 780 m/s²

16. A locust jumps by pushing with its hind legs over a distance of about 4 cm (0.04 m). If it starts from rest and pushes for a period of 0.02 s while moving this distance, what is its acceleration during this time?

Answer: 200 m/s² (20 times *g*)

17. If the mass of the locust in Problem 16 is 3 g (0.003 kg), with what average force does it push during the jump? (Notice that the locust pushes on the ground. The ground pushes back with a reaction force that accelerates the locust.)

Answer: 0.6 N (20 times its weight)

18. A swimmer hits the water 1.0 s after diving from a diving platform. With what speed does she hit the water?

Answer: 9.8 m/s (35 km/h)

19. How high is the diving platform in Problem 18?

Answer: 4.9 m

20. Wonder Woman jumps from the ground to a tenth-story window, 30 m high. How long is it from the time she leaves the ground until she gets to the window if she jumps straight up? (The time to go up is the same as the time it would take her to come down if she jumped from the window to the ground.) With what speed does she leave the ground?

Answer: 2.47 s; 24 m/s

21. A runner moving at a speed of 10 m/s, passes a parked Cadillac. At the instant he passes, the Cadillac starts to accelerate with a constant acceleration of 8 (km/h)/s, or 2.2 m/s². How long does it take for the car to catch the runner? How far does it travel before it catches him?

Answer: 9.1 s; 91 m

Home Experiments

1. Hook a strong rubber band to the handle of a toy wagon or other vehicle you can pull. On a smooth level surface (like the hallway of the physics building) give a friend a ride on the wagon by pulling on the rubber band so that it always stretches the same amount. The constant force will provide a constant acceleration and you will find yourself unable to keep up indefinitely. (You have to pull with a little more force than "just enough" to overcome friction.)

2. Tie the string of a helium-filled balloon to the floor of a car so that the balloon is held just below the ceiling. Does the balloon move in the direction you expect when you accelerate, brake, or round a curve? Explain.

3. Estimate your reaction time by having a friend hold a ruler just above your curved hand and then release it (Fig. 2.11). The distance it falls before you can grab it will tell you, after a simple calculation, your reaction time.

4. At some street intersections the yellow light is on for such a short time that a driver within a certain distance of the intersection can neither stop nor make it through before the light turns red. For a few intersections, measure the time the light is yellow and step off the distance across the intersecting street. Using these data and the stopping distances in Figure 4.21, find out if there is such a "dilemma zone" for these intersections.

Figure 2.11 A way to estimate reaction time.

References for Further Reading

Frohlich, C. "Resource Letter PS–1: Physics of Sports," *American Journal of Physics*, July 1986, p. 590. This guide to the literature on physics of sports gives a broad range of references from introductory to advanced treatments.

McMahon, T. A., and P. R. Greene, "Fast Running Tracks," *Scientific American*, Dec. 1978, p. 148. The Harvard University track was designed for faster running by analyzing the mechanics of human running.

Walker, J. *The Flying Circus of Physics with Answers* (John Wiley & Sons, Inc., New York, 1977). Hundreds of fascinating examples of how "Physics and physics problems are in the real, everyday world that we live, work, love, and die in" (quotation from preface of the book).

G R A V I T A T I O N   A N D   M O T I O N   A L O N G   A   C U R V E

Picture this in your mind: You're strolling hand-in-hand with someone on a beautiful autumn night. The light of the full moon is interrupted only by the scattered trees whose patchwork of shadows on the path add to the illusion that you're walking on air. Being caught up in the excitement of what is happening between you and your partner, you fail to notice the rock that has been left in your path. Suddenly, in an ill-timed step that yanks you back to reality, you find yourself lying facedown on the ground.

Should you be even the most dedicated of physics students and caught in this situation, it's unlikely that you would pause to notice that the same force that held the moon in its place and that held you to the path so that you could walk, is the one that also slammed you to the ground. The realization that there is one universal gravitational force that keeps planets in their orbits, that causes apples to fall, and that acts between *all* masses, was another of the major contributions of Sir Isaac Newton. Our first job in this chapter is to consider Newton's law of *universal gravitation*.

We have already studied the falling of bodies under the action of the gravitational force—their weight. But most objects that move as they're acted on by this force do not move downward in straight lines. They move in *curved* paths. If you pick up that rock and throw it, it becomes a *projectile* that moves along a curved trajectory. Planets and satellites move in circles or ellipses around the sun or earth. *Most* motion is along curved paths.

In this chapter we learn how to treat certain types of curvilinear motion. We first consider paths followed by projectiles. We then investigate the characteristics of circular motion and apply that knowledge to understanding the movement of satellites and planets in their orbits.

Notice that we'll be doing two kinds of things in this chapter. First, as we investigate the law of gravitation, we'll be considering one of the *fundamental forces* of nature—one of the basic ways in which one object exerts a force on another. We'll then use this force in Newton's laws of motion to describe the behavior of bodies acted on by it.

3.1 THE GRAVITATIONAL FORCE: ONE OF NATURE'S FUNDAMENTAL FORCES

We have already discussed one example of a gravitational force, the *weight* of objects on earth. However, there is a gravitational force between *any* two masses in the universe, no matter how small the masses or how far apart they are. Isaac Newton discovered the law that describes this force in quantitative mathematical language.

In his three laws of motion Newton clearly restated, using precise definitions and mathematical language, ideas already known to Galileo and others. But his theory of gravitation was a major contribution that simultaneously explained both the weights of ordinary objects and the motions of celestial bodies. To verify the correctness of his theory, Newton had to invent the branch of mathematics we call calculus.

Newton's **law of gravitation** may be stated as follows: *There is a force of attraction between any two masses that is directly proportional to each mass and inversely proportional to the square of the distance between them.* Consider the situation presented in Figure 3.1—two masses m_1 and m_2 separated by a distance d. In mathematical language, this law gives the force F as

$$F = G\frac{m_1 m_2}{d^2}$$

Figure 3.1 Any two masses attract each other with a gravitational force F.

The G is a proportionality constant called the universal gravitational constant, and has the value 6.67×10^{-11} N·m²/kg². Its units make the force come out in newtons when SI units are used for mass and distance.

This law is strictly true only if the size of the mass distribution is very small compared with the distance between the masses. Newton needed

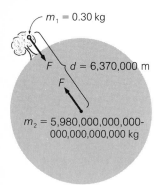

$m_1 = 0.30$ kg

F $d = 6,370,000$ m

F

$m_2 = 5,980,000,000,000-$
$000,000,000,000$ kg

Figure 3.2 A 0.30-kg apple
near the earth's surface is
attracted by a force—its
weight—of 2.9 N.

calculus to show that the same relationship holds for the force if the masses have a spherical shape and d is the distance between the centers of the spheres.

Newton reportedly started wondering about the gravitational force as he thought about why an apple always falls *downward* from a tree. Let's use the law to calculate the force on a 0.30-kg apple hanging on a tree, as shown in Figure 3.2 (with the tree and apple exaggerated in size relative to the earth). This force is, by definition, the *weight* of the apple. Since the earth is nearly spherical, the relevant distance d in our problem is the earth's radius, 6370 km. We'll arbitrarily take m_1 to be the mass of the apple and m_2 to be that of the earth, 5.98×10^{24} kg. Then

$$F = G\frac{m_1m_2}{d^2}$$

$$= 6.67 \times 10^{-11} \text{ N·m}^2/\text{kg}^2 \times \frac{0.30 \text{ kg} \times 5.98 \times 10^{24} \text{ kg}}{(6.37 \times 10^6 \text{ m})^2}$$

$$= 2.9 \text{ N (or } 0.65 \text{ lb)}$$

a good weight for a plump juicy apple.

Notice the law says that the two masses attract *each other*. This means that the apple exerts an equal force on the earth. You could turn the emphasis around somewhat and say that, near the surface of a 0.30-kg apple, the earth weighs 2.9 N.

• •

EXERCISE   Calculate the gravitational force the earth (mass $= 6.0 \times 10^{24}$ kg) exerts on the moon (mass $= 7.4 \times 10^{22}$ kg). How does that compare with the force the moon exerts on the earth? The average distance from earth to moon is 3.8×10^8 m. What would this force be if the earth and moon were touching each other? (radius of earth $= 6.4 \times 10^6$ m; radius of moon $= 1.7 \times 10^6$ m). (See Appendix II–B for a review of exponential notation.)   *Answer*: 2.0×10^{20} N; same; 4.5×10^{23} N

The gravitational force is very weak between objects of ordinary size. The force becomes significant only when at least one of the objects is extremely massive, like the earth.

• •

EXERCISE   Calculate the gravitational force between two 1-kg books $\frac{1}{2}$ m apart, and show that the gravitational attraction between them is negligible for practical purposes.   *Answer*: 2.7×10^{-10} N

$W = 32 \text{ N} = \frac{1}{25} W_e$

$W = 50 \text{ N} = \frac{1}{16} W_e$

$W = 89 \text{ N} = \frac{1}{9} W_e$

$W = 200 \text{ N} = \frac{1}{4} W_e$

$W = 800 \text{ N}$
$= W_e$

Earth

Figure 3.3 An astronaut experiences a gravitational force that decreases as the square of the distance from earth. The spaceship is shown at distances of one, two, three, four, and five times the radius of the earth. W_e is the astronaut's weight on earth—800 N.

Inverse-Square Laws

The gravitational force varies inversely as the square of the distance between the masses. Suppose, for example, an astronaut weighs 800 N at the surface of the earth, a distance r from the center of the earth in Figure 3.3. His weight is shown at various other heights as he recedes from the earth. The law of gravitation is one of many "inverse-square" laws found in nature. We'll consider others as we explore different topics in this book.

g versus G

From the law of gravitation we found the weight of a 0.30-kg apple to be 2.9 N. However, in Section 2.5 we found the weight of a body to be related to its mass by $W = mg$, where g is the acceleration due to gravity (9.8 m/s²). That equation also gives the apple's weight as 2.9 N. Some relationship must exist between these two equations.

Let's recast the law of gravitation, pulling m_1 out in front and grouping the other quantities together:

$$F = m_1 \left[G \frac{m_2}{d_2} \right]$$

We can now evaluate the quantity in brackets, taking m_2 as the mass of the earth and d as the distance from the earth's center to the surface—its radius. The value of the quantity in brackets is 9.8 m/s², just the value of g. Therefore, we get the same results as we do when using $W = mg$.

The quantity $[G \, m_2/d^2]$ gives the acceleration due to gravity at any point, not just on the *surface* of the earth, if you use the appropriate value of d. This quantity shows why g is very nearly constant everywhere on the earth's surface: A variation of several kilometers from the earth's center as you move from point to point on the earth's surface, makes little difference out of a total distance of over 6000 km. The bracketed quantity also gives the acceleration due to gravity on other celestial bodies, such as the moon, provided you use those values of m_2 and d that apply.

The Law of Gravitation versus the Laws of Motion

Newton's law of gravitation differs in character from the laws of motion. The law of gravitation describes one of the fundamental ways in which matter can exert a force on other matter—namely the gravitational interaction that depends on the masses and the distance between them. It tells us that, if two masses exist, they exert forces on each other. And it specifies the strength of those forces.

Newton's laws of motion, however, do not depend on the nature or the source of the force. They describe the way matter changes its state of motion when acted on by *any* force, regardless of its origin.

Both types of laws—the law of gravitation and the laws of motion—are fundamental. But they are fundamentally different in the kinds of things they tell us about our world.

Aside from the gravitational force, there are three other fundamental types of interactions that we'll consider in detail later in this book: the electrical (or electromagnetic) force and the weak and strong forces associated with the nuclei of atoms. Every force we know about fits into one of these four categories. What about "contact" forces that we mentioned in the introduction to Chapter 2? How do they fit into this scheme? Contact between bodies comes about as the atoms on the surface of one body interact with those on the surface of the other. And atoms interact through electrical forces. At that level there is no meaning to the concept of contact. But that's another story for another chapter.

· ·

APPLICATION   **BLACK HOLES***†

The notion of a *black hole*, a gravitational field so strong that it literally tears a hole in space, has been around since 1798 when French physicist Pierre Laplace discussed the phenomenon in the theoretical framework of Newton's mechanics and gravity. Shortly after Albert Einstein formulated his general relativity theory in 1916, the German astrophysicist Karl Schwartzschild reformulated the phenomenon in terms of this new theory of gravity. Real interest in black holes was generally lacking, however, until the late 1960s when astronomical observations of pulsars (rotating neutron stars) showed that black holes might indeed exist in nature and not just in our theoretical constructs.

Perhaps the easiest way to imagine a black hole is to consider (as Laplace did) the escape velocity, v_e—the velocity one body needs to escape from the gravitational field of another. Newtonian physics shows that for a spherical planet of mass m and radius r, $v_e = \sqrt{2\,Gm/r}$, where G is the universal gravitational constant. For earth, $v_e = 11.2$ km/s (about 7 mi/s). For example, this is the speed rockets need to break the bonds of earth's gravity. As either the mass increases or the radius decreases, the escape velocity increases; it has a limit, though—the speed of light, c. When the escape velocity equals the speed of light we have a black

**Written by John Broderick, Department of Physics, Virginia Polytechnic Institute and State University, Blacksburg, Virginia. Reprinted by permission.*

†**To the student**: *In several places throughout the book you will find interesting, detailed applications of the principles being discussed at that point.*

hole. Note that the radius r_s of the black hole depends only on its mass, $r_s = 2Gm/c^2$, called the Schwartzschild radius. Using r_s in the expression for escape velocity, and setting this velocity to c, we see that the sun would become a black hole if it were shrunk to a diameter of 6 km. The earth would be a black hole if it were about 2 cm in diameter instead of 13,000 km. Since the escape velocity at the Schwartzschild radius equals the speed of light, nothing, not even light, can escape the strong gravity and, for this reason, we call the object a black hole. No information about what is inside the black hole can ever come out; all we know of what's inside is the mass, not what the mass consists of (e.g., matter, anti-matter, or the type of particles that make up the matter). Actually there are two more pieces of information about the black hole that we can know; these are its electrical charge (Chapter 14) and its angular momentum (Chapter 6). Otherwise we have no knowledge of its interior.

Space-time in the vicinity of a black hole is severely distorted, so much so that time is stopped and space is ripped open, for we can't cross the radius successfully. The effects on the temporal dimension of space-time can be illustrated by considering a journey into a black hole. You and your robot (the one who'll make the journey which you observe from afar) synchronize your watches before beginning; you monitor the journey on your robot's watch, which you see through your telescope. As your robot falls toward the black hole you notice that the light emitted by its watch is redshifted—it appears more red than it did before the robot left. This occurs not only because it is accelerating to high speeds as it falls in but also because the intense gravitational field induces a redshift on the light. Even if the robot were to somehow stop in its tracks, the light from it would be redshifted by the gravity. Furthermore the time from its clock, as read by you, is earlier than what is given by your watch. In other words, time for the robot has slowed down as you see it. As it gets closer to the black hole the slowdown is greater, so much so that, if noon were the time you calculated that the robot would reach the Schwartzschild radius, then you would see its watch approach noon but never reach it, just as you would notice the robot approach the Schwartzschild radius but never get there.

For the robot, things are different; as it falls toward the black hole it notices the horizon (the boundary of the black hole as observed by the robot) rises up into the sky and shrinks into a smaller and smaller circle overhead until it finally vanishes when the robot reaches the Schwartzschild radius (when its clock reads noontime). Now the robot sees nothing of the outside world; for this reason the Schwartzschild radius is also called the "event horizon." What happens inside the black hole is anybody's guess and is not properly describable by someone outside. Perhaps the robot falls immediately to the center and is crushed by the infinite gravity, perhaps it comes out into another universe where robots aren't subjected to such cruel and unusual experimentation; perhaps something else. Notice that you, on the outside, never see the robot pass the

Schwartzschild radius—just approach it—but the robot notices that it fell in; this is an example of the severe distortion of space-time, quite literally a hole in space.

Black holes are thought to form in nature when a star of at least about 30 solar masses finishes its nuclear burning. (A *solar mass* is a unit equal to the mass of our sun.) No outward pressure mechanism is able to overcome the inward gravitational pull and much of the core of the star—probably at least 2 to 5 solar masses—collapses into a black hole while most of the outer parts of the star get blown into space in a supernova explosion. A few examples of such black holes seem to have been detected by telltale signs of matter stripped off nearby companion stars falling into the black hole, and on the way in, becoming very hot and luminous. When black holes are situated at the center of a galaxy, they can continue to grow, fed by material in an orbit through the galaxy center, reaching masses of perhaps a billion solar masses. Such supermassive black holes born approximately 10 billion years ago may be responsible for many of the explosive phenomena we observe occurring at the center of our own Milky Way and of other galaxies. However, the case for these phenomena being black holes is by no means proven and is one of the current frontiers of astrophysics.

3.2 PROJECTILES

Most falling bodies don't fall straight down in the way we discussed in Section 2.7. Instead they often move horizontally as they are falling. Some examples are a rock thrown outward from a cliff, a pitched baseball, a kicked soccer ball, a bullet fired from a gun, emergency food supplies dropped from an airplane. Probably you don't think of the ball or bullet in the examples as falling at all. We will see that they really do fall, just as if they had been dropped from your hand.

For the sake of discussion, let's focus on the motion of a baseball after it has been thrown. However, the following remarks are valid for any *projectile*—that is, any object given some initial velocity and allowed to move freely under the gravitational force of the earth. For now we will neglect the air resistance opposing the motion of the projectile, so that the only force on the projectile is the *constant downward* force of gravity—the projectile's weight.

Figure 3.4 shows this single force that acts on the ball as it is in flight. Since the acceleration of the ball must be in the direction of the net force on it, the only acceleration of the ball is the downward acceleration due to gravity, g.

Suppose a player throws the ball horizontally, exactly in the direction of another player's glove (Fig. 3.5). If there were no gravitational force,

$W = mg$

Figure 3.4 Neglecting air resistance, the only force acting on a baseball in flight is the downward force of gravity, its weight.

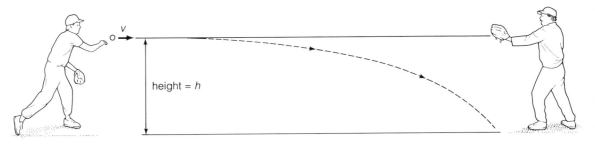

Figure 3.5 As the ball moves horizontally, in a given time it falls the same distance from a straight line as it would if it had been dropped vertically.

the ball would move at a constant speed along a straight line to the other player's glove. However, once the ball leaves the thrower's hand, it falls with acceleration g toward the ground as it moves at a *constant horizontal* speed.

For concreteness, let's add one more condition: Let's give the label t to the time it would take for the ball to fall to the ground if *dropped* from the height h shown. Then let the ball be thrown at a speed such that, if it continued along the horizontal line, it would get to the other player's glove in the same amount of time, t. Under that condition, as the ball moves the horizontal distance to the glove, it falls a distance h and hits the ground directly below the glove. (If thrown at a higher speed, it would get there sooner and, thus, not fall as far.)

The overall path of the ball results from a combination of constant velocity in the horizontal direction and constantly accelerated motion in the vertical direction. We call such a path, shown by the dashed line in Figure 3.5, a *parabola*. The downward curvature increases as the ball moves along because its vertical speed increases (recall Fig. 2.9). If the ball were thrown horizontally from a high cliff, its vertical speed would eventually increase so much that it may appear to move straight down. However, in the absence of air resistance it would maintain its constant speed horizontally.

· ·

QUESTION   Suppose you fire a high-speed rifle aimed exactly horizontally over a large flat area and, at the same instant, drop a bullet from the same height as the rifle. Which bullet would hit the ground first? Would it matter if the bullets were of a different size and mass?

Based on the discussion of the baseball, I hope you answered that they would both hit at the same time, independently of the mass.

If the horizontally aimed baseball hits below the targeted level, how *does* the shortstop get the ball to the first baseman so that it arrives at the same level as when thrown? It needs to be thrown at an angle above the horizontal (Fig. 3.6). If the ball would fall a distance d in the time

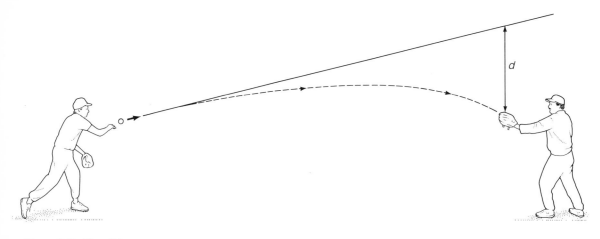

Figure 3.6 The distance the ball falls below the straight line is the same distance it would fall in that amount of time if dropped vertically.

needed to travel the horizontal distance to first base, the direction of the initial motion must be along a line that passes a distance d above the point at which it will be caught. The ball will then fall this distance d below the straight line as it travels, and thus end up at the glove as shown. (Baseball players instinctively aim the ball to take this fall into account.)

· ·

QUESTION   A teenager aims his slingshot at the last apple on the tree. The pebble leaves the slingshot headed directly for the apple, but the apple happens to start to fall from the tree just at the instant he fires. Will the pebble hit the apple?

So far we have neglected air resistance in our discussion of projectiles. A real projectile in the earth's atmosphere would be slowed somewhat by air resistance and therefore not travel as far horizontally as it would without air present.

3.3 BODIES IN CIRCULAR MOTION

We all spend our time going around in circles. We rotate with the earth every 24 hours. This motion is superimposed on the much larger path of the earth's revolution around the sun. We ride on highways whose curves are portions of circles. Our toys and our tools abound with wheels; the parts of those wheels move in circles.

With so much circular motion in our world, we need to understand what it takes to keep a body moving in a circle. In particular, we'll consider *uniform* circular motion, in which the body moves in a circular path with *constant speed*. Once we understand circular motion, we can apply our knowledge to gain an understanding of the orbits of celestial bodies.

Centripetal Acceleration and Centripetal Force

We know from Newton's first law that the natural tendency of matter in motion is to move in a straight line with a constant speed. For it to do otherwise, there must be a force on it. Any deviation from a constant straight-line velocity involves an acceleration. From Newton's second law, any acceleration requires a force. A body moving in a circle is changing direction continuously and, therefore, is accelerating continuously. Since we are talking about a change in direction and not a change in speed, the force that accelerates the body is perpendicular to its motion and is directed toward the center of the circle. As we said in connection with Figure 2.2, a rock twirling on a string moves in a circular path only as long as the string exerts a force on it toward the center of the circle (Fig. 3.7).

The force acting on a body that keeps it moving in a circle is called **centripetal force**. The name itself describes the force—"centri" meaning "center," and "-petal" meaning "to go toward" from the Latin *petere*. The centripetal force always acts *toward* the *center* of curvature. This term is just a name given to whatever force acts on the body to keep it in circular motion. For the rock on the string, the centripetal force is the pull of the string. For the stunt-riding skateboarder on the nearly vertical walls of a bowl-shaped track (Fig. 3.8), the walls exert the centripetal force. For the communications satellite in circular orbit around the earth, the gravitational force is the centripetal force.

The acceleration that accompanies the force is called a **centripetal acceleration** because it, too, is directed toward the center of curvature.

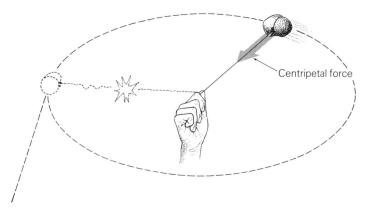

Centripetal force

Figure 3.7 The string provides the centripetal force that keeps the rock moving in a circle. When the string breaks, there is no force and the rock moves in a straight line.

Figure 3.8 The nearly vertical walls of the track provide this skateboarder with a centripetal force.

Figure 3.9 Notice the distortion of the right front tire. (Reprinted from *Popular Science* with permission. © 1983, Times Mirror Magazine, Inc.)

• •

QUESTION   An automobile rounding a curve is moving in a circular path. What provides the centripetal force for this motion? (Hint: See Fig. 3.9.) Why do highway builders usually "bank" sharp curves—that is, slope them downward toward the inside of the curve?

The centripetal force needed to keep a body in a circular path increases with an increase in its speed, and decreases as the radius of its path increases. That's why you need to slow down for a sharp curve (short radius of curvature) in the road. Otherwise, even dry pavement can't provide enough frictional force for the needed centripetal force.

Centrifugal Force

You have certainly had the experience of riding in a curved path and having to lean toward the center to keep from falling over or falling off. This may have been on a merry-go-round, a car moving around a curve, a bicycle, or a skateboard, for example. We will choose the merry-go-round to illustrate our point, but if you identify better with one of the other machines, the ideas still work. As the person in Figure 3.10 rides on the merry-go-round, he is thinking about staying on the horse. (We'll assume the horse only moves around in a circle and does not move up and down.) But the rider feels what appears to be a force F_{away} pulling him off the horse.

In reality, is there such a force? If you were watching from outside, you would certainly answer "No." The rider's body, having inertia, tends to move in a straight line. At the instant shown he is moving to the right. Without an outside force on him, he would continue in a straight line. But the horse he's on is moving in a circle and so curves to the left of that straight line. That is, the horse tries to move from under the rider. As it does, it exerts a force F_{toward} on the rider's leg. From your point of view, that is the only horizontal force on the rider. It provides the centripetal force needed for the rider to continue to move in the circle.

$F_{away}?$

F_{toward}

Center

Figure 3.10 A person on a merry-go-round feels a force F_{away} pulling him away from the center of his motion. To an outside observer, there is only a centripetal force F_{toward} pulling him toward the center.

Direction of rotation

Then what is this force the rider feels trying to pull him off the horse? The rider considers himself to be part of a rotating system; physicists would say that he is in the *reference frame* of the merry-go-round—a rotating reference frame. Because of that, he feels a very real force away from the center. We call this a **centrifugal** (center-fleeing) **force**. To you, an outside observer, the force doesn't exist: It is fictitious. The effect you observe is due entirely to the rider's tendency to move in a straight line.

The next time you get slammed against the car door because the driver went around a curve too fast, remember that the force pushing you there is purely fictitious from the reference frame of those of us sitting still in our chairs. It is no more real than the force that seems to throw you against the dash when the driver slams on the brakes. Both effects occur because your body obeys Newton's first law of motion.

Planets and Satellites

A planet moves in a curved path around the sun. Satellites move in curved paths around the earth. Keeping a satellite in a curved path requires a force toward the center of curvature. The reason we have said "curved" rather than "circular" is that most planets and satellites move in elliptical rather than circular orbits. Since the same basic ideas apply in either case, we can assume circular orbits.

The centripetal force needed for the circular motion comes from the gravitational attraction between the satellite and the earth, or between the planet and the sun (Fig. 3.11). The sun exerts a gravitational force on the earth, and this force provides the centripetal force. The earth exerts an equal reaction force on the sun. Comparing this with the rock on a string, you could say that the gravitational force acts as the "string" that transmits force between the sun and the earth.

Weightlessness.   In this age of space travel we hear lots of talk about weightlessness. Those of us who have not had the privilege of traveling in space have probably seen films of weightless astronauts floating and frolicking around in their capsules. What does it mean to be weightless?

The weight of an object is defined as the gravitational force on the object from the earth or whatever body it's on. An astronaut and his spaceship in orbit around the earth are still acted on by the gravitational force of the earth. Otherwise, his spaceship would not stay in orbit. Then how can he be weightless? If he were on the way to the moon he would pass a point where his weight due to the moon would exactly balance his weight due to the earth. (Would this point be halfway?) He would be weightless there, but beyond that point would have weight directed toward the moon. Only if he were far out in free space, a great distance from any other body, would his weight be nearly zero. Then why do we say that a space traveler near the earth or moon is weightless?

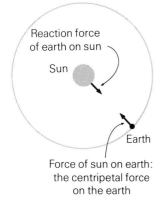

Reaction force of earth on sun

Sun

Earth

Force of sun on earth: the centripetal force on the earth

Figure 3.11   The centripetal force that keeps the earth in orbit around the sun is the gravitational attraction.

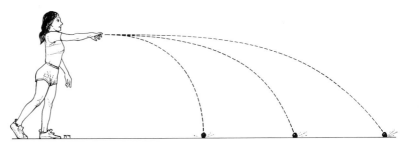

Figure 3.12 The harder you throw the rock, the farther it goes before hitting the ground.

We'll answer that question by considering the way a satellite is placed in orbit. Consider first a special satellite launched from a special launch-pad—a rock launched from your hand.

Figure 3.12 shows the paths taken by identical rocks thrown horizontally at different speeds on level ground. As you throw harder, the rock travels farther horizontally before hitting the ground. But the earth pulls downward with the same force on each rock, so that the time required for the rock to fall to the ground is the same for each throw. It goes farther when you throw harder because its faster horizontal speed covers a larger horizontal distance in the same amount of time.

But suppose we look at you from much farther away as you throw the rock. We'll retreat far enough to see the entire earth, and draw you large enough that we can see you (Fig. 3.13). As you throw harder and harder, the rock travels farther and farther around the earth as it falls to the ground. (We're assuming the earth to be smooth—no mountains—and neglecting air resistance.) When you throw the rock hard enough (path O), its curved path as it falls is exactly parallel to the surface of the earth. The rock continues to fall completely around the earth to the starting point. It would then keep on going indefinitely. For the rock to follow this path, the gravitational force between the earth and the rock must provide the *exact* centripetal force necessary for the circular motion. The centripetal force needed depends on the speed, and would equal the gravitational force when the speed is about 28,000 km/h (18,000 mi/h). To completely escape the gravitational field of the earth (path E), the rock must have a speed of at least 40,000 km/h.

The same ideas apply to the launching and orbiting of any space satellite. The force to launch the satellite usually comes from the thrust of a rocket. The satellite in orbit is continuously falling, but moves with the proper horizontal speed to stay in orbit.

Now let's go back to our question concerning the meaning of weightlessness. Suppose an astronaut is in a space shuttle orbiting the earth. Both shuttle and astronaut are continuously falling around the earth, as we just said. If he were to stand on scales inside the shuttle, he would exert no force on them because the shuttle and scales are falling along with him. He *appears* to be weightless because he exerts no force on the scales or anything else in the shuttle. The effect inside the spacecraft is the same as if there were no gravitational force. But viewed from the

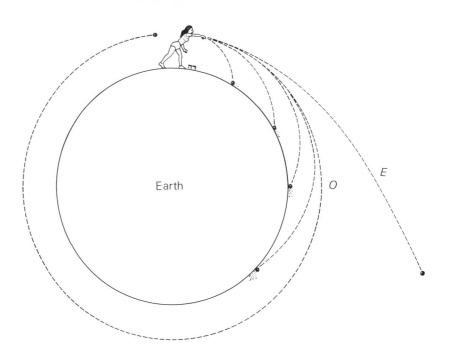

Figure 3.13 If you throw hard enough, the rock falls completely around the earth.

earth, each object has a weight and falls continuously because of that weight.

You would experience the same effect if you were inside an elevator and the supporting cable broke. The elevator and all its contents, including you, would be in free fall and therefore "weightless" until you hit the bottom. In the meantime, you could float weightlessly about the elevator.

Satellite descent. Our discussion so far has implied that, once a satellite is in orbit, it stays there forever, always "falling" around the earth unless intentionally brought back by firing its rockets. Orbiting artificial satellites eventually return to earth because, at their altitudes above earth, there is *some* atmosphere that supplies a frictional "drag."

This force slows the satellite, causing it to move to a lower altitude. As the altitude decreases, the surrounding atmosphere is denser and therefore the friction increases. This increased friction speeds the descent even more.

• •

Key Concepts

Newton's **law of gravitation** is one of nature's fundamental force laws. It can be stated as follows: There is a force of attraction between any two masses that is directly proportional to each mass and inversely proportional to the square of the distance between them. In symbols,

$$F = G\frac{m_1 m_2}{d^2}$$

The gravitational force is an example of an *inverse-square* law. The *weight* of an object on earth is the gravitational force between the object and the earth.

Projectile motion is described by a combination of constant velocity in the horizontal direction and constantly accelerated motion in the vertical direction. In a given amount of time, a projectile falls as far below the straight line of its initial direction as it would fall in that amount of time if dropped straight downward.

A body in circular motion must be acted on by a **centripetal force** directed toward the center of curvature. The corresponding acceleration is the **centripetal acceleration**. A **centrifugal force** is experienced by a body in a rotating reference system but is fictitious from the standpoint of an outside observer.

The centripetal force that keeps planets and satellites in orbit is the gravitational force on them. Persons in orbit appear *weightless* because they, as well as everything in their spacecraft, are falling at the same rate.

Questions

The Gravitational Force

1. Was Neil Armstrong's mass the same on the moon as on the earth? How about his weight?

2. Does it mean anything to talk about the *weight* of the earth? If so, what does it mean?

3. Would you expect there to be a place between the earth and sun where an object would experience no net gravitational force?

4. What are some reasons why the acceleration due to gravity g might differ a little from point to point on the surface of the earth?

5. Does the law of gravitation discovered by Newton tell us *why* we have weight?

6. The sun is 330,000 times more massive than the earth. On the surface of the earth, you are only 23,000 times farther from the sun than from the center of the earth. Why doesn't the sun pull you off the earth into itself?

Projectiles

7. When a baseball batter hits a hard line drive, does the ball travel in a straight line until caught by the shortstop?

8. You drop a rock straight down from a steep cliff into the river below. It takes 2.0 s to hit the water. If you throw the rock *horizontally* at

a speed of 10 m/s out over the river, how long will it take to hit the water?

9. How would you expect air resistance to affect the distance a real projectile will travel? Would it go as far as in a vacuum?

10. When you hold the nozzle of a garden hose horizontal and squirt water out in a narrow stream, the path of the water curves more sharply near the ground than near the nozzle. Why?

Bodies in Circular Motion

11. Why does a bicycle rider lean toward the inside in going around a curve?

12. Is a bug riding on the rim of a $33\frac{1}{3}$ RPM phonograph record accelerating?

13. If Santa Claus became the Arctic ambassador to Ecuador, would his accurate bathroom scales read the same when he stood on them in the embassy in Quito (latitude 0°) as they did at his home?

14. During the spin cycle of a washing machine's operation, water is separated from the clothes as the tub rotates (Fig. 3.14). Would you say that water is thrown from the clothes, or that

Figure 3.14 A spinning washing machine tub.

the clothes are pulled away from the water? Your answer may depend on whether you consider yourself to be outside the machine or inside it, spinning with the clothes.

15. Why does mud on the tires of a truck fly off as the truck picks up speed? What is the direction of motion of a hunk of mud relative to the tire?

16. Why do riders not fall out as they go through the loop-the-loop of the "Loch Ness Monster" (Fig. 3.15)?

17. In a treatment for certain immune disorders, the person's blood is circulated through a rotating bowl that spins it at rotational speeds of up to 1800 revolutions per minute. The blood separates into red cells, white cells, and plasma. Do the heavier red cells move to the outside, or stay near the center of rotation?

18. In the future, large numbers of people may live in space colonies—large self-sufficient space stations that produce "artificial gravity" by slow rotation of the entire station. Why would this rotation produce the same sensation as gravity to a person living on the *inner* surface of a shell surrounding the station?

19. Would you weigh more if the earth were not rotating?

Figure 3.15 The "Loch Ness Monster."

20. Is it correct to say that the earth is continuously falling around the sun?

21. If you jump from a 2-m-high wall, are you weightless until you hit the ground?

22. Is a skydiver weightless after jumping from an airplane, but before opening the parachute?

Problems

1. Calculate the gravitational force on you from the sun. (The sun is 1.49×10^{11} m away and has a mass of 1.99×10^{30} kg.)

    Answer: 0.359 N for 60 kg

2. Calculate the gravitational force acting between two apples of mass 0.25 kg hanging 2 m apart on a tree.

    Answer: 1.0×10^{-12} N

3. Captain Kirk of the Starship *Enterprise* has a mass of 70 kg and a weight of 700 N at the surface of Planet Cantaloupius. What would be his mass and weight when he is at a height equal to the planet's radius above its surface?

    Answer: 70 kg, 175 N

4. How high a flag pole would you have to climb to reduce your weight by 10 percent? (Assume a spherical earth and neglect the effect of the mass of the pole itself.)

Answer: 340 km

5. With what gravitational force does the earth attract the moon? (The moon is 3.84×10^8 m away and has a mass of 7.36×10^{22} kg.)

Answer: 1.99×10^{20} N

6. Using the answer to the preceding problem and Newton's second law of motion, what is the centripetal acceleration of the moon?

Answer: 0.0027 m/s^2

Home Experiments

1. Experimentally determine the answer to Question 2. Turn your bathroom scale upside down, stand on it, and weigh the earth. You can read the dial if you let it hang over the edge of a step and put a mirror under it. Are you surprised that the earth weighs the *same as you do*?

2. Check your answer to Question 8. Lay an unsharpened pencil or other light stick 8″ to 12″ long across the corner of a table as shown in Figure 3.16, and place a coin on each end. With your finger, flick one coin and the end of the stick in the direction indicated (or opposite to it). Hit the coin so that it moves off with a horizontal initial direction, and the other one falls to the floor. You should hear both hit the floor at the same time.

Flick in this direction 2–3 m.

Figure 3.16 Be sure you hit the coin so that its initial direction is horizontal and the other end of the stick just moves out from under the coin.

References for Further Reading

Edelson, E., "Saving Skylab—the Untold Story," *Popular Science*, Jan. 1979, p. 64. The story of the mid-1978 earth-bound efforts to revive the sleeping *Skylab* and save it in orbit until the arrival of the Space Shuttle.

Giancoli, D. C., *Physics: Fundamentals with Applications*, Second Ed. (Prentice-Hall, Inc., Englewood Cliffs, N.J., 1985). This book is among the many good algebra/trigonometry-based general physics books that give a more thorough coverage of the topics in this chapter.

C H A P T E R   4

WORK AND ENERGY

Work is a four-letter word, and most of us try to avoid it. In fact, one mark of a good physics student is a healthy dose of laziness. That's laziness of the right kind, of course—the kind that always looks for the easiest solution to a problem or that looks for the simplest law to explain the broadest spectrum of phenomena. So if you want to be a good physics student, work hard to avoid work. We're accustomed to using the word *work* to indicate how much effort we put into some project. However, in this chapter we're going to use a more restricted and precise definition of that word—one that is only indirectly related to effort expended.

In the last two chapters we looked at forces and their effect on motion. In this chapter we will combine both force and motion into one concept, that of work. Work is done whenever a force moves something through a distance. A force alone does not give us work; that force *must* act through a distance (Fig. 4.1).

A discussion of work leads us to the concept of *energy*. From just

Figure 4.1 The force on the car does no work if the car doesn't move. From the looks of the front tire, was *some* work done?

reading the newspaper we know that energy is very crucial to our lives; our society could not function without an ample supply of it. But just what *is* energy? We will answer that question in this chapter. Furthermore, we will consider the various forms in which energy exists.

The heart of this chapter is the principle of *conservation of energy*. Because of the way we define energy, it is an entity whose total amount stays constant, no matter what we do. This principle will provide us with a new and often easier way of analyzing the motion of objects. In fact, the principle can be used to understand a vast array of physical situations, and we will make use of it many times throughout the remainder of this book.

A closely related but clearly distinct concept is *power*. We'll close this chapter by defining power and showing its usefulness.

4.1 WORK

Suppose you're planning to hang up your grandmother's picture with a thumbtack. Your objective is to *push* on the tack and have it *move* into the wall (Fig. 4.2). That push, whether or not the tack moves, is a *force*. If that force also causes the tack to move into the wall, the force of your thumb does *work* on the tack. **Work** is done when a force *moves* something through a *distance*. When you pull on a desk drawer, you exert a force on it; if the drawer moves you do work on it.

Now, let's develop a more quantitative definition of work. Consider a situation in which the force acts in the *same direction* as (or parallel to) the motion. An example would be the work done by the force of a towrope on a water-skier when both ends of the rope are at the same height above water (Fig. 4.3). The force of the rope then does work in moving the skier along the water.

Figure 4.2 Your thumb does work on the tack *only* if the force moves it into the wall.

Work is defined as the product of the force parallel to the motion and the distance through which the force acts:

work = (force parallel to motion) × (distance the force acts)

Force

Figure 4.3 The force from the towrope does work in keeping the skier moving.

Gas prices expected to exceed
50 cents per gallon*

WASHINGTON, D.C. (DECEMBER 30, 1973)—President Nixon's new Federal Energy Administrator announced that he expects gasoline prices to rise from the present national average of 41.3 cents per gallon to more than 50 cents by February 1. This news is an added blow to many drivers already spending hours in line to get only a few gallons of gasoline at a time.

As the energy crisis worsens, the nation seems headed for the first peacetime rationing of gasoline in its history. A number of states have imposed their own measures for easing the crisis, including the odd/even system under which drivers of cars with odd license numbers buy gas only on odd days of the month, while drivers of cars with even license numbers get gas on even days. President Nixon has ordered thermostats to be adjusted downward in government buildings, imposed a 50-mile-per-hour speed limit on government cars, reduced unnecessary outdoor lighting, and encouraged similar voluntary actions in the private sector. But despite rising pressure to do so, the president has been reluctant to put a national rationing plan into action.

The fuel shortage reached crisis proportions in September when Libya's strongman Muammar Gadhafi nationalized all foreign oil companies operating in his country. And there is no immediate end in sight. Even if the Arab oil-producing nations lifted October's total embargo on shipments of oil to the U.S., it would take 45 days for the first tankers to reach this country.

We can write this definition symbolically as

$$\text{work} = F_p \times d$$

We put a subscript p on the force symbol as a reminder that we're considering a case where the force is *parallel* to the direction of motion.

If the boat pulls our skier 200 m in a straight line by exerting a continuous force of 100 N, the work done by this force is then

$$\text{work} = F_p \times d$$
$$= 100 \text{ N} \times 200 \text{ m}$$
$$= 20{,}000 \text{ N·m}$$

This calculation shows the unit of work to be the newton-meter, the result of multiplying force in newtons by distance in meters. This unit of work is called the **joule** (J), the SI unit of work. In other words, the joule is the amount of work done when a force of 1 N acts through a distance of 1 m in the direction of the force.

***To the student**: These simulated news releases scattered throughout the book bring out some historical sidelights.*

EXERCISE

Since a Big Mac weighs about 1 N (and therefore needs a force of about 1 N to lift it), and since a typical table is *roughly* 1 m tall, about how much work does it take to lift your Big Mac back to the table if you drop it onto the floor? You might use this result to gain some intuition about how "big" a joule is.   *Answer*: 1 J

The joule is named for James P. Joule, a nineteenth-century English physicist noted for helping to establish the relationship between work and heat. From time to time, it will be convenient to use other units of work. For example, in the English system of units, the unit of work is the *foot-pound*, the work done when a force of 1 lb acts through a distance of 1 ft. We'll define the other units of work as they are needed.

For many jobs, the force that does the work is *not* exerted parallel to the motion. For example, if you push a lawn mower, the direction of the force you apply is parallel to the handle, *not* parallel to the motion (Fig. 4.4). Thus, only a fraction of the force is effective in doing work. The smaller the angle between force and motion, the larger the fraction of the force that is effective. This description probably agrees with your experience. With thin, short grass, you can push the mower easily at almost any angle. But with tall tough grass, you keep the handle low to apply the force as nearly as possible in the direction of motion. If you haven't had the pleasure of mowing grass this way, you might have pulled a kid brother in his little red wagon. With only your brother in the wagon, the angle of pull doesn't matter much. When four friends get in with him, however, you may need to decrease the angle to make the same force do more work.

If the force and the motion are perpendicular, the force does *no* work. For example, when you ride in a car, your weight causes you to push downward on the seat. Since the motion is perpendicular to this direction, the force does *no* work. For most of the situations we'll consider, the force will be either parallel or perpendicular to the motion.

Figure 4.4 If the force and motion are in different directions, only a fraction of the force *is* effective in doing work.

Direction of force

Direction of motion

QUESTION   Assume that the earth is in a *circular* orbit around the sun. How much work does the gravitational force on the earth do in one revolution?

EXERCISE   If it takes a force of 160 N to slide your dresser across the floor, how much work do you do in sliding it 3.0 m?   *Answer*: 480 J

4.2 ENERGY

Now let's return to grandmother's picture (Fig. 4.2), which has fallen down because the tack wasn't strong enough to hold it up. You decide to put a nail, rather than a thumbtack, into the wall. Pushing with a thumb can't handle this project; the job calls for a hammer. Why not just hold the hammer against the nail and push? Having a handle to hold, you can exert much more force than by pushing directly with your thumb. But it still isn't enough; you have to *swing* the hammer. The *motion* of the hammer gives it the ability to exert a force on the nail and move it through a distance. That is, the hammer's *motion* gives it the ability to do work.

This *ability to do work* is our *starting* definition of **energy**. An object or group of objects that have the ability to do work—that is, the ability to exert a force through a distance—have *energy*.

The hammer's energy in our example comes from the fact that it's moving. We call that form of energy—energy resulting from motion—**kinetic energy**. You can probably think of some other everyday uses of kinetic energy, such as the work the force of a tennis racket can do on a ball because the racket is moving.

To see how to make use of the idea of energy, we need to establish some mathematical definitions of it. We'll find that energy exists in a number of different forms. The first form we consider is kinetic energy.

(a)

(b)

Figure 4.5   (a) A hammer of mass m striking a nail can be approximated by (b) a uniform mass m striking the nail.

4.3 KINETIC ENERGY

Let's find out how much work the hammer can do on the nail because of its motion. This will serve as a specific example that will illustrate how to calculate the kinetic energy of *any* object. Figure 4.5(a) shows the hammer acting on the nail. Since the hand has little effect during the impact with the nail, we'll neglect the hand. Furthermore, since most of the mass of the hammer is concentrated in its head, we can neglect the handle. In other words, the situation is essentially as pictured in Figure 4.5(b), a mass m moving horizontally at speed v and striking the nail. How much work can it do as it stops?

The work the mass does equals the force it exerts multiplied by the distance it travels in stopping:

$$work = force \times distance$$
$$= F \times d$$

The force F that the mass exerts on the nail is equal to the force the nail exerts on the mass (but in the opposite direction). Assuming this force to be constant, we can use Newton's second law to show that* this work is

$$work = \tfrac{1}{2}mv^2$$

Since the kinetic energy is the work the object can do, we equate the kinetic energy to the work done:

$$KE = \tfrac{1}{2}mv^2$$

In words,

$$kinetic\ energy = \tfrac{1}{2} \times mass \times (velocity)^2$$

This equation is the *definition of kinetic energy*. It applies to all bodies moving at speed v, not just the one we have considered. It says, for example, that if two bodies have the same speed, but one has twice the mass of the other, the more massive one has twice the kinetic energy. However, doubling the speed of a given body quadruples its kinetic energy; tripling its speed multiplies its kinetic energy by nine.

• •

EXAMPLE 4.1   A 0.20-kg arrow moving at a speed of 20 m/s pierces a 40-cm diameter pumpkin and stops just as the tip of the arrow reaches its far side. (a) What is the kinetic energy of the arrow before it hits the pumpkin? (b) How much work can the arrow do on the pumpkin? (c) What is the average force resisting the arrow's penetration?

(a) From the definition of kinetic energy,

$$KE = \tfrac{1}{2}mv^2$$
$$= \tfrac{1}{2}(0.20\ kg)(20\ m/s)^2 = 40\ J$$

*Using Newton's second law, we get: Work = $F \times d = (ma) \times d$. From the definition of acceleration, we get: $a = v/t$, where t is the stopping time. And since the average speed is $\tfrac{1}{2}$ the initial speed, we have: $d = \tfrac{1}{2}vt$. Therefore, work = $(mv/t) \times (\tfrac{1}{2}vt) = \tfrac{1}{2}mv^2$.

(b) The work the arrow can do equals its kinetic energy while moving—40 J.

(c) The definition of work, work = $F \times d$, gives

$$F = \frac{\text{work}}{d}$$

$$= \frac{40 \text{ J}}{0.40 \text{ m}} = 100 \text{ N}$$

This example illustrates the units we use for energy. Since the energy a body has equals the work it can do, *we express energy in the same units as we do work.*

The Work-Energy Theorem

We have seen that the work a moving body can do—its kinetic energy—is given by $\frac{1}{2}mv^2$. But how did that body *get* its kinetic energy if it started from rest? It got its energy by work being done on it. By a derivation similar to the one we've just discussed, we can see that the work needed to *accelerate* a body from rest to speed v is $\frac{1}{2}mv^2$. We can conclude, therefore, that the work done by the *net* (unbalanced) force on an object equals its *change* in kinetic energy. Calling this work the *net work*, we have

net work = change in kinetic energy

This statement is called the *work-energy theorem*. It applies whether the object is speeding up or slowing down. Since the force *on the object* by an external agent equals the force of the object *on the external agent* (Newton's third law), the work done on the object equals the work the object does on the external agent. For example, when the hammer hits the nail, the work done by the nail in stopping the hammer *equals* the work done by the hammer on the nail.

• •

EXAMPLE 4.2   Suppose you're in a head-on automobile collision while traveling at a velocity of 25 mi/h (11 m/s). Compare the average force on your head if it is stopped by the car dashboard in a distance of 1.5 cm with the average force on your head if it is stopped by an inflatable air bag in a distance of 25 cm.

We know that the work done by the force that stops your head must equal the *change* in your head's kinetic energy. Since *all* the kinetic energy is lost,

$$F \times d = \tfrac{1}{2}mv^2$$

Taking the mass of the head to be 2.0 kg, we calculate the force F_d of the dashboard from

$$F_d \times 0.015 \text{ m} = \tfrac{1}{2}(2.0 \text{ kg})(11 \text{ m/s})^2$$

which gives a force of 8100 N or 1800 lb. We get the force F_{ab} of the air bag from

$$F_{ab} \times 0.25 \text{ m} = \tfrac{1}{2}(2.0 \text{ kg})(11 \text{ m/s})^2$$

which gives 480 N or 108 lb.

· ·

QUESTION Why does the average braking distance for an automobile go from 10 to 40 m as its speed goes from 25 mi/h to 50 mi/h?

· ·

EXERCISE Find the kinetic energy of a 42-m/s fastball (mass = 0.15 kg). How much work does the ball do on the glove of the catcher?

Answer: 130 J, 130 J

Figure 4.6 As the weights in a grandfather clock descend, their gravitational potential energy does the work of turning the hands of the clock. (Courtesy Pulaski Furniture Corporation.)

4.4 POTENTIAL ENERGY

What provides the energy to do the work of running a grandfather clock? Certainly it doesn't take much work to rotate the hands and the internal gears for a day; but some work is needed and that work must come from some source of energy. If you're familiar with these clocks, you know that you "wind" them by raising heavy weights inside the clock. The weights then gradually descend and operate the clock for several days (Fig. 4.6). But what kind of energy is this? Even though the weights move slowly, the energy to run the clock is not kinetic.

The weights exert a downward force on their supporting chains because of their own *weight*—the force on them due to the earth's gravitational field. As they descend, this force does work. In other words, these weights have energy—they can do work—because they are in the earth's gravity. This energy is an example of what we call *potential energy*: in this case, *gravitational potential energy*.

Potential energy results either from the position of an object relative to some other object, or from the deformation of a given object. Grav-

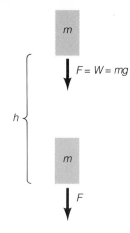

Figure 4.7 A mass *m* can exert a force *F* equal to its weight *mg* when in the earth's gravitational field. It does work *mgh* in moving downward from height *h*.

Figure 4.8 To lift the mass at constant speed, a force *F* equal to the weight must be applied.

Figure 4.9 The value of the potential energy has to be specified relative to some chosen level.

itational potential energy is of the first type; it comes from the position of some object relative to the earth (or another mass). *Elastic potential energy* results from the deformation of a given object. An example would be the energy available due to the stretching of a spring.

Gravitational Potential Energy

A mass *m* can exert a downward force equal to its weight *mg*. As such a mass moves downward from a height *h* (Fig. 4.7), it does an amount of work equal to the force times the distance:

$$\text{work} = F \times d = mg \times h = mgh$$

But the work it can do is its *gravitational potential energy*. Therefore, in general

$$GPE = mgh$$

Now reverse the situation and lift the mass through height *h* (Fig. 4.8). If it moves at a constant speed (without acceleration), the force needed equals its weight *mg*. Therefore, lifting it through height *h*

Figure 4.10 Thomas Jefferson's "seven-day" clock mounted over the door at his home, *Monticello*. The weights are groups of steel balls hanging from cables in each corner. The dark strips on the wall behind the right-hand cable have the days of the week written on them; Friday and Saturday are in the basement. (Photograph courtesy of Thomas Jefferson Memorial Foundation.)

requires an amount of work *mgh*—just the amount of potential energy the mass gains. In other words, giving a mass a certain amount of potential energy requires *exactly* that amount of work being done on it.

As you can see from the way we derived our potential energy equation, it really expresses the potential energy *change*, and it assumes the value of potential energy to be *zero* at an arbitrary point, often taken as the lowest point the mass reaches in a given problem. In other words, our equation gives the potential energy relative to whatever level at which we have taken its value to be zero.

If, as in Figure 4.9, a 9-kg mass is 2 m above the first floor, its potential energy relative to the first floor is 9 kg × 10 m/s^2 × 2 m, or 180 J. (We've used our approximate value for *g*.) But, if the floors are 3 m apart, the potential energy relative to the basement is 450 J and, relative to the second floor, the potential energy is −90 J.

· ·

APPLICATION   ## JEFFERSON'S "SEVEN-DAY" CLOCK

If you ever visited Thomas Jefferson's home, *Monticello*, you probably saw an excellent illustration of the fact that the value of potential energy depends on the level chosen for its value of zero. Among Jefferson's many ingenious inventions is his "seven-day" clock mounted in the entrance hallway (Fig. 4.10).

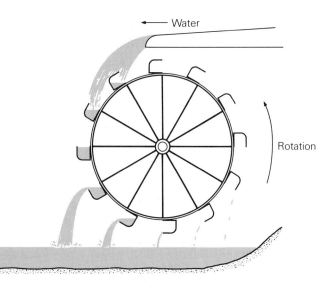

Water

Rotation

Figure 4.11 The gravitational potential energy of the water does the work of turning a waterwheel.

Mr. Jefferson wanted a clock that would run for a week. He suspended two sets of heavy weights from cables, ran the cables over pulleys in a corner of the room, and wound the other end of the cables around wheels inside the clock. As the weights gradually descended, they gave up their potential energy in doing the work of running the clock. But Jefferson ran into a snag when the weights hit the floor and the clock stopped before the end of the seven days.

Rather than increase the weight and redesign the clock, Jefferson cut holes in the floor to let the weights pass into the room below. He then had available the potential energy of the weights descending through two stories of height rather than just one. That got his clock through the seven days without anyone having to rewind the weights to the ceiling. In other words, by cutting holes in the floor he was choosing a different reference level for measuring the potential energy of his weights

Using water's potential energy. For hundreds of years the gravitational potential energy of water has been used to do practical work. The simplest example is the old-fashioned water wheel illustrated in Figure 4.11. Water flowing onto the wheel from above fills the upturned cups. The weight of the water in these cups causes the wheel to rotate, thereby operating the mill attached to the wheel. In other words, the potential energy of the water is used to do the work of turning the wheel that runs the mill.

In the modern version of the water wheel, large quantities of water flowing from a massive dam turn a turbine—the modern version of a water wheel—that, in turn, turns a generator that generates electrical energy. This energy is transmitted over wires to the users, both individuals and industries. (We'll deal with this topic in detail in Part Four.) Actually, we left out a step: The potential energy does the work of giving the water kinetic energy as it falls through large pipes leading to

Figure 4.12 In a hydroelectric power plant, gravitational potential energy of the water near the top of the dam does work that gives kinetic energy to the water that goes through the turbines. This kinetic energy does the work of turning the turbines. Wheeler Dam, shown here, on the Tennessee River has been producing electrical energy since 1936. (Courtesy Tennessee Valley Authority.)

the turbine (Fig. 4.12). This kinetic energy then turns the blades of the turbine in much the same way that air from your breath turns a pinwheel when you blow on it.

The ramp. Imagine that you're moving across town, and that you've packed all your books into a single box and rolled the box out to the street balanced on your roommate's skateboard. But somehow you're not especially inspired to lift the box onto the back of that rented pickup truck. An alternative approach, one that would require much less force, would be to lay a plank from the street to the truck and roll both box and skateboard up the plank (Fig. 4.13).

Suppose the truck floor is 90 cm from the street, and the box weighs 400 N (mass 40 kg). The work you need to do equals the change in potential energy, mgh or Wh. This is 400 N × 0.90 m = 360 J. The

Figure 4.13 The change in potential energy is the same whether the box is lifted directly onto the truck or rolled up the ramp. If friction is neglected, the work done is the same either way, but the force F needed to push the box up the plank is only $\frac{3}{10}$ as much.

Figure 4.14 As Jon stretches the slingshot, it stores elastic potential energy.

change in potential energy of the box is the same whether you lift it directly or roll it up the ramp. If we neglect the weight of the skateboard and the friction opposing its rolling, the work of pushing the box up the ramp is 360 J. Therefore,

$$\text{work} = F \times 3 \text{ m} = 360 \text{ J}$$

We can solve this equation for F and find that the force needed is 120 N, only 3/10 that needed to lift the box directly. In reality, the force would have to be stronger than this by the amount of frictional resistance plus the force needed to raise the skateboard itself.

· ·

QUESTION   Why are roads up steep mountains made to wind around rather than to go straight up the mountain?

Elastic Potential Energy

If you grew up in the country, you might have used a slingshot to acquire apples from an inviting tree (Fig. 4.14). What provides the energy that does work on the pebble to give it kinetic energy as the slingshot is fired? The answer to that question illustrates another form of potential energy:

elastic potential energy. Stretching the elastic bands stores energy as these bands are deformed. This energy is recovered as they are released to return to their equilibrium (unstretched) length.

Many useful devices make use of elastic potential energy stored in springs as they are compressed, stretched, twisted, or wound. The energy is then used to do useful work as the springs go back toward equilibrium. Examples include wind-up watches and clocks, window shades, screen-door closers, mouse traps, and heavy-duty staple guns. You can probably think of many other examples.

· ·

QUESTION   How do trampolinists and archers make use of elastic potential energy to achieve their goals?

The simplest device for storing elastic potential energy is a straight spring that can be compressed or stretched. The stretching (or compressing) force for such a spring is usually proportional to the distance stretched. Following the steps outlined in Problem 32 at the end of this chapter, you can show that the elastic potential energy of such a spring is given by $\frac{1}{2}kx^2$, where k—the "spring constant"—gives the force needed to stretch the spring one unit of length and x is the distance stretched or compressed from equilibrium.

4.5 ENERGY TRANSFORMATIONS

We started our discussion of energy with its simple definition as the ability to do work. Having looked at some specific forms of energy, now let's observe that our definition is not quite correct on two counts: First, keep in mind that it is actually the *force* that does work as it acts through a distance. Thus, a body or system with energy has the ability to exert a force that can do work. Second, there are some situations in which an object with energy cannot do any work because there is no way for it to give up its energy. An example is a body with gravitational potential energy but with no way to move to a lower level in order to give up this energy. Later on we'll encounter other examples of this second limitation.

Now notice that energy can equally well be defined as *the result of doing work*. As we saw in Section 4.3, objects with kinetic energy *get* their energy as work is done on them. Similarly, Section 4.4 showed that we increase a body's potential energy by doing work on it. *Any time a form of energy increases by some amount, it does so because exactly that same amount of work is done on it.*

Thus, we can conclude that as one form of energy does work, that work produces the *same amount* of another form of energy. In other words, when work is done, energy is *transformed* from one form to another, or perhaps to more than one other form.

Let's look at some illustrations of this point. An apple hanging on a tree has potential energy relative to the ground. If it falls from the tree, it loses potential energy as it falls; but the work done by this potential energy increases the kinetic energy by the same amount, as illustrated in Figure 4.15. As another example, consider a clock pendulum swinging back and forth. It continuously undergoes changes between potential and kinetic energy. At the highest point in its swing, where the pendulum reverses direction, it has no kinetic energy. As it swings downward, its potential energy decreases but its kinetic energy increases by the same amount.

In every case, *work transforms energy*. And in any such transformation, one form of energy *does* the work; the work done *increases* the other form of energy. But *the total amount of energy remains constant*.

· ·

QUESTION   Describe the energy transformations as a person bounces on a trampoline. For each energy interchange, describe the work done by one form of energy and how this work increases the other form or forms.

4.6 CONSERVATION OF ENERGY

The statement at the end of the last section—that the total amount of energy remains constant—is a statement of one of the most important principles in physics: the **principle of conservation of energy**. It can be stated in the following way: *Energy can be neither created nor destroyed; it can only be transformed from one form to another.*

Thus far we have considered only kinetic and potential energy. As we go along we'll discover that energy exists in a number of other forms. However, for now we'll concentrate on systems with only these two forms of energy, which are usually called *mechanical energy*. If no other forms of energy exist in a system, then total *mechanical* energy is *conserved*—that is, it remains constant.

If friction or air resistance exists in a system, then some of the mechanical energy may be used in doing work against friction. Therefore, we can assume conservation of *mechanical* energy only for situations in which frictional effects are small enough to be negligible.

Conservation of Mechanical Energy

In a system with only mechanical energy, we can write the principle of conservation of energy as kinetic energy + potential energy = constant

In symbols,

$$KE + PE = \text{constant}$$

Therefore, the total energy at one time must equal the total energy at some later time:

$$PE + KE \text{ at one instant} = PE + KE \text{ at later instant}$$

This principle lets us analyze the motion of objects whose total mechanical energy remains constant. Knowing the state of motion of the body at one instant of time, we can then determine its state at a later time when the body is at a different location. We'll illustrate such an analysis with two specific examples.

First, let's find the speed with which a diver hits the water, after diving from a 5-m-high platform (Fig. 4.16). We'll neglect any air resistance, so that total mechanical energy stays constant. That means the diver's

$$\text{total energy at platform} = \text{total energy at water}$$

or

$$(KE + PE) \text{ at platform} = (KE + PE) \text{ at water}$$

The *KE* is low enough to consider it zero at the platform, and we can choose the *PE* to be zero at the water. Then

Figure 4.15 The bar graph depicts the decrease in gravitational potential energy (*PE*) and increase in kinetic energy (*KE*) as the apple falls from the tree.

Figure 4.16 The diver's potential energy is measured relative to the water, meaning it is zero at water level. At the platform, the kinetic energy is zero.

$PE = 0; KE = \tfrac{1}{2}mv^2$

$h = 5\,\text{m}$

$$0 + mgh = \tfrac{1}{2}mv^2 + 0$$

where the left-hand side applies to the diver at the platform and the right-hand side at the water. We can divide both sides of the equation by m, and solve it for v to get

$$v = \sqrt{2gh} = \sqrt{2(10 \text{ m/s}^2)(5 \text{ m})}$$

or 10 m/s. (See Appendix I for a review of square roots.)

We could have solved *this* problem by the methods we previously used in studying free fall. The real advantage of the energy method appears in solving a problem when the direction of motion changes.

As our second example, consider a roller coaster traversing the path of Figure 4.17. If it starts at rest at point A, what is its speed when it gets to point B, assuming there is *no* friction? All we need to know is that energy is conserved: Total energy at point A equals total energy at point B. We don't care what happens in the middle. Since $KE = 0$ at point A and we can choose $PE = 0$ at point B, we have

$$PE \text{ at } A = KE \text{ at } B$$

or

$$mgh = \tfrac{1}{2}mv^2$$

Again, we get $v = \sqrt{2gh} = \sqrt{2(10 \text{ m/s}^2)(7.2 \text{ m})} = 12$ m/s. Using the energy method, we don't need to know any forces, heights, or directions at intermediate points.

$KE = 0;$
$PE = mgh$

A

7.2 m

$PE = 0$
$KE = \tfrac{1}{2}mv^2$

B

Figure 4.17 When there is no friction, the roller coaster's *total* energy at point B equals its *total* energy at point A. When there *is* friction, its total energy at B is less than that at A by the amount of heat generated as work is done against friction.

Notice that if we had chosen the zero of *PE* at any other point, the answer would have been the same. For instance, if we had taken the zero at the bottom of the pool in the first example, it would have added the same amount of potential energy to *both* sides of our equation, which would not have changed the result.

- -

EXERCISE Little Johnny slides down a sliding board whose top is 2.0 m off the ground. With what speed does Johnny hit the ground? Neglect the effect of any friction. *Answer:* 6.3 m/s

Friction

So far we have neglected any effect of friction or air resistance. Yet you know that friction plays a role in real situations. When one object rubs against another, there is always a force that we call *friction* that opposes this relative motion. If we could not neglect friction in the last exercise, the answer would be quite different. The smoother the sliding board (the lower the friction), the faster Johnny gets to the bottom. Or if Johnny is swinging and quits pumping, he gradually slows down and stops. That's because the friction in the swing and between Johnny and the air exerts forces on him and the swing. The energy in the system must do work to move the swing against these forces.

- -

QUESTION Why do you have to wind a pendulum clock? Note: The unwinding spring gives the pendulum a little "kick" during each swing. Compare this kick with the push somebody has to give you to keep you swinging at the same pace in a swing.

Any time work is done against frictional forces, the work done produces an amount of *heat* equal to the work done. Thus, heat is a form of energy. If you rub your hands together quickly, you can feel the heat produced.

- -

QUESTION Your car brakes work on the basis of friction between the brake pads and the disks that rotate with the wheels. Why will the brakes of your car overheat and fail if you keep them on constantly while driving down a high mountain?

If friction is present in a system, we need to modify our conservation-of-energy equation to take into account the mechanical energy that is converted to heat as work is done against friction. Our conservation of energy equation becomes

total energy at one instant

= total energy at later instant + work done against friction

We can apply this equation to the roller coaster problem (Fig. 4.17), now including friction. Suppose the coaster car and its occupants have a total mass of 1000 kg; the total path length from A to B is 50 m; and a continuous frictional force of 400 N opposes the motion at every point. The energy conservation equation becomes

PE at A = KE at B + work against friction

or

$$mgh = \tfrac{1}{2}mv^2 + (\text{frictional force} \times \text{distance})$$

or

$$(1000 \text{ kg})(10 \text{ m/s}^2)(7.2 \text{ m}) = \tfrac{1}{2}(1000 \text{ kg})(v^2) + (400 \text{ N})(50 \text{ m})$$

$$7.2 \times 10^4 \text{ kg·m}^2/\text{s}^2 = (500 \text{ kg})v^2 + (2.0 \times 10^4 \text{ kg·m}^2/\text{s}^2)$$

or

$$5.2 \times 10^4 \text{ kg·m}^2/\text{s}^2 = (500 \text{ kg})v^2$$

or

$$104 \text{ m}^2/\text{s}^2 = v^2$$

We can solve this equation for v to get a velocity of 10 m/s. As you would expect, this speed is a little lower than if there were no friction.

The energy conservation techniques that we've applied to these specific examples have general applicability, and are used routinely by scientists and engineers to solve problems of quite practical significance.

Notice that, when friction is not important, the speed of an object after it descends from some height does *not* depend on either its mass *or* the curvature of its path. The path the object takes determines its direction at the end, but not its speed. However, as the last example illustrates, when friction is important, we must know more about the system in order to determine the final state of motion.

Figure 4.18 As the wheel rotates, the chunk of matter labeled "x" moves in a circle shown by the dashed line.

Each part of this book will introduce new forms of energy. By way of a brief look ahead, we'll mention a few of these other important forms.

First notice that kinetic energy is associated with the rotation as well as with the straight-line motion of bodies. This *rotational kinetic energy* comes about because, as a body rotates, each piece of matter is moving in circular motion and, therefore, has kinetic energy associated with it. For example, if the wheel shown in Figure 4.18 is rotating, the chunk of matter marked "x" is moving and has kinetic energy. We lump the energy associated with all such chunks of matter together and say that the wheel has rotational kinetic energy.

Massive wheels for storing energy in this form are called *flywheels*. Flywheels can serve to store energy for later use, or they can serve to smooth out fluctuations in the rotational speed of devices such as automobile engines. Flywheels may someday play a significant role in reducing fuel consumption. Development work is being done on busses, subway trains, and even cars for city driving that are equipped with special flywheels. As the brakes are applied, the work done in stopping one of these vehicles goes toward increasing the rotational kinetic energy of the flywheels. This energy, that otherwise would have been wasted, is then used to get the vehicle back up to speed.

· ·

QUESTION Why are phonograph turntables usually rather massive? Hint: If they did not store significant amounts of energy while rotating, what would small forces due to the stylus or external vibrations do to the rotational speed?

Electrical energy plays a major role in our society. When we get to Part Four, we will gain insight into the two ways in which electrical energy shows itself: through the interaction of the electrical "charges" that exist on subatomic particles, and through the movement of these charges. *Chemical energy* that is released through the burning of fossil fuels, for example, is basically electrical energy of the first type.

Sound and *light* are two more important forms of energy. Try getting along without them! You might wonder how these fit our definition of energy as the ability to do work. Well, you know from biology that sound waves wiggle your eardrum when they impinge on it. That process involves doing work on your eardrum. Similarly, light waves do work in wiggling the atomic particles that make up the receptors in the back of your eyes.

Finally, there is *nuclear* energy, a topic of continuing controversy in our society. A particular type of force—called the nuclear force—binds the nuclei of atoms together. This force can do work on nuclear particles as it either rearranges them within the nucleus or ejects them from it. Thus, the rearrangement of nuclear particles to form different nuclei can result in the conversion of nuclear energy into other useful forms.

What about *mass*? Is it a form of energy? When we study Einstein's theory of relativity, we will discover that mass is not a *form* of energy, but rather that mass and energy are equivalent. They are related by Einstein's famous equation $E = mc^2$.

If we take into account all forms of energy and the mass-energy equivalence, we can state the *principle of conservation of energy* in its most general form: *In an isolated system* (one in which no energy enters or leaves), *the total amount of energy remains constant.* This principle will provide many useful insights as we study various topics in this book.

You may be thinking that there is something different about "biological" work. You *know* that you get tired holding up a stack of books, even if you are perfectly still and there is no motion. Or if you're in the situation shown in Figure 4.1, you become exhausted whether the car moves or not. Why should you get tired if no work is done?

No work is done by the force of your hand *on* the books or *on* the car. But that does *not* mean that no work is done by the muscles inside your body. Muscles do not "lock in" and become rigid. As the muscle pulls, its many fibers are constantly contracting, then relaxing slightly, then contracting again. The different fibers contract at different times so that the muscle as a whole seems to stay rigid. Each time a fiber contracts, it does real mechanical work—force × distance. Energy is needed for this work.

Where does the energy come from that provides the gravitational potential energy your body gains in walking up a flight of stairs? It comes from the *chemical* energy stored in food you have eaten. Your body is able to use this chemical energy to do work in walking. When walking upstairs, you have to do more work than on a level surface, because you have to increase your body's potential energy. Riding a bike uphill is harder than riding on level ground because additional work is needed to increase your, and the bike's, potential energy.

4.8 EFFICIENCY

Most of the time when we deliberately transform energy from one form to another, we don't get all of it converted into the form we want. For example, all the work you do on the pedals of a bicycle isn't converted into the work of moving the bike; some is used to overcome friction between the chain and sprocket wheels, and between the other moving parts of the bike. In the operation of a light bulb, some of the electrical energy used is given off as heat, not light. And when Mr. and Mrs. Johnson burn natural gas in their furnace, some of the heat produced gets lost from the ducts carrying hot air to their rooms; they do not benefit from *all* the energy in the fuel.

We use the term *efficiency* to describe how much we get out of a system relative to what we put into it. In particular, efficiency is the ratio of the output work, or energy in a useful form, to the input work or energy:

$$\text{efficiency} = \frac{\text{work (or energy) out}}{\text{work (or energy) in}} \times 100\%$$

If the Johnson's heating system is 60% efficient, then 60% of the chemical energy in the natural gas is delivered as heat to their house. If your bike is 90% efficient, 90% of the work you do on the pedals produces useful work in moving you along the street. The rest goes to heat resulting from friction between the moving parts of the bike.

• •

QUESTION   What does the energy conservation principle tell you about the maximum possible efficiency of *any* system?

The difference between energy (or work) input and the energy (or work) output is usually considered "lost" energy. It may be lost from doing useful work, but it must appear *somewhere* in *some* form. In a mechanical system where the loss is due to friction, the lost energy goes to heat.

In some situations the "lost" energy isn't lost from usefulness. An ordinary 100-watt light bulb is only 2.5% efficient at producing light. That is, only 2.5% of the electrical energy supplied to it gets converted to light. The rest goes to heat, most of which is radiated into the room. If it happens to be winter, the load on your heating system is reduced by the amount of heat supplied by the bulb.

4.9 POWER

Another quantity closely related to the concept of energy, is *power*. In fact, when you use the word power in ordinary conversation, its meaning is comparable to our definition of energy—it often means the ability to get things done. However, our definition of power is distinct from, though related to, energy.

Power *means the rate at which work is being done.* In other words, power is the amount of work done per unit of time.

Suppose you're spending the weekend with a friend who lives on the fourth floor, and you do 2400 joules (J) of work in carrying your suitcase up the three flights of stairs. You might do this work at a leisurely pace, taking 60 seconds (s) from bottom to top. But if you're in a hurry, you might make it in 20 s. The amount of work done on the suitcase is the same either way, but the rate of doing work, the power, is different.

In the 60-s case

$$power = \frac{work}{time} = \frac{2400 \text{ J}}{60 \text{ s}}$$

or 40 J/s, where J/s means joules per second. When you do the work in 20 s,

$$power = \frac{2400 \text{ J}}{20 \text{ s}}$$

or 120 J/s. The power is three times as much when you do the work in one-third the time.

We could express power in joules per second, but in the SI system the unit for power has the special name *watt* (W) after James Watt who, while not the inventor, made significant improvements to the steam engine. That is,

1 watt (W) = 1 joule/second (J/s)

The *horsepower* (hp) is a unit of power often used. When James Watt improved the steam engine, he wanted to be able to tell people how many horses his engines could replace. By suspending a weight in a deep well, and hooking a horse to a mechanism that could lift the weight, he and his coworkers estimated that a typical horse could do 746 joules of work per second, or 746 watts. Watt defined the horsepower to be that amount of power. The Watt engine of 1778 had 14 hp.

The horsepower of an automobile engine tells you the rate at which that engine is capable of doing work. If a 100-hp engine could accelerate a particular car from rest to a speed of 80 km/h in 20 s, a 200-hp engine could give the same car that speed in roughly 10 s. I say "roughly" because several factors could influence things somewhat: The capabilities of the two engines might vary differently with speed, the frictional effects might be different, and the more powerful engine would probably make the car weigh more. If all these extraneous influences were negligible, the time would be cut in half.

As we discussed earlier, any time work is done, some form of energy is used to do the work, and an equal amount of some *other* form of energy is created as the work is done. Doing work does *not* use up energy; the work transforms the energy into another form (or forms). Thus, we could also define *power as the rate at which some form of energy is used (or transformed):*

$$power = \frac{energy \text{ used (transformed)}}{time}$$

When we specify the "wattage" of a light bulb, for example, we specify the electrical power—that is, the rate at which the bulb consumes electrical energy. A 100-W bulb uses electrical energy at the rate of 100 joules per second. You would do work at exactly this rate if you lifted 120 5-kg (11-lb) bags of sugar from the floor onto a 1-m-high table in one minute. (See Problem 26 at the end of the chapter.)

Let's do one more thing with our definition of power. Since

$$\text{power} = \frac{\text{energy}}{\text{time}}$$

we can rewrite this equation as

$$\text{energy} = \text{power} \times \text{time}$$

In other words, if your electric heater uses a power of 1000 W, and you use it for 4 hours (h), the electrical energy used is

$$\text{energy} = 1000 \text{ W} \times 4 \text{ h}$$

$$= 4000 \text{ watt-hours}$$

We see from this example that the watt-hour (Wh) is an *energy* unit that results from multiplying the power in watts by the time in hours. Often we use a larger unit, the kilowatt-hour (kWh), which equals 1000 Wh or 3.6×10^6 J.

When you pay your electrical "power" bill, is your payment based on the power used or the energy used? This question is like asking whether you are paying for how much energy you use or for how fast you use it. Hint: look at the units used on your (or somebody else's) latest electric bill.

4.10 ENERGY AND SOCIETY

Energy makes our lives go. It moves us along roads; it scrambles our eggs in the morning; it warms our houses in winter; and it creates music to lift our spirits. Most of the energy that we use to do these jobs is stored in fuels that we burn. Having adequate energy supplies to keep our society functioning means having adequate supplies of fuel in a useable and economical form. Using energy for practical purposes means *transforming* energy from its form as stored in fuel to a form that can do the desired job. Let's look briefly at some of these transformation processes.

Figure 4.19 summarizes the major sources and final uses of energy in our modern society. The diagram also indicates the usual means by which energy is transformed from source to use. Under "uses," the mechanical

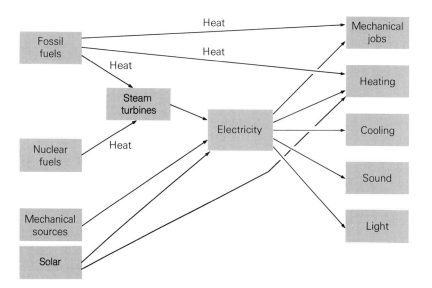

Figure 4.19 A schematic diagram showing the sources, intermediate stages, and final everyday uses of energy.

jobs include such applications as transportation, the operation of factories, and home projects like beating eggs and drilling holes. Heating includes space heating, cooking, water heating, and industrial uses of heat. Cooling includes refrigeration and air conditioning. Uses for sound and light need no explanation.

As the arrows in the diagram indicate, the first two categories of uses may get their energy directly from fuel. For example, we burn fuel in automobiles to release heat that the engine uses to propel the car. Or many of us heat our houses by the direct burning of some fossil fuel.

However, most of our everyday jobs use electrical energy. Some of this electrical energy comes from mechanical sources—hydroelectric plants and windmills—that use the kinetic energy of water or wind to turn electrical generators. More typically, fossil fuels are burned to release heat that produces steam, which operates a turbine to turn the electrical generator. In some plants, nuclear fuels provide the heat for this steam.

Finally, radiation from the sun provides direct solar energy for heat and, in limited situations, for the generation of electricity by way of solar cells.

In every case, whatever energy's source or final use, it is neither created nor destroyed; it is only transformed from the form of its initial source to one that will do the intended job. *Total energy is conserved.*

If energy is conserved anyway, what are we doing when we "conserve energy" by turning off unused lights, setting our thermostats low in winter, driving fuel-efficient cars, and minimizing our use of hot water? These practices do not conserve total energy; that happens whether we like it or not! Instead, they conserve energy in forms useable for practical purposes—forms that have a limited supply. In other words, "conservation" in the sense of not depleting energy supplies means saving useable *forms* of energy; "conservation" in the conservation-of-energy principle means that the total amount of energy remains constant.

. . . .

Key Concepts

Work is done when a force acts through a distance, and it is defined as the product of the force parallel to the motion and the distance through which it acts:

$$\text{work} = F_p \times d$$

Energy is the ability to do work and the result of doing work. **Kinetic energy**—energy by virtue of motion—is given by $\frac{1}{2} \times$ mass $\times$ (velocity)^2:

$$KE = \tfrac{1}{2}mv^2$$

The *work-energy theorem* states that the net work equals the change in kinetic energy.

Potential energy results from the position of one object relative to another, or from the deformation of a given object. **Gravitational potential energy** arises from the position of an object in the gravitational field of the earth. Its value at some height *h* relative to the defined zero is

$$GPE = mgh$$

Elastic potential energy results from the deformation of an elastic object.

The **principle of conservation of energy**: Energy can be neither created nor destroyed, only transformed; the total energy in an isolated system remains constant. Thus,

total energy at one instant

= total energy at a later instant + work done against friction

$$\textit{Efficiency} = \frac{\text{work (or energy) out}}{\text{work (or energy) in}} \times 100\%$$

Power is the time rate of doing work or of using (transforming) energy:

$$\text{power} = \frac{\text{work (or energy used)}}{\text{time}}$$

Questions

Work

1. The weightlifter of Figure 4.20 is holding the weights stationary over his head. Are his hands doing any work on the bar?

2. How much mechanical work is involved in
   a. lifting a 2-kg brick a height of 1m?
   b. holding the brick 1 m off the floor?
   c. slowly moving the brick a *horizontal* distance of 10 m?
   d. letting the brick fall a distance of 1 m?

3. On June 3, 1921, German strongman Herman Görner carried a 6423-N (1444-lb) piano a horizontal distance of 16 m (52$\frac{1}{2}$ ft). How much work did he do on the piano in carrying it this distance?

Figure 4.20 The weight lifter is holding the weights still.

Kinetic Energy

4. Do you have kinetic energy sitting still in your car while it's traveling at 85 km/h?

5. Why do experienced carpenters advise beginners to hold the hammer at the end of the handle, as far as possible from the hammer head?

6. We found that a baseball pitcher could give a 0.15-kg baseball a kinetic energy of 130 J. However, a javelin-thrower might give a 0.60-kg javelin a kinetic energy of 200 J. Do these facts necessarily mean that the javelin-thrower is stronger than the baseball pitcher?

7. Figure 4.21 is taken from a state driver's manual. Explain why reaction distance is proportional to speed, whereas braking distance increases much more significantly with increase in speed. Mathematically, how does braking distance vary with speed?

8. As you catch a fast-moving baseball, why is it wise to move your hand in the direction of the ball's motion?

9. If you're waiting for a stoplight to change, and you're about to be rammed from the rear, you'd probably prefer that it be by a compact Honda than by a loaded Mack truck. From an energy standpoint, why do you have this preference?

Potential Energy

10. You can easily catch a 2-kg brick dropped from a height of $\frac{1}{2}$ m. However, if the back of your hand is resting on a tabletop when the brick first contacts it, your hand may be injured. Why does catching the brick with your hand on a table increase the force on your hand so much? (Answer in terms of work and energy.)

11. Some people with fenced yards still use an ancient technique to close the gate (Fig. 4.22). The weight rises when the gate opens and pulls the gate shut later. What actually provides the energy to close the gate?

12. In treadmill tests for heart malfunction, the

Average reaction distance (ft)

Average braking distance (ft)

Figure 4.21 Reaction and braking distances in driving an automobile. Reaction distance is the distance traveled from the time you start to move your foot until you actually apply the brakes. Braking distance is the distance needed to stop once the brakes are applied.

patient walks on a moving inclined belt to simulate walking uphill. If the belt is moving at a speed of 2 m/s and is inclined at 10°, would the person do work at the same rate as if walking at 2 m/s up a hill inclined at 10°?

13. What are the similarities and the differences in principle between the devices shown in Figure 4.23 and the ramp shown in Figure 4.13?

Figure 4.22 A cheap but effective gate closer.

14. Many satellites move in elliptical rather than circular orbits. When the satellite is at a position such as point A in Figure 4.24, the gravitational force is acting in a direction to speed it up.

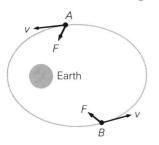

Figure 4.23 The wedge, the needle, the knife, and the screw.

When the satellite is at point *B*, the gravitational force slows it down. Describe how these changes are consistent with the principle of conservation of energy, since the potential energy decreases as the satellite gets closer to earth.

15. A stretched slingshot is capable of doing work on the pebble in it. What form of energy does the slingshot have?

16. Why does pulling the bow back farther cause an arrow to have a higher speed when released?

Figure 4.24 The arrows *v* and *F* show the velocity of and gravitational force on a satellite at two different points in an elliptical orbit.

17. When you push on the end of a retractable ball-point pen to project the point for writing, what form of energy are you storing in the pen for later use in retracting the point?

18. When you pull down a roll-up window shade, one form of energy decreases and another increases. What are the two forms? Which form of energy changes most? Which form does work later to raise the shade?

Energy Transformations

19. Pole vaulters now use flexible fiberglass poles (Fig. 4.25) rather than the rigid poles once used. Why does a flexible pole help them transform their energy to the kind they want in order to win? What kind of energy is that?

Figure 4.25 Modern pole vaulters use a flexible pole.

20. A roller coaster has no engine or motor inside, but starts its run after being pulled to a high point on the track. Can it ever get that high again on its own? What gives it the energy that makes it go?

21. Why do you need to wind a pendulum clock? Doesn't the pendulum keep it going?

22. What various forms of energy are present as you shoot an arrow with a bow?

23. What eventually happens to the electrical energy supplied to operate an egg beater? an electric fan? an electric shaver?

24. Describe the energy interchanges involved as you jump on a trampoline.

Conservation of Energy

25. Does "conserving energy" by turning off unnecessary lights and using less hot water have anything to do with the principle of conservation of energy?

26. To describe conservation practices such as those mentioned in Question 25, would it be better to use the term *fuel conservation* than *energy conservation*?

27. Space shuttles are equipped with heat shields to protect them from the tremendous amount of heat developed when they reenter the earth's atmosphere. Where does this heat come from?

28. Why does it "burn" to catch a fast-moving baseball?

29. When a train moving at a speed of 80 km/h stops quickly, what happens to the kinetic energy it had while moving?

30. A car initially moving at a speed of 80 km/h coasts on a perfectly level road until it stops. What happens to the kinetic energy it had?

31. When a baseball player slides into second base, what happens to his kinetic energy?

32. An apple hanging on a tree suddenly falls. Just before it hits the ground, what has happened to the potential energy it had while hanging on the tree? What has happened to it after it hits?

33. Can you explain *why* the total amount of energy in an isolated system remains constant?

Other Energy Forms

34. What forms of energy are associated with hammering a nail into a board?

35. Why do some A.A.U. track records say "wind-assisted"? From an energy/work viewpoint, why is it harder to run against a wind than to run when there is no wind? Why is it harder to run uphill than on level ground?

36. The heater on most cars works by circulating the water and antifreeze (that normally go to the radiator) through the heater to give up heat inside the car. Usually, a fan blows air through the heater to speed delivery of this heat to the passengers. Do you get lower gas mileage when you use the heater on such a car?

37. A car equipped with radial tires gets better gas mileage than a car that has bias ply tires. What does this say about the amount of friction that opposes the rolling of a radial tire compared with the rolling of a bias ply tire?

38. Why is the wheel of an old-fashioned spinning wheel (Fig. 4.26) so large? What form of energy is stored in this wheel as it turns?

39. Studies show that when porpoises swim faster than 5 m/s, they use less body energy by soaring through the air rather than staying in the water. What does this say about the work required to move a porpoise through air compared with that to move it through water?

40. If the engine dies in a helicopter, it can land safely anyway if it is *high enough* in the air. The pilot lets the copter fall, setting the blade rotating as the air moves through it (as when you blow on a pinwheel). By reversing the pitch of the rotating blades—the angle at which the blades slice the air—to provide lift at just the right time, the craft can be slowed for landing. What are the energy interchanges involved in this maneuver?

Figure 4.26 An old-fashioned spinning wheel.

Power

41. If you push a lawn mower across a yard in 10 s, how does the work done compare with pushing it across the same yard in 20 s? How does the power compare in the two cases?

42. Contrary to what you might predict, if you stop up the inlet hose of a working vacuum cleaner, it takes less power to keep the cleaner running. Why? Would the same be true if you stopped up the air-output end? (If you aren't convinced of this power reduction, try to persuade your instructor to connect a vacuum cleaner via an electrical current meter, and check it out.)

Problems*

1. Suppose that in throwing a ball you exert an average force of 50 N on it for a distance of 2 m. How much work do you do on the ball?

   Answer: 100 J

2. If you exert this force (Problem 1) for a time of 0.10 s, what is the power used on the ball? What is this in horsepower?

   Answer: 1.0 kW; 1.3 hp

3. Suppose you use a 200-N force to slide a large box 5 m across the floor in 20 s. You then use a 100-N force to slide a small box 20 m across the floor in 40 s. Which of these efforts takes more work?

   Answer: moving the smaller box (takes twice as much work)

4. Which of the jobs in Problem 3 takes more power?

   Answer: both jobs take the same amount

5. If you throw a ball and it leaves your hand with a kinetic energy of 100 J, how much work did the force of your hand do on the ball?

   Answer: 100 J

6. Based on the data in Figure 4.21, how much frictional force exists between the tires and the road when a 1000-kg car moving at a speed of 65 km/h (40 mi/h) is stopped in the shortest possible distance? (Note: 1 m = 3.3 ft.)

   Answer: 6000 N (1300 lb)

7. Riding in a car at 80 km/h (22 m/s), you collide head-on with an identical car moving in the opposite direction at the same speed. What was the kinetic energy of your body just before the collision? If it stops within a distance of 1 m during the collision, what average force will your body feel?

   Answer: (for a 60-kg body): 14,000 J; 14,000 N (3200 lb)

8. If your car traveling 50 km/h will stop in 15 m after you hit the brakes, what distance will it need to stop at 150 km/h?

   Answer: 135 m

9. Henri La Mothe holds the record for the highest shallow dive. He dived 12 m (40 ft) from the Flatiron Building in New York City into 0.32 m ($12\frac{1}{2}$ in) of water. What was his kinetic energy when he hit the water? What average net force was exerted on him as he stopped? (Assume his mass to be 70 kg.)

   Answer: 8400 J; 26,000 N (5900 lb)

10. A brick mason does 40 J of work in lifting a brick to chest height. How much does she increase the gravitational potential energy of the brick? If she then accidentally drops it on her toe, how much kinetic energy does it have when it hits her toe?

    Answer: 40 J; 40 J

11. If a sprinter running at a speed of 10 m/s could convert all his kinetic energy into upward motion, how high could he jump?

    Answer: 5.1 m

12. From what height would a car have to be dropped to gain the same kinetic energy it would have if it were moving horizontally at 95 km/h (59 mi/h)?

    Answer: 35 m

Several of these problems use metric prefixes (k, M, or G) with the units. Refer to Appendix III-A for a summary of these prefixes.

13. If you lose 1500 J of potential energy in sliding down a sliding board, and you hit the ground with a kinetic energy of 1000 J, how much heat was generated in your posterior and the sliding board?

    Answer: 500 J

14. A 50-kg stewardess slides down the emergency chute from a flaming airliner. If the top of the chute is 4 m above ground, and friction is negligible, with what speed does she hit the ground? Would her speed be different if she "weighed" 60 kg?

    Answer: 8.9 m/s; no

15. A 1000-kg car parked at the top of a hill loses its brakes and rolls down the hill. Assuming no friction or other opposing force, with what speed does the car get to the bottom of the hill, an elevation of 50 m lower than the top?

    Answer: 31 m/s (70 mi/h)

16. If the hill in Problem 15 is 150 m long as measured along the road, and an average frictional force of 1200 N opposes the motion, with what speed does the car reach the bottom of the hill?

    Answer: 25 m/s

17. Lucy Van Pelt slides down a sliding board whose top is 2 m from the ground. If her mass is 20 kg and she hits the ground at a speed of 5 m/s, how much heat is produced during her slide?

    Answer: 140 J

18. If Lucy's sliding board (Problem 17) is 3 m long, what is the average frictional force slowing her motion?

    Answer: 47 N

19. If the frictional force opposing the rolling of the box up the ramp of Figure 4.13 is 20 N, what is the efficiency of the ramp and rollers? Assume the 400 N includes the weight of the roller system, and recall that we found F to be 120 N with no friction.

    Answer: 86%

20. A liter of water loses 100 J of potential energy in falling 10 m. If 2000 liters of water pass over a 10-m-high dam and into a water turbine per second, what is the mechanical power output of the turbine? Assume the system to be 60% efficient at using the water's potential energy.

    Answer: 120 kW

21. Complete combustion of fuel oil releases about 40 MJ of energy per liter of oil. If the Jones family used 1200 liters of oil to supply 27 GJ of heat to their house last winter, what was the efficiency of their heating system?

    Answer: 56%

22. A particular power plant converts 35% of the chemical energy in the oil it burns to electrical energy delivered to a house. If this energy is used to heat the house with electrical heaters that are 98% efficient in converting electrical energy to heat, what is the overall efficiency of this method of burning oil to produce heat?

    Answer: 34%

23. A typical large city might produce 1 billion kg of garbage each year. If half of this is burned for producing heat to generate electricity, how much electrical energy in kWh is produced by a 32% efficient power plant if 1 kg of garbage releases 10 million J of energy when burned? If this energy is sold at 9¢ per kWh, what is the gross yearly income from the plant?

    Answer: 440 million kWh; $40 million

24. If running up a flight of stairs increases your gravitational potential energy by 2100 J, and you make it in 3 seconds, what is your horsepower?

    Answer: 0.94 hp

25. Show that 1 kWh = 3.6×10^6 J.

26. How many 5-kg (11-lb) bags of sugar would you have to lift from the floor onto a 1-m-high table in 1 minute to use energy at the same rate as it is used in a 100-W light bulb?

    Answer: 120

27. Lifting your suitcase to the top of the car for returning to campus might take about 300 J of work. If your body could use *all* the energy in

a piece of boysenberry pie (about 200 Calories*, or 800 kJ) to do this kind of work, how many suitcases could you lift to the top of the car on one piece of pie?

Answer: 2700

28. The average American eats about 3000 Calories* (12.5 MJ) in one day. If all this energy could be converted to mechanical work, how many 25-kg sacks of potatoes could this average person lift from the ground to a 1-m-high truck in a day? Since no one can do this much mechanical work, where does the rest of this energy go?

    Answer: 50,000; heat

29. A commuter train and 300 passengers might have a kinetic energy at full speed of 16 MJ ($M = 10^6$). If electrical energy to operate the train costs 9¢/kWh, how much does it cost to get this train up to full speed? Assume 100% efficiency in use of the electrical energy. (See Problem 25).

    Answer: 40¢

30. Suppose a train such as the one of Problem 29 could store in flywheels and reuse $\frac{1}{3}$ the energy dissipated in stopping. Estimate how much money New York City might spend equipping a train with flywheels and still save money. Assume, for example, the train lasts five years, and estimate the number of stops it might make per day.

    Answer: At 500 stops/day, $122,000

31. A truckload of eggs has 60 MJ more gravitational potential energy at the top of a 600-m-

high mountain than at the bottom. How much heat is generated in the brakes of the truck if it maintains a constant speed going down the mountain by using the brakes only? It takes about 500 kJ of energy to heat a gallon of water from 13°C (55°F) to 43°C (110°F). At 20 gallons per bath, how many baths could you take on the amount of heat generated in one trip down the mountain of the truckload of eggs?

    Answer: 60 MJ; 6

32. Springs and other elastic bodies return to their original shape when the distorting force is removed. The distance x that a spring stretches (or compresses), is proportional to the applied force F. Thus, $F = kx$, where k is called the spring constant. The *average* force needed to stretch (or compress) the spring a distance x is then $\frac{1}{2}$ the force *at* distance x. Show that the elastic potential energy of the spring, which equals the work done in deforming it, is $\frac{1}{2}kx^2$.

33. Use the result of Problem 32. A spring with spring constant 20,000 N/m is compressed a distance of 15 cm. How much work is done during this compression?

    Answer: 220 J

34. Suppose a mass of 1 kg is placed on the spring of Problem 33. If the spring is compressed vertically 10 cm (0.10 m) by an outside force and then released quickly, how far will the 1-kg mass fly above the point of release? (Hint: Use conservation of energy.)

    Answer: 10 m

Home Experiments

1. Have two friends hold a bed sheet in an approximately vertical plane by gripping the four corners. Stand in front of the sheet and throw a raw egg at it as hard as you can. In terms of the work-energy theorem, explain why the egg does not break when it hits the sheet. (Have

your friends hold the bottom corners up somewhat to form a valley in the sheet to catch the egg. Otherwise, the egg will fall to the floor and break.)

2. In a large empty parking lot or other area with no traffic, find out how a car's braking distance changes with speed. Start with a low speed, then double and triple it. Put on the brakes at

The nutritionist's Calorie (with capital C) is the kilocalorie (1000 cal) used by physical scientists, and is a unit of energy equivalent to 4190 J.

the same spot each time so that it will be easy to compare distances. Assuming the same stopping force each time, the work done in stopping the car is proportional to the stopping distance. Show that the kinetic energy is proportional to the square of the speed.

3. Drop a new tennis ball on various surfaces and measure the height of rebound relative to the initial height. What fraction of the energy goes to heat? Where is most of the heat produced— inside the ball or inside the surface material? Repeat using an old "dead" ball. Is the amount of heat produced *inside the ball* any different for the dead ball?

References for Further Reading

Johnston, W. D., Jr., "The Prospects of Photovoltaic Conversion," *American Scientist*, Nov.–Dec. 1977, p. 729. Discusses the factors involved in converting solar energy directly to electrical energy.

Lin, H., "Newtonian Mechanics and the Human Body: Some Estimates of Performance," *American Journal of Physics*, Jan., 1978, p. 15. On the basis of energy considerations and other basic physics, the author estimates the maximum human performance for field and track events, as well as certain everyday tasks. He compares these estimates with actual performance.

Marion, J. B., and M. L. Roush, *Energy in Perspective*, Second Ed. (Academic Press, New York, 1982). A discussion concerning all aspects of energy in useable form, including sources, relative consumption for various purposes, technology of supplying useable energy, problems, and future projections.

Post, R. F., and S. F. Post, "Flywheels," *Scientific American*, Dec. 1973, p. 17. Discusses the modern techniques used in one portable method of energy storage.

Romer, R. H., *Energy Facts and Figures* (Spring Street Press, Amherst, Mass., 1985).

Scott, D., "Flexible Rafts Harness Wave Power," *Popular Science*, Sept. 1978, p. 94. Describes an interesting approach to generating electricity by an unusual source. Every month for the past several years *Popular Science* has had at least one article concerning what individuals can do to conserve or provide useable energy, and usually at least one article on large-scale methods being used or tried to supply energy.

Starr, C., "Energy and Power," *Scientific American*, Sept. 1971, p. 36. A history of the technological use of energy and a discussion of present and future alternatives in this area.

MOMENTUM

When a fleet-footed running back pauses briefly, then sidesteps as the fast-moving linebacker approaches, the would-be tackler helplessiy passes on by, unable to change his course. Or, when you're driving your car along a gradual uphill slope and the engine dies, the car doesn't stop immediately—it coasts a distance up the hill. In either of these cases, is there a force acting on the body to keep it going? No. These bodies keep moving because they have what we call **momentum**. A body can change its momentum only when an external force acts on it.

Newton's first law of motion tells us that a body tends to continue in motion in the direction it is going. Momentum measures this tendency by taking into account the two properties that determine it—mass and velocity.

We define another quantity that is related to momentum—namely, *impulse*. We will see that impulse is especially useful in determining the effect of forces that act on a body for short periods of time.

In this chapter we introduce another important conservation principle: *the principle of conservation of momentum*. Finally, we'll use this principle to analyze the effects of collisions on bodies, to understand rocket propulsion, and to gain insight into a number of other practical situations.

5.1 WHAT IS MOMENTUM?

We define momentum as the *product of mass and velocity*:

$$\text{momentum} = \text{mass} \times \text{velocity}$$

$$= mv$$

A 7-kg bowling ball moving at a speed of 4 m/s has a momentum mv of 7 kg × 4 m/s, or 28 kg·m/s. But a 0.15-kg baseball would have to move at the unreasonably high speed of 187 m/s (418 mi/h) to have the same momentum (Fig. 5.1).

. . . .

EXERCISE   If you make a quick dash to get out of the rain at a speed of 9.0 m/s, what is your momentum?   *Answer*: (for 60 kg) 540 kg·m/s

Momentum = mv = 28 kg·m/s

m = 0.15 kg

v = 187 m/s

Momentum = mv = 28 kg·m/s

Figure 5.1   The lower-mass baseball must have an unreasonably high velocity to have the same momentum as a slowly moving bowling ball.

Momentum and Newton's Second Law

Our statement of Newton's second law of motion in Chapter 2 was not quite in the form that Newton himself used. Using our definition of momentum, Newton's original statement of the law was

net force = *rate of change* of momentum

$$F = \frac{\text{change in } mv}{t}$$

If m doesn't change, then

change in $mv = m \times$ *change* in v

The second law then becomes

$$F = \frac{m \times \text{change in } v}{t} = m \times \frac{\text{change in } v}{t} = m \times a$$

or

$$F = ma$$

as we stated the law previously. For most situations, the mass does not change, and thus, the two statements of the law are equivalent.

5.2 IMPULSE

Momentum tells us how much effort it takes to stop a moving body or to give it a certain velocity if it is at rest. That "effort" depends on two factors: How much force you apply and how long you apply it. Here's an example.

Suppose your car stalls on Main Street and you have to get out and push it down this level street to a parking area (Fig. 5.2–a). If you can push with a continuous 200-N force, you might get the car up to a speed of 10 km/h in a time of 30 s. However, if two of your friends come to your rescue (Fig. 5.2–b), you might expect that the three of you each pushing with a continuous 200-N force could get the car to the same speed in 10 s, or $\frac{1}{3}$ the time. The final momentum of the car is the same in both cases.

We can use Newton's second law, stated in terms of momentum, to show that the expectation that three times the force for $\frac{1}{3}$ as long does indeed give the same momentum. Multiplying both sides of the equation for this law by t gives

$Ft =$ *change* in *mv*

The product of the force applied and the time of application is what we call **impulse**. Thus

$$\text{impulse} = \text{force} \times \text{time}$$
$$= Ft$$

The equation $Ft =$ *change* in *mv* can then be stated in words:

$$\text{impulse} = \text{change in momentum}$$

Our car example (Fig. 5.2) is consistent with this equation: multiplying

$F = 200\,\text{N}$

(a) 10 km/h in 30 seconds

$F = 200\,\text{N}$

200 N

200 N

Figure 5.2 (a) If only one person pushes, it takes three times as long to reach some speed as (b) if three people push with the same force.

(b) 10 km/h in 10 seconds

the force by 3 means that the same change in momentum occurs in $\frac{1}{3}$ the time.

The analysis of a problem from the impulse-momentum viewpoint offers an alternative to applying Newton's second law. It is often easier to use the momentum approach for problems in which the force is applied for short periods of time, such as a ball being hit with a bat, two vehicles colliding, or a bullet being fired from a rifle.

In any case, you can achieve a given change in momentum with a smaller force by increasing the length of time it acts, and vice versa. You can increase the momentum change by increasing either the force, the time, or both. Here are a few everyday situations that you can analyze on the basis of impulse and momentum. You might at the same time analyze them from a work–energy viewpoint, and see if you reach the same conclusion.

1. Why are you less likely to do serious injury to yourself and your car if you crash into a collapsible guardrail rather than a concrete bridge support?

2. Whether you crash into the guardrail or concrete support, why are your survival chances better if you are wearing seat belts or if your car has inflatable air bags?

3. Why does a safety net under trapeze performers reduce the risk of injury in a fall?

4. If you jump from a high wall, why do you bend your knees when you land rather than hitting with your legs stiff?

5. Why can you safely jump from a much higher point into water than onto dry land?

· ·

EXAMPLE 5.1   A 70-kg diver hits the water with a downward velocity of 10 m/s, after diving from a 5-m-high platform. If his downward motion stops 0.7 s after hitting the water, what is the average net upward force on him as he is being stopped?

Impulse = change in momentum

or

Ft = change in mv

Since the diver's velocity goes to zero,

$$\text{change in momentum} = \text{initial momentum}$$

$$= mv$$

$$= (70 \text{ kg})(10 \text{ m/s}) = 700 \text{ kg·m/s}$$

Therefore,

$$F \times 0.7 \text{ s} = 700 \text{ kg·m/s}$$

or

$$F = 1000 \text{ N}$$

· ·

EXERCISE   A golfer hits the 46-g ball so that it leaves the club at a speed of 60 m/s. If the club and ball are in contact for 6.0×10^{-4} s, what is the average force the club exerts on the ball?   *Answer*: 4600 N

You may find the answer in this exercise surprisingly high. It's about 10,000 times the weight of the ball! The force must be so large because the contact time is extremely short. If you were to throw the ball instead—and you could easily throw it at about one-fourth this speed—you certainly would not exert anywhere near one-fourth this force. The impulse-momentum principle shows that you would then need a much smaller force because the time of contact is much longer.

5.3 CONSERVATION OF MOMENTUM

Imagine the following situation: Linda and Marvin are standing on roller skates facing each other. They put their hands together and push each other backwards (Fig. 5.3–a). Assume that the skates roll so easily that their resistance is negligible. Thus, the only horizontal forces on their bodies are from the push of their hands. Let's compare their individual momenta.

First notice that, from Newton's third law, the force that Marvin exerts on Linda, F_L, is exactly equal to the force Linda exerts on Marvin, F_M (though opposite in direction). Also, the time periods during which these two forces are exerted are exactly the same. Thus, the product *force × time*—the *impulse*—is exactly the same for both Linda and Marvin. From the impulse-momentum equation,

(a)

Linda   Marvin

(b)   Linda   Marvin

Figure 5.3 (a) Linda exerts force F_M on Marvin and he exerts force F_L on her. (b) As they move apart, Linda moves with speed v_L and Marvin, with speed v_M.

$$Ft = change \text{ in } mv$$

we see that the *change* in Linda's momentum must exactly equal the *change* in Marvin's. But since both started with zero momentum, they both must have the same *final* momentun after the push:

$$m_L v_L = m_M v_M$$

where m and v are the mass and velocity, and the subscripts refer to the particular person (Fig. 5.3–b).

If Linda's mass is 54 kg and Marvin's is 72 kg—hers, $\frac{3}{4}$ of his—then Marvin's speed is only $\frac{3}{4}$ of Linda's. The larger mass has the lower speed.

Momentum—a Vector Quantity

Velocity has a direction associated with it: An object cannot be moving without moving in *some* direction. Therefore momentum—mass × velocity—has a direction. Quantities that have both magnitude *and* direction are called **vectors**. We've encountered two other vector quantities so far: acceleration and force. Both of these have a direction as well as a magnitude.

When vectors act along a straight line, it's convenient to associate a positive value with the magnitude of the vector if it acts in one direction along the line, and a negative value if the vector acts in the other direction. For example, consider the force and velocity vectors in Figure 5.3. If we decide that the positive direction is to the right, then F_M and v_M have positive values, while F_L and v_L are negative.

This practice is very useful in analyzing momentum. Since v_L is in the negative direction, then Linda's momentum $m_L v_L$ is *negative*. Similarly, Marvin's momentum $m_M v_M$ is positive. But since these momenta are equal in size, the *total* momentum of Marvin and Linda adds to *zero*. Since their total momentum was zero before they pushed off—neither was moving—the total momentum has not changed. *The total momentum of the system has remained constant.*

The situation we have described for the system in Figure 5.3 holds for *any* system on which there is no net external force—that is, no net force on the system from outside. In fact, it is a general principle of physics *never* known to be violated: the **principle of conservation of momentum**. It says: *For a system on which no net external force acts, the total momentum remains constant.* Here "total" means the effective momentum of all the parts of the system. Notice that this principle is similar in wording to the conservation-of-energy principle. However, we are talking about a different concept here. Momentum and energy are clearly distinct quantities. Both turn out to be conserved in nature.

• •

EXAMPLE 5.2   Harold jumps from a stationary 35-kg canoe toward shore. If his mass is 70 kg and he leaves the canoe with a horizontal speed of 2.0 m/s relative to the water, with what speed does the canoe move away from shore? (see Fig. 5.4).

From conservation of momentum,

total momentum after jumping = total momentum before jumping

$$m_H v_H - m_c v_c = 0$$

$$(70 \text{ kg})(2.0 \text{ m/s}) - (35 \text{ kg})v_c = 0$$

Therefore,

$$v_c = 4.0 \text{ m/s}$$

Figure 5.4 As the person jumps one way, the canoe must move in the other direction in order to conserve momentum.

• •

QUESTION   If Harold in this example is accustomed to jumping from dry land, why might he be surprised not to make it to shore when he jumps?

___

5.4 APPLICATIONS OF MOMENTUM CONSERVATION

The conservation of momentum principle is often applied in comparing the state of motion of some system before and after a particular event in that system. Such an event might be a collision, an explosion, or a rapid expansion or contraction of the parts of the system. We can then restate the momentum conservation principle for such a situation: *The total momentum of the system after the event is equal to the total momentum before the event.*

We will consider a few examples of how the principle of conservation of momentum can be used to understand the motion of the parts of a system after some event occurs within it.

Billiard Ball Collisions

Suppose a moving billiard ball makes a direct head-on collision with another ball at rest on the table. If there is no english on the ball—that is, no special spin other than from rolling—what is the resulting motion of the balls? If you're a billiard player, you know that the first ball stops

Figure 5.5 A head-on collision between the cue ball with no "English" and the eight ball at rest.

completely and the second one moves off with the same speed the first one had. This situation is easy to understand from the conservation-of-momentum principle.

Figure 5.5 shows the cue and eight balls before and after a head-on collision. If the mass of each ball is m, the momentum before the collision is mv. Therefore, the momentum after the collision must be mv. This can happen if the cue ball stops, and the eight ball moves off with speed v, since the mass m of both balls is the same. If the ball has any spin other than from its normal rolling across the table, then other forces act during the collision, making the analysis more complicated.

You might have seen an apparatus such as the one in Figure 5.6 used either as a toy or as a physics classroom demonstration. The billiard ball collision we just discussed should give you some idea why it behaves as it does: If you pull one ball aside and release it, another ball will fly off from the other end after the collision. If you pull two aside and release them, two others will fly off from after the collision.

Both billiard balls and the hanging balls of this toy deform slightly during collision, but then spring back to their original shapes. As a result, any kinetic energy used to deform the balls when they come together is regained as they separate. Except for small frictional losses, the total kinetic energy after the collision equals the total kinetic energy before the collision. A collision that conserves kinetic energy, as well as momentum, is called an **elastic collision**.

Many types of bodies can have *nearly* elastic collisions: billiard balls, bowling balls, marbles, springs. Only atoms and subatomic particles can have truly elastic collisions. Our next example discusses an obviously *inelastic* collision.

Automobile Collisions

Suppose a 1000-kg car moving at 80 km/h (22 m/s) collides on slick ice with a stationary 10,000-kg truck and the two stick together. With what speed do they slide across the ice together after the collision (Fig. 5.7)? Friction is involved with the crunching of the vehicles' metal. Therefore, the total kinetic energy does not stay the same: The collision is inelastic. In fact, most of the kinetic energy is converted to heat during the collision. Since we can neglect the small frictional force exerted by the ice, momentum *is* conserved:

momentum before collision = momentum after collision

Figure 5.6 A demonstration of the conservation of momentum.

Before collision

After collision

Figure 5.7 An inelastic collision conserves momentum but not *kinetic* energy.

or

$$1000 \text{ kg} \times 22 \text{ m/s} = (1000 \text{ kg} + 10{,}000 \text{ kg}) \times v$$

where v is the final velocity of the stuck-together vehicles. We solve this equation and find this velocity to be 2.0 m/s (7.2 km/h). Calculate the kinetic energy of the wreckage immediately after the collision and that of the car before to prove that the kinetic energy decreases during the collision.

The Firing of a Gun

Consider the cannon shown in Figure 5.8. Before firing, the entire system—cannon, cannonball, and powder—are not moving and therefore have zero momentum. From conservation of momentum, we know that the total momentum after the firing must also be zero. That may seem strange, since you know the cannonball has momentum. The total momentum can still be zero because momentum has a direction associated with it—it is a vector. We associate a *positive* sign with momentum in one direction. Then momentum in the other direction is *negative*. Since the cannon recoils in the opposite direction to the motion of the cannonball, we must have

$$(\text{momentum of ball}) - (\text{momentum of cannon}) = 0$$

or

$$(\text{mass of ball}) \times (\text{ball velocity})$$
$$- (\text{mass of cannon}) \times (\text{recoil velocity}) = 0$$

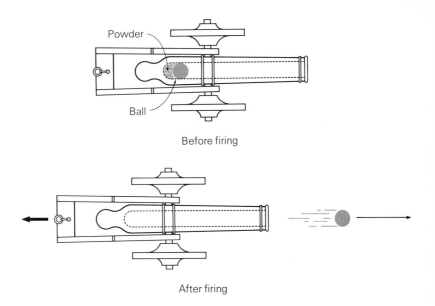

Powder

Ball

Before firing

After firing

Figure 5.8 The total momentum is zero both before and after the cannon is fired.

The mass of the cannon is greater than the mass of the ball; therefore the speed of the ball is greater than the recoil speed of the cannon.

The recoil effect just described for a cannon applies to the firing of any gun or rifle. For example, if you fire a high-powered rifle, you will feel a strong "kick" against your shoulder. This kick happens because the rifle must move backward with the same momentum the projectile has moving forward. The rifle, of course, has a much lower speed than the projectile because the rifle has much greater mass.

Let's look at this situation from the viewpoint of Newton's action-reaction law. The cannon and burned powder push forward on the ball, so the ball must push with the same force in the opposite direction on the cannon and burned powder. Since the force on each acts for the same period of time, the impulse-momentum relation tells us that the same amount of momentum is given to each, but in opposite directions.

Rocket Propulsion

Look again at the cannon of Figure 5.8. Suppose the cannon has a large supply of cannonballs stored inside it, and that it is capable of firing those balls in rapid succession, one immediately after the other. Each time a ball is fired to the right with a certain momentum, the cannon must gain that same amount of momentum to the left. If the balls are fired often enough and with enough velocity, the cannon must acquire a high velocity in the opposite direction. The cannon in effect becomes a rocket!

This action is exactly what happens in a real rocket. Figure 5.9 shows a simplified rocket diagram. Fuel is burned in the combustion chamber

Figure 5.9 The momentum of the rocket moving forward equals the momentum of the gases moving backward.

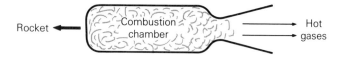

producing gases at extremely high pressure. These gases escape out the nozzle in the rear, so shaped that the gas particles exit with exceedingly high velocities. The momentum of the gas particles shooting out the rear (equivalent of cannon balls) must be matched exactly by the forward momentum of the rocket in the opposite direction. In this way, the rocket is propelled forward.

This process is also what happens when you blow up a balloon and turn it loose to fly around the room. The backward momentum of the air shooting out the rear must be matched by the gain in forward momentum of the balloon.

You may doubt that something as light as gas molecules could have enough momentum to balance the forward momentum of a massive rocket. Remember, though, that billions of billions of billions of gas molecules come out in a short time at speeds much higher than the speed of sound.

· ·

Key Concepts

Momentum is the product of mass and velocity:

Momentum = mv

Newton's second law in terms of momentum is:

$$F = \frac{change \text{ in } mv}{t}$$

Impulse = change in momentum

$Ft = change$ in mv

A *vector* is a quantity with both magnitude and direction. For a vector quantity, we associate a positive sign with its magnitude in one direction and a negative sign with its magnitude in the opposite direction. Examples of vectors are velocity, acceleration, force, and momentum.

The principle of **conservation of momentum**: *For a system on which no external force acts, the total momentum remains constant.*

Questions

Momentum

1. Which has more momentum: a 150-lb (mass 70 kg) sprinter running at 10 m/s or a 280-lb (mass 130 kg) tackle jogging at 5 m/s?

2. Doubling the speed of a car quadruples its kinetic energy. What does it do to the car's momentum?

3. Why does a firefighter need great strength to hold the nozzle of a hose squirting a high-speed stream of water?

Figure 5.10 The way to put a loose handle back into a hammer.

4. When the handle is coming out of your hammer, why should you hold it vertically as shown in Figure 5.10 and slam the handle on the floor rather than turning the handle up and slamming the handle into the hammer? How is your answer related to Newton's first law as well as to momentum?

5. Why does a long-jumper build up all the speed he can get before his jump, whereas a high-jumper runs fairly slowly?

Momentum and Impulse

6. Do you agree with the assertion in some physics books that "impulse = change in momentum" explains why follow-through is important in sports (hitting, kicking, or throwing a ball, for example)?

7. Some people claim that, in an accident, they are less likely to be injured if they are intoxicated than if they are sober. Can you think of a reason why this might be true?

8. Answer Questions 8 and 9 of Chapter 4 on the basis of impulse-momentum considerations.

Conservation of Momentum

9. If you drop a bowling ball, what happens to the momentum it had just before hitting the ground? Answer the same question for a Ping-Pong ball.

10. Charlie Brown is abandoned by his friends on perfectly slick ice (absolutely no friction). Is there any way for him to get off the ice?

11. Convince your roommate that having air behind a rocket or jet airplane for the expelled gases to push against has nothing to do with the propulsion. (For example, would the kick of a rifle be any different if it were fired in a vacuum?)

12. In 1854 an Englishman, J. Howard, long-jumped 29 ft, 7 in with the help of two 5-lb weights. He sprinted down the track with the weights held high, and threw them to the ground just as he launched himself. Explain what the weights did for him. (It took another 114 years before Bob Beamon broke the 29-ft mark without the aid of any weights.)

13. To avoid being knocked into the goal by a fast-moving soccer ball, goalkeepers need to jump slightly forward as they jump up to catch the ball. If ball and keeper are momentarily at rest after the catch, how must the horizontal momenta of the ball and keeper compare just before the catch?

14. If you're riding backwards on a skateboard and you jump off so that you hit the ground moving straight down, what must happen to the skateboard? How does its momentum compare before and after you jump?

15. A 7-kg bowling ball collides head on with the number-1 pin and gives it 10 kg·m/s of momentum. Can you tell from this information how much momentum the bowling ball had? Can you tell how much momentum the bowling ball lost during the collision?

16. In the movies, it is not unusual to see the bad guy knocked out a window as he is shot by the good guy with a hand gun. Yet the good guy never moves as he shoots. Are these effects consistent with the principle of momentum conservation?

Problems

1. What is the maximum momentum *you* can achieve in a 10-yard dash?

   Answer: (for 70 kg at 8 m/s) 560 kg·m/s

2. A cheetah, one of the world's fastest animals, might have a momentum at full speed of 800 kg·m/s. How fast would a 10,000-kg elephant have to run to have the same momentum?

   Answer: 0.08 m/s (or about 0.2 miles/hour)

3. A 0.60-kg basketball moving at a speed of 10 m/s collides with the rim of the basket and rebounds at 8 m/s in the opposite direction. If the ball is in contact with the rim for 0.02 s, what force does it exert on the rim?

   Answer: 540 N

4. A 20-kg child jumps from a third floor window 6.0 m above ground into her father's arms. At what speed does the child reach her father's arms, which are 2.0 m off the ground? If the father stops the child's motion in 0.20 s, what force does he exert on her (the net force on her plus her weight)?

   Answer: 8.9 m/s; 1080 N

5. A soccer player kicks the 0.40-kg ball with an average force of 1200 N and the ball leaves her foot at a speed of 25 m/s. How long is her foot in contact with the ball?

   Answer: 0.0083 s

6. Neglecting the impulse given *to* the leg by the player during this 0.008-s interval in Problem 5, how much momentum is lost by the player's foot and leg?

   Answer: 10 kg·m/s

7. You throw a 0.20-kg glob of putty with a speed of 20 m/s at a wall. If it stops in 0.0020 s and sticks to the wall, what average force does it exert on the wall? If, instead of the putty, you throw a very elastic ball of the same mass and it bounces back with the same speed, what force does it exert? Assume the same interaction time.

   Answer: 2000 N; 4000 N

8. Snoopy, whose mass is 5 kg, runs at a speed of 7 m/s and jumps onto his 2-kg skateboard. How fast do he and skateboard move?

   Answer: 5 m/s

9. A 1000-kg car traveling at 25 mi/h (11 m/s) collides from the rear with a 1500-kg car stopped on a level road without its brakes on. With what speed is the wreckage moving immediately after the collision?

   Answer: 4.4 m/s

10. You're stuck in the middle of a *perfectly frictionless* frozen pond 10 m across. You tilt your mouth upward, take a breath, and then blow it out horizontally. If you blow out 0.60 g of air at an average horizontal speed of 0.50 m/s, how long will it take you to reach the edge of the pond if no outside horizontal forces act on you?

    Answer: (for a 60-kg person) 280 h

11. In Question 13, suppose the 80-kg goalkeeper catches a 0.40-kg ball moving at 30 m/s. To meet the conditions described, what must be the keeper's forward speed?

    Answer: 0.15 m/s

12. Jesse James sets off an explosion between the last car of the train, which is carrying the gold, and the rest of the train. If the train is moving at 15 m/s and the explosion completely stops the last car, with what speed must the rest of the train be moving afterwards? Take the mass of the rest of the train to be 3 times that of the last car, and assume it stays intact.

    Answer: 20 m/s

13. A billiard ball collides head-on with another that is at rest. Momentum would be conserved if both balls moved off with $\frac{1}{2}$ the velocity of the incoming ball and in the same direction. Show that kinetic energy would *not* be conserved in this situation. You can assume a mass and velocity, or you can use symbols.

Home Experiments

1. Make a vertical stack of several identical coins on a smooth table. Lay another such coin on the table several centimeters away. Hit this extra coin with your finger so that it slides across and hits the bottom of the stack. If your aim is good, the extra coin will exactly replace the bottom one in the stack. Explain in terms of conservation of momentum.

2. The next time you go to a football game, study the effects of momentum conservation on the players and the ball.

Reference for Further Reading

Field, M. S., R. E. McNair, and S. R. Wilk, "The Physics of Karate," *Scientific American*, Vol. 240, 1979, p. 150.

R O T A T I O N A L E F F E C T S

Since the invention of the wheel, rotation has had a major impact on humankind. Think about the different wheels that influence your life. Probably your first thoughts turn to transportation. But if you were to take the cover off the working parts of your analog watch, your refrigerator, your hair drier, or your favorite stereo sound system—to name just a few—you would find an abundance of wheels, pulleys, shafts, and disks that turn as the device operates. Understanding rotation is an important part of understanding our world and our technological society.

Rotation is not limited to wheels, disks, and other wheel-like elements. Almost all tools—from pliers to fingernail clippers—operate as the parts rotate about some pivot point.

As we look at rotational effects in this chapter, we'll focus on two main categories: machines whose parts function through rotation and rigid bodies that rotate freely under the action of turning effects called *torques*.

6.1 SIMPLE MACHINES

In today's language the word *machine* can mean anything from a garden hoe to a digital computer. We will talk in this chapter only about mechanical machines, but even these can be as simple as a screwdriver or as complex as an automobile. Most mechanical devices are based on the

Figure 6.1 The work done by the plank on the piano equals the work done by the person on the plank.

idea that a small force moving through a large distance can do the same amount of work as a large force moving through a small distance. In this section we'll see how this application of the principle of conservation of energy makes some kinds of work much easier.

Did you ever have to move a piano? If so, you probably needed to lift it first, in order to put casters under its legs. If you couldn't get help, you might have raised one end at a time by using a plank as a lever (Fig. 6.1). As you exerted a fairly small force downward on the plank, it exerted a much larger force upward on the piano. Thus, the plank served as a "force magnifier." This might seem to violate the principle of conservation of energy but, as we'll see, it does not.

Let's look at the situation more closely with the help of Figure 6.2. You exert a force f downward at the right-hand end, which moves a distance D. The force F upward at the left-hand end moves only a distance d, much shorter than D. Since the lever is not a source of energy, the work done by f on the right must equal the work done by F on the left. (We are neglecting the small amount of friction between the lever and the objects it touches.) Thus,

work on left = work on right

From the defintion of work—force parallel to the motion times the distance through which it acts—we have

$$F \times d = f \times D$$

Since d is much less than D, F must be much greater than f for this equation to hold. Therefore, the "gain" in force results from a "loss" in distance—the larger force moves through a much smaller distance. The closer you move the fulcrum, or pivot point, to the left-hand end, the larger the force F you can exert for a given applied force f; however, the distance d decreases accordingly.

For example, if the piano weighs 2000 N (200-kg mass), you need to exert a force of 1000 N to lift one end. Suppose your plank is 240 cm long and you place the pivot 40 cm from the piano. The length of plank to the right of the pivot is five times that to the left; therefore, for each

Figure 6.2 Forces on, and distances moved by, each end of a lever.

Fulcrum

centimeter moved by the left-hand end, the right-hand end moves 5 cm. By balancing the work on the two ends, we obtain

$$1000 \text{ N} \times 0.01 \text{ m} = f \times 0.05 \text{ m}$$

Solving for f, you find that the downward force you need is 200 N (a force equal to the weight of 20 kg), only $\frac{1}{5}$ the weight of one end of the piano.

In summary, the plank functions in two ways. First, it serves as a "transfer agent" that transfers work from your hands to the piano. Second, it gives you an advantage, letting you raise the piano with a much smaller force than you would need if you had to lift the piano directly. You pay for this smaller force by having to move it a larger distance. If you don't need to move a piano, try lifting some small objects using a meter stick or other short stick. Check out the effect of different locations and different heights for the support point that serves as a fulcrum.

In this discussion we have used the symbol f for the weaker force and F for the stronger force, d for the shorter distance and D for the longer distance. *Throughout this chapter*, whenever we contrast smaller and larger values of the same physical property, we will use capital letters for the larger values and lowercase letters for the smaller values.

You have probably used a claw hammer to pull out a nail. The head serves as a built-in fulcrum (Fig. 6.3). A small force f from your hand causes the hammer to pull with a much larger force F on the nail. You, of course, have to move your hand much farther than the nail moves. Figure 6.4 shows several other examples of ordinary tools that magnify

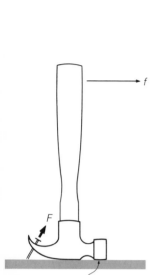

Figure 6.3 The claw hammer is just a lever when it's used to pull a nail.

Figure 6.4 Devices that exert a large force *F* over a small distance as the result of applying a small force *f* over a long distance: (a) nutcracker, (b) pliers, (c) wrench, (d) wheelbarrow, (e) bottle opener.

force at the expense of a larger distance. For each one, f represents the smaller force applied by the hand, whereas F represents the larger force exerted by the device. The fulcrum, or pivot point, which is shown for each device, is often—(a), (d), and (e)—at the end of the "lever." The force F acts always closer to the fulcrum than f, so that the distance F moves is less than the distance f moves. The nutcracker has a fulcrum for each arm.

Analyze each of these devices in terms of the applied and resulting forces and the work done by each. The wrench of Figure 6.4(c) may seem different from the other machines. In order to tighten the nut, you need to apply forces F to turn it as shown. These forces move in a circular path with a radius approximately equal to the "average radius" of the nut. Since the hand pushing with force f moves in a much larger circle in doing the same amount of work, the hand's force can be much smaller than that required at the nut itself.

In all our examples so far, the device supplies a resulting force larger

Figure 6.5 Devices that exert a small force f over a long distance as the result of applying a large force F over a short distance: (a) shovel, (b) oar of rowboat, (c) human forearm.

than the force the user applies to it. Sometimes, however, the need is not so much for a *larger* force but for one that can move a *longer distance*. If that's true, we may be willing to apply more force to the device than we get out of it in return for being able to move the weaker force over a longer distance. Examples of this are certain uses of the shovel and the oar of a rowboat. (See Figs. 6.5–a and 6.5–b.)

A person shoveling dirt from a loose pile often holds the end of the shovel handle almost stationary while exerting a force about midway up the handle. This grasp pivots the shovel about the end of the handle. The force on the dirt is less than the force you exert on the handle, but acts over a larger distance, thereby moving the dirt farther and faster than by hand. The work done on the dirt by the force *f* is equal to the work done by the shoveler's hand exerting force *F* on the handle. The rower of a boat wants the end of the oar in the water to move farther and faster than he can move his hand. This process can happen only if the force *f* on the water is less than the force *F* with which the rower pulls on the oar. Thus, more oar must be outside the boat than inside.

An example even closer to home is your own forearm (Fig. 6.5–c). The biceps muscle is attached typically about 5 cm from the elbow. If the palm of the hand is 35 cm from the elbow, the biceps muscle must apply a force of 154 N (35 lb) to lift a 22-N (5-lb) bag of sugar.

We often intuitively do things in a way that uses as little force as possible. We pay for this small force by a longer distance of movement. For example, when you open a heavy department store door, you normally would exert a force such as *f* in Figure 6.6 near the edge *opposite* the hinge. The door could also be opened by pushing with a force *F* closer to the hinge. The same amount of work is done (or energy expended) by either *f* or *F*, but *F* does this work in moving the shorter distance *d*; thus, *F* must be larger than *f*. Specifically, if *f* is exerted near the edge, but *F* near the midpoint, then *F* must be twice as large as *f*.

Figure 6.6 If you push on the door in the middle to open it, you have to push twice as hard as on the edge opposite the hinge.

6.2 TORQUE

Another way of describing the action of the simple lever-type machines we have talked about is to use a quantity called **torque**. We can use the lever shown in Figures 6.1 and 6.2 to define this term. We'll redraw this diagram (as in Fig. 6.7) to indicate the perpendicular distances from the fulcrum to the lines along which the forces act—their *lines of action*.

The symbols *l* and *L* represent the smaller and larger of these distances, respectively. Each such perpendicular distance from the line of action of the force to the fulcrum, or pivot point, is defined as the **lever arm** for that force. (This distance will actually vary somewhat as the lever tilts. However, if the lever doesn't rotate far from a horizontal position, this change is relatively small and negligible for our purposes.)

Figure 6.7 The lever of Figure 6.2, showing the lever arm for each force.

We can now define torque. *Torque is the product of a force and its lever arm.* That is,

torque = force × (lever arm)

= force × (perpendicular distance from fulcrum to line of action)

For the two forces shown in Figure 6.7,

torque on left = $F \times l$

and

torque on right = $f \times L$

Figure 6.8 The force *F* applied at point *e* has the largest turning effect because it has the longest lever arm. The lever arms for points *a* and *b* are zero. Those for *c*, *d*, and *e* are shown as l_c, l_d, and l_e.

You may ask: Why do we need to define a quantity such as torque anyway? To answer, suppose that you're using a wrench to loosen a very tight nut, as in Figure 6.8. You certainly must apply a force to the wrench. But there are many positions and directions to apply it, a few of which are shown. If you apply it as shown at *a* and *b*, you will certainly not make any headway. Applying the same force at *c* or *d* is somewhat better, but position *e* is the most likely to be successful. Why? The force is the same every time—the lever arm is what makes the difference. The lever arm is zero for positions *a* and *b*, but for position *e*, its value l_e is larger than for any other points and lines of force shown. For position *d*, the force is applied farther down the wrench than for *e*, but because of the angle of the line of force, its lever arm is about the same as for *c*.

The ability to turn the nut depends equally on the force and the lever arm. That is, it depends on the *torque*. Torque, then, is a measure of the **turning effect** of the applied force. The greater the torque, the greater the turning effect.

To start anything rotating, you need to apply a torque to it: to turn a doorknob, you apply a torque about its axis; to dial a telephone, you apply a torque around the center of the dial; to spin a plate on his finger, a juggler applies a torque about his finger supporting the plate by exerting a force at the rim. (See Fig. 6.9.)

Now let's go back to the torques on the lever of Figure 6.7. Is there some relationship between the torque on the left and that on the right? We pointed out earlier that the work done by the two forces *F* and *f* must be the same, or

work = $F \times d = f \times D$

Figure 6.9 A juggler starts a plate rotating by applying a torque to it about his finger.

But a definite relationship exists between the lever arms and the distances the ends of the lever move. (We used this relationship in finding the force necessary to lift the piano.) We can prove geometrically that the distance moved by the large force is in the same ratio to its lever arm as the distance moved by the small force is to its lever arm. That is,

$$\frac{\text{distance large force acts}}{\text{its lever arm}} = \frac{\text{distance small force acts}}{\text{its lever arm}}$$

or

$$\frac{d}{l} = \frac{D}{L}$$

(You can also test the equation by measuring the four distances on the figure.)

By using this relationship in our equation for the work on each end of the lever, we get, after a few algebraic manipulations,*

$$F \times l = f \times L$$

In words, this equation is:

(large force) × (its lever arm)

= (small force) × (its lever arm)

But from our defintion of torque,

torque = force × (lever arm)

which means that

torque on left = torque on right

This relationship, which comes from the equality of work done by each force, is sometimes easier to use than the work relationship. The reason is that the lever arm is known from the shape and size of the device, but the distances moved depend on the way the device is used.

All the devices discussed in the last section (Figs. 6.1 through 6.6) may be described just as well in terms of torques as in terms of work. As you analyze each device, keep in mind that the lever arm for each force is the perpendicular distance from the line of action of that force to the pivot point. Then the torque that the user applies to the device must equal the torque of the resulting force. Our piano example can be

---

Divide $F \times d$ by d/l and divide $f \times D$ by D/L (which is equal to d/l). Then d cancels on the left and D cancels on the right; l and L come to the numerator to give the desired equation.

solved directly in terms of torques by equating the torques on the two ends of the plank:

$$1000 \text{ N} \times 0.40 \text{ m} = f \times 2.0 \text{ m}$$

When we solve this equation for f, we find, as before, that a 200-N force is needed at the right end of the plank.

Balanced Torques

Some devices, for example, the juggler's plate (Fig. 6.9), rotate freely when a torque is applied and rotate faster and faster as long as the torque acts. For the lever devices we've been considering, there are balancing torques so that the amount of actual rotation is either limited or non-existent. When you lift the piano and hold it still (Fig. 6.1), you apply a clockwise torque to the lever. The piano in turn applies an equal torque *to the lever* in a counterclockwise sense. As a result, there is no net rotation of the lever. There is a slight unbalance in the torques as the motion starts and stops, but this unbalance is negligible in analyzing the forces and torques.

If you use a wheelbarrow (Fig. 6.4–d), the counterclockwise torque you apply is balanced by one in the clockwise sense due to the weight of the load. For any device of the type we've been discussing, the sum of the counterclockwise torques on the device must equal the sum of the clockwise torques.

For smooth seesawing, the torques from the weights of the two see-sawers must balance. If you want to seesaw with someone twice your weight, you need to be twice as far from the pivot point as that heavier person in order for the torques on the two ends to balance (Fig. 6.10).

Figure 6.10 For proper seesawing, the torque on both ends should be the same size, but in the oppostie sense.

$W = 2 \times w$

l

$L = 2 \times l$

EXERCISE

(a)

(b)

Figure 6.11 (a) The center of gravity of a symmetrical body is at its geometric center. (b) For an asymmetrical body, the center of gravity is closer to the heavier part.

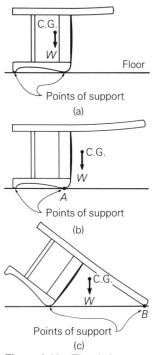

(a)

(b)

(c)

Figure 6.12 The chairs are stable in parts (a) and (c), but not in (b). C.G. is the center of gravity and W is the weight of the chair.

If one seesawer 2 m (6.5 ft) from the pivot point "weighs" 60 kg (132 lb) and the other "weighs" 80 kg (176 lb), how far should the heavier person sit from the pivot? What can you say about the work done by each in moving up and down? What about their changes in potential energy? Why is the word *weighs* in quotation marks?   *Answer*: 1.5 m

Center of Gravity

If, using one hand, you pick up a barbell with equal weights on each end, you will have to lift at the center if the barbell is to stay balanced (Fig. 6.11–a). If there is more weight on one end than on the other (Fig. 6.11–b), you will need to lift closer to the heavier end for balance. In either case there is one point, called the **center of gravity**, through which all the weight may be considered to act. When you lift at that point, the object is balanced, whatever its orientation, and it will not tend to rotate when you lift.

The center of gravity of a symmetrical object—a sphere, a cube, a barbell with identical weights at each end—is always at the geometric center. For an asymmetrical body—a chair, an axe, an unbalanced barbell—the center of gravity is near the heavier parts.

We may also define a *center of mass* for a body as the point at which all the mass can be considered concentrated. If you apply a force to an irregularly shaped body, parts of it may follow a wiggly path, but the center of mass will move according to Newton's second law of motion. Except for very tall objects, the center of mass and center of gravity are at the same point. Why should these two points ever be different?

Torques due to gravity. Several of the torques we described earlier depend on the force of gravity: the weight of the piano, the weight of the dirt in the wheelbarrow, the weight of the seesawers. When the weight of a body is the force providing a torque, the line of action of the force is downward through the center of gravity. Sometimes the weight of the lever itself provides enough torque to influence the other forces needed.

The location of the center of gravity relative to where a body is supported determines whether or not a body can rest stably without toppling over. In Figure 6.12(a), the center of gravity of the chair is located so that the weight acts downward between the points of support. There is no net torque tending to rotate the chair. For the high-backed chair of Figure 6.12(b), the center of gravity is located so that the weight acts outside the points of support. The weight gives a net torque that rotates the chair about point *A*. When the chair back hits the floor, as in Figure 6.12(c), the force from the floor at point *B* provides a torque that prevents further rotation.

Figure 6.13 (a) The equal arm analytical balance; (b) the unequal arm balance.

(a)

(b)

QUESTION If you stand with your heels against a vertical wall, why can you not touch your toes, no matter how flexible you are? Try it!

Figure 6.13 shows a couple of devices you might have used (maybe in chemistry lab) that depend for their operation on equal torques in opposite senses. Platform scales used by many doctors are basically the same as in (b). The interconnecting linkages, however, are somewhat more complicated.

6.3 APPLICATIONS TO COMPLEX MACHINES

Figure 6.14 The teeth of the right-hand gear exert a force F downward on the teeth of the left-hand gear. The left-hand teeth exert an equal force upward on the right-hand ones.

The principles in Sections 6.1 and 6.2 lay the foundation for all machines that transmit mechanical motion from one location to another, and that transform this motion into its most useful form. Examples of some complex machines are: the chain-and-sprocket system of a bicycle, the can opener, the egg beater, the door latch, and the automobile transmission. The principles we have described for the single-component machines apply to each component of a complex machine.

Many such devices use *gears* to transmit work. A gear is a wheel bearing teeth designed to mesh with the teeth of another gear. (See Fig. 6.14.) A gear is in reality nothing more than an odd-shaped lever. If the gear on the right in the figure turns counterclockwise, it exerts a force F on the tooth of the left gear. That force, acting a distance R from the axis of the left-hand gear, provides a torque that tends to rotate the gear. Since the force F on each gear is the same strength, the torques on each are proportional to the lever arms—the radii of the gears.

If you're a bicycle rider and not a mechanic, you might be more familiar with gears that use wrap-around chains rather than those that mesh directly.

QUESTION   For a ten-speed bicycle, explain on the basis of forces and torques why it's easier to climb a steep hill with the chain on the smaller front chain wheel. Why is hill-climbing easier with the chain on the largest rear sprocket wheel?

6.4 ROTATING RIGID BODIES

In Section 3.3 we studied circular motion, which sometimes comes about as a body rotates about an axis. Each particle of such a body moves in a circle about the axis. When we have this situation, it's usually convenient to analyze it from the viewpoint of the rotating body as a whole, rather than looking at the motion of the individual particles making up the body. That's our goal in this section.

In getting a handle on rotational motion, it is useful to compare rotational effects with motion along a line—often called either *linear* or *translational* motion for distinction. For each of the quantities that describe or influence some aspect of linear motion, there is a corresponding quantity for rotational motion. We'll use this correspondence to promote our understanding of rotation.

In Chapters 2 and 3 we found that *force* is the important consideration in determining translational motion. A body continues in its state of motion—at rest or with constant velocity—unless a net outside force acts on it. When a net force *does* act on the body, the body accelerates in proportion to and in the direction of the force. Newton's first and second laws describe these effects.

We noted in Section 6.2 that *torque* is a turning effect, that a torque is needed to start something rotating. In fact, *torque does for rotation what force does for linear motion.* A body continues in its state of rotation—either at rest or rotating at a constant rate—unless there is a net outside torque on it. If there *is* a torque, the rate of rotation changes. A rotational version of Newton's second law describes the rate of this change.

Table 6.1 summarizes the various quantities that describe and influence motion, and it compares the corresponding properties for linear and rotational motion. You're already familiar with the items listed under "Linear Motion." We'll look at each of the entries under "Rotational Motion" and see how to understand it in terms of its linear analogue.

The first three quantities describe rotational motion, just as speed, velocity, and acceleration describe linear motion. The angle of rotation corresponds to the distance; the angular velocity (the number of revolutions per second) replaces the velocity; and the angular acceleration tells us the rate of change of angular velocity. The other entries in Table 6.1 need somewhat more detailed comment.

Table 6.1 Comparison of linear and rotational motion

Linear Motion	*Rotational Motion*
distance	angle of rotation
velocity	angular velocity: number of revolutions per second
acceleration	angular acceleration: rate of change of angular velocity
force	torque
mass	rotational inertia

Newton's second law:
  force = mass × acceleration
	Newton's second law for rotation:
	  torque = (rotational inertia) ×
	  (angular acceleration)

translational kinetic energy:
  $\frac{1}{2}$ × mass × (velocity)^2
	rotational kinetic energy:
	  $\frac{1}{2}$ × (rotational inertia) ×
	  (angular velocity)^2

momentum:
  mass × velocity
	angular momentum:
	  (rotational inertia) ×
	  (angular velocity)

Newton's Second Law for Rotation

Just as Newton's second law—force = mass × acceleration—describes how forces influence linear motion, a variation of this law describes how torques influence rotational motion. This variation has exactly the same form as the second law:

torque = (rotational inertia) × (angular acceleration)

As you would expect, a greater torque on a particular body gives it a greater angular acceleration. But the quantity that measures the resistance to change in the rotational state is not mass, but rather something we call **rotational inertia**. This term is usually called *moment of inertia*, but *rotational inertia* describes its role better.

Rotational inertia depends on mass: It's certainly easier to spin a wooden block on a table than it is a lead brick of the same size. But rotational inertia also depends on how that mass is distributed relative to the axis of rotation. Let's illustrate. If you let a baseball bat hang in your hand, as in Figure 6.15(a), you can easily spin it back and forth about a vertical axis. But hold it as in Figure 6.15(b), and it's much harder—takes more torque—to spin it back and forth. The mass is the same both ways, but in Figure 6.15(b) it is distributed farther from the axis of rotation. The farther the mass is distributed from the axis, the greater its rotational inertia, and thus the greater the torque needed to give it a specific angular acceleration.

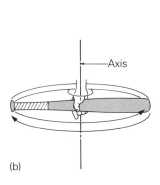

—Axis

—Axis

(a)

(b)

Figure 6.15 The rotational inertia of a body depends on how the mass is distributed relative to the axis.

(a)

(b)

Figure 6.16 Maximizing rotational inertia as (a) Kristen learns to walk and (b) a tight-rope walker balances.

· ·

QUESTION   (1) Why do children just learning to walk often hold their arms up or out as they walk (Fig. 6.16–a)? (2) Why does a tightrope walker carry a long skinny pole rather than a short fat one (Fig. 6.16–b)?

Rotational Kinetic Energy and Angular Momentum

Based on our analogy with linear motion, the definitions of **rotational kinetic energy** and **angular momentum** in Table 6.1 should be reasonable to you. Both depend on rotational inertia in the same way that translational kinetic energy and linear momentum depend on mass.

Figure 6.17 A skinny skater with fat hands can easily reduce his rotational inertia by pulling in his hands.

(a)   (b)

· ·

QUESTION

Why are flywheels for storing rotational kinetic energy often made with most of their mass near the outer radius of the wheel rather than near the axis?

Rotational kinetic energy, like any other energy form, must comply with the principle of conservation of energy. Any time the rotational kinetic energy of a body increases, that increase must be at the expense of a decrease in some other form or forms of energy, and vice versa.

A conservation principle also exists for angular momentum: The total angular momentum of a body, or system of bodies, remains constant if there is no torque on the body or bodies from outside.

Have you ever wondered how figure-skaters can rotate slowly and then suddenly, as if by invisible forces, begin to spin rapidly? How they pull this off can illustrate both conservation of angular momentum and conservation of energy in a rotating body.

Imagine a very skinny skater with fat hands rotating slowly with arms extended, as in Figure 6.17(a). What happens if he pulls his hands in close to his body, as in Figure 6.17(b)? Because much of his mass is then closer to the axis of rotation, his rotational inertia is *less* than before. Let's see what the two conservation principles, angular momentum and energy, tell us about what *should* happen.

Conservation of angular momentum says that the total angular momentum after the skater pulls in his hands must be the same as it was before, assuming we can neglect the small torques from friction in the skates and the air. That is,

angular momentum after = angular momentum before

or

(rotational inertia × angular velocity) after
= (rotational inertia × angular velocity) before

Figure 6.18 A skater uses the principles of energy and momentum conservation to increase rotational speed.

When the skater pulls in his arms and decreases his rotational inertia, his angular velocity must *increase* so that the product (rotational inertia × angular velocity) remains constant. Reextending the arms reverses the process and slows the rotation.

Thus, the skater spins much faster with hands pulled in and much slower with hands extended. A real skater with normal body and hands accomplishes the same result by pulling in arms and hands, plus maybe an extended leg (Fig. 6.18).

Although the explanation is somewhat more complicated, we can reach the same conclusion on the basis of energy conservation. Since rotational kinetic energy $= \frac{1}{2} \times$ (rotational inertia) × (angular velocity)^2, the skater's angular velocity must increase to compensate for a decrease in rotational inertia. In addition, chemical energy stored in her body lets her muscles do the work of pulling in her hands, thereby further increasing the rotational kinetic energy and adding to her rotational speed.

If you find the idea of rotational inertia somewhat hard to swallow, we can also explain what happens in terms of translational kinetic energy. The skater's hands have translational kinetic energy associated with their motion along a circular path. The direction of motion is continuously changing, but that doesn't change the kinetic energy. When the skater pulls in her hands, the radius of the circular path decreases, which decreases the circumference of the path. Since a hand then goes a shorter distance in one revolution, it must make more revolutions per second to keep the same speed and therefore the same kinetic energy. As before, the kinetic energy also increases by the amount of work done in pulling in the hands. Therefore, pulling in her hands increases the skater's rotational speed.

· ·

Key Concepts

If only two forces act on a lever, and we neglect friction, both forces do the same amount of work. This fact is the basis of mechanical machines, both simple and complex. The use of a machine can reduce the external *force* needed to do a certain amount of work, but the distance the force acts must increase correspondingly.

Torque = force × lever arm

where the **lever arm** is the perpendicular distance from the force to its line of action. Torque is a **turning effect**. When a weight is the force providing a torque, its line of action is downward through the **center of gravity**.

A body that is either not rotating or rotating at a given rate continues to do so unless there is a net outside torque on it. The rate of change of angular velocity, the angular acceleration, is directly proportional to the net applied torque and inversely proportional to the body's **rotational inertia**. Rotational inertia depends on mass and on how that mass is distributed relative to the axis of rotation; rotational inertia is greater the farther the mass is distributed from the axis.

Rotational kinetic energy and *angular momentum* depend on rotational inertia and angular velocity in the same way that translational kinetic energy and linear momentum depend on mass and linear velocity, respectively. The angular momentum of a system on which there are no external torques remains constant.

Questions

Simple Machines and Torque

1. How many machines that serve as force magnifiers (such as those of Fig. 6.4) have you used in the past week? How about those that exert a weaker-than-applied force, but over a larger distance (as in Fig. 6.5)?

2. In Figure 6.5(b), what exerts the force that makes the boat go forward? Why does the boat go in a circle if you row from only one side?

3. Suppose you use a wrench to loosen a spark plug on your car. If you do 10 J of work on your end of the wrench, about how much work does the other end of the wrench do on the spark plug?

4. Why do the block-and-tackle systems of Figure 6.19 result in more load W being lifted than force F exerted? Estimate the ratio W/f for each system. (Hint: Consider the relative distance of travel by W and f.)

5. When a threaded pipe is especially hard to unscrew, plumbers sometimes put a piece of pipe over the handle of their pipe-wrench to extend the length of the handle. How does this help?

6. In sweeping the floor, you usually hold the top of the broom handle with one hand. Compare the relative merits of different positions along the broom handle for the other hand.

7. For aesthetic reasons, doorknobs are sometimes put in the center of a door. What are the disadvantages of this arrangement?

8. Where should a doorstop be located relative to the door in order to minimize the force on the hinge screws when the door gets stopped?

9. It's easy for the coattail of a coat hanging in a closet to get caught between the closet door and the door casing on the hinged side of the door. Why will this easily damage the hinges, even though you don't push very hard to close the door?

Figure 6.19 The block and tackle.

10. If an arm of a mobile suspended from the ceiling is slightly out of balance, the arm tilts. What keeps it from continuing to tilt more and more until the arm becomes vertical? Hint: If the arm is perfectly straight, it *will* tilt until vertical.

11. Where does the torque come from that causes the arms of a windmill to rotate?

12. What torques act on a diving board to keep it from rotating when you stand on the end of the board? What are the senses (clockwise or counterclockwise) of these torques relative to the one you exert as you stand?

13. Archimedes, the noted Greek mathematician of the third century B.C., reportedly said that if he had a suitable fulcrum on which to rest his lever, he could lift the world. In what ways is his boast meangingless? (See Chapter 3, Question 2.)

14. Suppose you make a cylindrically shaped hunk of Jell-O and put it on a plate as in Figure 6.20(a). If it gradually distorts as shown from (a) to (b) to (c), draw its configuration when it first starts to topple over.

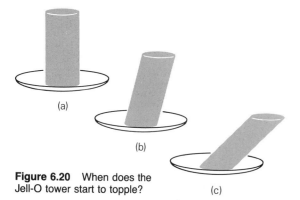

Figure 6.20 When does the Jell-O tower start to topple?

15. Freddie the Wino had four pallbearers at his funeral. If the front two hold the casket $\frac{1}{4}$ its length from the end, and the rear two hold it at the end, compare the relative loads carried by the pallbearers.

16. The curvy piece of wood in Figure 6.21(a), sometimes called a "sky hook," would obviously not stay in a horizontal position if you supported it only at the tip. Yet, when a belt is inserted in it as in Figure 6.21(b), you *can* support it that way. Why?

17. Some high-jumpers claim that, as they go over the bar, they bend their bodies in such a way that their centers of mass go *under* the bar while they go over the bar. Do you believe this is possible? (See Fig. 6.22.)

18. For maximum ease in pedaling uphill, should the front chain wheel of a bicycle be large or small? What about the rear one?

Rotating Rigid Bodies

19. Why does choking the bat (grasping it a short distance from the end, rather than at the end) make it easier for a baseball batter to swing?

20. In terms of rotational inertia, explain why: (a) short-legged people step more quickly than long-legged ones; (b) it's hard to walk without bending your knees; (c) you hold your arms straight out to the side when you're walking on a log across a creek; and (d) you can swing a short bat faster than a long one of the same weight.

Figure 6.21 The "sky hook": (a) shown, (b) in action.

21. Why does a helicopter (Fig. 6.23) need a second set of rotor blades on its tail?

22. You can throw a dart harder by holding the feathered end and flipping it than by holding it along its sides. Why? Why doesn't it continue to flip end-over-end?

23. Why do you swing your arms when you walk?

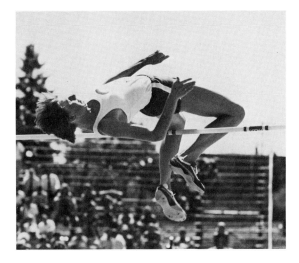

Figure 6.22 Joni Huntley as she clears the bar in a high jump.

Figure 6.23 A helicopter needs two sets of rotor blades.

Figure 6.24 Some typical dimensions for the forearm.

Problems

1. With what force must you push down on the plank of Figure 6.1 if the plank is 1 m long, if the piano weighs 1600 N, and if the fulcrum is placed 20 cm from the piano?

   Answer: 200 N

2. In Figure 6.24 what force F is exerted by the biceps muscle on the forearm as the ball of mass 15 kg is held in the position shown? (Remember to distinguish between mass and weight.)

   Answer: 1300 N (290 lb)

3. Suppose there is a pecan 2 cm from the fulcrums in the nutcracker of Figure 6.4(a). If you squeeze with a 40-N force 8 cm beyond the nut, with what force do the jaws push on the nut?

   Answer: 200 N

4. If the wrench of Figure 6.4(c) is 20-cm long, what torque can you exert on the nut with a force f of 150 N?

   Answer: 30 N·m

5. On the wheelbarrow of Figure 6.4(d) the distance from the wheel axle to the center of the body is 2 ft, and the distance from the wheel axle to the end of the handles is 5 ft. About how much force is needed on the handles to lift 200 lb of dirt? (Neglect the weight of the wheelbarrow.)

   Answer: 80 lb

6. Superman is shoveling coal with a supershovel (Fig. 6.5–a). He has 900 kg (mass) of coal in the shovel as he holds it in a horizontal position. If his left hand is halfway between the coal and his right hand, what force does his left hand exert? If we consider his left hand as the fulcrum, what is the direction and strength of the force exerted by his right hand?

   Answer: 18,000 N; 9000 N downward

7. The beautiful princess is rowing toward the charming prince in a boat with oars 2-m long (Fig. 6.5–b). Where must the oar's fulcrum be if she wants the end in the water to move three times as far as the end inside the boat? If she

pulls with a force of 200 N on her end of the oar, what force does the other end exert on the water?

Answer: 50 cm from inside end; 67 N

8. If you are seesawing with a friend and you weigh only $\frac{2}{3}$ as much as your friend, where should he sit if you sit on one end of the seesaw? Assume the pivot point to be at the center.

    Answer: $\frac{2}{3}$ as far from the pivot

9. In the unequal arm balance of Figure 6.13(b), suppose the empty pan exactly balances the empty beam with the movable weight w removed. The distance l is 3 cm, the distance from the fulcrum to the first notch is 2 cm, and the notches are 2 cm apart. Putting w in the first notch balances a 10-N weight in the pan. Moving w each additional notch to the right balances an additional 10-N weight. What is the weight of w?

    Answer: 15 N

10. A continuous torque of 50 N·m brings a fly-wheel from rest to an angular velocity of 6000 revolutions/min in 30 s. How long will it take a continuous torque of 150 N·m to bring it to the same angular velocity?

    Answer: 10 s

Home Experiments

1. With a meter stick or similar bar as a lever, try lifting objects of different weights. Compare the effect of different positions and different heights for the fulcrum.

2. Tie a rope around your waist and throw the free end over the limb of a tree. Pull yourself up to the limb by pulling down on the free end. Compare the strength needed in your arms this way with that needed to climb straight up a single length of rope tied to the limb.

3. Analyze the motion of a rocking chair. When you rock backwards and lift your feet, what provides the torque that rocks you forward again? What are the energy transformations as you rock all the way back and then stay still and let the oscillations die out?

References for Further Reading

Koff, R. M., *How Does It Work?* (Signet, New York, 1961). Simple but clear discussion of the workings of many everyday gadgets.

The Way Things Work, An Illustrated Encyclopedia of Technology, V. I and V. II (Simon and Schuster, New York, 1967). Detailed mechanisms as well as some of the physics of hundreds of everyday and industrial devices.

F L U I D S

You, I, and all other forms of life live immersed in a sea of fluid. For fish, that sea is the water in their ocean, river, lake, or stream. For you and me, and for most of the life forms around us, that "sea" is the air we breathe. Fluids influence our lives in countless ways: we travel on them in airplanes and boats; we burn them as fuel in our furnaces and ranges; we take baths in them; we drink them; and we depend on them to carry nutrients through arteries to all parts of our bodies.

Even though we often take the word *fluid* to mean liquid, a **fluid** is any substance that *can flow*. This category includes both *liquids* and *gases*. The molecules of a liquid can move around, letting the liquid take on the shape of its container. Yet these molecules stay about the same distance apart, making the liquid essentially *incompressible*. That means a given mass of liquid keeps the same volume, aside from minute changes. For a gas, however, the molecules bounce around randomly, filling up the complete volume of the container. A gas is therefore *compressible* and behaves differently from a liquid in certain ways. In spite of the differences, many properties of a fluid are independent of whether that fluid is a liquid or a gas.

In this chapter we examine some of the many aspects of the behavior of fluids. Before we can begin, however, we need to know the meaning of two important concepts that apply to solids as well as to fluids: namely, density and pressure. We can then discuss atmospheric pressure and the way pressure varies throughout any fluid. This pressure variation within a fluid is what accounts for the buoyant forces that make objects float on, or seem lighter in, fluids.

Practical use of fluids often depends on our ability to move them—to pump them—from one place to another. After looking at the physical bases for pumping, we'll consider a few properties of fluids in motion.

We can explain our final topics, *surface tension* and *capillary action* in liquids, only on the basis of the interactions between individual molecules of the liquid.

7.1 DENSITY AND SPECIFIC GRAVITY

Suppose you have three bricks, all of the same shape and size, but made of different materials: one wood, one concrete, and one lead. You know already that each of these has a different weight, and therefore a different mass. Substances with the same volume but different mass differ in their *densities.*

Density is the mass of a unit of volume of a substance. Thus,

$$\text{density} = \frac{\text{mass}}{\text{volume}}$$

In symbols,

$$D = \frac{m}{V}$$

Density may vary widely from substance to substance. For a gas, which is compressible, the density can vary over a wide range, depending on how much the given gas is compressed.

If we have a block of aluminum 20 cm × 10 cm × 5.0 cm and we measure its mass to be 2.7 kg, we can calculate its density by

$$\text{density} = \frac{\text{mass}}{\text{volume}} = \frac{2.7 \text{ kg}}{0.20 \text{ m} \times 0.10 \text{ m} \times 0.050 \text{ m}}$$

or 2700 kg/m^3.

Table 7.1 gives the densities of a few ordinary substances. Rather than using the SI units kg/m^3 as in Table 7.1, it's sometimes easier to work with densities in the smaller metric unit g/cm^3. You can get the density in g/cm^3 by dividing the values in this table by 1000. Why?

· ·

EXERCISE   What is the mass of the mercury that fills a 1-liter bucket? (1 liter = 1000 cm^3 = 0.001 m^3)   *Answer:* 13.6 kg

Are you surprised that the water has such a convenient density as 1000 kg/mg^3? This number is not an accident of nature, but comes from the

Table 7.1 Densities of selected substances

Substance	Density (kg/m^3)	Substance	Density (kg/m^3)
air (sea level, 20°C)	1.20	concrete	2800
pine wood	500	diamond	3300
butter	860	steel	7830
paper	920	lead	11,300
ice	920	mercury	13,600
water	1000	gold	19,300

original definition of the gram as the mass of 1 cm³ of water.

Since water is such an omnipresent substance, the densities of other substances are sometimes given as the ratio of *their* densities to that of water. This "relative density," called **specific gravity**, is defined as:

$$\text{specific gravity} = \frac{\text{density of the substance}}{\text{density of water}}$$

Being a ratio, specific gravity has no units.

· ·

QUESTION   Why does the specific gravity of a substance have the same numerical value as its density in g/cm³, 1/1000 the numerical value of its density in kg/m³?

We sometimes use the quantity **weight density**:

$$\text{weight density} = \frac{\text{weight}}{\text{volume}}$$

$$D_w = \frac{W}{V}$$

with SI units N/m³. (The British units are lb/ft³.) The subscript "w" distinguishes weight density from mass density. Since the weight of an object equals its mass multiplied by the acceleration due to gravity ($W = mg$, Chapter 2),

weight density = mass density × acceleration due to gravity

= mass density × g

or

$$D_w = Dg$$

For water, the weight density is 9800 N/m³ (62.4 lb/ft³) or roughly 10,000 N/m³. When we use only the word **density**, we'll mean *mass* density.

Knowing the density of a substance, we can get the mass of a certain volume of it from

mass = density × volume

Using symbols for the quantities in this equation, we have

$m = DV$

Similarly, we get the weight from multiplying weight density by volume.

7.2 PRESSURE

Hold a sharp pencil between your fingers and push as shown in Figure 7.1. You can push hard and the eraser end won't hurt. But the finger touching the pointed end hurts! You're applying the same force, but the point concentrates it in a much smaller area. The difference is in the pressure. **Pressure** is defined as the force per unit of area:

$$\text{pressure} = \frac{\text{force}}{\text{area}}$$

If the force you exert with the pencil is 25 N and it is distributed over the eraser of area 32 mm² or 3.2×10^{-5} m², then

$$\text{pressure} = \frac{25 \text{ N}}{3.2 \times 10^{-5} \text{ m}^2} = 7.8 \times 10^5 \text{ N/m}^2$$

Estimate the pressure you expect from the same force applied to the point. If the point area is 0.20 mm², you can calculate the pressure to be 1.25×10^8 N/m², or 160 times as much as with the eraser.

Many footballs are marked "inflate to 13 lb." But I doubt seriously that the one your favorite quarterback passed for the last game-winning touchdown *weighed* 13 lb. However, it probably was inflated to 13 *lb/in²* of pressure—a force of 13 lb on each square inch of area inside the ball. Similarly, inflating your tires to 29 "pounds" means 29 lb/in² of pressure.

QUESTION   Why is a period typed with a typewriter harder to erase than an *m*?

Figure 7.1 The same force on both fingers gives a much higher pressure on the right than on the left because the area of contact is much smaller on the right.

In the SI system of units, the newton/meter² is defined as the pascal (Pa). Thus, 1 Pa = 1 N/m². It may be a good idea to train your mind to *think* "newton/meter²" when you read or use the word *pascal*. That way you are reminded of the meaning of pressure.

EXERCISE   Convert a pressure of 29 lb/in² to pascals.
Answer: 200 kN/m² or 200 kPa (k = 1000)

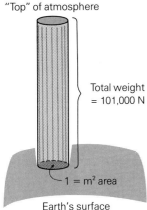

"Top" of atmosphere

Total weight = 101,000 N

1 = m² area

Earth's surface

Figure 7.2 Atmospheric pressure is caused by the weight of the air above the area. A column of air of 1 m² in cross-sectional area from sea level up weighs 101,000 N.

Atmospheric Pressure

The sea of air that we live in—the atmosphere—exerts a pressure on us and on everything else submerged in it. This pressure comes from the weight of the air itself as it pushes down on the earth and its occupants.

Figure 7.2 illustrates this point. Consider a square meter of area on the earth and the cylinder of air above it 1 m² in cross section extending to the "top" of the atmosphere. If the surface area is at sea level, the amount of air in that cylinder weighs 101,000 N. This means the **atmospheric pressure** at sea level is 101,000 N/m² (101 kPa), or 14.7 lb/in². At higher altitudes there is not as much air above a given area, and therefore the atmospheric pressure is lower. That's why your ears "pop" as you go up or down a tall mountain.

Atmospheric pressure not only pushes down on the top of horizontal surfaces, but upward on their bottoms and sideways on their sides. It pushes perpendicular to *any* surface, whatever its orientation. Since the air is free to flow, it exerts equal pressure in all directions at a given location. This condition holds true for any fluid under pressure. When you dive deep into water, you feel the pressure in all directions on your body.

EXERCISE   Estimate the force with which the atmosphere pushes on your chest.
Answer: (for 0.06-m² chest): 6100 N (1400 lb)

We don't feel the roughly 10 N on each square centimeter of our bodies (15 lb on each square inch) because we are used to it, and our

Figure 7.3 When the air is pumped from an empty can, the outside atmospheric pressure easily crushes it.

bodies push outward with an equal but opposite pressure all over.

Atmospheric pressure exerts a force of about 3700 N (820 lb) on the side of the empty can shown in Figure 7.3(a). Why doesn't it crush the can? You know the answer is that air is also inside the can pushing out. But if we connect a hose from the can to a vacuum pump and pump out the air inside, atmospheric pressure quickly crushes the can into the distorted mess shown in Figure 7.3(b). It's almost as entertaining to see the can come back nearly to its original shape when you pump in air at a pressure a little higher than atmospheric.

We often use sea–level atmospheric pressure as a *unit* of pressure. For example, we say that a pressure of 250 kPa is 2.5 atmospheres, since this pressure is 2.5 times "standard" atmospheric pressure.

The 90 kPa (13 lb/in²) of pressure in your football and 200 kPa (29 lb/in²) in your car tires are actually the difference in pressure between the inside of the given inflatable object and the atmosphere. The usual way of measuring pressure is just that: the difference between the internal pressure and the external atmospheric pressure. The total air pressure inside the tire is, therefore, 301 kPa.

Pressure in Liquids

When we spoke earlier of the "top" of the atmosphere, we put the word *top* in quotes because the atmosphere doesn't actually have a top, but gradually it thins out from about 6×10^{20} molecules/cm³ at sea level to about 1 molecule/cm³ in interstellar space. The density at any particular point depends on the pressure at that point.

Liquids, however, are relatively incompressible: Their densities stay almost the same for large variations in pressure. For that reason, we can easily find the pressure at a given depth caused by the liquid above it.

Consider the square area on top of the fish in the tank in Figure 7.4. The downward force on this area is caused by the weight of the liquid in the column directly above: its density $\times$ g $\times$ its volume. Since the

h = depth of fish
   = height of water
     column above fish

Figure 7.4 The pressure due to the water on the fish at depth h below the surface is determined by the weight of the column per unit of area. This weight depends only on the density of the liquid and the depth.

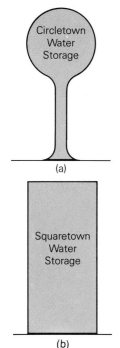

(a)

Squaretown
Water
Storage

(b)

Figure 7.5 The pressure at the bottom of both tanks is the same since the heights of both are the same.

volume of the column is its area × its height, we can calculate the pressure from

$$\text{pressure} = \frac{\text{force}}{\text{area}} = \frac{\text{weight}}{\text{area}} = \frac{\text{density} \times g \times \text{area} \times \text{height}}{\text{area}}$$

$$= \text{density} \times g \times \text{height}$$

In symbols, this equation is

$$p = Dgh$$

Notice that the height of the water column is the *depth* under water of the point where we're finding the pressure.

The pressure is directly proportional to the depth in the liquid, and it is independent of the volume or shape of the container holding the liquid. You've probably seen water storage tanks shaped like the one in Figure 7.5(a). The pressure at the bottom is the same for both tanks (a) and (b). For (a), the tank walls push up on the lower sloping walls of water to partly support its weight.

If, as usual, there is atmospheric pressure on the surface of a liquid, the total pressure at some depth would be the atmospheric pressure *plus* that due to the liquid. We'll illustrate with an example.

EXAMPLE 7.1

Compare the pressure 50 cm below the water surface with that 3.0 m down. The pressure due to water only is given by

$$p = Dgh$$

At 0.50 m,

$$p = 1000 \text{ kg/m}^3 \times 9.8 \text{ m/s}^2 \times 0.50 \text{ m} = 4900 \text{ Pa}$$

Thus, if p_a is atmospheric pressure, then

$$\text{total pressure} = p + p_a = 4.9 \text{ kPa} + 101 \text{ kPa} = 106 \text{ kPa}$$

At $h = 3.0$ m, a similar calculation gives the pressure of the water only to be 29 kPa, and a total pressure of 130 kPa: 23% more than near the surface. Does that explain why your ears may hurt when you go deep into the water after a high dive?

QUESTION

Did the little Dutch boy who held back the whole North Sea with one finger have superhuman strength?

Figure 7.6 Sea level atmospheric pressure pushing down on the open surface of mercury holds the mercury 76 cm up in the tube. This principle is the basis of the mercury barometer.

Measuring Atmospheric Pressure

Imagine that you take a glass tube about a meter long that is closed at one end, hold it with the open end up, and fill it with mercury. Then carefully invert it in a bowl so that the open end stays immersed in the mercury that runs out. You would find that not all the mercury would run out, but only enough to leave about a 76-cm-high column of mercury in the tube, as shown in Figure 7.6.

To find out why, let's calculate the pressure at a depth of 76.0 cm below the surface of mercury, which has a density of 13,600 kg/m³. Since there is vacuum (nothing) in the tube above the mercury, the pressure at the bottom is due entirely to the mercury:

$$p = Dgh = 13,600 \text{ kg/m}^3 \times 9.80 \text{ m/s}^2 \times 0.760 \text{ m}$$
$$= 101,000 \text{ N/m}^2 \text{ (or 101 kPa)}$$

But this value is exactly sea-level atmospheric pressure. The atmosphere pushing down on the open surface of the mercury pushes it up into the tube until the pressure at the surface level caused by the weight of the mercury column exactly balances the atmospheric pressure.

This principle is used in making a mercury *barometer*—a device for measuring atmospheric pressure. Atmospheric (*barometric*) pressure varies somewhat from place to place, and from time to time at a given place. The height of the mercury column supported in this way is a measure of the atmospheric pressure. In weather reports, this pressure is usually reported directly as the height of the mercury column—for example "760 mm" or "29.92 inches."

· ·

QUESTION   If we increased the size of the tube holding the mercury, would the height of the mercury column change?

· ·

EXERCISE   Could you make a barometer using water instead of mercury? How tall would the water column need to be?   *Answer*: 10.3 m (33.8 ft)

7.3 PRESSURE DISTRIBUTION IN A FLUID

Blaise Pascal, a seventeenth-century French theologian and scientist, discovered that if you apply a pressure to one part of an enclosed fluid, the

Figure 7.7 In a hydraulic press, the pressure is the same on both pistons. The force is therefore proportional to the area. The smaller piston must move farther in the same proportion.

Figure 7.8 Hydraulic disc brakes for automobiles. When you push on the brake pedal, the increased pressure is transmitted throughout the brake fluid, causing the piston to push the pad against the disc. The caliper slides so that both pads clamp on the disc, with the resulting frictional force opposing the rotation.

pressure *increases* by that same amount everywhere in the fluid. We call this effect **Pascal's principle**.

In the "hydraulic press" of Figure 7.7, the force on either movable piston is the product of the pressure and the area of the piston. Thus, the force F must be as much greater than f as the area A is larger than a, since the pressure is the same. In other words, the press acts as a force magnifier in the same manner as the levers of Chapter 6. Just as in the case of the levers, the smaller force must act over a proportionately longer distance. This principle is used in many hydraulic devices, including jacks and car lifts used by service stations, dentist and barber chairs, and brakes in cars and trucks (Fig. 7.8). For the pressure and areas shown in Figure 7.7, verify that the forces shown on the pistons are correct. Also verify that the relative distances, d and D, are valid examples.

7.4 BUOYANCY

You have probably found that it's easy to lift a person who is almost completely under water. Also, when you've been in the water for a long time and climb out, you feel unusually heavy as you get out because you've gotten used to being partly held up by the water. Evidently, as these two examples hint, there is an upward force on objects immersed in a liquid. That force is strong enough that some objects—boats, rafts, persons with life preservers, beach balls—float on top of liquids. This effect is not just confined to liquids. Helium-filled balloons and the Goodyear blimp float in air. Any object totally or partially submerged in a fluid experiences a *buoyant force* that is responsible for all the effects we've mentioned.

The buoyant force exists simply because the pressure increases with depth in the fluid. Consider the fish relaxing in the sun in Figure 7.9. Because its bottom is a little lower than its top, the pressure is a little greater on its bottom. The water, therefore, pushes up on it somewhat harder than it pushes down. The difference is just enough to hold the fish up so that it doesn't sink. The lead weight on the fishing line, being denser than the water, would sink if dropped. But the weight feels lighter to us under water than out because the pressure on its bottom is a little greater than on its top.

In the third century B.C. Archimedes observed a simple but useful fact: *The buoyant force pushing up on a totally or partly submerged object equals the weight of the fluid it displaces.* We call this statement **Archimedes' principle**. You can understand the reason for this principle by looking at the fish or the weight in Figure 7.9. If the object were not there, water

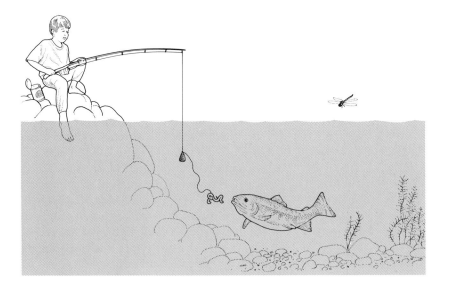

Figure 7.9 The buoyant force on the fish and on the weight comes from the pressure being higher below than above them.

Indecent exposure charged to Archimedes

SYRACUSE (JANUARY 30, 222 BC)—Several people in Syracuse reported seeing a strange-looking man wearing only a towel and running through the streets shouting, "Eureka! Eureka!" He had vanished, however, before police arrived on the scene to investigate.

Unidentified reliable sources report the person to have been Archimedes, the mathematician who has recently been working on a problem of great concern to Emperor Hiero: whether or not his crown is pure gold. According to these unnamed sources, Archimedes realized while taking a bath that he could measure the exact volume of the crown by finding the volume of liquid it displaces. By weighing it, he could then determine the density, and thus the purity, of the metal. This discovery allegedly produced the uncontrollable outburst of joy displayed in the streets.

would be in its place. The force pushing up on the water would hold it up, and therefore would have to equal its weight.

When you're underwater, you displace a volume of water equal to your body's volume. Since your body is mostly water, its weight is about the same as that of an equal volume of water: Its density is about the same as that of water. The buoyant force is therefore about equal to your weight so that it takes little or no additional force to hold you up. If your head is out of the water, you displace less water, and some force is needed. The farther you get out of the water, the less water is displaced, and the more upward force you must supply to keep yourself up.

A floating object, like a boat, sinks far enough to displace a weight of water equal to that of the boat and its contents. If a 700-N person (mass about 70 kg) steps onto the boat, it sinks a little bit—far enough to displace 700 N. We'll illustrate the use of Archimedes' principle with one numerical example.

- -

EXAMPLE 7.2

How many 70-kg people could get into a 15-ft canoe in still water without sinking it if everybody stands perfectly still? To make calculation easier, assume that the volume of the parts of the canoe lower than point M (the first point to go under) is the same as that of the rectangle shown in dashed lines in Figure 7.10—0.61 m³. When the canoe is about to go under,

$$\text{weight of water displaced} = \text{density of water} \times g \times \text{volume}$$
$$= 1000 \text{ kg/m}^3 \times 9.8 \text{ m/s}^2 \times 0.61 \text{ m}^3$$
$$= 6000 \text{ N}$$

Figure 7.10 Estimate the volume of the canoe by that of the dashed rectangle.

Point *M*

30 cm

45 cm

4.5 m

The buoyant force can support this total weight, which allows for a 340-N (76-lb) canoe and eight 700-N (157-lb) people.

· ·

QUESTION In terms of Archimedes' principle, explain why an upright steel ship will float but a capsized one will not.

· ·

EXERCISE In lifesaving practice Tricia, who weighs 490 N (110 lb), has the job of bringing Don, who weighs 980 N (220 lb), to the surface of the pool. How much upward force does Tricia need to exert on Don to get only his head out of the water? (Assume Don's average body density is the same as that of water, and his head occupies 10% of his body's volume.)   *Answer*: 98 N (22 lb)

Some liquids will float or sink in other liquids if the two do not combine chemically when mixed. The less-dense liquid is pushed to the top by the denser one. That's why oil spills stay on the top of water. Did you ever drink nonhomogenized milk and have to shake it to mix the cream and milk? If not, then maybe you've had to shake the oil-and-vinegar dressing before putting it on your salad.

7.5 PUMPS

Fluids that we must use do not do us any good until we get them to the place where we need them: blood must be pumped to our brains, water to our bathtubs, natural gas to our furnaces. In this section we look at how some of the principles we've discussed apply to this practical job of pumping. You won't find any new physical principles here, but you'll see how several of the old ones work together.

Breathing

A pump extremely important to your life is the one that pumps air into and out of your lungs many times a minute. This pump is simple in principle. As your diaphragm (the muscle that separates your chest and abdominal cavities) moves down, the chest cavity becomes larger. This decreases the pressure inside your chest, making the pressure in your lungs lower than atmospheric. The atmosphere then pushes air into your lungs. When your diaphragm moves upward, the higher pressure inside the lungs pushes air back outside.

Even when you open your mouth and pull in a breath quickly, you're not really *pulling* it in. You're merely lowering the diaphragm quickly and *letting* the air rush in faster.

Drinking from a Straw

You probably don't think of a straw as a pump, but you and the straw together do make up one. When you start to suck on the straw, you lower the pressure in the top of the straw and in your mouth, letting atmospheric pressure *push* the drink into your mouth. You don't *pull* it in at all!

The Siphon

A siphon is not a pump in the usual sense, but it depends on some of the same principles. Figure 7.11 shows two possible situations that might exist if you try siphoning gasoline from your car tank to use in your lawn mower. In part (a) the gasoline level outside is higher than that in the tank; in part (b) it's lower. Based on what we've already said, explain why in (a) the gas will run back into the tank, but in (b) it will run out. Three hints: (1) the same atmospheric pressure acts down on the gas in the tank and up on the exposed gas surface at the other end; (2) the pressure difference between two heights in a liquid column is proportional to the height difference; and (3) at the highest point, look at the pressure trying to push liquid to the left and the pressure trying to push it to the right.

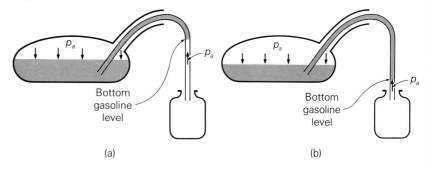

Figure 7.11 (a) The outside level of the liquid in the siphon is higher than that in the tank. (b) Here, the outside level of the liquid is lower. Why does the liquid run out only in (b)?

Figure 7.12 A force pump.

The Force Pump

The pump shown schematically in Figure 7.12 may seem more appropriate as a machine of your grandmother's youth. However, this "force pump" serves as a good introduction to the pump we'll discuss next—one that will certainly not be outdated in the near future.

The main body of the force pump is a cylinder containing a piston that moves up and down as the handle is moved. A pipe extends from this cylinder down into the water. As the piston moves up, atmospheric pressure pushing down on the water in the well *pushes* water up into the additional space in the cylinder. As the piston moves down, the water pressure from above closes check valve *A*, forcing the water to go up through check valve *B* into chamber *C*. Some of this water compresses the air in the top of the chamber; some goes out the spout. When the piston stops and starts to move up again, the pressure in Chamber *C* closes valve *B*, and the expanding air in the chamber continues to push water out the spout. The cycle continues in this way, with the compressing and expanding of the air in the chamber providing a fairly steady flow of water.

QUESTION Explain why such a pump cannot operate on the moon, and why there is a limit on earth to how high the pump cylinder can be above the water in the well. Hint: What role does the atmosphere play?

APPLICATION

THE HEART

Your heart is basically a force pump that pushes blood through the arteries, capillaries, and veins in the different parts of your body. Rather than having a piston that moves up and down, the heart has muscular walls that contract and relax, changing the volume of the heart's chambers as it "beats."

Figure 7.13 Pumping by the human heart. (a) Low pressure in the veins forces blood into the relaxed auricles and ventricles. The output valves are closed by the pressure in the arteries. (b) Contraction of the ventricles—the heart beat—pushes blood into the arteries. Pressure within the ventricles closes the input valves between the auricles and ventricles.

Output valves
Veins
Right auricle
Input valves
Left ventricle
Right ventricle
Arteries

(a)   (b)

Chapter 7   Fluids

Figure 7.13 shows the main steps in the pumping action. Notice the action of the valves that do the same job as the check valves in the force pump. The arteries are very elastic—a feature that smooths the blood flow just as the air chamber in the pump smooths the water flow. When the heart beats and pumps blood into the arteries, they expand. Their elastic contraction keeps the blood flowing as the heart relaxes for the next beat. The two sides of the heart are separate pumps. The right side pumps blood to the lungs and back to the left side, where it is pumped to the rest of the body.

· ·

QUESTION   Blood pressures are reported as two numbers—for example, 120/80, which reads "120 over 80"—that correspond to the highest (systolic) and lowest (diastolic) pressures in the artery of the arm during each pumping cycle. The units are millimeters of mercury. Why does the pressure vary up and down during each beat cycle? How would you expect this ratio to change when a person has arteriosclerosis (or hardening of the arteries, a condition where artery walls become rigid and inflexible)?

Notice that in our discussion of pumping we have made *no* mention (which is often made) of creating a partial vacuum that "pulls" the fluid into some area of the pump. Pumps do create a partial vacuum—a pressure lower than atmospheric. But *a vacuum can do no pulling!* It merely lets the higher outside pressure *push* the fluid in.

7.6 FLUIDS IN MOTION—FLYING AIRPLANES AND CURVING BASEBALLS

When fluids are moving, the rules are different than when they're at rest. In looking at some of these different rules, we'll start by considering a fluid confined inside a pipe, where we can account for all of it. We can use what we learn about that situation to describe what happens when unconfined fluids move past solid objects.

Suppose Uncle Charlie plants his petunias so far from the faucet that his hose won't reach to water them. We'll assume for now that the connection between his hose and the smaller one he borrows from his neighbor to reach his petunia patch is very smooth, as shown in Figure 7.14. Because the water is incompressible, the volume of a given mass doesn't change with pressure. For a steady rate of flow out the end, the same volume of water must pass both points *B* and *A* each second. However, the area is smaller at point *B*, so that the water must move faster at that point to get the same amount through per second. The speed must be exactly inversely proportional to the cross-sectional area of the pipe. That is, if the area is $\frac{1}{2}$ as large, the speed is doubled.

Figure 7.14 Since the same volume of liquid passes *B* and *A* in a given time interval, the speed is larger at *B*. The energy conservation principle shows that the pressure is lower at *B* where the speed is higher.

Compare this with the situation where two lanes of traffic merge into one to cross a bridge. If you're caught in traffic that is backed up and moving slowly in the two-lane section, then you move about twice as fast once you get into the one-lane part. Assuming that fluid is incompressible is like assuming the cars stay the same distance apart in both one- and two-lane sections.

If the fluid is compressible, as air or some other gas would be, the inverse proportionality doesn't hold exactly, but the speed does increase with decreasing pipe size.

The lines with arrowheads, called *streamlines*, show the direction of flow at any point, and they help in visualizing the flow pattern. Since the streamlines are continuous, they are *closer together* in a region of *higher speed*. If the junction between the large and small regions of hose is abrupt rather than smooth, there may be some turbulence in the junction region instead of the smooth lines of flow we've shown.

QUESTION   With a garden hose, why can you shoot a stream of water much farther if you have a nozzle with a small opening than if you just use the end of the hose? Almost the same *amount* of water comes out per second either way. Does partially covering the end of the hose with your finger make it squirt farther for the same reason?

Next consider what happens to the *pressure* as the speed of the fluid particles changes. Looking again at Figure 7.14, you might guess that the pressure in the pipe at point B is larger than at point A, since the fluid is being pushed through a smaller pipe. But if you measured the pressure at both points, you'd find just the opposite case: It's *lower* at B!

The reason for the lower pressure at B is the higher speed of the fluid particles, as we can see from conservation of energy. Concentrate on the fluid between A and B. For the fluid particles to gain kinetic energy in going from A to B, work has to be done on them. This work requires a net force in the direction of the increased speed. Therefore, the pressure pushing on the fluid between A and B from the left of A has to be larger than that pushing on it from the right of B. This means that the pressure in the fluid at B has to be less than that in the fluid at A.

If the fluid flow is from the smaller to the larger region, the pressure is still higher in the larger region. The higher pressure then slows the fluid particles to give the lower speed consistent with a larger hose.

Even though we've shown the relationship between pressure and speed for the special case of a liquid in a pipe of varying size, it is generally true that *the pressure in a fluid decreases as the velocity increases*, and increases as the velocity decreases. Scientists call the relationship between the pressure and energy of a fluid in motion **Bernoulli's principle**.

Figure 7.15 You can demonstrate Bernoulli's principle by blowing across the top of a piece of paper that is hanging down.

Figure 7.16 The higher pressure in the low-speed region of air pushes the ball toward the center line of the air coming out the hose.

Figure 7.17 As the air is pushed down, the reaction force pushes up on the airplane wing.

We can demonstrate Bernoulli's principle by some simple experiments in which air flows over solid objects. One way is by holding a piece of paper as shown in Figure 7.15 and blowing across the top. The lower pressure of the moving air on top lets the higher pressure of the still air under the paper push it up. Or you can suspend a light ball—Ping-Pong or beach ball—in midair by hooking a hose to the air-output end of a vacuum cleaner, and aiming it up at the ball as in Figure 7.16. The bombardment by the air pushes the ball upward. As the ball falls below the centerline of air flow, it feels a lower pressure near the centerline because of the higher speed of the air there. The ball is therefore pushed back toward the centerline by the higher pressure of the slower air.

We'll now discuss two additional questions related to the motion of fluids: What holds an airplane up in flight, and what makes a baseball curve? Viewed from the ground, both of these involve the motion of an object *through* the air. But they are both easier to understand as viewed from the reference frame of the object itself. There, the object is stationary and the air rushes by. The force causing these effects is often explained in terms of Bernoulli's principle, but it's easier to understand in terms of Newton's third law of motion—the *action-reaction law.*

First, let's examine the force, or "lift," that holds the airplane up. If you were to cut off an airplane wing in its middle, the end where you made the cut would have a shape something like that shown in Figure 7.17. From the reference frame of the airplane, air is passing the wing as shown by the streamlines. The wing is shaped and oriented so that the air, after passing the wing, has a partial *downward* motion. If the wing pushes the air *downward,* the action-reaction law tells us the air must push on the wing upward. This upward reaction force provides the lift. Except for low-speed flight, the wing would not be tilted as much as we've shown. Wing designs often have the lower surface fairly flat and the upper one curved. Air curving over the upper surface leaves the wing with the needed downward motion.

The effect described by Bernoulli's principle also provides lift. With the upper wing surface curved and the lower one flat, the air moving over the wing has farther to go and, therefore, must move faster than that moving under. Thus, the pressure is higher below the wing, giving a net upward force to the wing.

QUESTION In terms of the number of air particles passing the wing, explain why the angle of tilt can be decreased as the speed increases.

Understanding why a baseball curves is a little more complicated. From the ball's frame of reference, air passes a *nonspinning* ball as viewed from above the ball in Figure 7.18(a). The air cannot follow the sharp curvature of the ball, and it separates, leaving a turbulent area behind the ball. Since the flow is the same around both sides, there is no net sideways force and the air moves off in its original direction as shown by the arrow at the bottom. But for a curve ball, Figure 7.18(b) shows what happens. The pitcher puts a fast spin on the ball. As it spins, the seams drag a thin layer of air around with it. The air spinning with the ball on the left side opposes the overall flow there, causing an early separation of air from the ball. On the right side, the spinning air and passing air move in the same general direction, letting the air stick to the ball longer and thus leave in a slight leftward direction. The reaction to this push to the left on the air pushes the ball to the right, making it curve to the right.

Bernoulli's principle also contributes to the sideways force on the ball. Concentrate on the left-hand side of the ball in Figure 7.18(b). Since the surface of the ball drags air around with it, the air very near the left-hand surface is moving upward in the figure. That farther out on the left is moving downward. There must, therefore, be a point in between where the air is stationary relative to the ball. However, on the right, the air moves only downward relative to the ball. Bernoulli's principle, then, tells us that the pressure is higher on the left, giving a force to the right on the ball. The same ideas apply, of course, to a spinning tennis ball or golf ball that has a surface rough enough to drag air around with it.

· ·

QUESTION   In what direction must a ball be spinning when the pitcher throws a curve that curves downward?

7.7 MOLECULAR FORCES AND SURFACE TENSION IN LIQUIDS

So far, we haven't paid any attention to the fact that fluids are made up of molecules. However, to explain an important characteristic of liquids—their surface tension—we have to look at interactions between individual molecules. Strongly dependent on surface tension is the phenomenon called **capillary action**, a process that nature uses in many practical ways.

Surface Tension

You know that a solid chunk of steel won't float in water. To make it float (as a steel ship floats), you have to shape the steel so that it has a

Airflow

No spin

(a)

Airflow

Spin

Force

(b)

Figure 7.18 (a) When a baseball is not spinning, the air flow is the same on both sides and there is no net force. (b) When a "curve" is pitched, the spinning ball drags air around with it, giving the air a slight sideways motion as it passes. The reaction force pushes the ball in the other direction.

hollow region that will displace a weight of water equal to the weight of the steel. This holds for a chunk of any size—large or small. If you drop a needle or razor blade into a glass of water, either will certainly sink.

Now, suppose you don't *drop* these small items, but instead lay them gently on the water surface with their long or flat side parallel to the water. Figure 7.19 shows what happens: They stay on top of the water. The surface molecules of the water are attracted to their neighbors on the surface with a force called **surface tension**. The force is strong enough that the weight of small thin objects cannot break the surface tension and get through to the interior of the liquid. Once a corner or edge breaks through the surface, the object immediately sinks.

Let's see why surface tension exists. In Figure 7.20 the circles represent the centers of the molecules of liquid. Each molecule, such as the one shown in black and labeled *A*, exerts a force on its nearby neighbors. The force on any one molecule acts roughly as it would if all the molecules were rubber balls embedded in bubble gum. There would be a given distance between a molecule and its neighbor at which there would be no force. If the two molecules got closer they would be repelled, with the repulsion increasing drastically as the distance decreased (the rubber balls would repel each other as their sides started to touch). As the molecules move apart from the no-force distance, they are attracted together with a force that first increases, but then decreases with increasing separation (the bubble gum would stretch giving a strong force at first but then a weaker one as it stretched farther and thinned down).

Our rubber ball and bubble gum analogy breaks down, in that the molecules are free to move about throughout the liquid. As molecule *A* moves around, it feels the same forces, whatever its neighbors. Averaged over time, there is no *net* force on it.

Figure 7.19 Laid carefully on the surface, a needle or razor blade will stay on top of the water.

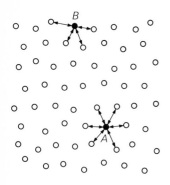

Figure 7.20 A molecule in the interior of a liquid is attracted in all directions by its neighbors. One on the surface is attracted only sideways and downward. This accounts for the surface tension of liquids.

Surface tension minimizes surface area, keeping water out of crevices

Figure 7.21 Dirt is not washed away by pure water because surface tension keeps water from seeping into the cracks between dirt and fabric.

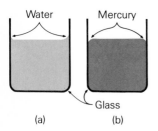

Figure 7.22 (a) Adhesion between water and glass is stronger than cohesion in the water, causing the contact region to curve up. (b) For mercury, cohesion in the liquid is stronger than adhesion to the glass, causing the contact region to curve down.

But what about molecule *B*, which is on the surface? It has no molecules above it that exert forces on it, so it's pulled and pushed only by its neighbors beside and below it. If something tries to increase the amount of surface, either by penetrating it or stretching it, the net downward and sideways attractive forces resist this increase. This tight binding along the surface makes up the *surface tension*.

Surface tension not only resists the penetration of a liquid surface, but causes it to take on a shape that minimizes the surface area. Drops of liquid that do not stick to a surface (for example, drops of water on a greasy pan) or falling drops of liquid are roughly spherical because a sphere gives the smallest surface area for a given volume. The supporting surface or the air resistance distorts the drop from being perfectly spherical.

Detergents

If you pour a few drops of dishwashing detergent in the water of Figure 7.19, the needle and razor blade immediately sink. Evidently, the detergent reduces the surface tension of the water. In fact, reducing surface tension is one of the main jobs of a detergent.

What does reduced surface tension have to do with cleaning? Most "dirt"—either from clothes or dishes—is greasy or oily material that doesn't mix with water. As Figure 7.21 shows, surface tension keeps a smooth surface on the water, not letting it get under the dirt to wash it away. Detergent reduces the surface tension, thus allowing the water to get between dirt and fabric. Detergent makes the water "wetter."

In hot water, the molecules bounce around faster than in cold. They are, therefore, not bound as tightly and thus hot water has lower surface tension than cold. That's one reason hot water washes better than cold, although many modern detergents are effective even in cold water.

Capillary Action

If you look closely at the water in a clean glass, you will see that the surface curls up at the edges where it meets the glass (Fig. 7.22-a). This curl happens because water molecules are attracted more strongly to glass molecules than to each other. This attraction between unlike molecules is called *adhesion*, whereas the attraction between like molecules is called *cohesion*. Because of the stronger adhesive forces, water is pulled a small distance up the glass surface. The travel stops when the adhesive forces between glass and water, the cohesive forces in the water, and the weight of the small elevated ring of water are in balance.

Between certain other substances, for example mercury and glass, the adhesive forces are much weaker than the cohesive forces within the liquid. In this kind of situation, the surface tension in the liquid causes it to curve downward (see Fig. 7.22-b).

Figure 7.23 In a small glass tube, water will rise above the level of the water in the bowl. The smaller the tube, the farther the water rises.

Suppose one end of a small glass tube is put into a bowl of water, as in Figure 7.23. The level of water in the tube will be higher than in the bowl. Also (as the figure illustrates) the smaller the tube, the farther the water will rise. In a 0.2-mm diameter tube, water will rise about 15 cm; in a 0.02-mm tube, it will rise about 150 cm. This rising of liquid in such a tube is called **capillary action**, or *capillarity*.

Water rises in the tube because adhesion pulls water up along the glass walls. Surface tension in the liquid then pulls the center up more nearly level with the edges. This allows the edges to move up more, with cohesion then pulling the center up. This upward motion continues until the forces pulling the liquid up balance the weight of the liquid in the tube. Since, for a given height, the weight is less for a smaller tube, the liquid can rise higher in the smaller tube.

Capillarity is responsible for such phenomena as the rising of melted wax in candle wicks and the traveling of water through a towel, even though only an edge of the towel may be in the liquid. But nature uses capillarity in many important ways. These include bringing water through the soil to the roots of plants, assisting other mechanisms in raising sap in the plant stems, and transferring blood through capillaries (very small blood vessels in the human body).

· ·

Key Concepts

A **fluid** is a substance that can flow—either a liquid or a gas. Liquids are incompressible, gases compressible.

Density is the mass per unit of volume. **Weight density** is the weight per unit of volume. The **specific gravity** of a substance is the ratio of its density to that of water.

Pressure is the force per unit of area. Sea level *atmospheric pressure* is 10^1 kN/m², which can support a column of mercury 76 cm high. The pressure at any depth in a liquid caused by the liquid itself is given by density $\times$ g $\times$ depth.

Pascal's principle states that increasing the pressure at one place in a stationary fluid increases the pressure by the same amount throughout. Hydraulic devices operate on this basis.

Archimedes' principle states that a body partially or totally immersed in a fluid is buoyed up by a force equal to the weight of the fluid it displaces.

Bernoulli's principle expresses conservation of energy for a moving fluid. Regions of lower velocity of fluid particles have higher pressure.

A molecule in a liquid experiences a force from its nearby neighbors. The force repels it if it gets too close, attracts it as it moves away. At the surface of the liquid, molecules are bound together with a **surface tension** that tries to minimize the surface area.

Capillary action causes a liquid to move through small tubes or between the fibers making up a material. It raises the liquid to a height such that the capillary lift equals the weight of the liquid raised.

Questions

Density and Specific Gravity

1. A mixture of 50% water and 50% antifreeze (enough antifreeze to lower the freezing point to −30°C) has a specific gravity of 1.070. The acid solution in a fully charged car battery has a specific gravity of about 1.28, and a "dead" battery has a specific gravity of about 1.13.

Why do service station attendants use a hydrometer—an instrument for measuring specific gravity—to check your antifreeze and to find out if your battery needs charging? (See Problem 16.)

Pressure

2. Why are water beds sometimes recommended for eliminating bed sores on bedridden persons? (Bed sores usually occur at the points of strongest contact between the body and the bed.)

3. Figure 7.24 shows how a tire flattens out on the road under the weight of a vehicle. Why do larger tractor tires require only low air pressure (about 80 kPa), whereas bicycle tires require high pressure (around 500 kPa)? (See Problem 8.)

Figure 7.24 A tire flattens until the total force on the road from the tire pressure equals the weight supported by the tire.

4. Cheap soccer balls are sometimes labeled "Inflate to 0.6 kg or 8 lb." What is missing from these units? What is especially bad about this use of the kilogram?

5. The total downward force from the atmosphere on top of a Ping-Pong table is 421,000 N (47 tons). Why do you not have to lift that hard to lift the table?

6. How do suction cups work? (No, vacuum does not hold them on.)

7. With what force does each square centimeter of your body push outward on the atmosphere?

8. Even though it is seemingly not a popular practice, some women have been known to supplement their own anatomies by wearing inflatable bras. Comment on the hazards of this practice when flying in high-altitude airplanes, especially if the cabin suddenly depressurizes.

9. Is the pressure on your body when you are 2 m below the surface of the water in a swimming pool twice what it is when you are 1 m down?

10. Some large coffee urns have a vertical glass tube along the outside to show the coffee level. Why is the level in the tube the same as that in the pot?

11. True or false: When you get into a half-filled swimming pool, the pressure all over the bottom of the pool increases a small amount.

12. True or false: A mercury barometer works because there is a vacuum in the top of the tube that pulls up on the mercury.

13. When you're washing dishes, and you raise a submerged glass out of the water with its *bottom up,* why does no water leave the glass until the glass is completely out of the water?

14. You operate a plunger for unclogging drains by covering the drain with the plunger, pushing down slowly, and then *quickly* pulling up. The *upward* motion is supposed to shake loose the clog, which then gets washed down by the water. Is the clog *pulled* up or *pushed* up when you pull up?

Pascal's Law

15. In a hydraulic lift (Fig. 7.7) you get a larger force at the larger piston, since the pressure is the same at both. Does this mean you get more work done *by* the larger piston than you do *on* the smaller one?

16. Which of the three systems shown in Figure 7.25 is balanced? In each case the small piston has an area of 10 cm², the large one, 100 cm².

Buoyancy

17. Most people just barely float in fresh water, although some cannot. Estimate the average density of the human body.

18. Which, if either, is easier to float in: deep water or shallow? (Assume you don't touch the bottom either way.) Does the size of the pool make a difference?

19. Crocodiles have an average density somewhat less than 1000 kg/m³. They sometimes swallow stones, and this habit assists them in sneaking up on their prey in water. Why?

20. Sea water has a specific gravity of about 1.025. Why is it easier to float in the Atlantic Ocean than in the Mississippi River?

21. Since ice is less dense than water, it floats in water with a small fraction of the ice sticking out of the water. If a glass is filled to the very top with ice and water, will some water run out as the ice melts?

Pumps

22. The nurse tells Grandpa that he has hypertension (high blood pressure). He thinks that means he's unusually healthy because his heart is pumping at a higher pressure. Actually, his circulatory system has tightened up, increasing the resistance to blood flow. How would you expect this condition to influence the load on

Figure 7.25 The three hydraulic systems contain movable pistons. Which system is in balance?

Grandpa's heart and the likelihood of heart problems?

23. A giraffe has a blood pressure of about 260 mm of mercury when standing, but only about 120 mm when lying down. A dog's blood pressure is about 110 mm either way. Explain why these differences are appropriate.

Fluids in Motion

24. A pipe of 2-cm diameter branches off to four pipes of 1-cm diameter that supply water to four faucets. If *one* faucet is turned on, how does the number of liters of water leaving the faucet per minute compare with the number of liters passing some point in the 2-cm pipe? How does the speed of water in the 1-cm pipe supplying the open faucet compare with that in the 2-cm pipe?

25. Why does a steady stream of water from a faucet get narrower, the farther it gets from the faucet? Does this narrowing process continue, no matter how far it falls?

26. When a convertible travels down a highway, why does the top bulge upward as far as possible?

27. The effect of the Bernoulli principle is one reason a shower curtain pulls in at the bottom when you turn the water on. Explain.

28. When the roof of a house comes off during a tornado, the reason is usually not that the wind blows it off, but rather that the air pressure inside the house pushes it up and then the wind blows it away. Explain. Would it help to open the windows?

Surface Tension and Capillary Action

29. If you had a body of water densely populated with hexopuses all holding hands and pulling on each other, would there be a surface tension around the outer surface of the hexopuses? (Note: A hexopus can be created by surgically removing two tentacles from an octopus.)

30. Why are drops of mercury on a table and drops of water on a greasy pan nearly spherical, except where they are supported by the table or pan?

31. If you dip a dry, soft-bristled paint brush with spread-out bristles into water and remove it, the bristles all stick together. Why?

32. Why is the sand at the beach wet considerably beyond the highest point the water reaches?

33. Where power lines are connected to a house, "drip loops" are provided to keep rainwater dripping outside, rather than running down the entrance line (see Fig. 7.26). Discuss the action of adhesive and cohesive forces in making the water run both ways toward the sag, and drip only near the bottom of the sag.

Figure 7.26 "Drip loops" are provided where power lines connect to the side of the house, to prevent water running down the entrance wire.

Problems

1. If you have 4.3 g of a substance that has a volume of 50 cm³, and you know that it is one of the substances in Table 7.1, which substance is it?

   Answer: butter

2. Superman compresses a rock the size of a tennis ball (volume 160 cm³) with a density of 2.5 g/cm³ into the size of a marble (volume 1.1 cm³). What is the new density?

   Answer: 360 g/cm³

3. What is the specific gravity of each of the substances listed in Table 7.1? What is the weight density of each in N/m³?

   Answer: (for steel) 7.83; 76,700 N/m³

4. A freezer rests on four legs, each 4.5 cm in diameter. If the food and freezer together weigh 3500 N (790 ib), what is the pressure under the legs if the load is equally distributed among them?

   Answer: 550 kPa (80 lb/in²)

5. Women's shoe fashions fluctuate from year to year, but heel sizes reached a minimum in the early 1960s. If we *guess* that then–First Lady Jackie Kennedy's mass was 50 kg, and assume

that $\frac{1}{2}$ her weight was supported by her two heels, then how much pressure did her 1-cm-diameter heels exert on the White House floor?

   Answer: 1600 kPa (230 lb/in²)

6. Find the pressure on a phonograph record from a needle that has a circular point 0.025 mm in diameter and "tracks at 2 grams."

   Answer: 40 × 10⁶ Pa (5800 lb/in²)

7. A queen-size water bed typically has a mass, including the frame, of about 800 kg and typically is supported on a pedestal shaped as illustrated in Figure 7.27. A conventional bed typically has a mass of about 90 kg and is supported on four posts. If the water bed pedestal

Figure 7.27 A queen-size water bed pedestal.

is made of 2-cm-thick lumber and the conventional bed posts are 4 cm in diameter, compare the pressure on the floor for the two beds when each holds two people totaling 140 kg. (Caution: In considering the feasibility of a water bed, you also have to consider whether or not the floor structure can support the *entire* weight.)

Answer: water bed—39 kPa; conventional bed—450 kPa

8. Suppose a car and its contents have a mass of 1400 kg, and the air pressure in the tires is 200 kPa. How much tire area of each tire is in contact with the road? (See Question 3.)

Answer: 0.017 m^2 (170 cm^2)

9. The water pressure in city water lines is typically about 40 lb/in^2. Express this pressure in pascals and atmospheres.

Answer: 280 kPa; 2.7 atm

10. What is the force from the atmosphere on an eardrum of area 0.60 cm^2? A 100-m increase in altitude above sea level changes the air pressure about 1%. If, while ascending this distance in an elevator, the pressure of the fluid inside the ear does not change, what is the net outward force on the eardrum?

Answer: 6.1 N; 0.06 N

11. Swimming pools are sometimes built with observation windows in the side for underwater viewing. What is the total net force on a window 2.0 m by 1.0 m and centered 2.0 m below the surface? Why do you not need to consider atmospheric pressure?

Answer: 39,000 N

12. Pressure gauges on older boilers for hot-water home-heating systems usually show the pressure in the boiler as "feet of water." If the gauge reads "14 ft," what is the pressure in pascals and atmospheres?

Answer: 42 kPa; 0.41 atm

13. A hydraulic automobile lift has a 30-cm diameter piston. What pressure is needed in the fluid to lift a 1200-kg car?

Answer: 170 kPa (1.6 atm)

14. How many 70-kg people can float on an empty, leakless 160-liter (42-gal) barrel without submerging it, if the barrel's mass is 12 kg?

Answer: 2

15. Consider a uniform solid cylinder submerged upright in a liquid. Show, from the difference in pressure on the top and bottom, that the net upward force on the cylinder equals the weight of the water it displaces.

16. A hydrometer for measuring the specific gravity of antifreeze solution or battery acid is made as shown in Figure 7.28. (See Question 1.) Liquid is drawn into the large tube by squeezing and releasing the bulb at top while the tube is in the liquid. Suppose the float rests as shown, and displaces exactly 100 cm^3 of pure water. If a mixture of 50% antifreeze and 50% water with a specific gravity of 1.07 is drawn in, how much higher will the float rest if its stem is 1.0 cm in diameter?

Answer: 8.3 cm

Figure 7.28 A hydrometer for measuring the specific gravity of a liquid.

17. If the atmospheric pressure above a 100-m^2 roof is reduced 1% during a hurricane (see Question 28), what is the net atmospheric force acting upward on the roof?

Answer: 100,000 N (11 tons)

Home Experiments

1. The next time you go "tube-ing," either down the river or down the hill on snow, blow the inner tube up with your breath. You may be surprised how tight you can get it that way. Then estimate the pressure with which you blew as follows: Lay the tube in sand or snow and put a piece of plywood or other stiff board big enough to cover most of the tube on top of it. Stand on the middle of the board so that it stays level. Then get off, measure the approximate diameter and width of the indentation left by the tube, and calculate the pressure from your weight and the indentation area. This value is roughly the pressure with which you blew. (You can do this experiment in your room by sprinkling powder or flour and then measuring the area flattened by the tube. I found I could blow to a pressure of a little over 6 kPa.)

2. If you can get a hot water bottle, a tube for connection to it that is a little longer than your roommate is tall, and a funnel to fit in the tube, ask your roommate to stand on a book on the water bottle. Hold the tube up and pour in water until your roommate is lifted. Calculate and compare the pressure from your roommate's weight and that from the water column. Notice that the water column weighs much, much less than your roommate. Does it matter what size the tube is?

3. You can carry out the can crushing of Figure 7.3 without having a vacuum pump. Get a rectangular can of about 4-liter volume—such as a paint thinner can—with a tight-fitting top. Put about 1 cm of water in the bottom and put the can *with the top off* on the surface element of a range or hotplate. When the water starts to boil, remove the can from the heat, put the top on tightly, and let it cool. Running cold water on it speeds the cooling. The steam in the can condenses, greatly reducing the pressure inside, and letting the atmospheric pressure crush the can.

4. Fill a glass with water, put a piece of paper or cardboard tightly on top, turn the glass upside down while holding the paper in place, then release the paper. Why does the water not run out? You can also do it with a partially filled glass, but you may need to push up a little on the center of the paper before releasing it.

5. Carefully put a fresh egg in a glass of water and notice that it sinks. Remove the egg, stir in as much salt as will dissolve, and put the egg back. It will now float. What can you say about the relative densities of water, saltwater, and fresh egg?

6. You can demonstrate Bernoulli's principle with several simple experiments. (1) Lay a Ping-Pong ball on the edge of a table and put your mouth down where you can blow along the tabletop and beside the ball. A quick puff will make the ball roll toward the blown air, not away. (2) Put a straw in your mouth and aim it upward at a slight angle to the vertical direction. Hold a Ping-Pong ball above the end of the straw and blow with a steady even breath. As you start to blow, release the ball. With practice you can make the ball hang there a few seconds. Experiment with different angles and different starting ball positions. (3) Hang two apples from strings tied to their stems so that they are a few centimeters apart. Blow between them. Are you surprised by the direction of their motion? (4) Blow between the flames of two candles placed a few centimeters apart. Which way does the flame tilt? (5) Hold a straw vertically in a glass of water and blow with a quick gust *across* the top of the straw. With the right technique, water will come up the straw and spray out with your air. (You, the glass, and the straw will have become an atomizer.)

7. You can simulate the curving of a golf ball, tennis ball, or baseball by tying a string around a Ping-Pong ball and hanging it from about one meter of string. Pull it aside and let it swing as a pendulum. It swings in a straight line. Then twist the string about 50 turns and again pull it aside to swing back and forth. The direction of oscillation gradually rotates. Does it curve the right way?

8. Carefully lay a needle, a razor blade, or a piece of wire screen on the surface of a bowl of cold water. (The first two will be easier to place if you rub them with a light coat of oil.) Now touch the surface with a bar of soap, or pour in a little liquid detergent to break the surface tension. Kerplunk!

9. Sprinkle pepper on the surface of a bowl of water. Then touch the center of the water surface with a piece of soap and watch the pepper "split."

10. Take two clean flat pieces of glass—windowpane or picture frame glass—hold two edges together, and leave a small gap between the opposing edges. Dip them vertically into a pan of water and notice how capillary action pulls the water up farther where the plates are closer together.

References for Further Reading

Boys, C. V., *Soap Bubbles* (Doubleday & Co., Garden City, New York, 1959). These three lectures given by Boys in 1889 and 1890 describe many interesting effects of surface tension and capillary action. Includes a list of experiments to demonstrate these effects.

Milne, L. J., and M. Milne, "Insects of the Water Surface," *Scientific American*, April 1978, p. 134. Four insects found all over the world depend on the surface tension of water for their existence.

Smith, N. J., "Bernoulli and Newton in Fluid Mechanics," *The Physics Teacher*, Nov. 1972, p. 451. Focuses on the explanation of airplane lift and baseball curves.

Strother, G. K., *Physics with Applications in the Life Sciences* (Houghton Mifflin Co., Boston, 1977). Chapter 11 includes many applications of fluid principles in biological systems.

Walker, J., *The Flying Circus of Physics* (John Wiley & Sons, New York, 1977). See especially Chapter 4. Considers many interesting effects in fluids and gives references to literature that discusses in detail those things that we could only cover briefly here.

PART TWO

HEAT

I . . . am persuaded, that a habit of keeping the eyes open to everything that is going on in the ordinary course of the business of life has oftener led, as it were by accident, or in the playful excursions of the imagination, . . . to useful doubts and sensible schemes for . . . improvement, than all the more intense meditations of philosophers, in the hours expressly set apart for study.

COUNT RUMFORD

Benjamin Thompson, born in Woburn, now Concord, Massachusetts, in 1753, was an opportunist active in political life during the time of the American Revolution. Even though he was in general loyal to the British side, he was careful always to be identified with the winners. His scientific and technological contributions included major breakthroughs in understanding the nature of heat, improvement of the fireplace, and invention of the drip coffeemaker. He made many of his discoveries during the course of his everyday routine. King George III conferred knighthood on him and he was made Count of the Holy Roman Empire, choosing for his title the name Rumford.

H E A T   E N E R G Y

How many roles does heat play in your life? Some of these roles come to mind at once: its effects on climate, personal comfort, cooking, and the warming of houses. But heat also influences our lives in subtle ways that we don't usually think about.

Take another look at Figure 4.19 of Chapter 4 and notice that heat was often the intermediary between source and use of energy. Almost all electricity is generated from heat: Some form of fuel is burned to release heat that produces steam that turns a turbine that runs a generator that generates electricity. Automobile, train, and airplane engines burn fuel that releases heat to make the engines go. Heat is involved in the manufacture of thousands of useful everyday products: steel and aluminum production requires large amounts of it; it is used to dry paint on bicycles and to cure colors in fabrics; and it's needed to put the "memory" into permanent press fabrics. Your own body depends on heat for proper functioning. Almost all the energy we use is in the form of heat in one stage of its exploitation.

In this chapter and in the next three we'll see how heat performs its many jobs. We'll find that when heat is applied to a substance, it can have either or both of two effects: Heat can change the temperature of the substance, or it can change the substance's state of matter. We'll consider the subject of "change of state" in Chapter 9. Here, we will define temperature and heat and describe how they are measured. Then we'll discuss specifically how heat and temperature relate to each other. We'll consider sources of heat and, finally, we will examine an effect that usually accompanies a temperature change, an effect of tremendous practical importance: expansion or contraction. Chapter 10 considers how heat moves from one location to another; that is, how heat is transferred. Finally, Chapter 11 discusses how we can use heat to do work and how we can pump heat from one area to another of higher temperature.

8.1 TEMPERATURE VERSUS HEAT

When we begin to discuss any new topic, the first step is to define the terms involved. This topic is no exception. Even though the terms *temperature* and *heat* are both familiar, we must carefully distinguish between them.

Temperature

When people use the word heat, they often mean **temperature**. For example, you may say, "Turn up the heat, I'm freezing." But what you are really saying is, "Set the thermostat so that the *temperature* will get higher and I will be warmer." When you are heating water for that 3:00 A.M. cup of coffee while you study for an English test, you know that the water boils when it gets to a certain *temperature*. When it's "104 in the shade," it's the *temperature* that gets to you. When you are sick, your body *temperature* goes up.

From an everyday point of view, *temperature is a measure of the "hotness" or "coldness" of an object or region.* Soon we'll see how this relates to what goes on inside the object.

Heat

Knowing that we want a certain temperature doesn't tell us what to do to *change* the temperature. You know that it would take longer to bring a large potful of water to the boiling point than it would a small one placed on the same burner. Yet, either one will burn your tongue just as quickly because both boil at the same temperature. For similar reasons it takes a much larger furnace to heat a 15-room house than a neighboring 3-room house constructed of the same materials.

To see the why, consider an analogy. Suppose you have a pint jar and a bathtub side by side, each containing water up to a level of 5 cm from their bottoms. If you then add a cup of water to each, what does this additional amount do to the water levels? In the pint jar it about doubles the level. In the bathtub you probably don't see any change. Yet the same amount of water was added to each container. In summary, we can change the water level in either container by adding water to it. But to change the *level* by the same amount in both containers, we must add more water to the larger-volume container.

Figure 8.1 The larger letters symbolize the larger quantity of heat needed to change the temperature of the larger pan of water the same amount.

The relationship between heat and temperature is the same as the relationship between water added and water level in a container. To change the temperature of something, you have to add heat to it. The more of the "something" you have, the more heat it takes to change its temperature a certain amount. Figure 8.1 symbolizes the situation for the large and small pots of water. More heat is needed to bring a large pot of water to a boil than a small one, even though both start and stop at the same temperature. Similarly, more heat is needed to change the temperature of the air and contents of the 15-room house a fixed amount than to do the same for the 3-room one.

One way of defining heat is to say what it does. *Heat raises the temperature of matter to which it is added and lowers the temperature of matter from which it is removed*, provided the matter does not change state—such as from solid to liquid or from liquid to gas. (As mentioned, we'll consider that possibility in Chapter 9.)

Internal Energy versus Heat

We have said what heat *does*, but not what it *is*. In order to say what it is, we need to consider the fact that all matter is made up of molecules— either single atoms or combinations of atoms. There are 33 million billion billion of them, for example, in a liter of water. These molecules continuously jiggle around at random.

In gases at ordinary (near atmospheric) pressures, the molecules bounce around freely, interacting little except during collisions. In liquids and solids the molecules continuously push and pull on each other by means of electrical forces. In the case of solids, these electrical forces hold the molecules in the same average position relative to each other: The molecules behave *as if* they were held to each other by tiny springs, as shown in Figure 8.2 for only four molecules.

As a molecule jiggles around, its total energy consists of its kinetic energy and its share of the potential energy of the compressed or stretched springlike forces. Thus, energy is stored within the substance as kinetic and potential energy of the molecules. We refer to this energy as **internal**

Figure 8.2 A model that shows how the electrical force between molecules in a solid acts as if the molecules were tied together by tiny springs. The circles represent molecules. The springs do not exist, of course, but the force between molecules varies with separation much like the force of a spring.

energy. Another term sometimes used in such a discussion is *thermal energy*. This is a general term that refers to the energy associated with the random motion of the molecules within a substance.

In forming a mental image of what's going on inside a substance, it helps to concentrate on the kinetic energy associated with the jiggling molecules. Keep in mind, however, that potential energy is also present. The molecular energies vary over a wide range within the substance but, at any instant, have some *average* value.

When we add or remove internal energy, we say we have added or removed **heat**. Thus, *heat is internal energy that is added to or removed from matter*. In other words, if we refer to the energy stored in the molecules of a substance, we are talking about internal energy. But if we discuss adding or removing internal energy to or from the body, we are talking about heat. Heat, then, is not really a new form of energy. It merely describes mechanical energy transfer associated with individual molecules.

The principle of conservation of energy still applies, but now it must include changes in the internal energy of the body. When this principle does in fact include the effect of internal energy changes, we usually call it the *first law of thermodynamics*. We'll say more about this special form of the energy conservation principle in Chapter 11.

But why would heat enter or leave a body? Consider the simple practice of heating a pan of water on an electric range. Figure 8.3 shows an imaginary slice through an electric heating element and a pan. The molecules of the heater and the pan are jiggling around with a certain amount of thermal energy. Those molecules on the heater surface in contact with the pan collide with the molecules on the pan's surface. Similarly, those of the pan collide with the molecules of the heater. If, on the average, the molecules of the heater have more energy than those of the pan, the heater molecules will give up some of their energy as they collide. That way, the heater will lose some internal energy, and the pan will gain some. This process means that heat has been transferred from the heater to the pan. Heat is transferred *through* the pan in a similar way.

As long as the heating element is operating, electricity supplies the heater with internal energy that is transferred to the heater's surface by molecular activity. When you turn the heater off, energy transfer continues until the *average* energy of the molecules in both heater and pan is the same. But you already know that, with the heater off, both it and the pan eventually come to the same temperature. Any time a hot body and a cold body are put in contact. the hot one cools and the cold one warms until they both reach the same final temperature. If we combine this fact with what we just said about the heater and pan molecules, we can conclude that when two bodies are at the same temperature, the average thermal energy of the molecules of the bodies is the same. In other words, *temperature is a measure of the average thermal energy* of the

Figure 8.3 Heat is transferred from the range-heating element as molecules of the heater give up thermal energy to molecules of the pan.

molecules of the substance. As the *average* molecular energy goes up, the temperature goes up. As the *average* molecular energy goes down, the temperature goes down.

Two bodies are at the same temperature when: (1) the average thermal energy of their molecules is the same; and (2) no heat flows from one to the other if they are placed in contact. Actually (1) and (2) are equivalent— if the average molecular energy of one were higher than that of the other, energy would be transferred. If two bodies in contact are at different temperatures, *heat always flows from the hotter to the colder*, and never the reverse.

Let's summarize the terms we have defined.

1. *Temperature:* the hotness or coldness of the material, as determined by the *average* thermal energy of the individual molecules.

2. *Internal energy:* energy stored within a substance as potential and kinetic energy associated with the molecules.

3. *Heat:* internal energy being added to or removed from the substance.

We often use the word *heat* rather loosely when we are really talking about internal energy. You can usually tell from the context which idea is intended.

To illustrate the difference in terms, let's suppose the small pan in Figure 8.1 contains 1×10^{25} molecules and the large pan, 3×10^{25} molecules. If both pans are at room temperature, the total *internal energy* of all the molecules in the large pan is three times as much as that in the small pan. But because they are both at the same temperature, the *average thermal energy* of the molecules is the same in each. If we put them both on a stove and raise their temperatures to the boiling point, we must add three times as much heat to the large one, since it has three times as many molecules. But once both are at the temperature corresponding to the boiling point, the average thermal energy is the same in each pan. This average is, of course, higher than it was at the lower temperature.

8.2 MEASURING TEMPERATURE AND HEAT

Now that we know what heat and temperature mean, we can describe how to measure them.

Temperature Measurements

We call a device for measuring temperature a *thermometer*. Thermometers work by using some property of matter that changes with temperature. Properties often used are: the expansion of a material with increasing

Celsius   Fahrenheit

100°C   212°F

60°C   140°F

37°C   98.6°F

20°C   68°F

0°C   32°F

−18°C   0°F

−20°C   −4°F

100 C deg

180 F deg

Figure 8.4 A liquid-in-glass thermometer, comparing the Celsius and Fahrenheit temperature scales.

temperature and contraction with decreasing temperature, the pressure of a gas that changes with temperature, the electrical resistance (a property discussed in Chapter 15), and the color of the light given off by a body that is hot enough to glow.

The liquid-in-glass thermometer (Fig. 8.4) is probably the most familiar type. As the temperature rises, the liquid in the bulb at the bottom—usually mercury or alcohol—expands more than the glass. The only place for expansion is up the tiny tube. As the temperature drops, the liquid contracts, moving down the tube. The top level of the liquid tells the temperature.

Temperature Scales

To assign a specific value to a temperature, we need some **temperature scale**. We can establish such a scale by assigning a number to any two points on the scale whose temperature can be reliably reproduced. Thus, we usually assign one number to the temperature at which water freezes and another number to the temperature at which it boils. (The temperatures of these changes of state depend on the atmospheric pressure. In *this* chapter whenever we refer to the freezing or boiling points, we will be assuming sea-level atmospheric pressure.)

Most of the world uses the Celsius scale, named after the Swedish astronomer Anders Celsius, who introduced it in 1742. On this scale, the freezing point of water is assigned the temperature 0 degrees (0°), and the boiling point, 100 degrees (100°).* For many years this was called the centigrade scale—"centi" standing for the 100 divisions between the two fixed points.

Until recently, we in the United States have used the Fahrenheit scale for all except scientific purposes. On this scale, the freezing point of water is 32°, and its boiling point is 212°. In setting up his temperature scale in about 1717, the German instrument-maker Gabriel Fahrenheit chose one fixed point as the lowest temperature he could produce in his laboratory with a freezing mixture of ice and salt, and called it 0°. The other point was the normal temperature of the human body, to which he assigned the number 96°. The freezing and boiling points of water then came out to be 32°F and 212°F. When these two points were accepted as standards and thermometers were made more accurately, the normal body temperature turned out to be 98.6°F. Fahrenheit also invented the mercury-in-glass thermometer and used it in defining his temperature scales.

Figure 8.4 shows a comparison between the Fahrenheit and Celsius temperature scales for some familiar temperatures.† Many of us in the

*Celsius first proposed a scale in which these two numbers were reversed—0° for boiling and 100° for freezing.

†The relationship between these two scales is given by the equation $C = \frac{5}{9}(F - 32)$, where C and F are the temperatures in degrees Celsius and degrees Fahrenheit, respectively.

United States are just learning to "think Celsius." One way to make approximate conversions near familiar temperatures is to notice that 100 C deg spans the same interval as 180 F deg. This span means that a change of 1 C deg is about 2 F deg (actually 1.8 F deg). If you then remember a few Celsius-Fahrenheit equivalents, you can make conversions near these temperatures in your head. For example, you might remember that 68°F is the same as 20°C. Then each 2 F deg change near there gives a change of about 1 C deg. For instance, to the nearest degree, 70°F is 21°C. After a 10-F-deg variation from the exact equivalent, your estimate is wrong by about $\frac{1}{2}$°C.

[Note: To avoid confusion, we are following the common practice of writing Celsius degrees (C deg) when we are talking about a temperature *interval*, and degrees Celsius (°C) when talking about an actual temperature.]

Absolute Temperature

In laboratories that have facilities for taking substances to extremely low temperatures, experimenters have found that the closer they get a substance to the temperature of −273.16°C, the more difficult it becomes to keep lowering its temperature. In fact, the temperature of −273.16°C has never been reached. Current theory predicts this value to be the lower limit of temperature, and thus not reachable at all.

It is often useful to have a temperature scale for which the lowest possible temperature is defined as 0°. We call such a scale an *absolute* temperature scale, with the lower limit being called *absolute zero*. The *Kelvin* temperature scale is an absolute scale that uses the same size degree as the Celsius scale. The melting and boiling points of water are then 273 K and 373 K, respectively. The kelvin (symbolized K) is the unit of temperature in the SI system.

Heat Measurements

Since heat is a form of energy, quantities of heat can be measured in any energy units—joules, foot-pounds, kilowatt hours, for example. In the SI system, the basic unit of heat is the joule. However, separate units of heat were defined before anyone knew that heat was a form of energy. These special heat units are still used, both because of tradition and because they are about the right size for many heat measurements.

The traditional metric unit of heat is the calorie (cal) or kilocalorie (kcal; 1000 cal). The kilocalorie is now defined as 4190 J. Originally it was defined as follows: *1 kcal is the heat needed to raise the temperature of 1 kg of water 1 C deg.* (One liter—just over a quart—of water has a mass of 1 kg.) Because of tradition and the intuitive definition of the kilocalorie, we will use it in our discussion of heat. However, keep in mind that we are talking about *energy*, and could just as well use joules as the unit of measure.

Confusion sometimes occurs because the Calorie (with capital C), used to give the energy value of foods, is the equivalent of the kilocalorie, or 1000 cal. To avoid such confusion, many nutritionists now use the term *kilocalorie* rather than *Calorie*.

In the British system of units, the heat unit is the British thermal unit (Btu), defined as: *1 Btu is the heat needed to raise the temperature of 1 lb of water 1 F deg.* Even after the American switch to metric units is complete, older heating and air conditioning systems rated in Btu will be in use for some time.

Table III.3 of Appendix III gives a comparison between these and other energy units. From this comparison, we see that 1 kcal = 4190 J = 1.16 Wh = 0.00116 kWh = 3.97 Btu.

· ·

EXAMPLE 8.1   How much energy is needed to heat the water to take a bath? Assume you use 100 liters (about 26 gal) of water that enter your house at 13°C (55°F), and you like your bath water at 43°C (110°F). If your water is heated with electricity costing 9¢ per kWh, how much does it cost to heat this water?

Since 1 kcal raises the temperature of 1 kg of water (which has a volume of 1 liter) 1 C deg, the total heat needed is found by multiplying this number by the mass of water and by the temperature change:

$$\text{heat needed} = 1 \text{ kcal/kg·C deg} \times \text{mass of water} \times \text{temperature change}$$

$$= 1 \text{ kcal/kg·C deg} \times 100 \text{ kg} \times 30 \text{ C deg} = 3000 \text{ kcal}$$

We can convert this to kilowatt hours:

$$\text{heat needed} = 3000 \text{ kcal} \times 0.00116 \text{ kWh/kcal} = 3.5 \text{ kWh}$$

At 9¢ per kWh,

$$\text{cost} = 3.5 \text{ kWh} \times 9¢/\text{kWh} = 31¢$$

8.3 CHANGE OF TEMPERATURE AND SPECIFIC HEAT

Let's step back briefly and look at some of what we have said so far: (1) temperature, the hotness or coldness as determined by the average molecular energy, is changed by adding or removing energy in the form of heat; (2) temperature is measured on an arbitrary scale defined by

Table 8.1 Specific heats of ordinary substances

Material	Specific Heat (kcal /kg·C deg)	Material	Specific Heat (kcal /kg·C deg)
water	1.00	table salt	0.20
ice	0.50	hamburger	0.75
air	0.24	mercury	0.033
wood	0.42	steel	0.11
glass	0.20	aluminum	0.21
sand	0.20	gold	0.032

assigning numbers for the temperatures of certain fixed points; and (3) heat can be measured in any energy units, but is conveniently measured in units such as kilocalories that are defined in terms of the heat needed to raise the temperature of water.

Because of the way we defined the heat units, we know how much heat it takes to warm a certain amount of water (or how much heat has to be removed to cool it) a fixed temperature difference. But what about heating and cooling other materials? For example, why is it that you can pick up bare-handed a thin aluminum pie pan that has just come out of the oven, but you better not try doing so with a steel one? Or, why does your freezer run longer (indicating more heat removal) after you plug it in if it's full of room-temperature food rather than full of room-temperature air? The answers to these questions are related to the property of matter called **specific heat**.

Specific Heat

The specific heat of a substance is the amount of heat needed to raise the temperature of a unit mass (for example, 1 kg) *of the substance 1 degree.* The values of specific heat vary widely from substance to substance. Table 8.1 presents some specific heat values for a few ordinary materials, given in kcal per kg per C deg.

Knowing the specific heat and how much of a substance we have, we can calculate how much heat we must add to or remove from the substance to change its temperature a fixed amount. Since specific heat is the heat needed *per unit of mass* and *per degree of temperature change,* we multiply it by both of these factors to get the total heat added or removed:

heat to change temperature = specific heat × mass × temperature change

You'll notice that we have in effect already used this relationship in working Example 8.1, but we'll illustrate it with another example.

Count Rumford boils water without fire

MUNICH, BAVARIA (OCTOBER 3, 1794)—Count Rumford, Inspector General of Artillery for the Army of Bavaria, took advantage of the beautiful autumn afternoon to hold another of his public demonstrations on boiling water without using fire. The Count had one of his machines for boring cannon immersed in a large tank of water. He astonished onlookers by showing that, just by turning the boring tool inside the cannon, he could heat the water to boiling and continue it boiling no matter how long he turned the tool.

In a news conference after the demonstration, Count Rumford explained that "being engaged lately in superintending the boring of cannon in the workshops of the military arsenal at Munich, I was struck with the considerable heat which a brass cannon acquires in a short time in being bored." He went on to explain that his demonstration here today cast grave doubt on the views long held by philosophers that rubbing objects together heats them because it squeezes caloric fluid out of the objects. "How," he asks, "can you explain the inexhaustable supply of caloric in the cannon that is evidenced by the continued heat production, no matter how long I turn the boring tool?"

Count Rumford, born Benjamin Thompson in the British Colony of Massachusetts, entered the service of Elector Karl Theodor ten years ago. He was elevated to the rank of Count of the Holy Roman Empire seven years later. The count chose for his title the name Rumford, the original name for Concord, New Hampshire, where he married the wealthy widow of Colonel Benjamin Rolfe in 1772.

Rumford concluded his news conference by pointing out that this way of producing heat is not very efficient. Conceding that a great deal of heat is produced by the borer, he explained that if you burn the fodder fed to the horses turning the drill, the resulting fire could heat much more water.

· ·

EXAMPLE 8.2   How much heat has to be removed from 50 kg of hamburger placed in a freezer at 20°C to lower the temperature to 0°C? Compare this with the heat removed to change the temperature of the same volume of air by the same amount. (The mass of this much air is 0.072 kg.)

heat removed = (specific heat) × mass × (temperature change)

$$= 0.75 \text{ kcal/kg·C deg} \times 50 \text{ kg} \times 20 \text{ C deg} = 750 \text{ kcal}$$

For air,

heat removed = 0.24 kg/kcal·C deg × 0.072 kg × 20 C deg

$$= 0.35 \text{ kcal}$$

In Example 8.2 notice that the specific heat of hamburger is about three times that of air. At first glance you might expect the hamburger to take only three times as much heat. But since the heat needed depends on the mass (which is much higher for the hamburger), the heat needed is over 2000 times as much.

. . . .

QUESTION   From the specific heats in Table 8.1, it might first appear that the aluminum pie plate mentioned near the beginning of this section might store more thermal energy and therefore inflict a worse burn on your hand. What other factor must be considered in providing an explanation?

_____

Water and Its Specific Heat

The specific heat of water is exactly 1.00 kcal per kg per C deg. This handy figure is not just a fluke of nature but comes from the way we define the kilocalorie. From the definition of this unit, water must have this specific heat. This means that the numerical value of the specific heat of any substance is given by the ratio:

$$\frac{\text{heat needed to change temperature of substance 1 degree}}{\text{heat needed to change temperature of same mass of water 1 degree}}$$

(Compare this ratio with the way we defined specific gravity. Since we find water almost everywhere, it is often easy to compare a certain property of other substances with the same property of water and report the result as a ratio.)

The fact that the specific heat of this ever-present substance, water, is higher than for most ordinary materials is an important part of nature's design, and it has a profound effect on your life. Some examples:

1. The surface of the earth is dominated by water. Water's high specific heat means that the earth can absorb or give off large quantities of heat without much temperature change, making the temperature difference between winter and summer acceptable for life.

2. Your body, which is largely water, can also absorb or give off a lot of heat without much temperature change.

3. Water, which can carry the most heat of any ordinary material, is readily available for carrying heat from furnace to user in applications such as home heating.

4. Areas near large bodies of water tend to have a more nearly constant year-round temperature. This is evident in coastal areas, particularly the west coast of the United States, where the prevailing wind is from sea to land.

5. This high specific heat has the disadvantage that heating water for your shower takes lots of fuel or electrical energy, most of which goes down the drain.

Energy Conservation

The energy conservation principle tells us that, when a hot and a cold substance are combined, the heat lost by the hot substance must be gained by the cold one. The use of this idea in studying specific heats (or the amount of hot or cold substance needed) is often called *calorimetry*. An everyday example will illustrate the principle.

· ·

EXAMPLE 8.3 How much hot water do you have to mix with 1 liter of cold water (mass = 1 kg) in your wash basin to make it comfortable for washing your face on a cold day? Take the cold water to be fresh from the supply lines at 8°C (46°F) and the hot water to be at 60°C (140°F).

Since both hot and cold water come to the same final temperature, which we'll take as 43°C, we have:

heat lost by hot water = heat gained by cold water

(specific heat) × (mass of hot water) ×
    (temperature change of hot water) = (specific heat) × (mass of cold water) × (temperature change of cold water)

(1 kcal/kg·C deg) × (mass of hot water) (60°C − 43°C)
   = (1 kcal/kg·C deg) × (1 kg) × (43°C − 8°C)

We can solve for the mass of hot water, which comes out to 2 kg, or 2 liters of hot water.

· ·

EXERCISE How much energy is needed to heat these 2 liters of water to 60°C if they enter the water heater at 8°C? Compare this answer with the typical 200 kcal in a piece of boysenberry pie. *Answer:* $\frac{1}{2}$ piece of pie

---

8.4 SOURCES OF THERMAL ENERGY

Most of the thermal energy we use comes from burning some kind of fuel. In heating our homes, many of us burn natural gas, fuel oil, coal, or wood. These fuels release chemical energy as they burn. During the

Table 8.2 Energy content of certain fuels

Fuel	Heat of Combustion (kcal/kg)	Fuel	Heat of Combustion (kcal/kg)
gasoline	11,500	sirloin steak	3100
fuel oil	10,600	ice cream	2100
natural gas	9500*	apple	580
coal	7100	peanut butter	5800
charcoal	7300	lemon meringue pie	2600
oak wood	4000	chocolate candy	5000

*In kcal/m³ at standard atmospheric pressure.

burning (oxidation) process, the original chemical form of the fuel is changed to one with lower chemical energy. The released chemical energy shows up as kinetic energy of the molecules—thermal energy. This energy heats the air and other surrounding materials.

Automobile engines burn gasoline to release heat that operates the engine. Most electrical power plants burn coal, gas, or oil to heat water and make steam for turning the turbines. Therefore, even homes or industries that use electrical energy for heating obtain their energy, in reality, from burning fuel.

The food we eat basically serves as "fuel" for our bodies. The nutrients in the food are distributed to all parts of the body, where they are oxidized—burned—to release energy. This energy is used to keep the body itself functioning, to do mechanical work, and to produce heat to keep the body warm. A healthy human body uses about 70 kcal/h even when sleeping. The same body when running fast uses about 600 kcal/h.

- -

EXAMPLE 8.4   If you eat a piece of a cheesecake containing 400 kcal of chemical energy, how far do you have to run if you want to exercise this energy off?

If you run fast enough to use 500 kcal/h, you need to run for $\frac{4}{5}$h. At a speed of 10 km/h (about half the speed a long-distance runner could run for that time), you need to run about 10 km/h $\times \frac{4}{5}$h = 8 km, or 5 miles.

The energy released by completely burning a unit of mass of a substance is called the **heat of combustion**. Table 8.2 gives the heat of combustion of some ordinary fuels. Included in Table 8.2 are some foods that are usually "burned" in a living body rather than in a furnace.

- -

EXAMPLE 8.5   How much heat is released in the complete burning of 900 kg (a pickup truck load) of oak firewood in a wood stove?

From Table 8.2, the energy content per kilogram of oak wood is 4000 kcal/kg. Thus:

$$\text{heat released} = \text{heat of combustion} \times \text{mass}$$

$$= 4000 \text{ kcal/kg} \times 900 \text{ kg} = 3.6 \times 10^6 \text{ kcal}$$

· ·

EXERCISE   How much fuel oil, completely burned, would release the energy found in Example 8.5?   *Answer*: 340 kg (95 gal)

8.5 EXPANSION AND CONTRACTION WITH TEMPERATURE CHANGE

Why does your toast pop out of the toaster when it's done? Why does your circuit breaker automatically turn off when you plug too many electrical appliances into the receptacle? Why do some concrete highways have those annoying cracks every few car lengths that go thump, thump as you ride down the road? The answer to all these questions is related to the fact that most substances expand when their temperature rises and contract when their temperature falls. We have already used this property in talking about liquid-in-glass thermometers—both the liquid and glass expand with increasing temperature, but the liquid expands more than the glass.

One prominent exception to this behavior occurs in water when its temperature is below 4°C. Water then *contracts* with temperature rise, and expands with temperature drop. This effect has an important impact on life on earth. However, since the reason for this behavior is closely related to the melting and freezing of water, we will discuss it later. In this chapter we will consider only those substances that expand with increase, and contract with decrease, in temperature.

When the temperature of a substance rises, the average kinetic energy of the molecules increases. Since the molecules then collide harder with each other, they spread out, causing the material as a whole to expand. (You can see a similar effect on a crowded dance floor when the tempo changes from a slow to a fast beat.) This effect holds in solids, liquids, and gases. A uniform cube of a solid expands equally in all directions, but if a solid object is longer in one particular dimension, the overall expansion in that direction is greater than in the others. The amount of expansion or contraction for each degree of temperature change varies from substance to substance.

QUESTION

Figure 8.5 Does the hole in the washer get bigger or smaller as the temperature rises? Hint: Is the longer dimension parallel to, or perpendicular to, the dashed line?

For a flat washer (a flat piece of metal with a hole in the center) such as is shown in Figure 8.5, does the hole get bigger or smaller as the temperature rises? Your instructor may need to demonstrate this effect before you are convinced of the right answer.

In the case of a liquid, only volume changes have meaning, since a liquid takes the shape of its container. In the liquid-in-glass thermometer, the liquid is confined to a tiny tube, so that the change is mostly in one direction.

Thermal expansion is both very useful and very much a nuisance. We will mention a few examples on each side—first, the useful.

Bimetallic Thermometers and Thermostats

(a)

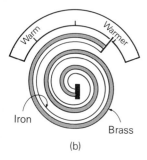

(b)

Figure 8.6 A bimetallic strip bends away from the metal that expands most as the temperature rises.

Figure 8.7 A bimetallic strip is used as a thermostat to turn heating and cooling systems on and off.

Suppose we have thin strips of two different metals welded together as in Figure 8.6(a). If one of these metals expands more for a given temperature change than the other, bending occurs as illustrated in the figure. For example, brass expands about 50% more than iron for a given temperature increase. When the temperature is low, the bimetallic (two-metal) strip bends to the left; when the temperature rises, the brass expands more, causing the strip to bend to the right. This device could be a crude thermometer.

To get a lot of bending from this simple strip, you need a large temperature change. The effect can be amplified by shaping the strip into a coil as in Figure 8.6(b). Then, a fairly small temperature change can produce a measurable movement of the pointer. Most thermometers now used on control units for heating and air conditioning systems have a coiled bimetallic strip.

Most *thermostats*, whether on home heating and air conditioning systems, electric blankets, or pop-up toasters, use some variation of the bimetallic strip. A simplified version of a thermostat used to control an electrical heater is shown in Figure 8.7. As the temperature around the thermostat drops, the bimetallic strip bends downward, making contact with the terminal below. Electrical current can then flow from the power source, through the strip, through the heater, and back to the source. When the temperature rises, the strip bends upward, breaking contact so that no current can flow to the heater. For a large heating system, the current flowing through the thermostat when contact is made usually operates another switch (called a relay) that turns the system on. For controlling an air conditioner, the contact is on the other side of the bimetallic strip, so that contact is made and current flows when the temperature *rises* above a certain point.

Real thermostats are usually more complicated than we've shown. There is usually an adjustment so that contact is made at whatever temperature you want. Often the strip is in the form of a coil, such as in the thermometer of Figure 8.6(b). In most cases, the thermostat merely turns the system on or off at some preset temperature rather than regulating the rate at which heat is produced or removed.

· ·

QUESTION   Suppose you come into a cold house and are in a hurry to get the temperature up to 20°C. Will the temperature get there faster if (a) you set the thermostat on 30°C and then set it back to 20°C when the room temperature gets to that value, than if (b) you set it on 20°C at the start?

Many household appliances use bimetallic-strip thermostats to turn the heating element on and off in order to control their temperatures within a given range. These include electric frying pans, percolators, hair driers (as a safety device to prevent too much heat), pop-up toasters (to activate an electromagnet that releases the racks), tabletop ovens, irons, and electric blankets.

Unwanted Thermal Expansion

We mentioned earlier that thermal expansion can be a nuisance. As Figure 8.8 shows, it also can be hazardous if not properly allowed for. In concrete

Figure 8.8   During the hot summer of 1966 in Stewartville, Minnesota, this buckling of the tracks occurred due to thermal expansion.

Figure 8.9 Expansion joints (such as this one from the George Washington Bridge) are provided at each section of a bridge. As the bridge span expands and contracts, fingers of the expansion joints come together and withdraw.

Figure 8.10 A gas expands when heated only if it is in a container whose volume increases.

highways or sidewalks, gaps must be left every few meters to allow for expansion and contraction as the temperature changes. Bridges also need room to expand and contract with temperature change (Fig. 8.9). Power lines, if stretched tightly in summer, would break in winter. A cold glass bowl may break when filled with hot water because the layer of glass that gets hot first expands faster than the cold layer below, resulting in fracture.

Expansion of Gases

Gas molecules bounce around freely and fill up the complete volume of their container. When a gas is heated, it expands only when its container increases in volume. If the volume of the container doesn't change, the gas cannot expand. Instead, the molecules of the gas merely pound harder against the walls of the container and against each other, causing an increase in the pressure of the gas.

Suppose a gas is confined in a cylinder with a movable piston (Fig. 8.10). The weight on the piston keeps constant pressure on the gas while letting it expand as heat is added. Work is done in increasing the potential energy of the piston and the 1-kg mass. If the piston were held fixed, the gas could not expand during heating, and the pressure would rise instead.

As an illustration, consider what happens to the air pressure inside an automobile tire when it turns on a hot highway. The air inside the tire absorbs heat from the road. Even if the road were not hot, the air would be heated from the friction of the continuous flexing of the tire, especially if the pressure were low. Since the tire stretches very little, the air pressure must increase. That's why tire manufacturers recommend checking the pressure when the tires are cold and *not* letting air out if the pressure goes up while on the road.

As a second example, consider this question: How do you keep the bounce in your tennis balls when they lie around unused for a long time? Most tennis balls come pressurized in a can to maintain high pressure inside the balls and give them a lively bounce. After the can is opened, the gas in the balls leaks out gradually, reducing the pressure inside so that the balls become "dead."

One way to prevent this leakage is to buy a gadget from the sporting goods supplier for storing the balls under pressure when not in use. There's a simpler way: Reduce the pressure *inside the balls* when they're not in use so that the air doesn't try to get out! That's really not so hard: Store the tennis balls in the freezer compartment of your refrigerator. As the temperature drops, the pressure drops accordingly. (The balls won't bounce well until they warm up after you take them out.)

Key Concepts

Temperature, a measure of the hotness or coldness of a substance, is determined by the *average* energy of the molecules of the substance. **Internal energy** is the energy stored within a substance as the kinetic and potential energy of its molecules. **Heat** is internal energy being added to or removed from a substance.

If two bodies at different temperatures are placed in contact, heat always flows from the warmer to the cooler body and never the reverse. Two bodies are at the same temperature if no heat flows from one to the other when they are placed in contact. Adding heat to a substance raises its temperature; removing heat lowers its temperature (if the substance does not change its state).

A **temperature scale** is defined by assigning a number (temperature) to two fixed points and dividing the interval into equal parts. Heat, a form of energy, can be measured in any energy units, but is often measured in kilocalories—the amount of heat needed to raise the temperature of 1 kg of water 1 C deg.

The **specific heat** of a substance is the heat needed to raise the temperature of 1 kg of the substance 1 C deg. The **heat of combustion** of a substance is the energy released in the burning of a unit of mass of the substance.

Most substances expand when heated and contract when cooled.

Questions

Temperature and Heat

1. When it's cold outside and your roommate leaves the apartment door open, does cold come in or does heat go out?

2. Which does an air conditioner remove from a room, temperature or heat?

3. If the nurse removes a thermometer from your mouth and it reads 37°C (98.6°F), do you have a temperature?

4. True or false: Any two pans of water at 50°C both contain the same amount of heat. Explain the reasons for your answer.

5. On a warm day, the water in the university president's outdoor pool is 20 C deg warmer than that in his indoor one. Is he correct in claiming that *all* his indoor molecules have more kinetic energy than his outdoor ones?

6. When two bodies at different temperatures are put in contact, why does the warmer one get cooler, and the cooler one get warmer? Explain the heat flow in terms of molecular activity.

7. Father Eskimo puts a hot 25-kg stone in bed with him to keep his feet warm; Baby Eskimo does the same with a 10-kg stone. As they heat the stones from −25°C to +45°C near the fire, which stone gains more heat? Which gains more internal energy? For which does the average molecular kinetic energy change the most?

8. After the stones of Question 7 have reached 45°C, which stone has more internal energy? For which is the average molecular kinetic energy higher?

Temperature and Heat Measurements

9. From the facts that 68°F is 20°C and a temperature difference of 2 F deg is roughly 1 C deg, what are the approximate temperatures in °C of 66°F, 70°F, and 72°F?

10. Dr. Frankenstein sets up a temperature scale on which he assigns the value of 0° to the temperature of a freezing human body and 100° to the temperature of boiling blood at sea-level

atmospheric pressure. Is his scale as legitimate as the scale of Celsius?

11. Why does the diameter of the bulb that holds most of the mercury at the bottom of a mercury-in-glass thermometer have to be larger than the tube running up the length of the thermometer?

12. Why do you have to shake a clinical thermometer (Fig. 8.11) to make the reading go down after it's used?

13. What are some advantages and disadvantages of defining units of heat on the basis of the temperature rise of water?

Specific Heat

14. If you keep your feet warm at night by taking a hot object to bed with you, would you be better off taking a steel cannon ball or a piece of wood the same size and at the same temperature? Hint: You've got to consider more than just specific heat.

15. Why does your freezer run longer after plugging it in if it is full of room-temperature hamburger than if it is full of room-temperature air?

16. Antifreeze has a specific heat of less than 1. What disadvantage would there be to using pure antifreeze rather than a mixture of antifreeze and water in your car's cooling system?

17. Why is water a good medium for carrying heat to baseboards or radiators in a home heating system?

18. Athletes use cold packs made of a jelly-like substance that can be cooled in the freezer and then used on injuries in place of an ice pack. What can you say about the specific heat of this substance, if it is to work well?

19. If Snoopy is to be warmest in his outdoor doghouse, should the house be large and roomy or just big enough for him to fit in comfortably? (The house is being heated by Snoopy's body heat only.)

20. Why does the temperature in Omaha vary so

Figure 8.11 Notice the wiggle in the tube of the clinical thermometer.

much more between winter and summer than the temperature in San Francisco?

21. Does a bridge built in January in San Francisco need as large a gap in the expansion joint as one built at the same time in Chicago?

Thermal Energy Sources

22. Estimate how far you would need to run to work off the energy in a 30-gram Hershey bar.

Thermal Expansion

23. Why can you loosen a "stuck" peanut butter jar lid by pouring hot water over the lid?

24. Which is more likely to crack if you pour hot water into it—a thick glass or a thin one?

25. Which, do you think, would expand more for each degree change in temperature—Pyrex or ordinary glass?

26. When you turn on some hot water faucets part way, the flow will slow down and sometimes almost stop after a brief time. This doesn't happen with the cold water. Why?

27. When a hot water heating system first comes on, why do you hear creaking in the pipes or in the radiators or baseboards?

28. A machinist wants to fit a round shaft into a hole the same size and have a tight fit so that the shaft will not slip out. How can he use thermal expansion to aid his assembly?

29. Since dentists successfully use gold for crowns and bridges on teeth, how would you expect the thermal expansion properties of gold to compare with those of tooth enamel?

30. Sections of automobile exhaust pipes are fitted together snugly by heating one piece before sliding it over the other. Contraction then gives a very tight fit. Why doesn't the connection loosen when the pipe gets hot during operation?

31. When you change the temperature setting of a surface heating unit on an electric range, you shift a movable electrical contact closer to or farther from the end of a bimetallic strip (Fig. 8.12). This strip turns the current on and off periodically, with the percent of ON time determining the temperature of the heater. For high heat, should the contact be closer to, or farther from, the end of the strip? Assume that high temperatures cause the strip to bend away from the contact. (Note: Some ranges use a different type of mechanism.)

32. Draw a design that you make up for the mechanism of a simple pop-up toaster, using a bimetallic strip to release the toast elevator. Include a darkness control.

33. When you fill your car tank with gasoline on a warm day, why does the gasoline overflow after sitting for a while, even though you didn't fill it to the top?

34. Why is the door of a freezer harder to open after it has been closed a few minutes than right after it's closed? Hint: The effect is more noticeable if the freezer is nearly empty.

35. Refrigerators, ovens, oil-fired water heaters, and various other devices use a bellows-type thermostat (Fig. 8.13). The bulb is in the region where the temperature is to be sensed, and it is connected by a long tube to the bellows. As the fluid in the bulb expands and contracts, the end of the bellows moves back and forth, opening and closing the contacts of an electrical switch. Is the thermostat shown in the figure intended for a refrigerator or for an oven? How would you modify the thermostat to make it work for the other device?

Figure 8.13 A bellows thermostat.

Figure 8.12 Electric current flows to the surface heating unit of an electric range only when the bimetallic strip touches the movable contact. The percentage of ON time is set by the distance of the contact from the strip.

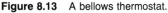

Problems

1. On January 28, 1977 the temperature in Cuck-oo, Virginia, dropped 40 deg from 48°F to 8°F in just a few hours' time. How much was this temperature *change* in C deg?

   Answer: 22 C deg

2. I like to shower in water at a temperature of about 110°F. What is this temperature in °C?

   Answer: 43°C

3. Refer to Example 8.1. How does the energy needed to heat the water for a bath compare with the energy needed to pump this much water from a 60-m-deep well?

   Answer: 210 times as much to heat

4. The cook at the Denver Hilton boils potatoes at 95°C. How much heat is needed to get 40 liters of water from 10°C to this temperature?

   Answer: 3400 kcal

5. Assume that the range used by the cook in Problem 4 is 50% efficient in getting heat into water. How much does it cost to heat this water, at 9¢ per kWh?

   Answer: 71¢

6. If 3 liters of water are heated in a 1-kg aluminum pot from 10°C to 100°C, how much heat goes to the pot? How much goes to the water?

   Answer: 19 kcal; 270 kcal

7. The River City Boys Band has collected 2000 metric tons (1 metric ton = 1000 kg, and weighs 2200 lb) of old steel automobile engines for melting down and recycling. How many kilowatt-hours of energy are needed to get this much steel from 20°C to the temperature at which it starts to melt (1430°C)?

   Answer: 310,000,000 kcal or 360,000 kWh

8. An auditorium 60 m long by 25 m wide by 8 m high holds 3000 people. The average healthy person burns up about 100 kcal of food energy each hour while sitting still. If each person continued to produce heat at this rate, and no heat were removed or leaked out, how much would the temperature of the air in the audi-

torium rise in 2 hours? The density of air is 1.2 kg/m³.

   Answer: 170 C deg

9. For a properly operating fireplace, about 15 m³ per minute of air is pushed up the chimney. If this air filters in from all over the house and is heated by the central heating system from −10°C to 20°C before it gets to the fireplace, how much heat is wasted in 8 hours of heating this air? (Density of air = 1.2 kg/m³)

   Answer: 62,000 kcal (72 kWh)

10. If your bathroom is 10 m from your water heater, about 1.3 liter of hot water stays in the pipe and loses its heat after the faucet is turned off. If you do this 6 times per day, how much useable energy do you waste in a year? Assume the water heater heats the water from 12°C to 60°C.

    Answer: 160 kWh

11. Superman found that 26,000 kcal were needed to change the temperature of 20 kg of kryptonite from 0°C to 60°C. What is the specific heat of kryptonite?

    Answer: 22 kcal/kg·C deg

12. If 1% of the chemical energy stored in a $\frac{1}{2}$-kg sirloin steak could be converted to doing mechanical work, how many 40-kg sacks of mail could your mail carrier lift onto the 1-m-high mail truck after eating such a steak?

    Answer: 166

13. A Labrador retriever eats four 440-g cans of Ken-L-Ration a day. She lies in the sun all day and in her doghouse all night. Assume an energy content of 4000 kcal/kg and that her body converts 75% of what she eats to heat. How much heat is available to heat her doghouse during a 10-h night?

    Answer: 2200 kcal

14. A power company burns fuel oil and generates 720 MWh of electrical energy in a month. If the plant is 38% efficient, how much fuel is used in a month. At 25¢/liter, how much does

this fuel cost? (1 MW $= 10^6$ W; 1 liter of fuel oil has a mass of 0.94 kg.)

Answer: 1.54×10 kg; $41,000

15. If you leave the room briefly, should you turn the light off or leave it on? An ordinary incandescent light bulb uses about 10 times as much power at the instant it is turned on as it does after the filament is hot. Assume that during the heating process, a 60-W bulb generates heat at an *average* rate of five times its rated power, or 300 W. If the mass of the tungsten filament is 15 mg (1 mg $= 10^{-6}$ kg), how much heat is needed to raise the temperature of the filament from 20°C to 2520°C? If we neglect the energy radiated from the filament during this time, how long would it take to heat this filament? For what maximum length of time would you save electrical energy by leaving the light on rather than turning it off? The specific heat of tungsten is 0.037 kcal/kg·C deg.

Answer: 0.00139 kcal; 0.0194 s; 0.078 s

Home Experiments

1. Take two identical pans and put twice as much cold water in one as the other. Compare the time required for them to reach the boiling temperature when they are placed on identical heating units.

2. The next time you bake potatoes put a rock about the size of one of the potatoes in the oven at the same time. When the potatoes are done, compare the time needed for the rock and the potato to cool to about the same temperature. What can you say about the relative specific heats of the two? (The rock's mass will be about 2.5 times that of the same-sized potato.)

3. Find out how much heat you lose from your water heater, even if you use no hot water. At the nearest faucet to the heater, run the water until it is as hot as it will get. Using a thermometer that registers up to 100°C, measure the temperature of some of this water. When no hot water has been used for a while (so that the heater is completely full of hot water), turn off the heater for eight hours. Remeasure the temperature as before. From the temperature drop and the size of the heater tank, you can now calculate the heat loss. From the cost of your energy supply, you can then estimate how much money you can save by better insulating your water heater.

4. Contraction with *increasing* temperature can be observed by tying an ordinary rubber band to a weight that is heavy enough to stretch the band slightly. Move a burning match or candle up and down near the band. The band will contract and raise the weight a noticeable amount. Put a stationary object beside or just below the hanging object to make the movement easier to see.

References for Further Reading

Brown, S. C., *Count Rumford Physicist Extraordinary* (Anchor Books, Doubleday & Co., Garden City, N.Y., 1962). An account of a man who was a gifted experimenter, prolific inventor, and social reformer, but who at the same time was a scheming adventurer, cynical soldier of fortune, and spy.

Dyson, F. J., "What Is Heat?" *Scientific American*, Sept. 1954, p. 58.

CHANGE OF STATE

Snow melts; water in the ice tray freezes; puddles of water evaporate; frost appears on the grass on a cool autumn morning; and the water in the pot of beans boils away if we leave the pot on the stove too long—we're all familiar with water's changes of state.

But many other everyday substances also change state. Moth balls gradually vaporize and disappear, the vapors keeping the moths away. When we solder metal pieces together, we melt solder that later solidifies to hold the parts together. Industries melt iron and aluminum to form them into many of the mechanical gadgets we use. Refrigerators and air conditioners work by continuously evaporating and condensing refrigerant inside the coils. Our bodies are cooled on a hot day by the perspiration evaporating from our skin.

In this chapter we study how substances change state and how these changes affect our lives. After summarizing the properties of the states of matter, we'll discuss the specific changes that can occur between states and the energy transfers accompanying these changes. We'll find out why water expands rather than contracts when it freezes, and we'll consider some of the external influences on the temperature at which the state changes. Finally, we'll discuss a topic very relevant to our comfort: humidity.

9.1 THE STATES OF MATTER

You learned in elementary school that matter at ordinary temperatures exists in one of three states—solid, liquid, or gas. Before we discuss

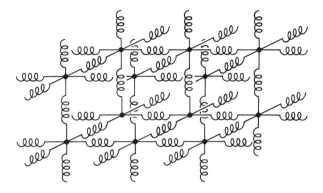

Figure 9.1 The molecules in a crystalline solid are arranged in a regular pattern. The forces holding them together behave like tiny springs. (The springs, of course, do not actually exist.)

what's needed for changing the states of matter, let's briefly explore the properties of each of the states and the ways in which they differ.

Solids

In **solids** the molecules are held close together by forces that keep any given molecule in the same place, on the average, relative to its neighbors. Most solids have a crystalline structure, in which the molecules are arranged in some regular pattern (Fig. 9.1). The molecules vibrate back and forth around their average positions, as if they were held together by tiny springs. As the temperature is raised, the molecules jiggle around faster and move farther from home in their back-and-forth motion, but nonetheless they keep their same average positions except for a slight overall expansion.

If you try to compress a solid, the springlike forces "give" some, allowing the body to compress slightly. Similarly, these forces let you stretch a solid a small amount without a permanent deformation or break. We mentioned this elasticity of solids—their ability to return to their original shape when the deforming force is removed—in connection with elastic potential energy in Chapter 4, and with elastic collisions in Chapter 5.

Liquids

Figure 9.2 In a liquid, the molecules bounce around at random within the bounds of the surface of the liquid and the container.

In **liquids** the molecules still exert fairly strong forces on each other, but not strong enough to hold them in fixed positions. The molecules bounce around freely in all directions as they collide with each other and with the walls of the container (Fig. 9.2). Even though the analogy is not perfect, the relationship between solid and liquid is similar to the relationship between zone and man-to-man defense in basketball. In zone defense (solid) each player (molecule) "bounces" around within a certain region as he interacts with other players. In man-to-man (liquid), the players move around freely throughout the court, interacting with other players but staying within the court boundary. High pressure can crowd the molecules of a liquid only slightly closer together. That's why we said in Chapter 7 that liquids are "almost" incompressible.

Gases

In **gases** the molecules are far apart so that the force between them is quite small except during collision. Gas molecules spread out to fill the entire volume of their container. Our basketball analogy is not quite as good here, but the gaseous state is somewhat like the end of the game, when the players spread out over the entire town, only occasionally interacting with each other.

If this were a detailed study of the gaseous state, we would distinguish a gas from a vapor. For our purposes, however, we will consider a gas and a vapor to be the same. For example, we'll consider steam and water vapor both to be the gaseous state of water.

9.2 THE PROCESSES OF CHANGING STATE

Materials such as glass are mixtures of different elements and do not have a crystalline structure with a regular ordering of molecules. As you heat glass, it gradually becomes softer and softer until it is a free-flowing liquid. There is no clear-cut condition for which we can say the glass has changed from the solid to the liquid state. Many plastics behave the same way. Such materials are said to be *amorphous*.

Most substances are crystalline in the solid state, and for these, a definite temperature exists at which the transition from one state to another occurs. In discussing change of state, let's focus our attention on water, since water is familiar to most of us in all three of its states. The change of state of other crystalline substances is similar to that of water in most respects. We're using the word *water* here in the general sense to include solid (ice), liquid, and vapor (steam).

Melting and Freezing

When ice or some other solid melts, the molecules must be pulled apart, freed from their characteristically regular pattern so that they can move around as a liquid. Work must be done in breaking down this regular structure, and therefore energy must be supplied. We usually provide this energy in the form of heat. Here, the heat does the work of overcoming the bonds that hold the molecules in the solid state, rather than changing the temperature of the material. For crystalline solids, this process happens at a particular temperature called the *melting point*. For ice at sea-level atmospheric pressure, the melting point is 0° C.

Suppose you have a well-stirred glass of ice water with lots of ice. As this water sits in a warm room, some heat will enter the glass from the room and melt a little of the ice. This heat goes to the work of changing some of the ice to water, *not* to changing its temperature. Thus, the temperature of the *water* after melting is the same as the temperature of the *ice* before melting.

Freezing is just the reverse of melting. Since ice absorbs energy when it melts, water gives up energy when it freezes. If we put our glass of ice water in a freezer, heat will leave the glass to the colder surroundings, causing some of the water to freeze as it gives up this energy.

When ice melts by using heat from its surroundings, the loss of heat cools the surroundings. On the other hand, freezing adds heat to the surroundings.

· ·

QUESTION   Why do people sometimes put a tub of water in their fruit cellar on bitterly cold nights?

The heat required to change 1 kg of solid into liquid is called the **latent heat of fusion**, or just the *heat of fusion*, of the substance. Exactly the same amount of heat is *released* as 1 kg of liquid freezes to form a solid. The heat of fusion of water is 80 kcal/kg. That is, it takes 80 kcal of heat to melt 1 kg of ice at a temperature of 0°C. If you want to know the total amount of heat needed to melt a certain mass of solid, multiply the total mass by the heat of fusion. Thus,

heat to melt = mass × heat of fusion

As the same quantity of liquid freezes, exactly the same amount of heat is given off and must be absorbed by the surroundings.

· ·

EXAMPLE 9.1   How much heat must be removed from the water in an ice tray to freeze it, once the temperature is at 0°C. A typical ice tray holds about $\frac{1}{2}$ liter (about a pint), or a mass of $\frac{1}{2}$ kg.

heat removed = mass × (heat of fusion)

$$= \tfrac{1}{2} \text{ kg} \times 80 \text{ kcal/kg} = 40 \text{ kcal}$$

· ·

EXERCISE   How much heat was removed from the water in the ice tray of Example 9.1 to get its temperature from 18°C to 0°C? Compare your answer with the result of Example 9.1.   *Answer*: 9 kcal

Figure 9.3 The molecules of water do not all have the same kinetic energy. The high-energy ones can escape through the surface to the air above and become vapor.

Evaporation

You know that a puddle of water slowly evaporates after a rainfall. So does the water in clothes hung on a clothesline, perspiration on your skin, and alcohol rubbed on your tired feet. The arrows in the water puddle of Figure 9.3 indicate the directions of motion of a group of molecules at any instant, the length of the arrows being proportional to their speeds. Not all molecules have the same speed—some move very slowly, some very fast, with a continuous distribution in between. Molecules near the surface feel a downward force from the other molecules in the water. But the faster-moving ones have enough kinetic energy to break through this surface layer and escape into the air above. Being widely separated from the others, these molecules are then in the vapor state.

Since only high-energy molecules escape, the *average* energy of those remaining is lower. This condition means that the temperature of the remaining water is lower. *Evaporation is*, therefore, *a cooling process*.

If you've ever spilled gasoline or rubbed alcohol on your skin, it feels very cool. Both of these liquids evaporate very rapidly, leaving the remaining liquid, and your skin, cooler. Your body uses this mechanism to eliminate heat. You are cooled by perspiring only when the perspiration evaporates.

Boiling

As the temperature of a liquid is raised, the average kinetic energy of the molecules increases. Therefore, the fraction of the molecules having enough kinetic energy to get through the surface increases, which in turn increases the rate of evaporation. When the temperature is raised until the average molecular energy is equal to the work needed to separate the molecules of the liquid from each other, *boiling* occurs. The temperature at which this happens is called the *boiling point*. For water at sea-level atmospheric pressure, the boiling point is 100°C.

Any heat added to the liquid at the boiling point produces vapor that forms bubbles throughout the liquid. The vapor molecules must be moving fast enough so that the pressure in a bubble exceeds the pressure from the atmosphere and the liquid above it. Otherwise, the bubble would collapse. Boiling starts when this condition is first met.

The heat needed to convert 1 kg of liquid to 1 kg of vapor at the same temperature is called the **heat of vaporization**. At sea-level atmospheric pressure, the heat of vaporization of water is 540 kcal/kg. This value is almost 7 times its heat of fusion and over 5 times as much heat as is needed to raise the temperature of the water from 0°C to 100°C. We can calculate total heat, using the heat of vaporization, in the same way we did using the heat of fusion.

Condensation

Condensation is the reverse of boiling. When the temperature of water vapor is at the boiling point and heat is removed, the vapor goes back to the liquid water state. In doing so, it gives up the heat of vaporization in exactly the same amount as was needed to boil it.

Condensation can also take place at a slower rate, the reverse of evaporation. When water vapor molecules above the liquid hit the liquid surface, they can be captured by this surface and returned into the liquid state. The more vapor there is above the liquid, the faster the rate of condensation.

· ·

EXAMPLE 9.2   A typical steam radiator can hold about 0.1 cubic meter of steam, which would have a mass of about 80 g. How much heat is given up to the room as this much steam condenses to water?

Heat loss = mass × (heat of vaporization)

= 0.080 kg × 540 kcal/kg = 43 kcal

We mentioned earlier that the molecules in a vapor (or gas) are quite far apart compared with their separation in the liquid state. This fact means that when a liquid is boiled, it expands tremendously. For example, for every liter of water boiled at 100°C, you have about 1700 liters of steam generated if the steam is confined and kept at atmospheric pressure. (About $\frac{2}{3}$ cup of water can produce the steam in the radiator of Example 9.2.)

· ·

QUESTION   Why does an old-fashioned steam heating system need only one pipe running to the radiators (see Fig. 9.4)?

Sublimation

Even in the solid state, a small fraction of the molecules of a substance have enough kinetic energy to break the bonds of the crystal and escape. Some of those near the surface may do so, thereby going directly from the solid to the vapor state. This process is called *sublimation*.

Sublimation is what happens to moth balls as they gradually disappear. Another example would be dry ice (solid carbon dioxide), which, at ordinary temperatures and pressures, sublimates directly to the vapor state. That's the reason it's called *dry* ice. And if you live in a climate

Figure 9.4 A diagram of a steam heating system such as those found in older houses. Notice that in this one-pipe system, the supply pipes have to slope downward from the radiators.

Radiator

Steam supply line
(sloped)

Boiler

Water return line

where it snows, you've probably noticed that snow gradually disappears even if the temperature never gets to the melting point. This is another example of sublimation.

QUESTION How would you expect the temperature of a moth ball to compare with the temperature of the surrounding air?

9.3 THE ENERGY TO CHANGE THE TEMPERATURE AND STATE OF A SUBSTANCE

We can now summarize the result of adding heat to a substance in any particular state. For convenience we will use water, but the basic ideas apply to most pure substances.

Suppose you have 1 kg of ice at a temperature of −40°C and you start adding heat to it slowly. The graph in Figure 9.5 shows the temperature after various amounts of heat have been added. From the heats of fusion and vaporization (80 and 540 kcal/kg, respectively), and from the specific heats of ice, water, and steam (0.5, 1.0, and 0.5 kcal/kg·C deg, respectively) you can verify the numbers on the graph. Notice some of the important features:

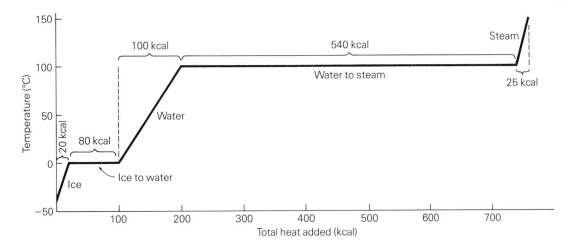

Figure 9.5 A graph showing the variation in temperature with total heat added to 1 kg of ice at −40°C until it is changed to steam at 150°C.

1. Once the melting or boiling point is reached, the temperature does not rise until all of the substance has changed state. The added energy goes to change the state.

2. The temperature of the ice just as it starts to melt, and the temperature of the water just after it has all melted, are the same—0°C. The same is true of the water and steam just before and just after boiling.

3. By far the greatest part—70%—of the energy is needed to change the state from liquid to vapor. Once the water has vaporized, the steam must be confined in a container in order to add more heat to it.

4. If heat is removed from 150°C steam, the graph looks the same. The substance goes from vapor to liquid to solid with exactly the same quantities of heat being removed as were added in going the other way.

· ·

QUESTION   When the outside temperature is dropping and gets to 0°C, why does it often stay at about this temperature for some time before getting any colder?

If a hot substance and a cold substance are mixed, we can use the principle of conservation of energy to determine the overall effect on the system. If no heat is lost to the outside, the heat given up by the hot substance equals the heat gained by the cold substance. Let's illustrate this concept with an example.

EXAMPLE 9.3

A large scoop of crushed ice at 0°C is added to $\frac{1}{2}$ glass (about 250 g) of tea at a temperature of 80°C. How much of the ice melts?

Since the tea is mostly water, its specific heat will be about the same as that of water. There is plenty of ice, so the tea will come to the temperature of the ice as the ice melts; that is, its temperature changes by 80 C deg.

$$\begin{aligned} \text{heat lost by tea} &= \text{(specific heat)} \times \text{(mass of tea)} \times \\ &\quad \text{(temperature change)} \\ &= 1 \text{ kcal/kg·C deg} \times 0.25 \text{ kg} \times 80 \text{ C deg} \\ &= 20 \text{ kcal} \end{aligned}$$

Assuming no heat lost to the outside, this amount of heat is absorbed by the ice as it melts. Thus,

$$\text{heat gained by ice} = \text{(mass of ice melted)} \times \text{(heat of fusion of ice)}$$

$$20 \text{ kcal} = \text{(mass of ice melted)} \times 80 \text{ kcal/kg}$$

Thus,

$$\text{mass of ice melted} = \frac{20 \text{ kcal}}{80 \text{ kcal/kg}} = \tfrac{1}{4} \text{ kg} = 250 \text{ g}$$

9.4 THE UNUSUAL EXPANSION OF FREEZING WATER

— Molecules

Figure 9.6 The molecules of ice are arranged in a hexagonal lattice that takes up more space per molecule than when the molecules bounce around at random in the liquid state.

When most substances freeze from liquid to solid, the molecules arrange themselves in a crystal lattice so that they are, on the average, closer together than in the liquid state. Such substances, therefore, contract when they freeze and expand when they melt. But for water, just the opposite happens: Water expands when it freezes and contracts when it melts.

If you live in a climate where the temperature goes much below 0°C, you don't leave water pipes exposed to outside temperatures—the water will freeze and burst the pipes. In whatever climate you live, you know that you do not put water or food (mostly water) in a freezer in a glass container—it will expand and break the glass.

Water expands when it freezes because of the way the molecules arrange themselves to form the ice crystal. The molecules of ice are arranged in a hexagonal lattice (Fig. 9.6). In this arrangement they are farther apart, on the average, than they are in the liquid state, where they bounce around freely. When ice melts, the molecules get closer together and the substance as a whole contracts. On freezing, the water expands as the molecules spread out and form the hexagonal lattice.

Chapter 9   Change of State

The questions at the end of the chapter point out some of the other problems caused by the expansion of freezing water. This behavior leads to an important benefit that we'll consider in Chapter 10: Deep lakes freeze only near the top, thus allowing marine life to continue to live in the water below.

9.5 CHANGING THE MELTING AND BOILING POINTS

Several times we have mentioned that the melting and boiling points of water are 0°C and 100°C *at sea-level atmospheric pressure*. As this remark implies, these melting and boiling points are different at other pressures. They also change if you add some impurity such as salt to the water.

The Effect of Atmospheric Pressure

Boiling occurs when the molecules of vapor in a bubble formed below the liquid surface have enough kinetic energy to make the pressure in the bubble at least as high as that in the surrounding liquid. As the outside pressure on the liquid increases, the temperature of the liquid must be higher in order to boil. That makes the boiling point increase with increasing pressure and decrease with decreasing pressure on the liquid.

At high altitudes, where the atmospheric pressure is low, water boils at a lower temperature. Since the rate at which foods cook depends on their temperature, cooking food in boiling water takes longer as you go higher in altitude. In Sopchoppy, Florida you can hard-boil an egg in 12 minutes. On Mt. Mitchell, North Carolina it takes about 20 minutes, and on Mt. Whitney, California, about 35 minutes. On Mt. McKinley, Alaska, you might as well give up and have corn flakes.

Furthermore, higher than normal atmospheric pressure increases the boiling point. The steam pressure in a steam heating system is usually at about $1\frac{1}{2}$ times atmospheric pressure, making its temperature about 110°C. A pressure cooker confines the steam until the internal pressure builds up to about twice atmospheric pressure. The water reaches approximately 120°C, thereby bringing the food to that temperature for faster cooking.

Since the energy needed to separate the molecules of a liquid increases with increasing pressure, the heat of vaporization also changes with change in pressure. But for pressures near standard atmospheric, the heat of vaporization varies only slightly.

The pressure on a substance also influences its melting point. As we saw in Section 9.4, water expands when it freezes. If high pressure is placed on ice, it can force the crystals to collapse and melt at a temperature below 0°C.

The reason your ice skates glide so easily over the ice is that the narrow blade puts enough pressure on the ice to melt it, letting the skate slide on a layer of water. Once the skate moves off, decreasing the pressure, the water refreezes.

Many of us who live in climates where snow comes periodically, but usually melts within a week or two, often complain that "people around here just don't know how to drive in snow. Up North they drive on a layer of snow all winter without a bit of trouble." The fault may not always be with the drivers. Driving over snow that is only slightly below the freezing point can cause the snow to melt from the pressure under the tires, forming a slick layer of water. This water then refreezes into solid ice as the tire moves off. The difficulty of removing this layer of packed snow and ice may have some bearing on another of our frequent complaints: "They just don't have the equipment to clear the roads around here like they do up North where it snows all the time." In colder temperatures, the snow remains powdery, which makes it both easier to remove and easier to drive over without slipping.

The Effect of Impurities

Salt or other chemicals are scattered on snow-covered highways to speed the melting of snow and ice. We put antifreeze in our car radiator and windshield washer tank to lower the freezing point and prevent freezing of the water. In general, adding impurities to water changes its melting point.

These impurities affect the freezing by mixing molecules of a different type with the water molecules, thereby impeding the formation of the hexagonal lattice. With additives, the water must be brought to a lower temperature before the crystal lattice forms.

Impurities such as salt or antifreeze also raise the boiling point of water. Water molecules are attracted to those of the salt, and therefore need a higher kinetic energy to escape the liquid.

• •

| APPLICATION | **THE ICE-CREAM FREEZER** |

The freezing of homemade ice cream excellently illustrates several of the processes we have talked about. The ice-cream freezer, (Fig. 9.7) consists of an inner metal container with a fairly large gap between it and the outer bucket. The ice-cream mixture is placed in the inner container and a mixture of ice and salt is placed in the outer.

For the ice cream to freeze, its temperature must be lowered to the freezing point (around 0°C), and then more heat must be removed to change its state. Pure ice in the outer container would lower the temperature to 0°C, but no more heat would leave the ice cream because it would have the same temperature as the ice. When salt is added to the

Figure 9.7 An ice-cream freezer illustrates many of the characteristics involved in changes of state between liquid and solid.

ice, however, it mixes with the outer layers of the ice chunks, lowering the ice's melting point and causing it to melt. But melting requires heat. This heat is taken from the remaining ice and surrounding water, which reduces their temperature. Heat then passes from the ice cream to this lower-temperature mixture of ice, water, and salt, thus causing the ice cream to freeze. Further melting of the ice absorbs this heat.

From the standpoint of overall energy transfer, the heat given up as the ice cream freezes is absorbed by the ice as it melts. Turning the crank merely turns the inner container while the dasher and outer bucket stay fixed, thereby keeping everything well mixed to speed up the heat transfer and give a smooth texture to the ice cream.

9.6 HUMIDITY

The earth's surface contains vast amounts of water, with some of it continually evaporating into the air. But the amount of water vapor in the air, the *humidity*, varies from time to time.

In a pan of water such as the one in Figure 9.8(a), high-energy molecules near the surface continually escape from the liquid into the air above it. At the same time, many vapor molecules hit the water from above, get captured, and become part of the water again.

If we put a lid on the pan, as in Figure 9.8(b), the vapor molecules will be trapped and will increase in density until the number leaving the water equals the number returning. We say that the air is then **saturated**—it holds all the water vapor it *can* hold. If we increase the water temperature, more molecules have enough kinetic energy to leave the liquid, and more water vapor will be in the air when it gets saturated.

The air temperature also limits the amount of water vapor the air can hold. When vapor molecules collide at low speeds, they may stick together forming tiny water droplets. The higher the air temperature, the higher the temperature of the droplet, and therefore, the greater the

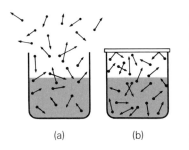

(a)   (b)

Figure 9.8 When the air above a liquid gets saturated, the same number of molecules return as leave the liquid per minute.

Table 9.1   Water vapor in saturated air

Temperature (°C)	Water Vapor (grams per cubic meter)
−8	2.74
0	4.84
8	8.21
16	13.50
18	15.70
20	17.12
24	21.54
32	33.45

chance it will evaporate. Thus, the higher the temperature, the higher the fraction of molecules *not* sticking together when they collide, and therefore the more water vapor the air can hold. Table 9.1 gives the amount of water vapor in saturated air at various temperatures.

Relative Humidity

The *absolute humidity* is the actual amount of water vapor in the air at any instant. The term used more often is **relative humidity**, the ratio of the amount of water vapor in the air to the amount it could hold if saturated at that temperature. For example, if the temperature is 16°C and the air contains 8.0 g/m³ of water vapor, the relative humidity is $(8.0 \text{ g/m}^3)/(13.50 \text{ g/m}^3)$, which is 0.59 or 59%. (The 13.50 g/m³ comes from Table 9.1.)

Relative humidity tells how fast you can expect evaporation to occur. For example, if the relative humidity is 100%, you know that the rate of condensation is exactly equal to the rate of evaporation, so that there is no *net* evaporation. However, if the relative humidity is quite low, say 10%, you can expect rapid evaporation. This evaporation rate has a pronounced effect on your comfort. If it's a hot day and the humidity is low, perspiration evaporates quickly from your body, using up body heat to change the state of the perspiration. You can then be quite comfortable at fairly high temperatures. But if it's a humid (muggy) day, even slightly high temperatures make you uncomfortable because little net evaporation of perspiration occurs. Perspiration may pour off your body, but unless it evaporates, it doesn't cool you. That's why an air conditioner needs to remove water vapor as well as heat from the room air.

· ·

QUESTION   Why does water drip from the outside end of a working room air conditioner, especially on a humid day?

Chapter 9   Change of State

In winter, the problem is usually too *little* humidity inside the house. If the humidity is low, any perspiration evaporates immediately, taking heat from your body. Therefore, a higher temperature is needed for comfort. But your comfort is affected in another important way by low humidity. Rapid evaporation can cause excessive drying of mucous membranes in the nose and throat. The resulting swelling and irritation can cause sore throats for anyone, and even worse difficulties for people with respiratory problems. That's why doctors recommend the use of humidifiers during certain illnesses and for people with allergy problems. Around 35% to 45% relative humidity is considered best.

But why *is* the inside humidity so low in winter? Except for moisture added inside the house, the actual amount of water vapor in a cubic meter of air inside is the same as that in a cubic meter of air outside. Even if the relative humidity is high outside, the higher inside temperature means a much lower relative humidity inside. For each 10 C deg we increase the air temperature by heating, the relative humidity is approximately cut in half. Let's illustrate with a specific example.

EXAMPLE 9.4 Suppose the outside temperature and relative humidity are $-8°C$ ($+18°F$) and 70% respectively. For the same absolute humidity, what is the relative humidity inside a house where the temperature is 20°C (68°F)?

Table 9.1 gives the water vapor in saturated air at $-8°C$ as 2.74 g/m^3. At 70% relative humidity,

$$\text{absolute humidity} = 0.70 \times 2.74 \text{ g/m}^3 = 1.92 \text{ g/m}^3$$

The absolute humidity inside is the same. But at 20°C, the inside air can hold 17.12 g/m^3 of vapor. Therefore, inside

$$\text{relative humidity} = \frac{\text{absolute humidity}}{\text{absolute humidity at saturation}}$$

$$= \frac{1.92 \text{ g/m}^3}{17.12 \text{ g/m}^3} = 0.11, \text{ or } 11\%$$

QUESTION Why does the use of a "cool mist" humidifier in a room cause the room to cool somewhat? (A cool-mist humidifier, the kind most doctors recommend, sprays tiny droplets of *room temperature water* into the air.)

Dew, Frost, and Fog

You are no doubt familiar with the dew often found on grass and elsewhere outdoors in the morning. Where does dew come from?

Suppose that during the day the outside temperature is 32°C and the relative humidity is 40%. The air then contains 13.4 g/m³ of water vapor. If this amount of water vapor does not change much as the temperature drops at night, the air will be completely saturated (100% relative humidity) when the temperature falls to 16°C. If the temperature drops more, the air must give up some of its humidity. Some of the water vapor condenses from the air that is near grass and other cool objects. The temperature at which dew starts to form is called the *dew point*. The dew point depends on the actual amount of water vapor in the air.

If the temperature of the cool surfaces is below the freezing point before the air gets saturated, the vapor sublimates out as *frost* rather than dew.

QUESTION   Why does frost build up on the inside of a freezer or the freezing compartment of a refrigerator? Does the rate of frost buildup have anything to do with how often the door is opened? (You don't see frost in a "frost-free" model because the coils are periodically heated enough to melt the frost, but not enough to thaw the food.)

When warm, moist air moves into a cool region, the air may be quickly cooled below its dew point. The vapor then condenses into little water droplets that we call *fog*. This process often happens as a warm breeze blows from a large body of water over the cooler land. If the cooling and condensation occur high in the earth's atmosphere, we call the result a *cloud*.

Key Concepts

In a **solid**, the molecules are held tightly in a fixed average position relative to each other. They vibrate back and forth around this average position. In a **liquid**, the force of attraction is weaker so that the molecules bounce around at random throughout the liquid. In a **gas** (vapor), the molecules are far apart and experience little force except during collisions with other molecules.

Energy must be added in the form of heat to change a solid to a liquid or a liquid to a vapor. The same amount of energy is removed in the reverse processes. The (**latent**) **heat of fusion** is the heat needed to melt 1 kg of a substance, and it is the heat given up as 1 kg of the substance freezes. The **heat**

of **vaporization** is the heat needed to change 1 kg of a liquid substance to vapor, and it is the amount of heat given up as 1 kg of the substance condenses.

Evaporation is a cooling process because high-energy molecules escape the liquid and become vapor molecules. **Boiling** occurs when the average molecular energy equals the work needed to separate the liquid molecules from each other. **Condensation** is a change from the vapor state to the liquid state. **Sublimation** is a change directly from the solid to the vapor state, or vice versa.

Water expands when it freezes and contracts when it melts.

Melting and boiling points depend on the pressure on the substance and on the amount of impurities in the substance. The higher the pressure, the higher the boiling point. For water, which expands as it freezes, increased pressure decreases the melting point.

Relative humidity is the ratio of the amount of water vapor in the air to the amount it can hold at saturation. The air is **saturated** when the rate at which molecules go from the liquid to the vapor state equals the rate at which they return to the liquid state. The amount of water vapor in the air at saturation increases with increasing air temperature.

Questions

Change of State

1. What happens to the energy that is added to ice to melt it?

2. Is energy absorbed or released as water freezes?

3. True or false: When we perspire, our bodies are cooled because the perspiration gives up heat as it evaporates.

4. Grandpa dips his shirt in the creek and wears it wet while working in the hot corn field. His "city slicker" son tells him he's fooling himself because the wet shirt just gets hot and doesn't keep him any cooler. Is the son right?

5. Why does a hog wallow in a water hole? Hint: to get wet.

6. Wet tennis courts will dry much quicker if you sweep the water out of the puddles and spread it out. Why?

7. Is boiling merely fast evaporation?

8. When water is boiling, does boiling it faster make it any hotter? Does this suggest a way to conserve useable energy while cooking?

9. What is the stuff you see rising from a pan of boiling water? (It is not steam; steam is invisible). Why does it disappear after rising a small distance?

10. Why does coffee rise in a percolator stem?

11. Why do orange growers spray their trees with water when there is a high probability of freezing weather?

12. A cold unfrozen Pepsi will sometimes freeze when you remove the top from the bottle. Why? Hint: notice the gases that escape when you open the bottle.

13. Why can you see your breath in cold weather? Why can you not see it in warm weather?

14. Does snow have to melt to go away after it falls on the ground?

15. Will clothes on a clothesline dry if the air temperature is below the melting point?

16. Why do moth balls get smaller with age?

Changes of Temperature and State

17. Why does it take so much longer to boil away a pan of water than to raise its temperature to the boiling point?

18. When you buy coffee-to-go at a restaurant, there's always a small hole in the top placed on the cup. Why?

19. A CIA agent is being tortured by a foreign government in an attmept to extract national secrets from him. He is given a choice of having his hand placed for 2 seconds in boiling water or in steam at the same temperature. Which would you advise him to choose and why? (One choice will be much worse.)

20. Some people claim that you get ice quicker if you start with hot water rather than cold. Do you believe this claim? (See the Kell reference at the end of this chapter.)

21. If you wash your car on a cold day with a bucket of *hot* water, the water will probably freeze on the car as you wash. Use *cold* water

from the hose and this probably won't happen. Is this inconsistent with Section 9.3?

22. Will your portable cooler keep your picnic food colder if you dump out the water from the melted ice or leave it in the cooler?

23. You can buy a container of reusable ice substitute to freeze in your freezer and use repeatedly in your portable cooler in place of ice. What should be the physical characterisitics of this substance?

24. Why is the sand at the beach quite cool near the water but often too hot to walk on several meters from the water?

Freezing of Water

25. Why is it important, when pouring the concrete footings making up the foundation of a house, that the footing bottoms be underground, below the frost line? (The frost line is the lowest depth at which water in the ground freezes.)

26. Swimming pool owners often put logs or tires in their pools in winter to prevent the cracking of the pool walls. Why does this help?

27. Why does ice float at the top of a glass of water?

Changing the Melting and Boiling Points

28. True or false: Food cooks faster in a pressure cooker because the increased pressure increases the speed of the cooking process.

29. Is it more important that you take your pressure cooker with you on a camping trip to the beach or to the high mountains?

30. Why is it possible to make a snowball? Why is it *not* possible to make a snowball if the snow is too cold?

31. When you put ice chunks in a towel and crack it by hitting it with a hammer, why does the ice stick to the towel?

32. Why is ice skating not possible in extremely cold temperatures?

33. Special waxes are put on snow skis to make them slide easier. Why is such waxing more important at low temperatures than when the temperature is near the melting point?

34. Is it inconsistent that we add salt to an ice cream freezer to make the ice cream freeze, but we put salt on the streets to make the ice melt?

35. I once tried to make homemade ice cream for some friends using ice taken out of my home freezer and used immediately. Shortly after putting in the ice, the freezer would no longer turn because the dasher could not move relative to the can containing the ice cream mixture. My claim was that the ice was *too cold*. Would you have joined my friends in ridiculing my explanation?

36. Why does antifreeze keep the water in a car radiator from freezing?

Humidity

37. Why is a humidifier needed so much more in the winter than in the summer?

38. If you take a shower in a chilly house, why are you warmer if you stay in the steamy bathroom to dry off than if you step into the next room, which is at the same temperature? Why do you *not* feel so much difference once you're dry?

39. Why does a cold glass of lemonade sweat?

40. Why do the insides of windows sometimes get wet in winter, especially if they are single thicknesses of glass without storm windows?

41. House insulation usually has a vapor barrier on one side that water vapor cannot penetrate. Why should this barrier be placed on the interior side of the insulation?

42. To get a cold, frosty glass for drinks, some people dip their glasses in hot water and then put them in the freezer for a while before using them. When does the frost form? Does the water for dipping need to be hot? Is dipping necessary?

43. The electric fan was still running in the first floor apartment when the landlady found the body. The police knew the murder occurred

either July 25 or July 26. Sherlock Holmes said that, because the humidity was 18% on July 25 and 98% on July 26, he was sure he knew the correct day. Does he have grounds for being so sure?

Problems

1. How much heat has to be removed from 50 kg of hamburger to freeze it once its temperature has reached the freezing point? The heat of fusion of hamburger is about 50 kcal/kg. Compare your result with the result of Example 8.2 for the heat removed in lowering the temperature to the freezing point.

   Answer: 2500 kcal

2. As much as 40 million metric tons of snow can fall in a snowstorm. How much energy is needed to melt this snow? If it takes 30,000 kWh of heat to heat a house in winter, how many such houses could be heated with this amount of energy? (1 metric ton = 1000 kg; 1 kcal = 0.00116 kWh.)

   Answer: 3.2×10^{12} kcal; 124,000

3. How much heat is absorbed as a 30-g ice cube melts?

   Answer: 2.4 kcal

4. How much heat is needed to melt the iron that goes into one 200-kg automobile engine, once the iron has reached the melting point? The heat of fusion of iron is 12 kcal/kg.

   Answer: 2400 kcal

5. How much heat is needed to boil away 1 pint of water (470 g), once the water has reached the boiling point?

   Answer: 250 kcal

6. When the baby fell off the bed, the harried father forgot the potatoes he had left boiling in a pan with 2 liters of water. If the stove supplies 30 kcal/min of heat, how long will it take the pan to boil dry?

   Answer: 36 min

7. Suppose 5 cm of rain falls in one week on the state of Rhode Island, an area of 3.1 billion square meters. How much energy is released in the clouds as water vapor condenses to form this rain? The people of Rhode Island consume about 100 billion kWh of useable energy for all purposes in one year. How does this compare with the energy released during the 5-cm rainfall?

   Answer: 8.4×10^{13} kcal or 97 billion kWh

8. During the night a "cool-mist" humidifier sprays 6 liters of water into the air in the form of tiny water droplets. How much heat is taken from the air as this water vaporizes? As a result, how much would the temperature of the air drop in a room 4 m by 4m by 2.5 m, *if no heat were added or removed* from the air? (The density and specific heat of air are 1.2 kg/m³ and 0.24 kcal/kg·C deg.)

   Answer: 3200 kcal; 280 C deg

9. How much heat is needed to change 40 kg of ice at $-20°C$ to steam at 130°C? (The specific heat of steam is 0.5 kcal/kg·C deg.) How much heat is removed in changing it back to ice at $-20°C$?

   Answer: 30,000 kcal; 30,000 kcal

10. Suppose that for your Fourth-of-July picnic with your friends, you put twenty-four 280-g bottles of Coke in an insulated tub and dump in lots of ice at 0°C. If the Coke temperature is 30°C, how much ice will be melted in bringing the Coke down to a temperature of 0°C? (Neglect the heat absorbed by the bottles and the tub, and take the specific heat of Coke to be about the same as that of water.)

    Answer: 2.5 kg

11. Assume a wooden tub in Problem 10 and make some reasonable assumptions about the mass of the tub and bottles. Then estimate their effect on the amount of ice melted. (See Table 8.1, Chapter 8.)

12. If, at a temperature of 24°C, the air contains 13.5 g/m³ of water vapor, what is the relative humidity? What will the relative humidity be if the temperature drops to 20°C? At what temperature will dew start to form?

Answer: 63%; 79%; 16°C

13. If a glass of iced lemonade is not to sweat, what is the highest possible value of the relative humidity when the temperature is 24°C?

Answer: 22%

14. If the outside relative humidity and temperature are 60% and −8°C, what is the relative humidity inside a house where the temperature is 24°C and no extra water vapor is added inside? What is the relative humidity inside if the temperature is 18°C?

Answer: 7.6%; 10.5%

15. The dew point is 18°C when the temperature is 32°C. What is the relative humidity?

Answer: 47%

Home Experiments

1. Pour some water into an ordinary paper cup and hold it over the flame of a candle. You can hold it there until the water boils and the cup will not burn. Why?

2. Tie a very thin wire around a block of ice and suspend the block by the wire. The wire will work its way through the ice, melting from the pressure above the wire, refreezing below the wire as the pressure is relieved. The wire will pass completely through the ice, leaving the block intact. (You can get a block of ice by freezing water in an old plastic or paper milk carton. Before starting the experiment leave the ice at room temperature until the outside starts to melt. This process will ensure that the ice is near the melting point, and the experiment will go faster.)

3. Place an ice cube in a glass of water. Wet one end of a thread or string and lay it across the ice cube. Sprinkle some salt on top of the cube. After a few seconds, lift the string by the end you did *not* wet and you will lift the ice cube out of the water. Why does this work?

References for Further Reading

Kell, G. S., "The Freezing of Hot and Cold Water," *American Journal of Physics*, May 1969, p. 564. A study of which freezes faster, hot or cold water.

Plumb, R. C., "Squeak, Skid, and Glide—The Unusual Properties of Snow and Ice," *Journal of Chemical Education*, March 1972, p. 179. Why snow and ice behave as they do when very cold.

Pounder, E. R., *The Physics of Ice* (Pergamon Press, New York, 1965).

Thompson, D. T., and R. O'Brien, *Weather* (Time-Life Books, New York, 1968). Beautifully illustrated with color photographs and overlays that supplement the description of all aspects of weather.

Tufty, B., *1001 Questions Answered about Storms and Other Natural Air Disasters* (Dodd, Mead, & Co., New York, 1970). Many of these questions involve everyday changes in the states of matter.

T R A N S F E R   O F   H E A T

When you stand in front of an open fireplace, you expect heat from the fireplace to warm your body. When you put a pot of beans on the stove to cook, you expect heat to go from the burner into the beans. When you sit in front of a fan, you expect to be cooled as heat leaves your body.

In many of our daily activities we expect heat to be transferred from one place to another. But how does it get transferred? What happens within objects and between objects that causes heat to move from one place to another?

Heat may be transferred by three different mechanisms that we call *conduction, convection,* and *radiation.* We will discuss each of these mechanisms separately, with examples from our everyday lives, and then show how, in most situations, heat travels by a combination of the three means.

10.1 CONDUCTION

If you've ever scrambled eggs in a frying pan with a metal handle, you have probably found that, by the time the eggs are done, the handle is too hot to hold in your bare hand. Or you may have held the end of a poker in a fire for a while and found that the end you were holding became hot. In both of these examples heat traveled through the metal by a method called **conduction**.

Figure 10.1 shows symbolically what happens when one end of an object is held in a hot region such as a flame. The molecules in the end

Figure 10.1 Symbolic representation of the motion of three molecules in the rod. The longer arrows indicate higher thermal energy for molecules near the heat source.

that is heated gain thermal energy and therefore vibrate more vigorously, as indicated by the longer arrows. Since each molecule exerts forces on its neighbors (recall Fig. 9.1), some of this excess energy is passed on to the neighboring molecules, which in turn pass it on to others in a "bucket brigade" fashion. In this way heat is transferred—conducted—from the warmer to the cooler part of the object. *Conduction is the transfer of heat from molecule to molecule in a substance by means of the interactions between neighboring molecules.* As long as one part of the object remains warmer, heat will be conducted to the cooler part.

In Chapter 8 when we described what happens in a situation such as the one in Figure 8.3, where two bodies at different temperatures are in contact, we were describing the process of conduction. Conduction can take place not only within a body whose parts are at different temperatures, but between two contacting bodies at different temperatures. The better the contact (that is, the greater the area actually touching), the more heat will be conducted.

The ability of materials to conduct heat varies widely from substance to substance. This ability is described by a property called **thermal conductivity**. The higher its thermal conductivity, the better a substance can conduct heat. Table 10.1 gives the thermal conductivities of a few ordinary materials. We'll discuss what the units mean shortly; for now, just compare the relative size of the thermal conductivities.

Generally, solids are better conductors than liquids or gases. The reason for this situation is that the molecules in solids are closer together and interact strongly with each other. Gases are poor conductors because the molecules are far apart and interact significantly only when they collide with each other.

As Table 10.1 shows, thermal conductivities vary widely, even among solids. In general, metals are better conductors than nonmetals because the atoms of metals have free (valence) electrons that can move easily throughout the body. Experiments show that the electrons rather than the molecules as a whole are responsible for most of the conductivity of metals.

Table 10.1 Thermal conductivities of materials

Material	Thermal conductivity, k $(kcal/m \cdot s \cdot C \ deg)$	Material	Thermal conductivity, k $(kcal/m \cdot s \cdot C \ deg)$
silver	0.10	brick	0.00015
copper	0.092	pine wood	0.000028
aluminum	0.051	plaster	0.0001
stainless steel	0.0033	ice	0.00052
ordinary steel	0.011	snow (compact)	0.000051
and cast iron		water	0.00014
glass (window)	0.00025	air	0.0000055
porcelain	0.00025		

QUESTION   If you were concerned only with getting heat most efficiently from the stove burner into the food, which of the materials in Table 10.1 would you choose for constructing a frying pan?

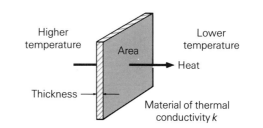

Figure 10.2 The rate of conduction of heat through a material depends on all the factors shown here.

The amount of heat conducted through a body depends on more than just its thermal conductivity. Figure 10.2 illustrates the factors that influence the rate of heat flow. By rate of heat flow, we mean the amount of heat flowing through the object per unit of time (for example, kcal/s). We are not talking about how fast the heat actually moves through the material, but *how much* heat gets through in a given amount of time. (For comparison: When you're running water in your bathtub, you don't worry about how fast the water molecules are moving through the pipe on their way to the faucet; you are only concerned with how much water comes out of the faucet per unit of time.)

In Figure 10.2 the slab of material of thermal conductivity k separates a high-temperature region from a low-temperature region. If you think of the conductor as a sort of pipe that carries heat, you can probably predict correctly how the amount of heat conducted depends on the factors shown in the figure. You would probably expect the rate to be higher for a bigger temperature difference across the slab, and for a larger area. Also, you would probably expect the rate to be lower the farther the heat must travel—meaning, the thicker the slab is. If so, your expectations are all correct and can be put in equation form as

$$\text{heat conducted per second} = \frac{k \times \text{area} \times (\text{temperature difference})}{\text{thickness}}$$

Convince yourself that increasing any of the factors on the right has the expected effect on the heat conduction.

An Illustration: Design of Cooking Utensils

Cooking utensils hold food during cooking and conduct heat from the stove burner into the food. The chemical changes involved in cooking take place only at high temperatures. We will pose a few questions for you to answer on the basis of what we have said about heat conduction.

1. How would doubling the thickness of the bottom of an aluminum pan influence the rate of heat transfer through the bottom?

2. Since stainless steel is much stronger than aluminum, stainless steel pans can be made thinner than aluminum ones. How much thicker than steel could you make an aluminum pan and still have the same rate of heat conduction through the bottom?

3. Most stove burners are hotter in some spots than others. To achieve uniform heat throughout the inside of the pot, should the bottom be thick or thin? Should the conductivity of the material be high or low?

4. Why do some manufacturers put a layer of copper over the bottoms of their stainless steel pans?

5. The specific heat (see Chapter 8) of aluminum is about twice that of cast iron, but cast iron is about three times as dense as aluminum. Which would stay hot longer—a cast iron or an aluminum pot?

6. If a pan is to be used on the perfectly flat heating element of an electric range, how will any nonflatness of the pan bottom affect the rate of heat transfer?

Calculations of Heat Conduction

We will work an example that illustrates the use of our heat conduction equation for calculating the rate of heat transfer. First, let's take another look at the meaning of the thermal conductivity k. We said that the rate of heat transfer is proportional to the thermal conductivity. In the units in Table 10.1, the thermal conductivity is the amount of heat in kcal conducted in 1 s through an area of 1 m^2 of 1-m-thick material across which there is a temperature difference of 1 C deg. To get the total heat transferred per second, we multiply the thermal conductivity by the total area and temperature difference, and divide by the thickness.

• •

EXAMPLE 10.1   A window 120 cm by 180 cm is made of glass 2.0 mm thick. If the inside surface of the window is at 20°C (68°F) and the outside surface is at −18°C (0°F), what is the rate of heat flow through the window?

$$\text{heat conducted per second} = \frac{k \times \text{area} \times (\text{temperature difference})}{\text{thickness}}$$

$$= \frac{0.00025 \text{ kcal/m·s·C deg} \times 1.2 \text{ m} \times 1.8 \text{ m} \times 38 \text{ C deg}}{0.0020 \text{ m}}$$

$$= 10 \text{ kcal/s}$$

Based on the answer in Example 10.1, this window would conduct about 290,000 kcal or 340 kWh of heat to the outside in 8 hours. At a price of 9¢/kWh, that would cost about $31 per window, or $620 per 8 hours in a house with 20 windows! Obviously, something is amiss in what we did.

Even though your hand is an unreliable thermometer, it's a good enough instrument to give you a hint at where we went astray. If you touch the inside of a single-glass window on a cold day, it is definitely much colder than the room temperature. In the example, our implicit assumption that the glass surfaces were at the same temperature as the inside and outside air is not valid. The air next to the window on both sides must transfer heat to and from the window, and air is a poor conductor of heat. The inside and outside glass surfaces differ little in temperature and are somewhere near the average of the inside and outside air temperatures. However, the glass temperatures and the overall heat transfer rate depend strongly on how much the air itself moves, bringing cold or warm air into the vicinity of the window. That brings us to our next method of heat transfer: *convection*.

Figure 10.3 Heat is transferred by convection when warmer air is forced up by the cooler air that is more dense.

10.2 CONVECTION

You have probably heard ever since you were in elementary school that "hot air rises." Let's examine this old wives tale and see why it's a distortion of the facts.

Consider a room that has a heater on one wall and a window on the opposite wall (Fig. 10.3). If the outside air is cold, heat will be conducted through the window, cooling the air nearby. As this air cools, it will contract and become more dense—a given volume will weigh more. This cooler, heavier air will push aside the warmer air below it and move down to exchange places with the warmer air. Thus, the warmer air will be *pushed* up by the cooler air.

On the other side of the room, where the heater is, the air near the heater is being warmed. Here, therefore, the air expands so that a given volume is lighter. This lighter air is then *pushed* up by the colder air above it that comes down to take its place. In both cases, the warmer air does *not rise* of its own accord. It still has weight and is being pulled toward the earth. But the cooler air has more weight per unit volume and bullies its way in to be closer to the earth. Hot air rises *only if* there is cooler air above it, and then only because it is pushed up by the cooler air.

As a result, there is a natural circulation of air in the direction of the arrows in Figure 10.3. The lower part of the room always contains cooler air, the upper part, warmer air. A room arranged this way can easily have a temperature difference of 10 C deg (18 F deg) between head and foot level.

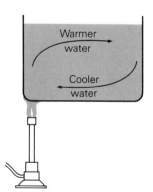

Figure 10.4 Any fluid—liquid or gas—can transfer heat by convection. Here the cooler, denser water pushes the warmer water to the top.

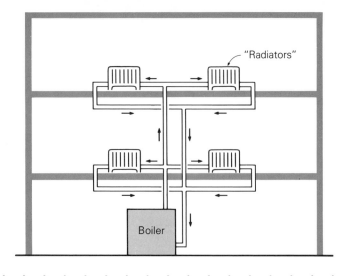

Figure 10.5 A gravity-fed hot water heating system. The arrows indicate the direction of water flow.

"Radiators"

Boiler

• •

QUESTION   Where should the heater be placed for the most even distribution of temperature in the room? Why? Hint: Where are heaters usually located?

The process of heat transfer we have described is an example of **convection**. Convection is a transfer of heat by movement of the heated medium itself—in this case, air.

A similar process happens in liquids. Suppose a flame is held under one side of a pan of water as in Figure 10.4. When the water directly above the flame is heated by conduction and expands, it is forced up by the cooler denser water. Thus, heat is transferred from the bottom to the top of the pan by convection.

Many houses built 30 to 40 years ago used this natural convection effect to distribute heat to "radiators" placed throughout the house. In Figure 10.5, water heated in the boiler is less dense and is forced up into the "radiators" by the denser water that has already conducted its heat to the air in the rooms. Heat is distributed throughout the room by convection in the room air.

Rather than relying purely on the natural convection from the expansion of heated fluid, heating systems now usually rely on forced convection. For example, hot water systems normally include a circulator to pump the water to, and through, the room heaters and back to the boiler for reheating. Hot-air central heating systems usually have a fan that blows air from the furnace through ducts to the rooms. Return ducts bring the cooled air back to the furnace.

Movement of air in the earth's atmosphere—winds—are merely convection currents. The sun heats the earth more in some places than in others. As the air above these places gets warmed (by conduction), convection causes a redistribution of air. The cool daytime breeze coming

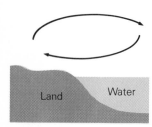

Figure 10.6 Convection currents between the warmer land and the cooler water cause the cool daytime breeze felt near large bodies of water.

Land   Water

in from large bodies of water is caused by convection. Since the land is warmer than the water, the air above the land gets warm. As Figure 10.6 shows, the cooler air above the water moves in and replaces this warmer air.

· ·

<blockquote>**APPLICATION**</blockquote>

THE YANKEES KNEW ABOUT CONVECTION—A CIVIL WAR EXAMPLE

In the Battle of Petersburg, Virginia—the last large battle of the Civil War before General Lee surrendered to General Grant at Appomattox—the Union and Confederate armies had been at a standoff for months. In an attempt to penetrate the Confederate lines, the Forty-eighth Pennsylvania Regiment, made up mainly of coal miners, dug a 511-foot tunnel that ended under the Confederate Pegram's Battery (Fig. 10.7). They also dug two 38-foot branches to the right and left at the end, and packed them with 8000 lb of powder. On July 30, 1864, 35 days after the project began, they exploded the powder, completely destroying Pegram's Battery. According to Col. Henry Pleasants, commander of the Forty-

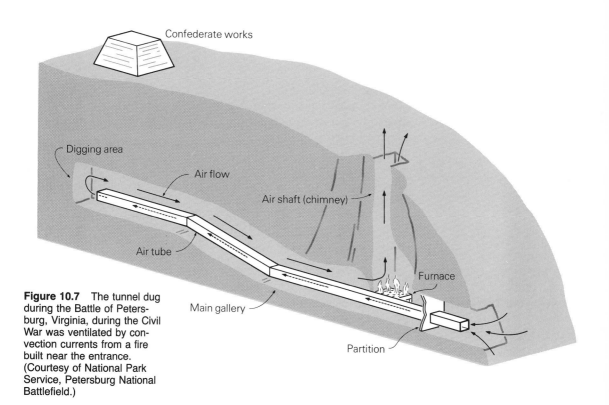

Figure 10.7 The tunnel dug during the Battle of Petersburg, Virginia, during the Civil War was ventilated by convection currents from a fire built near the entrance. (Courtesy of National Park Service, Petersburg National Battlefield.)

Figure 10.8 A fireplace needs convection currents for the fire to burn.

eighth, "the crater formed by the explosion was at least 200 feet long, 50 feet wide, and 25 feet deep."*

While digging the tunnel, the soldiers needed some means of getting fresh air to the digging area. To get fresh air, they ran an 8-inch square wooden tube from the entrance to near the end of the tunnel and extended it as the tunnel was lengthened. They sealed the entrance, except for the tube opening, with an airtight partition. Beyond this partition, they cut a vertical chimney to the surface, and kept a fire burning below the chimney (see Fig. 10.7). *Convection currents from the heated air forced fresh air through the tube to the far end of the tunnel, with air from there traveling back to the fire and up the chimney.*[†]

Convection currents make home fireplaces work. The fire gets fresh air needed for combustion from convection, which brings in new air to replace the heated air and combustion products that are pushed up the chimney (Fig. 10.8). A fireplace needs up to 17 cubic meters of air per minute—enough to empty a 5-m-by-6-m (about 15 ft by 20 ft) room in 5 minutes. Unless there is a special vent to the outside, this air must come from the room, which in turn must get its air from the outside through (usually small) cracks around windows, doors, and other places in the house. A fireplace could not operate in a completely leakproof house without vents to the outside.

QUESTION   Suppose a fireplace gets its air by infiltration from the outside through small cracks and openings throughout the house, as most do. If the

From the Official Records of the War of Rebellion, Vol. 40.
[†]*The results of the effort were disastrous for both sides. Not only were hundreds of Confederates killed during the explosion, but, during the melee that followed, masses of Union soldiers, horses, and equipment poured up the hill and tumbled into the crater left by the explosion, where they were either crushed by others falling in on top of them or later fell easy prey to Confederate guns.*

thermostat for the rest of the house is set to its usual value, so that the central heating system must heat this air before it gets to the room with the fireplace, is the homeowner likely to save on his fuel bill by using his fireplace? (Notice that the draft doesn't stop when the fire dies down. Much of it continues to flow as long as the fireplace damper is open.)

. . . .

EXERCISE   If you haven't worked Problem 9, Chapter 8, now would be a good time to work it.

10.3 COMBINATIONS OF CONDUCTION AND CONVECTION

Many important heat transfer processes are combinations of conduction and convection. We've already mentioned some of these in our comments on convection. The air touching a cold window is cooled by conduction of heat to the glass. In Figure 10.4, the heat passes through the container and into the water above by conduction. In the "radiators" of Figure 10.5, heat passes from the water in the radiator to the air outside by conduction. Warm earth heats the cooler air above it by conduction.

Once heat travels by conduction from a solid to a fluid, convection usually dominates the transmission in the fluid. But, as we discussed earlier, the rate of conduction between the solid and the fluid depends on the temperature difference between the two. Look again at Example 10.1 and the comments concerning why the answer for the heat transmitted through a window was much too high. Once the air near the window gets warmer, less heat will be conducted to it because of the smaller temperature difference between window and air. As that air moves away by convection, cooler air takes its place to again increase the rate of conduction.

This effect is greatly increased if the wind is blowing, because large quantities of cool air are brought into contact with the window to absorb its heat by conduction. Therefore, aside from the increase in the leaking of cold air through the cracks around doors and windows when the wind blows, more heat is also lost by conduction. Adding a storm window typically cuts the heat transfer rate about in half by providing a layer of air between the two sheets of glass that confines convection currents to that space. Conduction in the air is extremely slow.

A more personal example demonstrates this same effect. Your body is usually warmer than the surrounding air, so it loses heat by conduction to the air. If the air is still, a layer of warmed air surrounds your body and reduces the rate of conduction. In a breeze, this warm air is swept away and cool air brought in to pick up heat.

Table 10.2 Wind-chill factors (equivalent temperatures in °C)

Air Temperature (°C)	Wind Speed (km/h)				
	10	20	30	40	60
0	−3	−9	−13	−16	−19
−5	−8	−16	−20	−23	−27
−10	−13	−21	−26	−30	−34
−20	−25	−35	−42	−46	−59
−40	−47	−61	−69	−76	−81

The harder the wind blows, the more heat can be transmitted. Wind-chill factors, often given in the weather report in winter, are a measure of the effect we are describing. The *wind-chill factor* is the temperature in still air that has the same cooling effect on the body as that particular combination of wind and temperature. Table 10.2 gives some wind-chill factors for various temperatures and wind speeds. Wind speeds above 60 km/h have little additional effect.

- -

APPLICATION   **INSULATION**

Why does insulating a house result in less heat loss through the walls? Figure 10.9(a) shows a cross section of the wall of a house that is not insulated. For most houses, the walls have a space that is hollow except for studs placed every 16 or 24 inches. The air in this space next to the warmer side is heated; the air next to the cooler side is cooled. The result is a convection current in the direction of the arrows, carrying heat from the warmer to cooler side. The heat transfer from convection in this air space is many times more important than the transfer from conduction.

The purpose of insulation is to reduce convection. When the air space is insulated, it is broken up into small pockets, as shown in Figure 10.9(b). Heat transfer by convection is drastically reduced because it can occur only in small loops, as shown by the arrows near the top of the figure. The effect of conduction stays small because the insulation is fluffy, leaving the space still filled mostly with air, which has low conductivity.

Figure 10.9 Cross section of uninsulated and insulated walls. When insulation is placed in a wall, heat transfer by convection is greatly reduced because air can circulate only in small pockets.

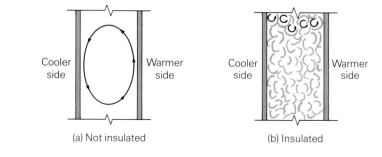

(a) Not insulated   (b) Insulated

Chapter 10   Transfer of Heat

Why does wearing thick clothes keep you warm?

Commercially available insulation is rated with an "R" value that stands for resistance to the flow of heat. For example, 6-inch-thick fiberglass is rated R-19, as marked on the bundle. The higher its R value, the more effective the insulation.

The R rating basically combines the thickness and thermal conductivity factors in our conduction equation. The effect of any convection is also lumped into the R value. The total rate of heat transfer is

$$\text{heat transfer rate} = \frac{\text{area} \times \text{temperature difference}}{\text{R rating}}$$

Commercial R ratings like the R-19 are in units of (ft$^2\cdot$F deg)/(Btu/h). To use R ratings directly, you need to express area in square feet and temperature difference in Fahrenheit degrees. The heat transfer rate then comes out in Btu/h.

Conduction, Convection, and Freezing Lakes

The low thermal conductivity of ice and water, and the natural convection in water, play an important role in the freezing, or lack of freezing, of deep lakes. To understand what happens, we first need to know the expansion properties of water. In Chapter 9 we discussed the fact that water expands when it freezes. But water *above* the freezing point has unusual expansion properties also.

Figure 10.10 shows how the volume of 1 kg of water changes with temperature between 0° and 100°C. The inset for the low-temperature region shows that the volume is least at 4°C. Above that temperature, water expands with temperature rise, as do most substances. But below 4°C, water expands with temperature *drop*. As water freezes, it expands even more, so that the volume of 1 kg of ice is larger than that of 1 kg of water at 0°C.

As we said in Chapter 9, the collapse of the hexagonal crystal structure of ice (Fig. 9.6) causes the contraction that water undergoes when it melts. But after melting, certain groups of molecules keep this hexagonal structure. As the temperature rises, more and more of the structures collapse. At 4°C the contracting effect of the collapsing crystals is balanced by the ordinary expansion due to higher molecular kinetic energy. Above this temperature the expansion has the larger effect.

If you were a fish, you would be extremely glad that water behaves as it does near the freezing point. Suppose the air above the deep lake in Figure 10.11(a) is warm—say 35°C—and that the temperature of the water near the top is only a few degrees lower. As the air temperature

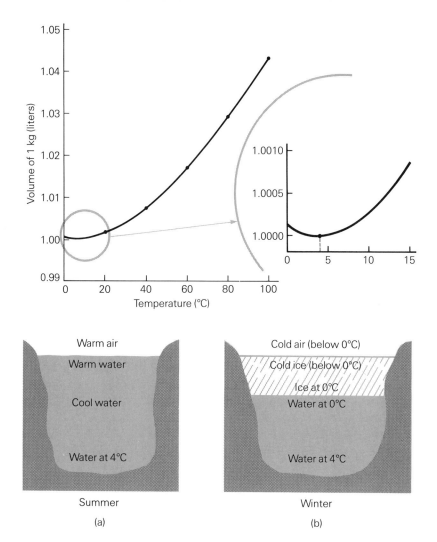

Figure 10.10 The variation in volume of 1 kg of water between 0° and 100°C. The low temperature region is shown in detail in the inset. The overall fractional change in volume between 0° and 100° is only about 4%.

Figure 10.11 Even in extremely cold climates, deep lakes do not freeze all the way to the bottom. The water in the lower part stays at 4°C year-round.

drops below the water temperature, the air cools the water by conduction. As the surface water cools, it contracts. Being more dense, it sinks, pushing the warmer water to the top where it gets cooled. As the air temperature falls to near 0°C, this convection process continues until the whole lake is at 4°C. Further cooling of the water at the top causes it to expand. The water is then less dense—that is, a given volume is lighter—than the water below it, so it stays on the top.

As the water temperature drops below its freezing point, the water at the surface freezes. In freezing it expands more, becoming less dense still and therefore staying on top as shown in Figure 10.11(b). The water just below the ice is colder than that further down. It, therefore, can freeze only by losing heat in conduction through the ice to the cold air above. But ice is a poor conductor of heat. Therefore, the thickness of the ice layer increases very slowly. If the lake is deep, the heat cannot transfer through the ice fast enough to freeze the entire lake before spring comes and the ice starts to melt. In fact, the bottom portion remains at 4°C, the temperature at which the water is most dense.

In summer, as the temperature of the water at the surface goes above 4°C, it expands and stays on top. To warm the lower water, heat must be conducted through this upper layer of water. For a deep lake the bottom then stays at 4°C year-round. Water either warmer or colder than this is less dense and therefore moves no heat downward by convection. Only conduction occurs, and water is an exceptionally poor conductor of heat.

· ·

QUESTIONS
1. If water continued to contract on down to the freezing point and also contracted on freezing, why would all lakes freeze solid once the temperature above them got below the freezing point?

2. What would have been the effect on the evolution of animal life if water had not been designed to have this anomalous expansion property?

10.4 RADIATION

The two methods of heat transfer we have talked about so far both involve a medium. In conduction, heat is transferred through a stationary medium by the collisions of the molecules. In convection, heat is transferred by movement of the heated medium itself.

But heat is transmitted sometimes without a medium. The most obvious example is heat from the sun. The sun emits **radiation** that travels millions of miles through empty space in getting to the earth. Once here, most of the radiation passes through the earth's atmosphere and is absorbed by the earth and objects on it. That way, heat is transferred from the sun to the earth, even though the radiation itself is not actually heat.

The radiation that transfers thermal energy is a form of what we call *electromagnetic waves*. In later chapters we'll say more about what these waves are. For now, we'll just notice that light, x rays, gamma rays, radio waves, and microwaves are also examples of electromagnetic waves. The only difference between them is in a property called *wavelength*. Figure 10.12 illustrates different wavelengths in a familiar type of wave. The wavelength is the distance from crest to crest, and it is much longer for the ocean wave than for the waves in the pond. Electromagnetic radiation with wavelengths between about 10^{-7} m and 10^{-4} m transfer thermal energy; we call this *infrared radiation*. Infrared's neighbor on the short wavelength side is visible light, the radiation to which the eye is sensitive.

Figure 10.12 The wavelength, the distance from crest to crest, is much longer for ocean waves than for waves in a small pond.

When you hold your hand near an ordinary incandescent light bulb, you feel heat. The radiant energy emitted includes not only visible light but infrared (heat) radiation as well. When you stand in front of an open fireplace, most of the heat you feel is from radiation. The heat from infrared radiation can relieve the pain of sprains, arthritis, bursitis, and various other aches.

All bodies radiate energy continuously. This fact means that, if the temperature of a body does not drop, it must also continuously absorb radiant energy. Some bodies are better emitters and absorbers than others. But we can show from the energy conservation principle that a good emitter is also a good absorber, and vice versa. To do this, we'll pick the example of a stove in an empty room.

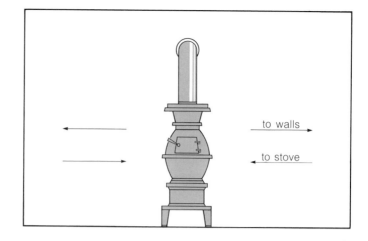

Figure 10.13 The stove emits radiation to, and absorbs radiation from, the walls, floor, and ceiling. The equal-length arrows symbolize that at equilibrium the rate of emission of radiation equals the rate of absorption.

Suppose the fire goes out in the stove of Figure 10.13 but the stove is still hot. (Imagine that the air has been pumped out of the room, so that we do not have to consider other forms of heat transfer.) The stove radiates heat to the walls, and the walls radiate to the stove. Since the stove is at a higher temperature than its surroundings, it emits faster than it absorbs radiation, so that its temperature drops. Once the stove and walls are *at the same temperature*, the stove must *emit and absorb radiation at the same rate*; otherwise, its temperature would change. This condition must hold whether the stove is a good or a poor emitter of radiation. Since energy is conserved and the temperature neither rises nor drops, we conclude that a good emitter must also be a good absorber.

Influences on the Rate of Emission and Absorption of Radiation

Several factors influence the rate at which a body emits or absorbs radiation. We'll consider the main ones.

Temperature. The rate of radiation from a body increases greatly with increasing temperature. In fact, this rate is proportional to the fourth power of the *absolute* temperature. This means that a body at 600 K (327°C) radiates at about twice the rate of one at 500 K (227°C). Doubling the absolute temperature gives 16 times as much heat radiated. Any body at a temperature above absolute zero (0 K), which means *every* body, radiates heat. Depending on the relative temperatures of a body and its surroundings, there may or may not be a *net* loss of heat.

Surface condition. A smooth, shiny surface reflects much of the radiation that hits it. On the other hand, a rough surface effectively has more surface area that can absorb radiation—the reasoning here is the same as that of my college friends from West Virginia who claimed that, if West Virginia were flattened out, it would be bigger than Texas. In other words, a good absorber has a coarse-textured surface rather than a shiny one. From our earlier arguments, a good emitter has the same characteristics.

Surface color.   You may have noticed that a dark-colored car gets much hotter when left in the summer sun than a light-colored one; a black asphalt driveway burns your feet worse in July than a concrete one; or various other evidences that dark objects get hotter when exposed to radiation. The natural conclusion to draw is that dark-colored objects are better absorbers than light-colored ones. However, that reverses the cause and effect. The object is dark-colored *because* it absorbs most of the radiation hitting it. A truly black object is one that absorbs all the radiation that hits it, thereby reflecting none. Physicists call such an object a *blackbody*, although a perfect one doesn't exist. The darker a body is, the closer it approximates a blackbody. By the reasoning we've used before, a dark-colored body is, therefore, one that is also a good emitter of radiation.

• •

QUESTION   Why will a black dog go outside and lie in the sun on a cold clear (even windy) day, rather than stay in its cozy insulated doghouse?

Temperature and Wavelength

All the radiation given off by a body is not of the same wavelength. The graph of Figure 10.14 shows that the radiation is spread out over a broad range of wavelengths. The vertical axis shows the relative amount of radiation emitted at various wavelengths along the horizontal axis. The vertical wiggly lines *symbolize* the increasing wavelength to the right.

Figure 10.14   A graph of the amount of radiation emitted by a body versus wavelength of the emitted radiation. The separation between wiggles in the vertical lines indicates the variation in wavelength. The dashed line is for a lower-temperature body than the solid line.

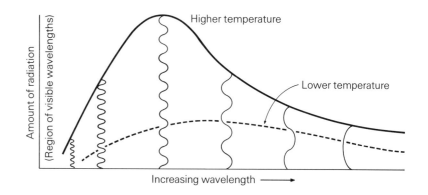

Chapter 10   Transfer of Heat

The peak in this curve corresponds to the wavelength at which the most radiation is emitted, although some radiation has very short, and some very long, wavelengths.

As the body's temperature decreases, the peak in the graph shifts to the right (toward longer wavelengths), and the total quantity of radiation decreases. The dashed line shows the emission curve for a body at a lower temperature. This curve peaks at a longer wavelength, and its overall height is lower.

A portion of the radiation on the short wavelength side has wavelengths in the range of visible light. As the body's temperature is raised, more and more of the radiation given off is visible. At 500°C, a significant amount is in the long-wavelength part of the visible region, which corresponds to red light. You see a dull red glow. As the temperature is raised, not only is more heat given off but more light of shorter wavelengths is emitted, making the light whiter. At 900°C the body is bright red and at 1500°C, white hot. You can see this effect as the heating element of an electric range or portable heater gets hot.

Figure 10.15 An infrared radiation thermogram can detect the slightly higher-than-normal temperature associated with breast cancer, even in its early stages. The light regions indicate the higher-temperature cancerous region.

Thermography. The infrared radiation given off by bodies can be a useful diagnostic tool in many areas. A device called an infrared scanner makes a *thermogram*, which is a picture based on the infrared radiation given off by a surface. Variations in temperature of small fractions of a C degree can be detected with sensitive scanners. An important application involves detecting cancers, which produce local "hot spots" in the skin temperature of as much as 1 or 2 C deg that can radiate at a 2% to 3% higher rate than normal (Fig. 10.15). Another use is the detection of areas of excessive heat loss from houses. The whole house is scanned on a cold day to detect spots where insulation is bad or where cracks exist. Electronic parts can be tested for flaws by looking for areas of excessive heat production.

Absorption of Radiation

We'll illustrate some of the important characteristics of the absorption of radiation by discussing some specific examples.

Figure 10.16 In a greenhouse, the short-wavelength radiation from the sun is transmitted, but the longer-wavelength radiation from the cooler objects inside is reflected.

Transparency and the greenhouse effect. Ordinary window glass is transparent—that is, light passes through it. At least, most of it does. But some light is absorbed, and the fraction absorbed depends on the wavelength of the radiation. Ordinary glass is also quite transparent to the short-wavelength infrared (heat) radiation emitted by very hot bodies. But glass absorbs most of the radiation from bodies at ordinary temperatures because most of it has much longer wavelength.

In a glass greenhouse (Fig. 10.16), the short-wavelength radiation from the sun is transmitted easily and absorbed by the contents of the greenhouse. The cooler objects inside emit mostly longer-wavelength rays that cannot pass through the glass, but are absorbed and reradiated, some back to the inside. The objects inside that are warmed by radiation conduct heat to the air confined inside the house, warming it. Convection distributes the heat throughout, but the relatively stationary air adjacent to the walls minimizes conduction to the outside.

The earth as a whole is a sort of greenhouse, housed not by glass but by the atmosphere. Most of the short-wavelength radiation of the sun is transmitted through the atmosphere and warms the earth. But much of the longer-wavelength rays reradiated by the earth do not get through the atmosphere. One of the potential hazards of air pollution (including excess carbon dioxide from the burning of fossil fuels) is that polluted air is less transparent to radiation from the earth. This fact might at first seem to have a beneficial effect—a warmer climate for all. One possible consequence, however, is that the polar ice caps could start to melt, flooding all our coastal cities.

MICROWAVE COOKING

Water is transparent to visible light and short-wavelength infrared rays, but paper is not. For different wavelengths of electromagnetic radiation, different materials are transparent. In fact, for *microwave* radiation such as police radar, just the opposite is true—paper is transparent but water is not. The radiation that serves as police radar and the radiation that does the heating in microwave ovens are basically the same—electromagnetic radiation with wavelengths of a few centimeters.

In microwave ovens (Fig. 10.17), the radiation passes unobstructed through paper, plastic, or glass containers, but it is absorbed in water and food. The absorbed radiant energy raises the temperature of the food and causes cooking. Some of the radiation penetrates several centimeters into the food before absorption, thus distributing the thermal energy throughout the food and giving quicker cooking than could be accomplished by conventional means.

Figure 10.17 A schematic diagram of the operation of a microwave oven. Microwaves travel down the waveguide and reflect from the stirrer that slowly rotates to scatter waves in all directions. The waves reflect back and forth from the walls until absorbed by the food.

　　　　Chapter 10　Transfer of Heat

In addition, microwaves reflect from metal. They bounce around inside the oven unattenuated until absorbed by the food. This fact means that almost all the radiant energy goes into the food. A conventional oven, however, heats both the air inside and the oven walls, which in turn heat the outer surface of the food. Heat is slowly transmitted by conduction to the interior. A large fraction of the heat is wasted by leaking through the walls of the oven into the room.

10.5 COMBINATIONS OF HEAT TRANSFER METHODS

When heat is transferred from one place to another, it usually moves by some combination of the three methods we have discussed. In all three methods, a major factor influencing the heat transfer rate is the temperature difference between the two regions. You can think of the temperature difference as the "driving effect" that causes heat to move from one place to another.

Isaac Newton, who probed into many different areas, noticed that *the rate of heat loss from a body by all methods is approximately proportional to the temperature difference between the body and its surroundings.* This fact is referred to as *Newton's law of cooling,* although it is not a fundamental law like Newton's laws of motion. It's only an approximate empirical law that holds under most circumstances.

• •

QUESTION   Based on Newton's law of cooling, what approximate relationship would you expect between the amount of fuel needed to heat your house (or apartment or room) during a particular season and the average temperature *difference* between the inside and outdoors?

Figure 10.18 A thermos bottle minimizes heat transfer by all three mechanisms.

A thermos illustrates the way different methods of heat transfer can interact. The bottle is designed to minimize heat transfer by any of the three methods. If the contents are hot, heat should be kept in; if they are cold, heat should be kept out.

Figure 10.18 shows how the thermos cuts down on each form of heat transmission. It consists of a double glass wall with silvered surfaces and a vacuum between the walls. Conduction and convection cannot occur across the vacuum. Conduction can occur along the glass-and-silvered surface, but the distance there is quite long. The silvered surface cuts down on radiation emission and absorption. The stopper is made of a material that is a poor conductor. (The bottle is housed in a metal or plastic can for protection.)

Key Concepts

There are three mechanisms of heat transfer—*conduction, convection,* and *radiation.*

Conduction is the transfer of heat from molecule to molecule within a substance. The **thermal conductivity** is the property of a substance that tells how effective it is in conducting heat.

Convection is the transfer of heat by movement of all or part of a heated fluid (liquid or gas). The circulation of fluid by means of a mechanical pump or fan is called forced convection. Conduction of heat from a body to an adjoining fluid is facilitated by convection (natural or forced), which brings more cool fluid into contact with the body.

Radiation is the transfer of thermal energy by electromagnetic waves. The rate of radiation of heat is determined by the temperature, color, and surface condition of the radiating body.

A body that is a good radiator is also a good absorber of radiation. All bodies radiate to, and absorb radiation from, their surroundings. A body at a higher temperature than its surroundings experiences a net loss of heat; one at a lower temperature, a net gain of heat. A body at the same temperature as its surroundings emits and absorbs radiation at the same rate.

The combined rate of heat transfer from a body to its surroundings by conduction, convection, and radiation is approximately proportional to the temperature difference between the body and its surroundings.

Questions

Conduction

1. Why are the handles of cooking utensils usually made of wood or plastic rather than metal?

2. When you pick up a cold hammer by the metal part, it *feels* much colder than when you pick it up by the wooden handle. Give two reasons why this is true.

3. If you sit in the bleachers during a soccer match in cold weather, why is your need for a cushion much more severe if the seats are aluminum than if they are wood?

4. Skis slide on snow that is not much colder than 0°C because friction melts the top layer of snow, letting the skis ride on water. Why do metal skis not slide as easily as wooden ones, even though they are smoother?

5. Suppose the burners on your electric stove are all 19 cm in diameter. Why does 1 liter of water heat quicker on one of these burners if you heat it in a 19-cm-diameter tea kettle than if you use a 14-cm-diameter one?

6. Discuss the electrical energy economy of using cooking utensils whose bottoms are larger than or smaller than the heating unit on which they are placed.

7. How much more heat will be lost through a window 3 m by 1 m than a window the same thickness but 2 m by 1 m if the temperature of the inside and outside surfaces of the glass are the same for both?

8. Aunt Jemima says that the pancake griddle is ready for cooking when "a small drop of water dances on it before evaporating." Why does the drop dance rather than quickly boil away? Why does the drop evaporate quicker when the griddle is cooler?

9. Certain circus performers and physics teachers have been known to plunge their *wet* hands briefly into molten lead without injury. Can you explain how this feat is possible?

10. If you put a mercury thermometer in something hot, the mercury moves up quickly at first, then slower and slower as it approaches the final reading. When you take it out, the same effect takes place—it goes down quickly

at first, then slower and slower as it settles down. Why?

11. If the nurse tells you to keep the thermometer in your mouth for three minutes, is it okay to take it out and look at it after half this time, as long as it stays a total of three minutes?

Convection

12. Why does the wind at the beach change direction at night?

13. What lifts a hot-air balloon into the air?

14. Why does it cost more in operational energy if you open your freezer when empty than when full? Hint: If the freezer is a chest type rather than upright, it probably doesn't.

15. Only about 10% of the heat produced in an ordinary fireplace gets into the room. Why is a fireplace so inefficient?

Combinations of Conduction and Convection

16. The difference between conduction and convection is sometimes pointed out by comparing a bucket brigade with a group of people all carrying individual buckets of water. Which mode of heat transfer would you associate with the bucket brigade and which with everybody having an individual bucket?

17. For a fireplace chimney, why can't you use a *bare* metal pipe running up the outside of your house?

18. When your skin temperature is warm, a large amount of blood flows through the small arteries near the outer layer of your skin. When the skin temperature drops below 37°C, your body compensates by constricting these blood vessels and limiting the flow of blood to the surface. How does this activity help to keep the inner part of your body warm? (The skin and fatty tissue underneath are poor conductors.)

19. Charlie Brown turns a fan on in his house and goes away for a while, expecting the house to be cooler when he gets back. Is he fooling himself?

Figure 10.19   A hot-water baseboard heater.

20. Baseboard hot water heaters are basically copper pipes with fins surrounded by a housing (see Fig. 10.19). What are the fins for?

21. The 110-lb ears of African elephants hold many blood vessels and are a great asset in keeping the elephant cool. What are the similarities between this effect and that of the baseboard fins discussed in the preceding question?

22. Suppose you are comfortable when the temperature is 0°C and the wind is blowing at a speed of 40 km/h. If the wind stopped blowing, how much colder could it get without your being uncomfortable in the same clothing?

23. A student claimed that his reason for missing physics class was that his car would not start when the temperature was *below* −5° F. Even though the temperature was 0° F that day, the wind-chill factor was −15° F. Should the professor accept his excuse?

24. Some states post signs on highway bridges that say, "Bridge freezes before road surface." Why does this happen?

25. Some oven manufacturers claim that self-cleaning ovens "save energy" and cost less to operate in the long run because they have more insulation in the oven walls. Is this claim reasonable?

26. How does heavy snowfall help plant roots and small animals that live underground to survive the winter?

27. Why does the hair on animals "stand up" in winter? Why do birds *look* fatter in winter?

28. Justify the statement made in the text that, if you were a fish, you would be very glad that water expands as the temperature drops below 4°C, and expands even more when it freezes.

29. What was the temperature at noon yesterday in the deep underwater den of Nessie, the Loch Ness monster? What was it at midnight last night? Barring unexpected developments, what will it be in exactly six months?

Radiation

30. Many houses are heated with electric heaters imbedded in the plaster in the ceiling. By which of the three mechanisms does heat warm the room? Why do your feet get cold if you are sitting with them under a table in such a room?

31. How does heat get from the heating element to the bread in a toaster?

32. Why do athletes put black under their eyes when performing in bright sunlight?

33. At our house one job we have to do before my mother-in-law comes to visit is polish the silver coffee pot she gave us. Aside from its appearance, is there a practical reason why this utensil will keep the coffee hot longer if kept shiny?

34. Explain why sprinkling soot over snow will cause it to melt more quickly.

35. Why do sailplanes experience strong upward air currents over freshly plowed fields or large paved areas?

36. If you want the snow to melt completely from your car as quickly as possible, should the car be a light or a dark color?.

37. If you live in a warm sunny climate, should the roof of your house be light-colored or dark? Should you buy a light- or dark-colored car if you want it to be cool?

38. Can the temperature of the water in the collector of a solar water heater ever get higher than that of the air outside?

Combinations of Heat Transfer

39. The waiter brings apple pie and coffee to your date, who does not want to drink the coffee until after finishing the pie. Knowing that you are taking physics, your date asks you for this advice, "If I want my coffee to be as hot as possible when I drink it, should I add the cream and sugar now, or wait until I finish my pie?" What is your answer?

40. Why do cookies near the edge of the baking sheet brown faster than those in the middle? Hint: On a flat sheet with no turned up sides, this does not happen.

41. Why do you save fuel in winter by keeping your thermostat set low?

42. Do you save fuel in winter by turning your thermostat down at night and when you are away in the daytime? Why so or why not? Base your answer on the principle of conservation of energy.

Problems

1. A store window is a single sheet of glass 3 m by 4 m by 5 mm thick. If the air temperature inside is at 20°C and that outside at −10°C, heat would typically transfer through this window at a rate of about 0.50 kcal/s. For this rate, what is the actual temperature difference across the glass only?

   Answer: 0.83 C deg

2. If the owner replaces the glass of Problem 1 with a sheet of 1-cm-thick pine plywood, what will be the rate of heat transfer if the temperature difference across the plywood is the same as it was across the glass? Would you expect this temperature difference to stay the same?

   Answer: 0.028 kcal/s

3. A copper rod 25 cm long and 2 cm in diameter has one end in boiling water and the other end in ice water. What is the rate of heat transfer down the rod? How much ice is melted per hour? (The rod and water are insulated so that

no heat is transferred except along the length of the rod.)

Answer: 0.0116 kcal/s; 520 g

4. A styrofoam ice chest 60 cm by 30 cm by 40 cm has 2-cm-thick walls. If it is filled with ice and put in a 30°C room, what is the rate of heat loss through the walls, assuming the outside walls of the chest to be 5 C deg lower in temperature than the room? How much ice melts in 5 hours? (The thermal conductivity of styrofoam is 0.00001 kcal/m·s·C deg.)

Answer: 0.0135 kcal/s; 3.0 kg

5. A wall 20 ft by 8 ft is insulated with R-11 insulation. Assuming this is the main resistance to heat flow, what is the rate of heat loss through the wall if the temperature difference is 70 F deg?

Answer: 1000 Btu/h

6. A room is heated by a steam radiator that gives up heat as steam condenses to water at a temperature of 100°C. An identical radiator contains hot water that heats a room as water cools from 90°C to 70°C. Compare the rate of heat transfer to the two rooms.

Answer: about 33% higher with steam

7. The *degree-day*, a unit used to determine heating requirements, is the difference between 18°C and the average temperature for the 24-hour day. (The value is taken as zero if the average is above 18°C.) In Atlanta the average number of C degree-days is 1600, and in Cleveland it is 3400. Approximately how would the fuel consumption for heating a particular house in Cleveland compare with that for an identical house in Atlanta?

Answer: about 2.1 times as much in Cleveland

8. The net rate of heat loss by radiation from an object is given by

$$R = \varepsilon\sigma A(T^4 - T_0^4)$$

where T_0 is the *absolute* temperature of the surroundings; T is the *absolute* temperature of the object; A is the area of the body; σ is a constant of value 5.67×10^{-8} W/m²·K⁴; and ε, the emmisivity, tells how effective the surface is as a radiator. The area of a human body is about 1.9 m² and ε is about 0.97. What is the rate of radiation of heat by such a body in a room at 20°C? The skin temperature is about 28°C. A person sitting still burns up about 100 kcal/h. What fraction of this energy is given off in radiation?

Answer: 88 W; 76%

Home Experiments

1. Check your answer to Question 11 by comparing your result from a continuous 3-min measurement of your body temperature with the result when you take the thermometer out once or twice during the measurement but still have a total in-time of 3 minutes.

2. Using a candy thermometer or other thermometer that measures temperature variations as small as 1 C deg, check your answer to Question 39. Pour two identical cups of coffee and add a measured amount of cream and sugar to one. After about 10 min, add the same amounts to the other. Measure the temperature of each. You should see 2 or 3 C deg difference. To avoid using heat from your experimental coffee to warm the thermometer, place the thermometer in a third cup of hot coffee first and then quickly transfer it to the one being measured. Any stirring should be done with a small stick or other low-heat-capacity device. Except for the time difference, be careful to do the same thing to each cup of coffee.

3. Extend Home Experiment 3 of Chapter 8 by estimating the energy savings by turning off the water heater for 8 hours. Measure the room temperature near the water heater. Subtract the room temperature from the highest water temperature. Subtract the room temperature from the *average* water temperature before and after the 8-h period. Based on Newton's law of cool-

ing, the ratio of these two temperature differences should be a good approximation to the ratio of heat loss for 8 hours with the heater on and with it turned off after the water is hot.

4. Bake cookies on a flat baking sheet and then on one with sides. Again use the flat sheet, but don't put the cookies near the edge. From a comparison of results, answer Question 40.

References for Further Reading

Anderson, B., *The Solar Home Book* (Cheshire Books, Harrisville, N.H., 1976). A beginner's book on solar energy.

Edelson, E., "Heat Pipes—New Ways to Transfer Energy," *Popular Science*, June 1974, p. 102. Shows how evaporation and condensation can be used as a very effective way to transfer heat.

Hodges, L., "Fuel Savings by Thermostat Setback: An Experimental Study," *The Physics Teacher*, Nov. 1977, p. 485. Experimental evidence that it does save fuel to set the thermostat back for part of the day.

Jones, T. H., "How to Figure the Best Places to Cut Heat Loss in Your House," *Popular Science*, Sept. 1975, p. 97. Shows how to calculate easily the rate of heat loss from a building.

Jordan, W., "The Bee Complex," *Science 84*, May 1984, p. 58. Shows the important role of heat transfer in the life of a bumble bee.

Plumb, R. C., "Sliding Friction and Skiing," *Journal of Chemical Education*, Dec. 1972, p. 830. Describes evidence based on thermal conductivity for the importance of friction in the easy sliding of skis.

USING AND PUMPING HEAT

At some time in your life, you've probably had a secret wish to get something for nothing. If that wish ever develops into an urge to try your hand at the dice, you might do well to remember what some people call the "Las Vegas laws":

1. You can't get something for nothing.

2. You can't even break even.

Using heat to do work is very much the same. As we discuss the first and second laws of thermodynamics and how they apply to engines that use heat to do work, you will see their similarity to the Las Vegas laws.*

Sometimes, rather than using heat to do work, we need to pump it from a cool place to a warmer one. That's what refrigerators, air conditioners, and less familiar devices *called* heat pumps do. We'll find that what happens in these devices is, in principle, just the reverse of what happens in an engine. We'll also find that for some of them, you get *more* than you pay for. As always, somebody has to pay the bill, just as when Mother takes you out to dinner. But here, we'll find that it's Mother Nature who foots the bill.

This chapter will draw on some of the ideas of the preceding three chapters. You need to recall: (1) that heat is internal energy added to or

---

The bit of folk wisdom incorporated in this terminology was passed on to me by Professor Jack M. Wilson.

removed from a substance (Chapter 8); (2) that heat is required to change a substance from solid to liquid and from liquid to vapor, but heat is given up in the reverse changes (Chapter 9); and (3) that heat flows naturally from a warmer to a nearby cooler region, never the reverse.

Before we study the laws of thermodynamics, we need to understand how the temperature of a gas changes as it expands or is compressed.

11.1 EXPANSION AND COMPRESSION OF GASES

In Chapter 8 we discussed the expansion that usually accompanies an increase in temperature and the contraction that usually accompanies a decrease. This effect is especially pronounced in gases, which expand or contract noticeably if kept at constant pressure and heated or cooled. (What happens to a balloon if you blow it up, tie it, and put it in a refrigerator?)

If we reverse cause and effect, you might expect that, by letting a gas such as air expand, it would become warmer. Instead, the opposite happens: A gas allowed to expand without adding outside heat gets cooler. For example, the air rushing out of a leaking tire is much cooler than it was when inside the tire. The air at high altitudes is cool because, as warm air rises, the pressure from other air above it decreases. The air expands, which cools it. Let's consider the conditions under which expansion of a gas causes it to cool.

The molecules of a gas have kinetic energy due to their continual motion. As gas molecules move, they collide with each other as well as with the walls of their container. When the gas is allowed to expand, it does work on the walls of its container as they move outward during this expansion. For example, as the piston in Figure 11.1 moves to the right, the molecules of the gas exert a force on the piston in the direction of its motion—that is, they do work on the piston. If no heat enters or leaves the gas, the energy used in doing this work comes from the kinetic energy of the molecules, causing their average kinetic energy to decrease. A decrease in average molecular kinetic energy means a decrease in temperature. Thus, if a gas does work as it expands, and no heat is added, this expansion is a cooling process.

Compressing the gas has the opposite effect. To compress a gas into a smaller volume, you have to do work. We can compress the gas in the cylinder of Figure 11.1 by moving the piston to the left. The gas molecules exert a force on the piston through their many collisions with it; from the action–reaction law, the piston has to exert an equal force back on the gas. As the piston moves to the left, this force does work on the gas. The work done increases the kinetic energy of the molecules, which means an increase in temperature. Thus, compressing a gas without adding or removing heat results in an increase in temperature.

Figure 11.1 The gas gives up energy as it does work on the piston moving to the right. The gas absorbs energy as the piston moving to the left does work on the gas.

We can describe the overall relationship between pressure, volume, and temperature of a gas by the equation

$$\frac{\text{pressure} \times \text{volume}}{\text{absolute temperature}} = \text{constant}$$

for a given mass of a gas. This equation, which we call the *ideal gas law*, is strictly correct only for an *ideal gas*—one whose molecules are volumeless points that interact only during collisions. Even though no such gas exists, most real gases at ordinary temperatures and pressures behave almost as an ideal gas would.

11.2 CONSERVATION OF ENERGY: THE FIRST LAW OF THERMODYNAMICS

If you have a fixed quantity of matter, you can treat it only four ways from an energy viewpoint: (1) you can add heat to it; (2) you can remove heat from it; (3) you can do work on it; or (4) you can let it do work. Applying this statement to the gas of Figure 11.1, you can add heat to or remove heat from the gas; you can do work on it by moving the piston to the left and compressing it; or you can let it do work by moving the piston to the right. Conservation of energy tells us that whatever energy or work you add to the substance must either stay there or come out. If it stays there, it increases the internal energy of the substance.

In summary,

net heat added = increase in internal energy + work done

where "work done" means *by* the substance. We call this way of writing the energy conservation principle the **first law of thermodynamics**. It basically says that, as far as energy is concerned, you can't get something for nothing. Notice that if heat is *removed*, or internal energy *decreases*, or work is done *on* the substance, that particular quantity goes into this equation as a *negative* value.

We'll use this law in discussing engines that use heat to do work and in discussing devices that pump heat from a cooler to a warmer region.

11.3 ENGINES THAT USE HEAT

Many different types of **heat engines** are available that can do everything from powering our cars to turning electrical generators. All of them do basically the same thing—add heat to a fluid that uses some of the heat to do mechanical work. To illustrate this process, we will describe the basic principles of an internal combustion engine, which is the type used in most automobiles as well as on most lawn mowers. We can then show how the same work and energy ideas apply to any heat engine.

THE INTERNAL COMBUSTION ENGINE

An ordinary gasoline engine consists of one or more cylinders, such as that shown diagrammatically in Figure 11.2(a). This figure shows the four strokes in a complete operating cycle of the engine. (Many small engines operate on a two-stroke cycle, but the basic ideas are the same.)

On the intake stroke (a), the intake valve opens, letting a mixture of fuel and air into the cylinder as the piston moves down. The intake valve closes, and this mixture is compressed during the compression stroke (b). A spark from the spark plug ignites the fuel, with the energy released during combustion causing a high rise in temperature. The resulting high pressure pushes the piston down for the power stroke (c). The exhaust valve then opens and the remaining gases are pushed out as the piston comes up during the exhaust stroke (d). The sequence then repeats.

Since only one of the four strokes is a power stroke, the engine has to be started by turning it with some external torque—either from a starter motor or a pull-rope, for example. A flywheel attached to the crankshaft stores some of the mechanical energy generated during the power stroke, and keeps the engine running until the next power stroke. The useful work output is done by whatever is attached to the rotating crankshaft. Automobile and other large engines usually have several cylinders of the type we've described.

Figure 11.2 The four strokes of the four-stroke internal combustion engine. (a) Intake stroke: The intake valve opens, admitting fuel and air as the piston moves down. (b) Compression stroke: The rising piston compresses the fuel-air mixture. (c) Power stroke: The fuel burns, increasing the temperature and pressure. The high pressure pushes the piston down. (d) Exhaust stroke: The exhaust valve opens, letting the gases out as the piston rises.

The heat released during combustion either does mechanical work or is rejected as heat outside the engine. This rejected heat either goes out the exhaust pipe with the combustion products or is conducted through the cylinder walls. The heat conducted through the cylinder walls is either given up directly to the outside air or to circulating water that later gives up its heat to air passing through the radiator.

Figure 11.3 The energy and work transfer in a heat engine.

General Energy Considerations

We can show the relationship between the overall energy transfer and the work done in an internal combustion engine, or any heat engine, in a diagram such as Figure 11.3. This diagram is purely symbolic and does not show the physical appearance of a real heat engine. The circle represents the engine. The "channel" on the left represents the heat input that, in the internal combustion engine, comes from burning fuel. The channel at the bottom represents the work output by the engine. The channel on the right represents the heat rejected to the outside.

At any point in the engine cycle, the internal energy of the fuel-air mixture is the same as at that point in the previous cycle. Thus, for a complete cycle, there is no net change in *internal* energy. The first law of thermodynamics then tells us that

heat in = heat out + work out

The widths of the channels in Figure 11.3 represent the amount of heat (or work) transfer in each region. The sum of the widths of the two output channels is equal to the width of the input channel. You might suspect that the width of the work channel is much too small compared with the heat output channel. Unfortunately, for most heat engines the widths on the channels in the figure are approximately correct in proportion. Most of the input heat is rejected to the surroundings and does no useful work.

The term *thermal efficiency* rates the performance of a heat engine and is defined as

$$\text{thermal efficiency} = \frac{\text{work out}}{\text{heat in}}$$

This equation means that, the greater the work output for a given heat input, the higher the thermal efficiency. However, the more heat it takes to do some amount of work, the lower the thermal efficiency. Figure 11.3 shows that we can also write

$$\text{thermal efficiency} = \frac{\text{heat in} - \text{heat out}}{\text{heat in}}$$

EXAMPLE 11.1   It typically takes about 600,000 J of work from the engine of a compact car to move the car 1 km along the highway. If the car gets 14 km/liter of gasoline (33 mi/gal), what is the thermal efficiency of the engine? The heat is supplied by burning gasoline that has a heat of combustion of 36×10^6 J/liter.

Basing our calculation on 14 km of travel,

$$\text{work out} = 60 \times 10^4 \text{ J/km} \times 14 \text{ km} = 84 \times 10^5 \text{ J}$$

$$\text{heat in} = 36 \times 10^6 \text{ J/liter} \times 1 \text{ liter} = 36 \times 10^6 \text{ J}$$

$$\text{thermal efficiency} = \frac{\text{work out}}{\text{heat in}} = \frac{84 \times 10^5 \text{ J}}{36 \times 10^6 \text{ J}}$$

$$= 0.23 \text{ or } 23\%$$

This efficiency is that of the engine alone. Multiply this by the approximately 40% combined efficiency of the transmission and rolling tires, and the overall efficiency is down to about 9%.

EXERCISE   A power plant burns coal at a rate of 200,000 kg/h and generates 600 megawatts (MW) of electricity. The heat of combustion of coal is 7100 kcal/kg (8.2 kWh/kg). What is the overall thermal efficiency of the plant?   *Answer*: 36%

11.4 YOU CAN'T BREAK EVEN: THE SECOND LAW OF THERMODYNAMICS

The nineteenth-century French physicist Sadi Carnot studied the cycles through which a substance can be taken in converting heat into work. He recognized that any time you operate a heat engine, there must always be some heat rejected *as heat* from that engine. The **second law of thermodynamics** describes this effect. You may see this law stated many different ways, but the form of the law we're concerned with here is: *No actual or ideal engine that operates on a cycle can convert all the heat supplied to it into work.* In other words, the "heat out" channel in Figure 11.3 must always be present. In any one expansion, all the heat can go to work. But for the complete cycle needed to *keep* an engine operating, heat must be rejected at some stage.

Carnot showed that for the most efficient possible engine you can have, the thermal efficiency is limited by the operating temperatures. If

T_{hot} is the *absolute* temperature at which heat is added during the cycle and T_{cold} is the *absolute* temperature at which it is removed, then

$$\text{maximum thermal efficiency} = \frac{T_{\text{hot}} - T_{\text{cold}}}{T_{\text{hot}}} = 1 - \frac{T_{\text{cold}}}{T_{\text{hot}}}$$

The higher you can make T_{hot} and the lower you make T_{cold}, the higher the theoretically possible efficiency.

A power plant using steam that enters its turbines at 873 K (600°C) and later condenses at 373 K (100°C) has a maximum theoretical efficiency of [1 − (373/873)], or 57%. That makes the 37% efficiency of a modern power plant seem not quite as bad.

You may wonder why a car with a diesel engine typically gets higher gas mileage than a similar car with a gasoline engine. The main reason is that the fuel in the diesel burns at a higher temperature and therefore gives the engine a higher thermal efficiency. A diesel has no spark plugs. Instead, the air in the cylinder is compressed during the compression stroke to three or four times the pressure in the cylinder of a gasoline engine. This compression raises the temperature high enough for the fuel to spontaneously ignite when injected into the cylinders.

11.5 PUMPING HEAT

Figure 11.4 The refrigeration unit of a refrigerator pumps heat from the food compartment to the room.

Heat flows naturally from a warmer to a nearby cooler region, just as water flows naturally downhill. However, not only can we pump water uphill by doing work on it, but we can pump heat from a lower to a higher temperature by doing work in just the right way. That's the purpose of any heat-pumping device, whether a freezer, refrigerator, air conditioner, or heat pump. Figure 11.4 illustrates this idea for a refrigerator. If its inside is to be cooled, and to stay cool when warm food is placed in the refrigerator, heat has to be removed by the refrigeration unit and dumped outside—usually at the bottom or at the back.

The refrigerator accomplishes this feat by pumping a cold fluid through tubes inside the food compartment. This fluid absorbs heat from the contents of the compartment. The fluid is then pumped outside the compartment and compressed so that its temperature rises enough to give up heat to the warm outside air. We'll take a look at this process in more detail.

The Refrigeration Cycle

Figure 11.5 shows the main components of an ordinary refrigerator. An air conditioner or heat pump has the same basic components, but the construction details differ.

The system contains coiled tubing that winds throughout the cold

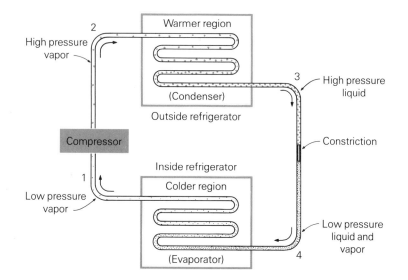

Figure 11.5 A schematic diagram for a compression refrigeration cycle.

region inside the refrigerator. Another coil in the warm region outside the refrigeration compartment usually has fins attached to help dissipate heat. Tubing joins the two regions to form a complete loop. An electrically operated compressor pumps a fluid called *refrigerant* (usually freon) around the loop in the direction of the arrows. Opposite the compressor is a constriction that restricts the flow of refrigerant. There may also be a fan to blow air over one or both coils to speed heat transfer.

Let's follow the refrigerant through its cycle. At point 1, it has just passed through the cold region and is in the vapor state at low temperature and pressure. The compressor compresses this vapor to a high pressure, causing its temperature to rise.

At point 2 the refrigerant is still a vapor, but at high temperature and pressure. As it passes through the coils outside the refrigerator—the *condenser*—its temperature is higher than that of the surrounding air. Therefore, heat flows from the refrigerant to the air, causing the refrigerant to condense.

At point 3, having given up heat and condensed, the refrigerant is a liquid at roughly the same temperature and pressure as at point 2. The refrigerant is next forced through the small constriction where it expands into region 4, which is at much lower pressure. The constriction resists the flow, and keeps a large pressure difference between 3 and 4. The drop in pressure lowers the boiling point, causing some of the liquid to vaporize. The heat needed for vaporization comes from the liquid itself, causing the temperature to drop to a low value.

At point 4 the refrigerant is mostly liquid, but some is vapor, at low temperature and pressure. As the fluid passes through the cold inside region—the *evaporator*—it is at a lower temperature than the surrounding medium so that heat is absorbed into the fluid from the surroundings. This heat evaporates the refrigerant to the vapor state, bringing it back to point 1 at low temperature and pressure, where it continues through the same cycle.

When the refrigerant is in the high-pressure part of the cycle, its temperature is that of its boiling point at that pressure. Similarly, the low-pressure side is at the temperature of the boiling point of the refrigerant for *that* pressure. The two pressures are chosen to produce a refrigerant temperature several degrees higher than the surrounding medium in the warmer region, and several degrees lower than the surrounding medium in the colder region.

The overall result of the cycle is to pump heat from the colder to the warmer region. Keep in mind, however, that in the condenser heat flows of its own accord *from* the *warm* coils *to* the *cooler* surroundings. In the evaporator, heat flows of its own accord *from* the *cold* region *to* the *colder* coils.

General Energy Considerations

We can describe the overall energy and work transfer in a refrigeration cycle by a simple symbolic diagram. In Figure 11.6, the circle represents the region where the refrigeration cycle operates. The region between the horizontal lines on the left symbolically represents a channel through which heat enters from the low-temperature region. Similarly, heat given up to the high-temperature region is represented by the channel on the right. The channel at the bottom represents the work done by the motor that runs the compressor.

The width of each of the channels is proportional to the amount of energy transferred in each process. Since there is no net change in internal energy for the complete cycle, the first law of thermodynamics tells us that

heat in + work in = heat out

The widths of the channels shown indicate that the heat input is about 2.5 times the work, and the heat output is about 3.5 times the work. These are typical values for real refrigeration systems. In other words,

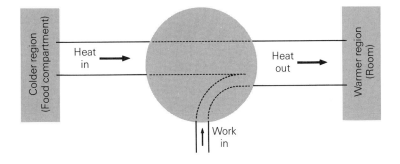

Figure 11.6 Energy and work transfer in a refrigerator.

Figure 11.7 (a) A heat pump serves as an air conditioner in the summer. (b) In winter it is reversed and pumps heat into the house from outside.

about 2.5 times as much energy is removed in the form of heat as the amount of work done in the compressor. Similarly, 3.5 times as much heat is delivered to the high-temperature region as is used in running the compressor.

Let's compare the energy transfer in an air conditioner and in a heat pump (Fig. 11.7). Figure 11.7(a) for an air conditioner is similar to the diagram for a refrigerator. Heat is removed from the cool inside and dumped outside the house. The total heat rejected equals the sum of the heat removed and the work done.

In winter you are concerned with adding heat *to* the inside. A heat pump (Fig. 11.7-b) is really an air conditioner in reverse. In a sense, you air condition the outside. Heat is removed from the cold outside and deposited in the house. Since the heat deposited is the sum of the heat removed from the outside and the work done, you get more heat inside than the energy you use in running the heat pump. The same piece of equipment can serve as both an air conditioner and a heat pump. A reversing valve reroutes the refrigerant so that the coils acting as an evaporator in summer become a condenser in winter.

Performance Ratings

Heating and cooling systems are rated in terms of how much heat they deliver or remove per unit of time. Most systems currently in use in the United States are rated in Btu/h, rather than in the metric units of energy or power that we have used in this book. For example, a 6000-Btu/h air conditioner (for a room) removes an average of 6000 Btu (1500 kcal or 1.8 kWh) of heat from the room each hour of operation.

It is also useful to know how much work is done (that is, how much electrical energy is needed) to add or remove the heat. Rather than defining an efficiency, we use the term *coefficient of performance* to rate refrigeration systems. The coefficient of performance is the ratio of the useful heat transfer to the work needed to transfer the heat. For a cooling system—refrigerator or air conditioner—

$$\text{coefficient of performance (cooling)} = \frac{\text{heat removed}}{\text{work done}}$$

In a heat pump used for heating,

$$\text{coefficient of performance (heating)} = \frac{\text{heat added}}{\text{work done}}$$

For an air conditioner or refrigerator the coefficient of performance is typically about 2.0 to 2.6, meaning that about 2.3 times as much heat is removed as work is done.

If you look at the sales information provided for air conditioners (see, for example, the Sears, Roebuck catalog), you will notice that it usually refers to something called the *energy efficiency ratio* (EER). The EER gives exactly the same information as the coefficient of performance. But the way the units are used in the EER does about the same thing to a physicist as scraping a fingernail across the blackboard. The definition of EER is:

$$\text{EER} = \frac{\text{heat removed in Btu/h}}{\text{power to the compressor in watts}}$$

An 8000-Btu/h air conditioner that uses 1000 W of electrical power has an EER of 8.

The EER presents problems for the physicist because its definition requires different types of units to be mixed in the same equation. By comparison, the coefficient of performance assumes that you use the same units for heat transfer as for work done. The heat transfer rate could be given in any *power* units, such as watts. Since 1 W is equivalent to 3.4 Btu/h, the 8000-Btu/h system removes heat at the rate of 8000/3.4 or 2400 W. For this air conditioner,

$$\text{coefficient of performance} = \frac{2400 \text{ W}}{1000 \text{ W}} = 2.4$$

(Why can we use power here rather than energy?)

Air conditioner EERs usually run between 5 and 12, with a value of 10 or above being excellent. For low power consumption, you obviously need an air conditioner with a high EER. If the EER is not given directly, you can calculate it by finding the heat removal rate and the electrical power used from the nameplate or the owner's manual.

EXERCISE   What is the EER of a 6000-Btu/h air conditioner that uses 1200 W of electrical power? What does it cost to operate this unit for 10 h if electrical energy costs 9¢ per kWh?   *Answer*: 5; $1.08

EXERCISE   What electrical power would be needed if this air conditioner had an EER of 10? What would it then cost for 10 h of operation?
Answer: 600 W; 54¢

The amount of work needed to pump a certain amount of heat depends on the temperature difference through which the heat is being pumped. This is like saying it takes more work to carry 5 gallons of water up Mt. Rainier (elevation 4392 m) than up to Clingman's Dome (elevation 2024 m). For air conditioners and refrigerators, the temperature difference between warmer and colder regions usually doesn't change much. But for heat pumps that take heat from outside air for which the temperature varies widely, the coefficient of performance can vary appreciably with temperature change.

For an outside temperature of 10°C, a typical heat pump has a coefficient of performance around 3. This fact means that for every 1 kWh of electrical energy delivered to the heat pump, 3 kWh of heat are delivered to the house. You get 2 of the 3 kWh free! Compare this with direct electrical resistance heaters such as baseboard or ceiling heaters. For these heaters, the coefficient of performance is 1: You get only as much as you supply in electrical energy. But the coefficient of performance of heat pumps drops drastically with temperature. At about −10°C, most heat pumps need to be assisted by another heat source in order to keep up with the demand. In moderate climates where the temperature does not stay below −10°C for extended periods in winter, and where a central air conditioning system would be installed anyway, a heat pump can save both money and usable energy in the long run.

An interesting variation of the "air-to-air" heat pump we've been discussing is a water-to-air system that takes advantage of the large heat of fusion of water. In winter the heat pump removes heat from a large tank of water, turning it into a block of ice as the heat of fusion from the water is pumped into the house. In summer the process is reversed—heat is pumped from the house into the block of ice, turning it into water. Such an approach is obviously best suited for a climate where the amount of heat removed from a house in summer is about equal to the amount added in winter. This system has not yet been widely used.

· ·

Key Concepts

With no heat added or taken away: Letting a gas expand as it does work cools the gas; compressing the gas warms it.

The **first law of thermodynamics**:

$$\text{net heat added} = \text{increase in internal energy} + \text{work done}$$

For a **heat engine**,

$$\text{heat in} = \text{heat out} + \text{work done}$$

$$\text{thermal efficiency} = \frac{\text{work out}}{\text{heat in}}$$

The **second law of thermodynamics**: No actual or ideal heat engine that operates on a cycle can convert all the heat supplied to it into work.

$$\text{maximum thermal efficiency} = 1 - \frac{T_{\text{cold}}}{T_{\text{hot}}}$$

where the Ts are *absolute* temperatures.

For the **refrigeration cycle**:

heat in + work in = heat out

coefficient of performance (cooling) =

$$\frac{\text{heat removed}}{\text{work done}}$$

coefficient of performance (heating) =

$$\frac{\text{heat added}}{\text{work done}}$$

The energy efficiency ratio (EER) gives the same information as the coefficient of performance, but it is defined in terms of a mixed set of units. The coefficient of performance decreases as the temperature difference between colder and warmer regions increases.

Questions

Expansion and Compression of Gases

1. Why is it usually cold on top of a tall mountain?

2. True or false: Since a gas expands as its temperature rises, if you allow a gas to expand, its temperature will rise.

3. Why does a tire valve get cool as you let the air out of the tire?

4. When you pump up a basketball with a hand-operated pump, the pump gets very warm. Is there anything causing this heating other than friction? If so, what? Why does a diesel engine not need spark plugs?

Heat Engines

5. Why do most automobiles have a radiator? Why can some get by without one?

6. Why do lawn mower engines have fins along the sides near the cylinder?

7. How does thermal efficiency compare with the efficiency in general as defined in Chapter 4?

8. Why does an automobile engine have a flywheel attached to its crankshaft?

Second Law of Thermodynamics

9. Could you run a ship by extracting heat from the water and converting it to work for propelling the ship? Would it matter if the water were very hot?

10. Suppose a power plant has an overall efficiency of 37%. Does this mean that if the plant were run properly, it could produce 2.7 times as much electrical energy for the same amount of fuel burned?

11. Power plants heat their steam to as high a temperature as the construction materials can withstand and still confine the steam. Why not just use more steam at a lower temperature, and that way have a safer plant?

Pumping Heat

12. Does heat ever flow naturally from a particular region to one at a higher temperature?

13. In the refrigeration cycle, how is it possible for the refrigerant to be at the boiling point in both the condenser and the evaporator?

14. Could a refrigerator or air conditioner be operated with water as a refrigerant?

15. Why does heat enter the refrigerant in the evaporator of a refrigerator? Why does heat leave the refrigerant in the condenser?

16. An enterprising freshman proposes to make a refrigerator in the following way: Draw room

air into a compressor and compress it until it cools below the temperature inside the refrigerator. Pass this cold air through coils inside the food compartment to pick up heat inside. Then let it pass through a tiny opening and expand back to atmospheric pressure and be released into the air. What is wrong with this idea?

17. Can you cool your kitchen by leaving the refrigerator door open?

18. If you heat your house with electricity (resistive heating such as baseboard or ceiling heaters rather than a heat pump) does it cost you anything to operate your refrigerator or freezer in the winter?

19. A handy way to dry kids' tennis shoes after they've walked in water is to leave them on the floor just in front of the refrigerator. Would putting them in front of the TV work just as well?

20. Devices are available that save electrical energy or fuel by using the heat output of an air conditioner to preheat water before it goes to a conventional water heater. Why does this actually save fuel or electrical energy rather than just increase the energy use by the air conditioner?

21. What is wrong with calling the coefficient of performance an efficiency?

22. In a heat pump you get more heat energy out than you put in as electrical energy. Does this fact violate the principle of conservation of energy?

Problems

1. If 0.30 kcal of heat is added to the gas in the cylinder of Figure 11.1, how much does the internal energy of the gas change if the piston does not move? If the piston then moves 20 cm to the right as the gas exerts an *average* 14,000-N force on it, what is the net change in internal energy for the two processes?
   Answer: 1260-J increase; 1540-J decrease

2. A particular gasoline engine has a thermal efficiency of 20%. How much work is done by the engine for each liter (0.74 kg) of fuel used? How much heat is released to the air? (The heat of combustion of gasoline is 11,500 kcal/kg.)
   Answer: 2.0 kWh; 7.9 kWh

3. What is the thermal efficiency of an engine that takes in 400 kcal of heat and does 500,000 J of work?
   Answer: 30%

4. The steam that turns a particular power plant generator goes into the turbine at 500°C and is condensed at 100°C. How much would the ideal maximum thermal efficiency be increased if the steam entered at 600°C?
   Answer: 5.5%

5. Suppose someone came to you suggesting that you invest in a company that will make a heat engine that, in each cycle, takes in 3000 J of heat at 400°C, performs 2000 J of useful work, and exhausts the remaining 1000 J of energy as heat at 0°C. Do you think that this would be a good investment? Explain why or why not.
   Answer: No; thermal efficiency would exceed the theoretical limit of 59%.

6. A 14,000-Btu/h air conditioner has an EER of 5.1. How much electrical power does it take to operate it? What does it cost per hour of operation if electrical energy costs 9¢ per kWh? What would be the power and cost if the EER were 10.1?
   Answer: 2750 W; 25¢; 1390 W; 12¢

7. What is the coefficient of performance of an air conditioner with an EER of 8.4?
   Answer: 2.5

8. The *ton*, sometimes used in rating air conditioners, is a unit of power equivalent to the rate of heat transfer needed to freeze 1 ton of water in 24 h. What is the ton's equivalent in watts?
   Answer: 3.51 kW

9. A typical house in Ohio needs about 20,000 kWh (68 × 10⁶ Btu) of heat during a winter. If electrical baseboard heaters are used, how much electrical energy is needed to heat the house? If a heat pump with an average coefficient of performance of 2 is used, how much electrical energy is needed?

Answer: 20,000 kWh; 10,000 kWh

Home Experiments

1. Hold the ends of a rubber band in your hands, with the middle held lightly between your lips. Alternately stretch and then release the tension in the rubber band, keeping the ends in your hands and the middle between your lips. Notice the change in temperature that you feel in your lips. Which way, stretching or compressing, is the band a heat engine, and which way a heat pump? Interpret the results in terms of the first law of thermodynamics.

2. If you have access to an air conditioner, determine its EER from the information given on the nameplate or in the owner's manual.

3. Observe the heat given off at the bottom or back of a working refrigerator, the outside section of a window air conditioner, and the outside unit of a central air conditioning system.

References for Further Reading

Derwin, R., "Heat Pumps—Cheapest Cooling and Heating for Your Home?" *Popular Science*, Sept. 1976, p. 92. Shows how the same heat pump is used for cooling and heating, and evaluates the economics of its use in various parts of the United States.

Fowler, J. M., *Energy and the Environment*, (McGraw-Hill Book Co., New York, 1975). See especially Chapter 4. Discusses efficiency of various types of heat engines and their uses in energy conversion.

Leff, H. S., and W. D. Teeters, "EER, COP, and the Second Law Efficiency for Air Conditioners," *American Journal of Physics*, Jan. 1978, p. 19. Compares actual coefficients of performance (or EERs) of air conditioners with their maximum values allowed by the second law of thermodynamics.

Walker, G., "The Stirling Engine," *Scientific American*, Aug. 1973, p.80. A heat engine with low noise and low pollution.

PART THREE

WAVES AND SOUND

Just because one loves to fiddle,
It's still not right to try to diddle
About your skills to other folks
Or you'll become the butt of jokes.
Still, the dilettante has the right
To scratch and scrape all through the night;
But so his neighbors do not mutter
He must kindly close the shutter.

A. EINSTEIN

Albert Einstein was born in 1879 in the former German state of Würtemberg. Even though his prep school teachers considered him uneducable, at age 17 he entered the Swiss Federal Polytechnic School after considerable difficulty in being admitted because of his inadequacy in mathematics. After graduation Einstein worked for eight years in a patent office in Bern, Switzerland, where his duties left him ample time to develop ideas that have revolutionized our understanding of our universe. Even though he is best known for his theory of relativity, Einstein won the 1921 Nobel Prize for his explanation, given in 1905, of the quantum nature of light. After holding several European professorships, he accepted a position at the Princeton Institute for Advanced Study in 1933, which he held until his death in 1955. Music, especially playing the violin, provided his greatest pleasure in life.

W A V E S

What image does the word *wave* conjure up in your mind? Perhaps you think of a friendly wave of the hand in good-bye, the shape of a friend's hair, or the breakers you enjoy at your favorite beach. But none of these is the kind of wave you encounter most often. You see this page by means of light waves that travel to your eyes. You hear your favorite music by means of sound waves that come to you through the air. Most of the sound you hear originates as waves in a vibrating medium.

In this chapter we study the general properties that apply to all waves. In the next chapter we consider sound waves in particular. When we get to Part Five, we will look at the wave behavior of light.

About the only kinds of waves that we often see—in the sense that you *see* this page—are water waves. You probably have some feeling for *their* physical properties from dropping pebbles into a stream, watching people in a pool, or playing in ocean waves at the beach. This chapter should help to extend your intuition to other types of waves.

12.1 WAVES ON A STRING

Take a long rope, tie one end to a post or doorknob, and hold the other end so that there is a slight tension in the rope. Give your end an upward jerk and quickly return it to where it was. You will see a *pulse*—a lump—in the rope (as in Fig. 12.1). This pulse travels from your hand to the other end.

Now suppose that, instead of just one wiggle, you move the end of

Figure 12.1 A traveling pulse in a rope. Successively lower lines show the pulse position as time goes on.

Figure 12.2 A traveling wave in a rope. Each line shows the wave at a later point in time.

the rope quickly but smoothly up and down, continuously. You then get a **traveling wave** (Fig. 12.2) that moves along the rope. Each successive line of the figure shows the waveform of the rope at a later point in time. When the wave gets to the end of the rope, it reflects back in the opposite direction. The general shape of the rope then gets messed up and no longer has the smooth form of Figure 12.2. We'll consider reflected waves a little later.

When you pluck a guitar string, waves like these travel back and forth in the string. The effect is not exaggerated as in our diagrams, but the overall behavior is the same. In describing this kind of wave, we usually refer to it as a wave on a *string*, even though that "string" may be a rope, a spring, or some other carrier of the wave. Compare the shape of the string with the shape of the water surface when you jiggle your finger in an otherwise-quiet pool.

The Language of Waves

Discussing waves calls for a new language. Even though the vocabulary is fairly brief, we need to understand it before we go further. Figure 12.3 shows three of the important terms. The **wavelength**, usually identified by λ (lambda), is the distance between any two points *in phase* with each other—the distance between two crests, two troughs, or between two cross-over points where the string is moving down, for example. The **displacement** of a point on the string, shown as y for one point, is the distance of the point from the equilibrium (central) position for that string particle. It varies from point to point along the string, and varies with time at any one point. The **amplitude**, represented by A in the figure, is the maximum value of the displacement.

Another important characteristic of a wave is its **frequency**. The *frequency* of a *sound* wave mainly determines its *pitch*. The *frequency* of a *light* wave determines its *color*. We define frequency in two equivalent ways: as the number of wave pulses passing some point in the string per unit of time, or as the number of complete oscillation cycles per unit of time of a point on the string. The usual unit of frequency is the hertz (Hz), defined as one cycle/second and named after Heinrich Hertz who, in 1886, first demonstrated the transmission of radio-frequency electromagnetic waves:

1 hertz (Hz) = 1 cycle/s

Even though it gives the same information as the frequency, the **period** T sometimes fills the bill better. The period is the time for one complete cycle. Since frequency is the number of cycles per second,

$$\text{period} = \frac{1}{\text{frequency}}$$

12.1   Waves on a String   **229**

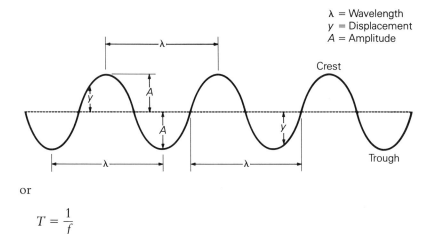

λ = Wavelength
y = Displacement
A = Amplitude

Crest

Trough

Figure 12.3 The wave-length, displacement, and amplitude of a wave. Wavelength and amplitude characterize the wave as a whole, but each particle has its own displacement, which varies with time between zero and the amplitude.

or

$$T = \frac{1}{f}$$

An analogy will show the relationship of these quantities to the speed with which the wave travels, the **wave velocity**. If you're waiting for a train to pass at a railroad crossing, you can find the train's speed by multiplying the number of train cars passing per minute by the length of one car (Fig. 12.4-a). For example, if 25 15-m-long cars pass per $\frac{1}{2}$ min—a frequency of 50 cars/min—the train speed is 50 cars/min × 15 m/car = 750 m/min or 45 km/h. Similarly, Figure 12.4(b) shows that the wave velocity v is the number of wave pulses passing a point per second f, multiplied by the length of each pulse λ. That is,

wave velocity = frequency × wavelength

$$v = f\lambda$$

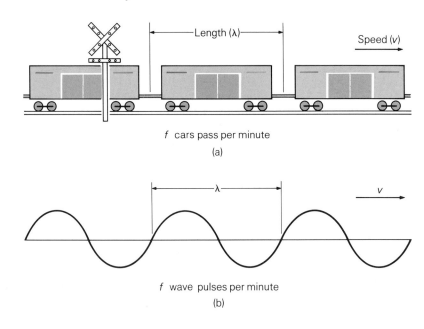

Length (λ)

Speed (v)

f cars pass per minute

(a)

λ

v

f wave pulses per minute

(b)

Figure 12.4 Whether you're talking about (a) a freight train or (b) a wave train, the speed *v* is the number of cars (pulses) passing in each unit of time multiplied by the length of each car (pulse).

EXAMPLE 12.1

The strings for middle C on a piano vibrate with a frequency of 262 Hz. If their wavelength is 2.00 m, what is the speed with which the waves move in the string? From the above equation,

$$\text{velocity} = \text{frequency} \times \text{wavelength}$$

$$= 262 \text{ Hz} \times 2.00 \text{ m} = 524 \text{ m/s} \ (1890 \text{ km/h})$$

• •

EXERCISE

CB radio waves are electromagnetic waves of frequency near 27 MHz (M = 10^6). If the speed of electromagnetic waves in air is 3.0×10^8 m/s, what is the CB wavelength?   *Answer*: 11 m

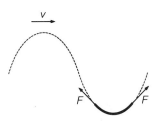

Figure 12.5 The pulse moves to the left as the string particles move up and down. The vertical arrows show the direction of motion of the string particles.

Figure 12.6 The tension provides force *F* acting on a segment of string at its lowest point. The more quickly the tension can move the segment back to equilibrium, the higher the wave velocity. This speed increases with increasing tension and decreases with increasing mass of the segment.

Which Way Does the String Move?

If you think again about the pulse you gave the string (Fig. 12.5), you realize that the string itself didn't move to the left; only the pulse did. As the pulse passed any particular little piece of the string, that piece merely moved up, and then back to its original position. For the continuous traveling wave of Figure 12.2, each piece of the string continually moves up and down with a periodic motion. The waveforms in Figures 12.2 and 12.3 describe the distance from equilibrium of any point along the string at some instant of time. But we could just as well think of them as graphs giving the distance of some point on the string from its equilibrium as time increases. Any point oscillates smoothly up and down with time, as shown by the graphs.

Wave Speed along a String

On a guitar, the bass note strings are more massive than those that produce a higher pitch. Tightening a particular string gives it a higher pitch. Both of these properties, mass and tension, influence the pitch by influencing the wave speed. As we've just seen, the frequency, which determines the pitch, is directly related to wave speed. Let's see how mass and tension determine the speed.

Figure 12.6 focuses on a short section of the string that, at the instant shown, is at the lowest point in its motion. This segment, shown as a solid line in the figure, needs to move upward as the wave moves to the right. The tension in the string—the force *F* stretching it—pulls on both ends of the segment. Since the tension slopes upward, it provides the string segment the force needed to accelerate it back toward equilibrium (recall Newton's second law, *F* = *ma*). The stronger the tension, the

12.1   Waves on a String   **231**

more quickly the segment gets moved back to equilibrium, and hence the faster the wave moves along the string. However, the more mass the segment has, the less its acceleration toward equilibrium and the slower the wave moves.

In summary, the stronger the tension, the higher the wave speed, and the greater the mass of a given length of string, the lower the wave speed. Does that fit with what we said about guitar strings?

Transverse Waves

The kind of waves we've been talking about are **transverse waves**: *Waves for which the particles of the medium vibrate in a direction perpendicular to the direction of wave propagation.* Electromagnetic waves such as light waves are also transverse. An important difference in their case is that they do *not* involve a transverse oscillation of a *medium*, but rather transverse variation of electric and magnetic fields.

12.2 LONGITUDINAL WAVES

A **longitudinal wave** is one is which *the particles of the medium oscillate in a direction parallel to the direction of wave propagation.* You may have played with a Slinky, a loosely wound springlike toy enjoyed by nine-year-olds and physics professors (Fig. 12.7). If you can get your professor to share one with you, have someone hold one end, and stretch the Slinky out across the floor (Fig. 12.8-a). Then give your end a quick push toward the other end, and then pull back (Fig. 12.8-b). A pulse made up of a compression followed by an expansion will travel down the Slinky as shown in Figure 12.8(c), (d), and (e).

If you then shake the end of the Slinky back and forth continuously, you will set up a traveling *compressional*, or *longitudinal*, wave (Fig. 12.8-f).

Even though the motion is parallel to the wave direction, we can still represent the situation by a wave diagram (Fig. 12.8-g). The crests then

Figure 12.7   A Slinky.

represent regions of compression, the troughs, regions of expansion. The wavelength λ is the distance between either. The other properties—frequency, period, velocity—have the same definition as for a transverse

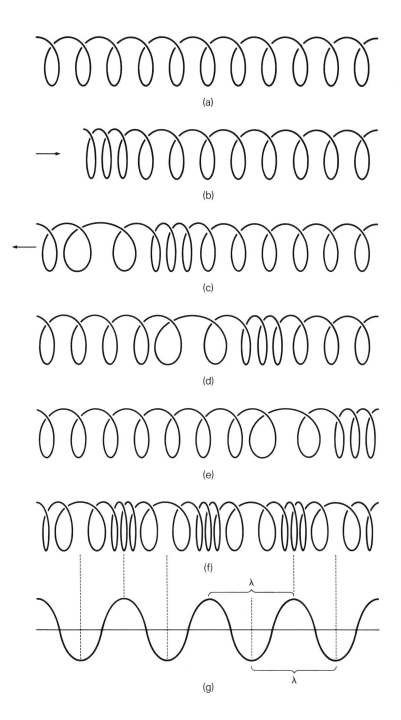

Figure 12.8 (a) The stretched Slinky. (b–e) By wiggling the left end one time you can set up a longitudinal pulse that travels from left to right, as in the successive lines. (f) A traveling longitudinal wave. (g) The waveform representation for this wave.

Figure 12.9 A longitudinal wave in a "line of molecules," and its waveform representation.

wave. A particle's displacement and amplitude are now parallel to the direction of wave motion.

Sound is a longitudinal wave. You can visualize how such a wave can exist in air by thinking of a long line of air molecules, as shown in Figure 12.9. As the wave passes along the line, there are regions of compression followed by regions of expansion. Any one molecule merely oscillates back and forth. In a real sound wave in the air, the effect is three dimensional, and the wave motion is superimposed on the random thermal motion of the molecules.

The Energy of a Wave

When you shake the end of a string to set up a wave, you do work on the particles in the end of the string. They in turn do work on the neighboring particles, which do work on those down the line. Thus, the work you do generates a form of energy that is transferred along the string.

We know that a wave carries energy because it can do work on—give energy to—an object with which it interacts. A water wave causes you to bob up and down as it passes you. A sound wave causes your eardrum to vibrate back and forth. Radiation from the sun heats bodies that it strikes.

We define the **intensity** of a wave as the energy transmitted by the wave per unit of time through a unit of area. Since power is the energy per unit of time, *intensity is the power transmitted through a unit of area.* You would probably guess that the intensity increases with increasing amplitude of the wave. It turns out that the intensity is directly proportional to the amplitude *squared*. Doubling the amplitude of a wave means it carries four times as much power.

No matter what the nature of the wave, it carries energy. Some of that energy is absorbed by whatever the wave interacts with.

12.3 INTERFERENCE

So far we have considered the situation in which only one wave existed in a given region. However, in most real situations two or more waves travel through the same region simultaneously. Let's look at the situation in which we have two *identical* waves in the same medium.

When the two waves are *in phase*—have their crests at the same point at any instant—the overall wave amplitude is the sum of the two individual amplitudes (Fig. 12.10–a). However, if two waves of the same amplitude are exactly out of phase, they cancel each other (Fig. 12.10–b). We call both of these effects **interference**; the first is *constructive* interference, the second, *destructive* interference. Intermediate phase differences give effects between these two extremes.

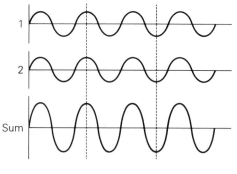

(a)   (b)

Figure 12.10 (a) Two waves undergo constructive interference if they are in phase. (b) They undergo destructive interference if they are exactly out of phase.

Interference is one of the most important effects in the behavior of waves. We'll discover its influence in a number of ways in both sound and light waves.

12.4 STANDING WAVES

In discussing Figure 12.2, we commented that when the wave gets to and reflects from the post, the shape of the string distorts from its regular wave form. However, if the frequency of the wave is just right, the reflected wave combines with the original wave to give a regular pattern of a different type. First we'll illustrate this effect, and then show how it comes about from the interference of two waves moving in opposite directions.

If your telephone has a stretchy coiled cord from the receiver to the body, hold the receiver with a slight tension in the cord and wiggle it back and forth. With a little experimenting, you can find the right frequency (probably around a cycle per second, depending on the nature of cord) that will make it vibrate between two extreme positions as in Figure 12.11(a). Notice that once you find the correct frequency, you need to move the end only slightly to have large vibrations, but you get only a jumble with a slightly different frequency or with an irregular frequency.

If you then wiggle it at about twice this frequency, you can get the pattern of Figure 12.11(b). At about three times the original frequency, you get pattern (c). If your cord is long and heavy enough, you might get 4, 5, and so on, segments for frequencies of 4, 5, and so on, times that needed for shape (a).

Notice that when one segment of the cord is up, the adjacent one is down. Figures 12.12(a) through (g) simulate "snapshots" of the way two neighboring segments would look at short separated time intervals.

A string vibrates in a regular pattern only at those frequencies that are some whole number, 1, 2, 3, 4, . . . , times the frequency that gives it a shape like that in Figure 12.11(a). We call the lowest vibrational

| (a) | (b) | (c) | (d) |

Figure 12.11 Photographs showing the vibration of a flexible "string" at its four lowest frequencies.

frequency the **fundamental** frequency. The whole series of frequencies, we call the **harmonics**—the fundamental being the *first harmonic*, the next highest frequency the *second harmonic*, and so on.

· ·

QUESTION   If the frequency of the second harmonic shown in Figure 12.11 is 4 Hz, what are the frequencies of the fundamental and the fourth harmonics?

At any instant the shape of our vibrating string looks suspiciously like its shape when a wave travels through it, and for good reason. The shape results from *two* waves in the string moving in opposite directions. There are certain points in the string, called *nodes*, that never move. Between these nodes the string moves up and down to form the pattern of Figure 12.11.

This pattern results from interference between the two waves moving in opposite directions. When the two waves are exactly in phase, as in Figure 12.10(a), the string takes on the shape of the solid line in Figure 12.13. After the waves have moved $\frac{1}{2}$ wavelength relative to each other, the two waves are exactly out of phase as in Figure 12.10(b)—resulting in a completely straight string. After another $\frac{1}{2}$ wavelength of relative

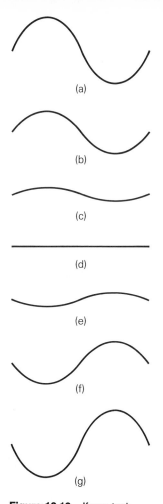

(a)

(b)

(c)

(d)

(e)

(f)

(g)

Figure 12.12 If you took snapshots of the cord of Fig. 12.11(b), successive shots would look as shown from (a) through (g).

motion, the waves are again in phase, but the crests and troughs have interchanged to give the dashed line of Figure 12.13.

To get the regular patterns of Figure 12.11, one more condition is necessary. The frequency and wave speed must be just right so that the positions of the maximum displacement always occur at the same points on the string. For a given wave speed, this happens for only certain discrete frequencies.

We then have positions in the string of no motion, called *nodes*, with points of maximum displacement, called *antinodes* between each pair of nodes. The overall pattern is called a standing wave—the terminology coming from the fact that the wave appears to stand still rather than move along the string. If you concentrate on either the solid or dashed line of Figure 12.13—the actual shape of the string at any instant—you see that the separation between nodes is $\frac{1}{2}$ wavelength.

We can now see why the harmonic frequencies are whole-number multiples of the fundamental. Figure 12.11 shows that the wavelength for the fundamental frequency is twice the length of the string—twice the distance between its nodes. The wavelength of the second harmonic is $\frac{1}{2}$ that of the fundamental, that of the third harmonic is $\frac{1}{3}$ that of the fundamental, and so on. From the relationship between wavelength and frequency,

$$\text{frequency} = \frac{\text{wave velocity}}{\text{wavelength}}$$

we see that reducing the wavelength to $\frac{1}{3}$ its fundamental value triples the frequency. Thus, the frequency of the third harmonic is three times the fundamental. A similar relationship holds for other harmonics.

When we study sound sources in the next chapter, we'll see that standing waves play a major role in the production of a broad range of types of sound.

12.5 RESONANCE

When you push somebody in a swing, you don't just push at random times or at some arbitrary frequency, you push at the natural frequency

Figure 12.13 A standing wave in a string fixed at one end and wiggled at the other. The string oscillates back and forth between the two "maximum" positions shown by the solid and dashed lines.

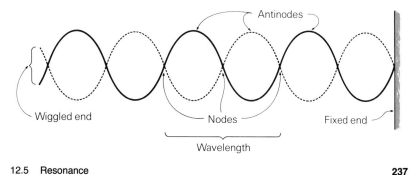

Antinodes

Wiggled end

Nodes

Fixed end

Wavelength

(a)

(b)

Figure 12.14 (a) Only four months after it opened in 1940, a mild gale set up resonant vibrations in the ½-mile-long center span of the Tacoma Narrows Bridge. (b) Because of the resonant vibrations, this bridge collapsed within a few hours.

of the swing. A small periodic push at that frequency makes the swing "go." To force it to swing at another frequency would take an extreme effort. Similarly, the pendulum of a grandfather clock and the balance wheel of a mechanical watch oscillate continually under small impulses from the spring mechanism at rates near their natural frequencies. We depend on this constant natural frequency to keep the watch or clock running at the right speed. When you jump up and down on a diving board, there is only one frequency at which you can jump and build up an appreciable height. If the tires of your car are out of balance, the vibration is strongest at certain distinct speeds. Even a graduate student on a pogo stick has no control over the frequency of his jumps; it depends entirely on his weight and the spring in the stick.

These examples illustrate that it is easy to excite large vibrations or oscillations in a body if the driving force has a frequency near a natural frequency of vibration or oscillation of the body. We call this phenomenon **resonance**, and the frequency or frequencies at which the vibration of the body is easily excited, we call **resonant frequencies**. When the body and driving force are in resonance, we can maintain a constant amplitude of vibration if the amount of work done each time we apply the driving force exactly equals the energy lost to friction in each cycle. Even the winds can excite resonant vibrations in large structures, as Figure 12.14 illustrates.

• •

QUESTION   A tall swing has a lower resonant frequency than a short one. By analogy, would you expect long-legged or short-legged people to take more steps per minute when they walk? Does your answer agree with your observations?

The standing waves discussed in the previous section are an important example of resonance. In that situation the strings have several resonant frequencies. Standing waves can be set up in a string *only* at one of the natural resonant frequencies of the string. Those frequencies depend on the wave velocity, which in turn, depends on the mass of and tension in the string, as described in Section 12.1. The resonant vibrations that wrecked the Tacoma Narrows Bridge (Fig. 12.14) were actually standing waves in the bridge structure.

One important application of resonance helps you to be sure you are never late for physics class. Wind-up watches keep time based on the oscillations of a mechanical balance wheel. Each time the balance wheel oscillates the watch ticks as the internal gears advance the hands a small amount. The accuracy of the watch depends on the stability of the resonant frequency of the wheel. This frequency is near 1 Hz, as you can tell by listening to the ticks. However, a "quartz" watch depends on the 33-kHz resonant vibrations of a quartz crystal. Shaking either watch at a frequency near its resonant value can easily disturb the vibration frequency and the corresponding accuracy of the watch. Which of these types of watch would you expect to be more sensitive to outside disturbances from ordinary movement of the watch? (Hint: Have you tried shaking your arm at 33-kHz lately?) Your answer provides the main explanation for the superior accuracy of "quartz" watches.

12.6 WAVE BEHAVIOR: REFLECTION, REFRACTION, DIFFRACTION

There are certain things that only waves can do. The names of some of these phenomena are *reflection, refraction,* and *diffraction.* We introduce these terms here because they apply to all waves. Their significance will be more obvious after we see in later chapters how they apply to sound and light waves.

Reflection

You might argue with the statement that only *waves* can reflect; and I wouldn't protest too strongly. Objects do bounce off others: basketballs off backboards, pool balls off the cushion. But we usually reserve the term *reflect* for *waves,* because they rebound in a particular way. Depending on how your opponent hits it, a racquetball may come off the wall at a weird angle. But waves from a smooth surface always reflect so that *the angle the reflected wave makes with the surface exactly equals the angle the incident wave makes with it.* This **law of reflection** applies to any kind of wave.

Figure 12.15 Waves incident from the upper left are reflected toward the lower left so that the angle of reflection equals the angle of incidence.

The waves we've discussed so far travel through media extended only in one dimension: transverse waves on a string and longitudinal waves in a Slinky, for example. Waves reflected from the end of such a medium have no choice but to go back in the opposite direction they came from. But water waves on a surface spread out in two dimensions. Sound and light waves spread in three dimensions and, in general, can travel in any direction in space.

When such waves strike a surface and reflect, they always obey the law of reflection. Figure 12.15 illustrates this behavior.

We're all familiar with the reflection of light from mirrors, from water surfaces, and from various other objects. The echo you hear when sound reflects from a distant hill is another example of this behavior. We'll investigate these effects in more detail later.

Refraction

Refraction is the bending of a wave's direction because it crosses from one medium, in which it has one speed, to another medium, in which it has a different speed, or because the wave speed is different in different parts of the same medium. The first condition is illustrated by the seeming bending of a straw in a glass of water, and by the ability of the lens of your eye to focus an image on its retina. The second condition would apply to sound if, for example, the temperature were different in different parts of the medium. It would also apply if parts of the medium itself were moving faster than others, since the overall wave speed is the speed of the wave through the medium plus (or minus) the speed of the medium.

To see why a speed difference causes a change in direction, imagine that you and a good friend are riding on your bikes side-by-side and holding hands. If your friend is on your right and moving just a little faster, the only way you can continue to hold hands indefinitely is for you both to curve gradually to the left. A similar effect occurs for the bending of a wave's direction across regions of differing wave speed.

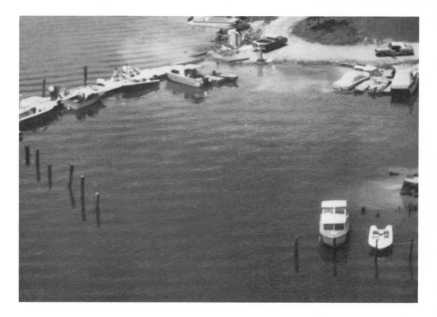

Figure 12.16 The posts cast sharp shadows in light waves but have little effect on the water waves, which have a much longer wavelength.

Diffraction

A tree or post casts a sharp shadow in sunlight. This is another way of saying that light waves do not bend around the object. But that same post in the path of water waves has almost no effect: The waves simply bend around the post and go on their way almost undisturbed (Fig. 12.16). The difference in behavior is caused *not* by the difference in the nature of light and water waves but by the *difference in wavelength relative to the size of the post.* When an object in the path of waves is much larger than the wavelength—as for the post in light—the waves do not bend around it but cast a sharp shadow. When the object is comparable in size or smaller than the wavelength—as for the post in water waves—the waves merely bend around it. This bending effect has the name **diffraction**.

The wavelength of most audible sound is between a few centimeters and a few meters. Sound, therefore, can diffract around ordinary-sized objects—people, trees, walls, and the like—and through ordinary-sized openings—doors, windows, and people's mouths.

· ·

Key Concepts

A **traveling wave** is a disturbance that propagates from one region to another in a medium. The distance of any point in the medium from its equilibrium position is its **displacement**. Wave properties include: (1) their **wavelength**, the distance between two points *in phase* with each other; (2) their **amplitude**, the maximum displacement; (3) their **frequency**, the number of wavelengths passing some point per unit of time; (4) their **period**, the time for one wavelength to pass; and (5) their **wave**

velocity, the speed with which the wave disturbance travels. In **transverse** waves the displacement is perpendicular to the direction of wave propagation; in **longitudinal** waves it is parallel to the propagation direction. For any type wave,

wave velocity = frequency × wavelength

The **intensity** of a wave is the power transmitted by the wave through a unit of area, and it is proportional to the square of the amplitude.

Interference between two waves is *constructive* if the waves are in phase and *destructive* if the waves are out of phase.

Standing waves occur when two identical waves move in opposite directions in a medium and have speed and frequency such that crests always overlap at the same point. The various *harmonic* frequencies at which this happens are all whole multiples of the lowest, or *fundamental*, frequency.

Resonance occurs when the periodic impulses that excite vibration or standing waves in a body are at a frequency near one of the natural, or **resonant**, frequencies of the body.

Some special areas of wave behavior include *reflection*, *refraction*, and *diffraction*.

Questions

Wave Properties

1. How does the wavelength of waves in the ocean compare with that of the waves you cause when you drop a pebble into a pond?

2. The low-pitch strings on some instruments consist of metal wire wrapped around catgut strings. What is the purpose of the wrapping? Hint: the higher the wave speed, the lower the frequency for waves on a string of fixed length.

3. Tightening the strings on an instrument makes them vibrate faster and give a higher pitch. If you hold the ends of a rubber band in each hand and stretch it while someone plucks it, there is almost no change in pitch as you increase the tension by stretching the band longer. Why?

4. How does a bottle bobbing up and down on a water wave demonstrate that the wave carries energy? How do you know electromagnetic waves from the sun carry energy?

5. If you set up a traveling wave in a rope by shaking its end, how does the frequency of the wave in the rope compare with the frequency of shaking?

6. In what ways can you discount the following statement: Since a longitudinal wave cannot be represented by Figure 12.3, there is no meaning to wavelengths for such waves, and the equation $v = f\lambda$ does not apply.

Interference

7. If you had two identical sound waves that were exactly out of phase as in Figure 12.10(b), what would you expect to hear if these waves arrive at your ear simultaneously?

8. If you moved one of the sources of sound described in the previous question $\frac{1}{2}$ wavelength farther from your ear, what effect would that have on the sound you hear?

Standing Waves

9. Suppose the person in Figure 12.11(c) is wiggling the end of the string at 3 cycles/s. At what frequency would he wiggle to get the other three waveforms shown?

10. The standing waves discussed in this chapter were transverse waves on a string. Describe, with the aid of a sketch, what would be happening to the various parts of a slinky in which there was a standing longitudinal wave.

11. If the lowest frequency that would give a standing wave in a particular string were 100 Hz, what would be the frequency of the fifth harmonic on this string? How many nodes would the string have in its fifth harmonic?

12. Looking at Figure 12.11, you might get the idea that the string "balloons" out at the antinodes

and compresses at the nodes, or at least splits into two strings at the antinodes. If that is not what is happening, what *is* happening in the picture and why does it look otherwise? If you look at a vibrating guitar string in a bright light, it might look similar. Why are looks deceiving there also?

Resonance

13. Why do you have to know about resonance to ride a pogo stick, even though you may not call it by that name?

14. What does resonance have to do with the following things you can observe riding in an automobile? (a) When your tires are out of balance, the vibration is much worse at certain speeds. (b) A bump in the road may cause the car to bounce wildly at one speed, but hardly be felt at a *higher* speed. (c) If your car is stuck in mud or snow, you can sometimes get it out by alternating between forward and reverse. The chance of success is better if you alternate quickly and rhythmically at a particular frequency. (d) Young drivers sometimes find it fun to rock the car when stopping, by applying and releasing the brakes at regular intervals. However, the frequency of brake application must be just right to get good bounce.

15. If you are trying to carry a pan of water without spilling it, why do some frequencies of shaking make the water spill so much more readily than others?

16. In some cars it's easy to drive with a cup of coffee on the dashboard and not spill it. In others it's impossible. This effect is not necessarily related to how smoothly the car rides. Explain.

Figure 12.17 An amplified voice can shatter a glass.

17. Certain singers have reportedly shattered crystal glasses by sounding a particular note, although this is normally possible only if the voice is amplified (Fig. 12.17). Is it only the intensity of the sound that counts?

Problems

1. Waves in a particular banjo string travel at a speed of 600 m/s. If the vibration frequency is 400 Hz, what is the wavelength of the wave in the string? What is the wavelength of the sound in air produced by this string? (The speed of sound in room temperature air is 344 m/s.)
   Answer: 1.5 m; 0.86 m

2. If you hear a clap of thunder 0.50 s after you see the lightning flash, how far away did the lightning strike? (Speed of light = 3.0×10^8 m/s, speed of sound = 340 m/s.)
   Answer: 172 m

3. A car with snow tires is traveling at a speed of 50 km/h. If the tread pattern on the tires repeats every 5.5 cm, what is the frequency of the sound wave that we interpret as hum from the tires?

Answer: 250 Hz

4. Longitudinal waves on a slinky have a frequency of 5 Hz. What is the period of these waves?

Answer: 0.2 s

5. The average intensity of the sound waves from a student's tape player is 0.05 W/m^2 near an open window. How much energy is transmitted by the sound through the 1-m-by-1.5-m window in 2 hours of operation?

Answer: 540 J

6. A guitar string carries a standing wave with a frequency of 500 Hz. If there are 3 nodes, including the ends, what is the frequency of the same string when there are 5 nodes?

Answer: 1000 Hz

7. The frequency of the third harmonic on a violin string is 660 Hz. What is the frequency of its seventh harmonic?

Answer: 1540 Hz

8. A 1-m-long string of a piano vibrates with a frequency of 756 Hz in its third harmonic. What is the velocity of the traveling waves that combine to form this standing wave? Recall that there is $\frac{1}{2}$ wavelength between nodes.

Answer: 504 m/s

Home Experiments

1. The next time you're bored with a telephone conversation, try some wave experiments. If the receiver is connected to the telephone with a coiled wire, notice that by varying the frequency with which you shake the receiver, you can set up various harmonics of standing waves in the cord. Estimate the frequency change for each different harmonic.

2. Try jumping on a diving board, pushing someone in a swing, rocking a boat, or riding a pogo stick at some frequency other than the resonant frequency.

3. Carefully observe a creek or stream where the water flows steadily over the rocks. Look for evidence of interference and standing waves—that is, look for points between ripples where there seems to be no wave motion.

References for Further Reading

Kock, W. E., *Radar, Sonar, and Holography, an Introduction* (Academic Press, New York, 1973). In addition to the topics in the title, the author discusses wave properties in general, including some interesting photographs of diffraction and other effects in sound.

Project Physics (Hold, Rinehart, & Winston, New York, 1975). Chapter 12 discusses and illustrates a broad range of wave behavior at a level consistent with this text.

S O U N D

When you think of sound, what comes to your mind? Is it the chirping of birds in spring? or maybe the rumble of city traffic? the band that's playing on campus next week, or an album of Beethoven's Fifth Symphony? the drone of your physics teacher's voice, or the bell that will end his or her monologue for another day? or maybe that one voice more special than all the others in the world?

Sound exists as a wave, and to understand the physical properties determining its behavior, we'll make use of our knowledge about waves from the preceding chapter. First we'll discuss the nature of sound waves and how they are produced. After briefly describing how the ear functions, we'll contrast the properties of sound that exist as measurable physical quantities with those that a listener perceives. We'll see how these ideas apply to sound that most of us like—the production of music—and then to sound that most of us do not like—noise. Finally, we'll consider some special effects that result from the wave nature of sound.

13.1 WHAT IS SOUND?

Sound is a longitudinal wave that travels through a material medium. We usually limit the word *sound* to mean those waves that have the right frequency to produce the sensation of hearing in the human ear. This **audible region** of sound includes frequencies between about 20 and 20,000 Hz, although it takes an extremely good ear to hear the lower and higher

Figure 13.1 The effect of a sound wave on air molecules. We have greatly exaggerated the difference in density.

frequency sounds within this range. Sounds with frequency below the audible region are called *infrasonic* and those above it, *ultrasonic*.

Figure 13.1 illustrates a sound wave passing through air molecules. The waveform at the bottom shows the pressure variation in the air. The peak-to-valley pressure difference in the faintest audible sound is only about 2×10^{-5} N/m² (3×10^{-9} lb/in²). For such a sound, a given molecule oscillates over a distance of only about 10^{-11} m: about 1/10 the diameter of a typical atom. These exceedingly small numbers show the incredible sensitivity of the ear!

Sound can travel through any material medium, but it usually gets to our ears by traveling through air. The speed of a sound wave depends on the density of the medium and on how easily the medium can be compressed. Its speed in air is 332 m/s at 0°C, and increases by about 0.6 m/s for each Celsius degree increase in air temperature. Thus, its speed in 20°C air is 344 m/s.

We can calculate the wavelength range *in air* of the audible frequencies from

$$\text{velocity} = \text{frequency} \times \text{wavelength}$$

At 20 Hz,

$$\text{wavelength} = \frac{\text{velocity}}{\text{frequency}} = \frac{344 \text{ m/s}}{20 \text{ Hz}} = 17 \text{ m}$$

At 20,000 Hz, the wavelength is 0.017 m, or 1.7 cm.

EXERCISE The speed of sound in water is about 1500 m/s. What is the range of audible wavelengths in water? *Answer*: 0.075 to 75 m

As sound travels from one medium to another, its frequency cannot change. (Reason out why that wouldn't be possible.) Since its speed does change, the wavelength of sound of a particular frequency depends on the medium.

13.2 SOURCES OF SOUND

Anything that vibrates with a frequency in the audible range can be a source of sound. That includes not only objects intended as sound sources, but things like electric shavers, automobile engines, boiling water, and wind "whistling" through the trees.

(a)

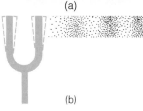

(b)

Figure 13.2 (a) A tuning fork. (b) The vibrating tuning fork is a source of single-frequency sound established by the vibrating arms.

Figure 13.3 In a loudspeaker, the vibrating cone-shaped diaphragm produces the sound. We show the spreading of sound from the source by wavefronts drawn one wavelength apart.

A tuning fork (Fig. 13.2-a) is a handy sound source for classroom demonstration because it vibrates with a single fixed audible frequency. As the sides vibrate, they set the surrounding air molecules in vibration at the same frequency (Fig. 13.2-b).

In a loudspeaker, sound is produced by a cone-shaped diaphragm that is set in vibration electrically (Fig. 13.3). We often represent sound, as well as other kinds of waves, emanating from a source by **wavefronts** as shown in the figure. We usually draw them one wavelength apart so that all points on all the wavefronts are in phase with each other: We can think of them as representing wave crests.

In most musical instruments, the sources of sound are either vibrating strings (in guitars, pianos, violins) or vibrating air columns (in pipe organs, trumpets, trombones, clarinets). In either strings or air columns, the vibration occurs at the resonant frequency that is characteristic of whatever is vibrating. We'll now look at the way strings and air columns vibrate.

Vibrating Strings

When you pluck a guitar string, for example, its vibrations set up sound waves that travel through the air to your ears. *The frequency of the sound is the same as the frequency of the vibrating string.* The vibrations in the guitar string occur at the resonant frequencies for the string; they are actually standing waves similar to those illustrated in Figure 12.11. As we discussed in Section 12.4, standing waves exist in a string when two identical waves travel in opposite directions in the string, and have the right wavelength so that the crests of the two waves always overlap at the same point in the string. In each section of Figure 12.11, the person is shaking the end of the string at just the resonant frequency to cause that particular harmonic in the string. (Remember, $\lambda = v/f$ tells you that choosing the frequency also determines the wavelength, since the velocity is fixed for a particular string under a given tension.)

For the guitar string, you don't shake the end; you merely pluck it at some point. That plucking is a disturbance that sends wave pulses of a broad range of frequencies (wavelengths) down the string. Only the resonant frequencies that meet the conditions for standing waves persist in the string. These standing waves gradually die out over a period of several seconds as their energy is used up sending out sound waves to the surroundings.

You may protest that you never saw anything like the waves of Figure 12.11 on a guitar string. They are the kind you get from shaking your telephone cord. However, in a guitar string, the tension and string mass are such that the waves typically travel at speeds of hundreds of meters per second—much too fast to be seen with the unaided eye. Furthermore, the amplitudes are only a few millimeters, at most.

Figure 13.4 The combination dash-dot lines show the extreme positions for the second harmonic on a string. The solid and dashed lines show the extremes for the fourth harmonic superimposed on the second. At one instant the string might appear as the solid line if these two harmonics were present simultaneously.

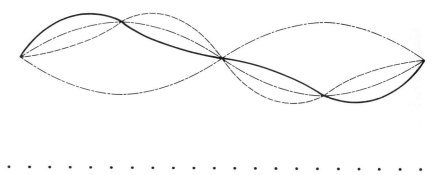

EXERCISE You can get a hint at what's happening on the guitar string by viewing the string in the light of an operating TV. The TV becomes a strobe light that flashes 60 times per second as the screen is scanned by the electron beam. (We'll discuss this process in Chapter 18.) If you tune the string to near a multiple of 60 Hz, you can see the distorted string appear to stand still or move slowly between distorted shapes.

If you are successful in this exercise, you will probably find that the string does not take on one of the smooth regular shapes of Figure 12.11. Instead, it may look more like the solid line of Figure 13.4. That's because several harmonics usually exist on the string at one time. Figure 13.4 shows only two harmonics superimposed on the string—the second and the fourth. The dot–dash lines show the extreme positions for the second harmonic alone. The dashed and solid lines show the extremes for the fourth harmonic superimposed on one of the extremes for the second. With several harmonics present, you can imagine that the string would take on a rather deformed shape.

On the string of an instrument many such harmonics usually exist simultaneously. The sound produced by such a string has the same frequencies as the waves present on the string do. In other words, the sound will have a frequency component corresponding to the fundamental frequency of the string, as well as components corresponding to whatever higher harmonics are present on the string. For example, if a guitar string vibrates with a 260-Hz fundamental frequency plus the third, sixth, and seventh harmonics, the sound will have frequency components of 260, 780, 1560, and 1820 Hz.

EXERCISE On a particular grand piano, the note A above middle C vibrates with a fundamental frequency of 440 Hz, and the second, third, fourth, fifth, and seventh harmonics are also present. What frequencies exist in the sound heard from the note?

Answer: 440; 880; 1320; 1760; 2200; 3080 Hz

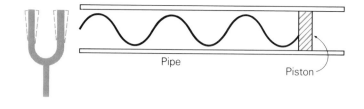

Figure 13.5 Sound waves travel through the air in the pipe and reflect from the moveable piston.

Pipe

Piston

There is another term applied to the higher harmonics, which is a carryover from its use in music. Harmonic frequencies above the fundamental are called **overtones**, the second harmonic being the first overtone, the third harmonic the second overtone, and so on.

Vibrating Air Columns

Suppose you put a vibrating tuning fork in front of a pipe that is closed at one end by a movable piston as in Figure 13.5. A sound wave will travel through the air in the pipe and be reflected from the closed end. Even though this wave is a longitudinal one, if the crests (high-pressure regions) and troughs (low-pressure regions) of the reflected wave always overlap the crests and troughs of the original wave at the same point along the air column, you get a standing wave in the air. Therefore, if you start with the piston near the tuning fork and move it to the right in the pipe, there will be certain lengths for the column of air at which it will be in resonance with the tuning fork. For these lengths, the vibrating tuning fork sets up standing waves in the air column that greatly increase the intensity of the sound emitted.

QUESTION   When the air column is the right length for standing waves, much more sound is given off by the system consisting of tuning fork and air column than at other lengths. From the energy conservation principle, how would you expect the length of time the tuning fork vibrates to be influenced by whether or not the column length is right for resonance?

If our air column were a fixed length and we could vary the frequency of the sound source, we would find increased sound intensity from standing waves only at certain distinct frequencies. The longer the pipe, the longer the wavelength for the lowest, or fundamental, resonant frequency. For a given pipe length, all other possible frequencies are harmonics of the fundamental: They are whole numbers multiplied by the fundamental frequency.

Vibrating air columns are the sound sources in pipe organs and in all other wind instruments.

EXERCISE The sound from a particular 25-cm-long organ pipe has frequencies of 344 Hz, 1032 Hz, 1720 Hz, and 2408 Hz. What harmonics are present in the standing waves in the pipe, assuming the lowest frequency given to be the fundamental. *Answer*: 1, 3, 5, and 7

EXERCISE Blow into an empty soft drink bottle so that you get a reasonably loud sound, indicating resonant vibrations (Fig. 13.6). Then add varying amounts of water and repeat each time, noticing how the pitch at resonance changes with height of *air* in the bottle. (We'll see later that higher pitch means higher frequency.)

Figure 13.6 With a little experimentation you can set up resonant vibrations in the air in the bottle by extending your upper lip over the bottle to direct air into it, as Doug is showing here.

The Human Voice

Figure 13.7 is an outline of the vocal tract. The vocal apparatus is essentially a wind instrument in which standing waves are set up in the air

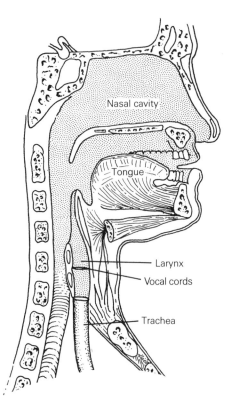

Figure 13.7 An outline of the human vocal tract.

by the *larynx* and various cavities of the mouth and head. Air forced up the trachea from the lungs has to pass through the opening between the *vocal cords*—two band-shaped membranes stretched across the larynx.

As air passes between the vocal cords, they vibrate back and forth, opening and closing the gap between them. We control the frequency of this vibration by controlling the tension in the cords. The frequencies in the sound emitted depend not only on the frequencies of the vibrating vocal cords but also on which of these frequencies resonate in the air within the throat, mouth, and nose. As we talk or sing, we simultaneously control the vocal cord tension and the shapes of the various air cavities, in order to produce the resonant frequencies needed for the sound we want at any instant.

13.3 THE HUMAN EAR

Now that we've discussed sources of sound, let's look at the other end of the picture—the ultimate *receiver* of sound, the human ear.

Figure 13.8 is a diagram of the human ear. Sound waves entering the auditory canal set up vibrations in the eardrum that are transmitted to, and through, the bones of the middle ear. A membrane called the oval window separates the middle and inner ear regions and is in contact with the fluid in the cochlea. Within the cochlea are hairlike cells that sense the vibrations and convert mechanical vibrations to electrical signals for transmission to the brain. The mechanism of this mechanical-to-electrical

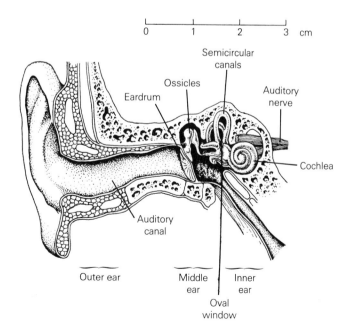

Figure 13.8 A diagram of the human ear.

energy conversion process is not well understood, but some of the theories concerning the process are discussed in the references at the end of this chapter.

13.4 THE OBJECTIVE AND SUBJECTIVE PROPERTIES OF SOUND

You might have heard someone claim: "If a tree falls in a forest and there is no person or animal nearby to hear it, there is no sound." The heated arguments this issue has been known to arouse center around the two different ways of defining sound: (1) the wave disturbance set up in a sound-transmitting medium; or (2) the sensation interpreted as sound by the brain. Whichever way you go, the question focuses on the two complementary aspects of sound that we'd like to compare and contrast in this section. One side of sound deals with its **objective** properties—those properties such as frequency and intensity, which are definite physical characteristics of the wave itself that can be measured with instruments and agreed on by all observers. The other aspect of sound deals with its **subjective** properties that are partly psychological and *depend on the responses of the particular person who hears the sound*. Properties in the second category include pitch, loudness, and quality.

For each *subjective* property, there is one *objective* property that mainly determines it, although the other objective properties have an influence. The situation is complicated because the relationship between subjective and objective properties is different for each listener. In this section we will look at each of the three subjective properties—loudness, pitch, and quality—and see what objective properties influence them.

Loudness and Intensity Level

You would probably guess that the **loudness** of sound would increase with an increase in the energy carried by the wave. That guess is correct, but loudness also depends on the sound's frequency. To see how this all fits together, we need to define a new term, **intensity level**, a term related to but distinct from *intensity*.

The *intensity* of a sound, like that of any traveling wave, is defined as the power transmitted by the sound through a unit of area. For example, if a rock band is playing in a room, and 0.30 J of sound energy emerge per second from an open 2.0-m-by-1.0-m door, the sound intensity is:

$$\text{intensity} = \frac{\text{power}}{\text{area}} = \frac{0.30 \text{ W}}{2.0 \text{ m}^2} = 0.15 \text{ W/m}^2$$

At a frequency of 1000 Hz, the faintest sound a normal ear can hear has an intensity of about 10^{-12} W/m². The loudest sound most of us can

Table 13.1 Intensity levels of certain sounds

Sound Source	Approximate Intensity Level (dB)
barely audible sounds	0
rustling leaves	10
whisper at about 1 m	20
city noise at night	40
ordinary conversation	60
noisy factory	85
Niagara Falls	90
riveter about 10 m away	100
rock band	110
sound becomes painful	120
jet plane take off, 30 m away	140
mechanical damage to ear	160
large rocket (nearby)	180

tolerate without pain has an intensity of about 1 W/m². In other words, our ears, if normal, are sensitive to sounds that vary in intensity by a factor of 1,000,000,000,000!

To avoid dealing with this broad range of numbers for intensity, we use the term *intensity level*, which is expressed in the unit *decibel* (1/10 bel), abbreviated dB. On this scale, we assign the intensity of 10^{-12} W/m² an *intensity level* of 0 dB. Then, each time the intensity is *multiplied* by a factor of 10, we *add* 10 dB to the intensity level. If the sound intensity is 10^{-11} W/m², that's 10 times that for 0 dB, or the intensity level is 10 dB. If the intensity is 10^{-9} W/m², that's 10×10, or 100, times that for 10 dB. The intensity level is then 10 dB + 20 dB, or 30 dB. If the intensity of a sound goes down from 0.37 W/m² to 0.037 W/m², the intensity level drops by 10 dB. Sounds with intensity level 120 dB have an intensity 10^{12} times those of intensity level 0 dB. The smallest *change* in intensity level that a person with normal hearing can detect is about 1 dB. Table 13.1 gives the average intensity level of a few well-known sounds.

EXERCISE Because of complaints from the old folks over 30 in the audience, a rock band turns the amplifiers down so that the sound energy output rate is only 1% of what it was. If the original intensity level were 110 dB, what is the new intensity level? *Answer:* 90 dB

For a given frequency, the loudness is roughly proportional to the intensity level. A sound of a particular intensity level is loudest at frequencies around 3000 to 4000 Hz. If you decrease or increase the frequency away from this region, the intensity level must increase somewhat to maintain the same loudness.

Pitch and Frequency

The **pitch** of a sound is its position on a musical scale, *as judged by a listener*. For example, a bass voice has a low pitch, a soprano voice, a high pitch.

Pitch depends *mostly*, but not entirely, on the frequency of the sound. A low-frequency sound has a low pitch, a high-frequency sound, a high pitch. Frequency is an objective, measurable property of the sound; it is the number of actual vibrations in the sound wave per second. But pitch is a subjective property that depends on the listener's response.

Pitch is influenced also by the loudness of the sound, particularly for low frequencies. In the frequency range of 200 to 300 Hz, an increase in intensity level of 40 dB can drop the pitch of a single-frequency sound by 10% for a sensitive listener.

Since most sounds are not *pure tones* of a single frequency, but are made up of a fundamental frequency and various overtones, you might wonder which frequency or frequencies determine the pitch. Experiments have shown that, even if one of the overtones is more intense than the fundamental, *the pitch depends mainly on the fundamental frequency*.

Our ears play a trick on us that sometimes turns out to be very handy. Suppose we hear a sound that has frequency components 400, 600, 800, 1000, and so on, Hz. These frequencies are the overtones for a sound of frequency 200 Hz. For this kind of situation, our ears often add the missing fundamental frequency, and we hear a pitch that corresponds to 200 Hz. (The term for this effect is *fundamental tracking*.)

One result of the production of these *subjective tones* by the ear is that the ear can partly compensate when we listen to inexpensive audio equipment. Small loudspeakers cannot produce low-frequency (below about middle C) sound. Yet we hear some bass from even the cheapest radios. Our ears use the high-frequency harmonics from the bass instrument to produce the sensation of bass even though it's not physically present in the sound hitting our ears.

Quality and Harmonic Content

If your mother, your rich aunt, and your closest female friend all hummed a note at exactly the pitch of middle C on the piano, you would have no trouble distinguishing one voice from the other, even though they sang with the same pitch and loudness. Also, you would have no trouble picking out the sounds of a piano, a trombone, and a factory whistle,

Figure 13.9 The waveforms of the sound of (a) a male voice singing the vowel ō, (b) a flute, and (c) a saxophone (from Hall, *Musical Acoustics*).

all sounding the same pitch at the same loudness. The main property that distinguishes two sounds of the same loudness and pitch, we call **quality**, or sometimes, **timbre**.

We can get a hint at the cause for differences in quality from Figure 13.9—the waveforms of the sound from three different sources. Each of these sounds has the same fundamental frequency, but the relative amounts of each of the higher harmonics are different. In general, *the relative amounts of the different harmonics present in the sound mainly determine its quality.* The difference in waveforms for different sound sources, as in Figure 13.9, arises from the combined effect of the harmonics.

Figure 13.10 shows bar graphs of the harmonic content of the sound from the two instruments of Figure 13.9. Considering the differences in harmonics present, it isn't surprising that the instruments have a different sound.

In addition to quality, another factor is important in distinguishing between instruments. It is the "attack" and "decay" characteristics of the sound—the way in which the sound intensity builds up and decreases regularly. For example, the intensity increases sharply as each note is struck on the piano. If that effect were removed, it would be much more difficult to distinguish between instruments.

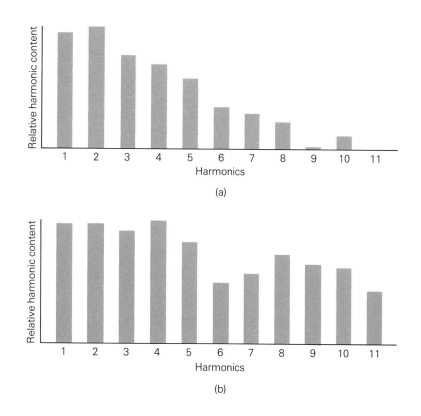

Figure 13.10 The relative harmonic content of sound from (a) a flute and (b) a saxophone as they produce the sound waves of Fig. 13.9 (based on Hall, *Musical Acoustics*).

APPLICATION

THE SOUND OF MUSIC

Music: The sound that really counts. We've already discussed the physical principles that form the basis for most musical instruments. We'll now point out in a general way how these principles apply to instruments.

Almost all the instruments that we ordinarily play or listen to are either stringed instruments or wind instruments. Some exceptions are those that use vibrating rods (the xylophone), those that use vibrating membranes (drums), those that use vibrating plates (cymbal and bell), and those in which the sound is generated electronically. We'll take a look at stringed and wind instruments separately.

The guitar, the violin, and the piano, as examples of stringed instruments, all use vibrating strings as their sound source. These three instruments illustrate the three ways that standing waves are set up in the strings: by plucking, by bowing, and by striking, respectively. (Have you ever looked inside a piano and noticed the little hammers that strike the strings when a key is pushed?) The method of exciting the vibrations and the location on the string at which the excitation is applied influence the quality of the sound: the relative harmonic content.

But if all you had were the strings, you wouldn't get much sound. In portable instruments, the vibrating strings set up standing waves in the body of the instrument and in the enclosed air volume. The air and the body together greatly amplify the sound and determine its harmonic content. In the piano, the strings are attached to a large, heavy wooden *sounding board* that resonates to a broad range of frequencies. (That's why the piano is so heavy!) The strings excite vibrations in the sounding board that supply the main part of the sound energy.

In principle, drums have a lot in common with strings. Rather than waves in a one-dimensional string, they have standing waves in a two-dimensional membrane that can have various harmonics present. Again, waves established in the enclosed air volume amplify the sound and influence which harmonics are present.

Wind instruments—for example, the pipe organ, the saxophone, and the tuba—all use standing waves in vibrating air "columns" as their main source of sound energy. When we considered standing waves in uniform air columns, we found that the length of the column determines the pitch of the sound. Wind instruments differ in the way we change the length of the air column. Pipe organs have different-length pipes for each note. But we change the *effective* length of portable instruments by opening or closing holes along the pipe—the clarinet—or by valves that add or remove sections of the column—the french horn. The particular shape of the instrument determines which harmonics exist and thus the quality of the sound: Notice, for example, the gradual increase in the size of the "pipe" that makes up a saxophone (Fig. 13.11).

Most wind instruments use one of two methods of setting up vibra-

Figure 13.11 A saxophone has a tapered, cone-shaped air column that allows for the existence of many harmonics.

tions in air columns: either *edge tones* or *reeds*. When air blows against a sharp edge, vortices are formed first on one side and then on the other. Their intermittent formation provides the vibration to set up standing waves in the air column.

In a reed-type instrument, air blowing across a thin flexible strip—a reed—causes it to vibrate and start standing waves in the air. For brass instruments, the player's lips form a pair of reeds that open and close quickly to set up the vibrations. The way the player blows can influence which overtones are set up.

Most of us would probably guess, just from playing a wind instrument or from watching it played, that the blowing of air through the body of the instrument and out the end is intimately connected with the sound given off—you sort of blow the sound out. This idea is probably reinforced in the type of instrument on which you open and close holes along the side, seeming to let the air out in different places. But *now you know* that the blowing merely starts the vibration by using reeds or edge tones, and that the main sound source is the standing waves in the air. These waves are not influenced by whether or not air is moving through the instrument. Opening and closing holes along the side changes the standing wave pattern by changing the effective length of the air column, but letting the air out has little, if any, effect.

The references at the end of the chapter will give you more details on the effects we have just mentioned, and they will show you how the various principles apply in specific instruments.

13.5 NOISE

Figure 13.12 The waveform of one example of noise. No regular repeating pattern exists. (*Courtesy Grayson Wright.*)

There is little doubt that what is music to one person is mere *noise* to another. You've probably experienced this difference of opinion between yourself and your parents. The discrepancy fits with one definition of noise: *any* unwanted sound.

Technically, however, we usually define **noise** *as sound having random frequencies mixed in random amounts*. The various frequencies are *not* harmonics of some fundamental frequency; rather, they have no particular relationship to one another. Notice that the noise waveform in Figure 13.12 has no regular repeating pattern, unlike those for the musical sounds of Figure 13.9. Noise produces no particular pitch or quality sensations.

Noise is not necessarily bad. It is important in the formation of consonants in speech. The sound from certain instruments—snare drums and cymbals, for example—is mostly noise. And experiments have shown that the noise burst during the first fraction of a second following the sound of a piano or harpsichord note is very important in our recognition of the sound.

You might be surprised to learn that noise has *color. White noise* consists of a fairly uniform mixture of frequencies throughout the audible range.

The term is borrowed from its use with light: *White light* is light such as sunlight, which is a mixture of colors. Just as pink light is light of mostly lower frequency and longer wavelength, *pink noise* is low-frequency noise; *blue noise* has the higher frequencies emphasized.

The concept of noise carries over into other fields. For example, if you are observing an electric current or voltage that varies smoothly with time, there may be random wiggles in the signal—called *noise*—superimposed on the smooth variation. For instance, in amplifying the small voltages measured across nerve cell membranes, random electrical activity in the amplifier might produce noise with larger variations than the signal being measured. For reliable measurements, it's important to have a large *signal-to-noise ratio*. The same is true if you're listening to a recording. You want the signal-to-noise ratio (music-to-scratches ratio) to be as large as possible.

In our increasingly technological society, *noise pollution* is becoming an ever more important consideration. The more gadgets we have, the more noises there are to pollute and mask the sounds we enjoy hearing. The psychological effects of noise pollution are yet to be thoroughly understood, but prolonged exposure to noise levels above about 85 dB often leads to permanent hearing damage. Rock musicians have been known to develop "boilermaker's ear," a name for a hearing loss whose origin you can easily imagine.

13.6 SPECIAL EFFECTS IN SOUND WAVES

In Section 12.6, we considered some special types of wave behavior—reflection, refraction, and diffraction. We'll now see how each of these effects influences the sounds that we hear. I suggest that you take another look at that section in which these terms were defined and described. In addition, we'll consider the *Doppler effect*—something that is common to all types of waves but that we most often observe in sound waves. And finally, we'll look at what happens to sound when its source moves faster than the sound itself.

Reflection

Take another look at Figure 12.15 and recall that it is illustrating the *law of reflection* for all waves: *The angle that the reflected wave makes with the reflecting surface exactly equals the angle the incident wave makes with it.* As you analyze the reflection of sound, keep in mind that this law always applies.

The most familiar example of sound wave reflection is the echo you hear when sound bounces off a distant large object. Reflection is important for proper acoustics in an auditorium. You want some reflection from the walls to make the sound "live"; too much reflection makes the auditorium like an echo chamber.

Sonar uses the reflection of sound waves (usually at ultrasonic frequencies) from underwater objects to determine their depth. Knowing the speed of sound in water and measuring the time between the emission and return of a sound pulse, you can calculate how far away the object is.

• •

EXERCISE

A ship sends out a pulse of sound of frequency 50,000 Hz that is reflected from a submarine and returns 1.00 second later. If sound travels at a speed of 1450 m/s in water, how far away is the submarine?

Answer: 725 m

Figure 13.13 One use of ultrasound in medicine. Reflected ultrasonic waves show a 16-week fetus inside the mother's uterus. (*Courtesy Joan K. Simons, R. T. (R) RDMS.*)

You may have been accused of being "blind as a bat." You also may not be as skillful as the bat in making up for visual deficiencies. Bats detect prey and obstacles in their paths by sending out short bursts of ultrasonic sound and detecting the echo reflected from the object. They can tell how far away their target is by timing the interval between the emission and return of the ultrasonic pulse.

Medical applications of ultrasonic sound, or *ultrasound*, already abound and are continuing to increase. One example is shown in Figure 13.13. Ultrasonic waves reflected from boundaries between two internal media can show the exact location of different organs, of abnormalities, or of "foreign" objects such as that in the picture. Ultrasonic techniques are sometimes not only more sensitive than x rays to variations in density of the tissue but also do not produce the harmful effects associated with x rays. Many interesting applications are discussed in the end-of-chapter references.

Refraction

In the last chapter we defined refraction as the bending of a wave's direction because it crosses a boundary between two media in which the wave travels at different speeds, or because the wave speed is different in different parts of the same medium. The second condition is the one that most often influences the behavior of everyday sounds. The following example illustrates such a case.

Suppose a sound wave, shown by the wave fronts in Figure 13.14, is moving to the right. If, as in (Fig. 13.14-a), the sound speed is higher closer to the ground, the wave will gradually bend upward, and a listener a moderate distance away won't hear it. If the speed is slower near the ground (Fig. 13.14-b), the sound wave will tend to bend downward, reflect from the ground, and travel a long way near the surface.

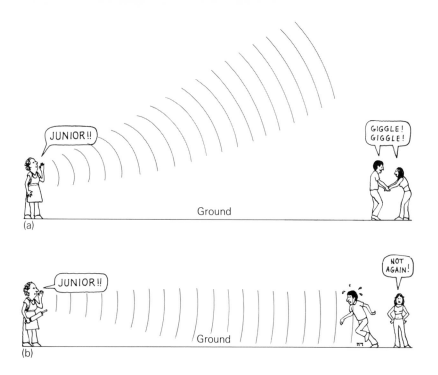

Figure 13.14 (a) If the wave speed is higher near the ground, the wave bends upward. (b) If it's lower near the ground, the wave bends downward and travels farther along the ground.

But what would cause the sound to travel faster or slower above ground level? One factor is a temperature difference. Sound speed increases with temperature increase. On a hot day when the ground is warm, the air near the ground is likely to be warmer than it is up a short distance; the sound curves up. On a cold day, or even better, a cold night, the opposite is true. Do these conclusions agree with your observations of how far away you can hear particular sounds on cold nights and warm days?

The wind has a similar effect. The speed of sound relative to the ground is its speed in air plus that of the wind (minus that of the wind if it and the sound are moving in opposite directions). Suppose in Figure 13.14 the wind is blowing from Junior to Mother. When he doesn't hear, she assumes the wind has held back the sound. In reality, the air is moving slower near the ground because of the frictional resistance of the ground. Thus, the net sound speed is a little higher near the ground. The sound curves up as in Figure 13.14(a). With the wind blowing from Mother to Junior, the net sound speed is slower near the ground, and the sound tries to curve down as in Figure 13.14(b).

Diffraction

In Section 12.6 we discussed the fact that an object casts a sharp shadow only if it is large compared with the wavelength of the waves impinging on it. If the object is about the size of or smaller than the wavelength, the waves merely *diffract* around the object and travel on almost undisturbed. The amount of disturbance depends on the ratio of object size to wavelength—being negligible for sizes much less than the wavelength and gradually increasing in effect on the waves as the relative object size increases.

As we calculated in Section 13.1, the wavelengths of audible sounds in air are between about 2 cm and about 20 m, with the longer wavelengths corresponding to lower frequencies. Thus, most of the sounds we hear have wavelengths comparable in size to the ordinary objects around us. That's why sound waves bend around ordinary objects rather than casting a shadow as light waves do.

. . . .

QUESTION   Why do you hear echoes only from large buildings or cliffs rather than from ordinary-sized walls? Why can you hear around a corner but not see around it?

Try to imagine how different our lives would be if sound had a much shorter wavelength so that things around us caused a shadow in the sounds we expect to hear. Just to name a few inconveniences, we would need to be sure no one walked between us and someone we were trying to hear; sound wouldn't travel very uniformly from room to room; and the loudness of the sounds around us would vary drastically as we walked in and out of the many sound shadows that would exist.

The Doppler Effect

If you've ever heard the "VAAROOOM!!" of a passing racing car, you've maybe noticed that the "ROOM" part has a lower pitch than the "VAA" part. Or if you've been passed by a train blowing its whistle, you may have heard the drop in pitch of the whistle as it passed. Turning the situation around, if you drive past a parked car blowing its horn, the pitch of the horn drops as you go by. All these examples illustrate the **Doppler effect**.

Figure 13.15 shows a situation like that of the parked car blowing its horn as you pass: A listener moving past a stationary sound source. The circles are wavefronts of the sound emanating from the source at their center. A traveler moving along the straight line meets the waves coming toward him as he *approaches* the source. He's going away from the waves that pass him as he *leaves* the source. Therefore, wavefronts pass him more often—he hears a higher pitch—as he approaches, and they pass him less often—he hears a lower pitch—as he leaves.

For a moving sound source—such as the passing racer—the effect is

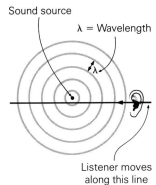

Figure 13.15 A traveler moving toward a stationary source intercepts waves at a faster rate than the source frequency. Going away he intercepts them at a slower rate.

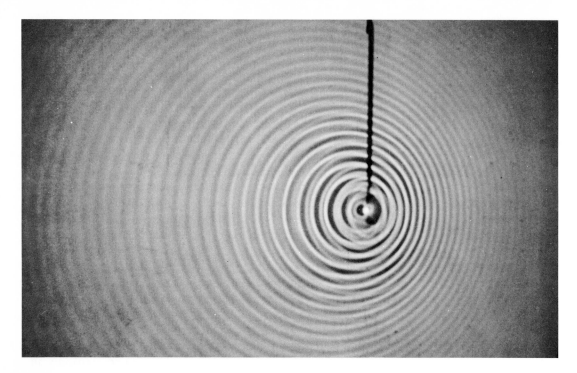

Figure 13.16 When the source is moving, the waves bunch up in front of the source and spread out behind it. This gives a shorter wavelength (higher frequency) in front, and longer wavelength (lower frequency) in back. The photograph is of water waves, but it illustrates what happens for a moving source of sound.

similar but the cause is different. Figure 13.16 shows what happens. The wave source at the center moves to the right with a speed lower than the wave speed. This movement causes the wavefronts to be crowded together with a shorter wavelength to the right of the source, and to be spread out with a longer wavelength to the left. Because the wave speed is the same on both sides, a listener hears a higher frequency (higher pitch) on the right than on the left. As the source passes, the pitch drops.

Waves reflected from a moving object have a frequency different from that of the incident waves because the moving object is actually the moving source for the reflected waves. This *Doppler shift* in frequency between emitted and reflected waves has a large number of useful applications. For example, sonar (which we discussed in connection with reflection) can do more than just tell the distance away of an underwater object. By measuring the Doppler shift in frequency between the emitted and returning pulse, you can tell the speed of the underwater object that reflects the sound. Police radar uses the Doppler shift of 16-GHz electromagnetic waves reflected from your car to monitor your speed.

Medical uses of ultrasonic sound in mapping internal organs can be enhanced appreciably by observing the Doppler shift in the ultrasound reflected from within the body. For example, the shift in frequency of waves reflected from moving blood cells can determine the speed of blood flow through a blood vessel or the heart, thus giving information on the condition of the organ.

EXAMPLE 13.1   Since blood cells are about 10 μm (micrometers; 10^{-6} meters) in diameter, what must be the approximate minimum frequency for ultrasonic sound if it is to reflect from individual blood cells? Take the speed of sound in the body to be the same as in water—1450 m/s.

If waves are to reflect appreciably rather than diffract around the cells, the wavelength must be *about* the size of the cell or shorter. Taking the maximum wavelength as the cell size,

$$f = \frac{v}{\lambda} = \frac{1450 \text{ m/s}}{0.00001 \text{ m}} = 145 \text{ MHz}$$

or around 100 MHz.

Shock Waves and Sonic Booms

The age of supersonic transport (SST) has brought livened interest in sonic booms, which were previously a concern only with military aircraft.

Figure 13.17(a) shows the wavefronts for a sound source moving to the right with speed lower than the speed of sound—the situation we described in the last section. Circles F_1, F_2, F_3, and so on are the wavefronts of the sound emitted when the source was at points 1, 2, 3, and so on, respectively. Point 6 is the "present" source position. The waves bunch together on the right. The faster the source speed, the closer the bunching. If the source speed exactly equals the wave speed, the wavefronts all crowd on top of each other at the source, and can never spread out in front of it (Fig. 13.17-b). When the source speed exceeds the wave speed, the individual circular waves spread out behind the source in a cone-shaped region, as in Figure 13.17(c). This cone-shaped wavefront on which the individual waves pile up on each other is called a *shock wave*.

When a supersonic airplane passes by, this conical shock wave spreads out from the plane. There are actually two shock waves, one from the front of the plane—a bow wave—and one from the back—a tail wave.

Figure 13.17 Wavefronts when the source is moving with speed (a) less than the wave speed, (b) equal to the wave speed, and (c) higher than the wave speed. The circles F_1, F_2, etc., are the wavefronts of the sound emitted when the source was at points 1, 2, etc. Point 6 is the *present* position of the source.

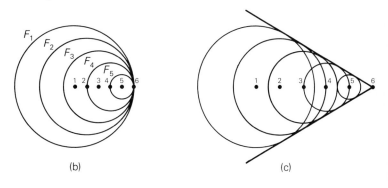

(a)   (b)   (c)

As the shock waves pass, there is a rapid rise and then fall in the air pressure—typically, a variation on the order of 100 N/m² above and below atmospheric for the *Concorde* SST. We hear a loud double "crack" that we call a **sonic boom**. For the *Concorde*, the perceived noise level for the boom is usually equivalent to about 120 dB.

· ·

QUESTION A lot of people erroneously think a sonic boom occurs only as an aircraft "breaks the sound barrier," not later. Explain why this belief is wrong.

There are other shock waves that we sometimes observe. As a high-speed bullet passes near you, the "crack" you hear is a shock wave. The V-shaped bow wave following a boat that is moving faster than the waves *in water* is an easily visible shock wave of the type we've described. You can even make them in your bathtub by running your finger along the surface of the water.

· ·

Key Concepts

Sound is a longitudinal wave disturbance in a material medium. The **audible** range of frequencies—the frequencies to which the normal human ear responds—is between about 20 and 20,000 Hz.

Any body that vibrates at audible frequencies can be a *source* of sound. Most musical instruments, including the voice, depend on standing waves in strings or air columns or both. The *harmonics* present in these standing waves form the frequencies present in the sound. Harmonic frequencies above the fundamental are referred to as **overtones**.

The **subjective** properties (those that depend on the response of the listener) of sound—**loudness**, **pitch**, **quality**—depend mostly on the **objective** properties (those that are physically measurable): **intensity level**, *frequency*, and *harmonic content*,

respectively. The intensity level of the faintest audible 1000-Hz sound is assigned a value of 0 decibel (dB), and the loudest 1000-Hz sound that does not cause pain has an intensity level of 120 dB. When the intensity is *multiplied* by 10, 10 dB are *added* to the intensity level.

Noise is sound having random frequencies mixed in random amounts. Its waveform has no repeating pattern.

Sound waves, as well as other kinds of waves, can *reflect*, *refract*, and *diffract*. The **Doppler effect** is the change in frequency resulting from the relative motion of the wave source and observer. *Shock waves* occur when the wave source travels at a faster speed than the wave speed in the medium. Shock waves from a sound source give rise to *sonic booms*.

Questions

Sound Waves

1. I was taught in my high school biology class that if a tree falls in a forest and there is no person or animal nearby to hear it, there is no sound. Was I taught correctly?

2. Suppose Neil Armstrong had decided to take a nap on the moon's surface and had laid his alarm clock on the ground beside him. Do you think he would have made his next appointment?

3. Sound will travel not only through air but through any material medium—for example, bone. What does that have to do with the reason *your* voice sounds so horrible on a tape recorder, while everybody else's sounds fine?

4. Research has shown that three-day-old babies can distinguish their mothers' voices from other sounds. How do you expect this result to relate to the ability of sound to travel through parts of the human body?

5. The speed of sound in water is about four times what it is in air. A waterproof bell ringing in air has a sound frequency of 400 Hz. What frequency will you hear if both you and the bell are underwater? Does your answer relate in any way to the observation described in Question 4?

6. Bowing a violin string produces sound as the bow alternately grabs and then slips a little. Describe how this same grab-and-slip process can produce some other types of sounds: chalk on the blackboard, squealing tires, creaking doors.

Sound Sources

7. Why does pushing a guitar string down against a fret increase the pitch?

8. On a given violin string you can play a note of any pitch within a certain broad range. On a guitar string, you play only discrete pitches within a certain broad range. Why?

9. The highest piano note has a pitch corresponding to a frequency of 4186 Hz. The upper limit of other ordinary instruments is lower than this. Does this mean that the only value in having a sound system that can reproduce frequencies higher than 4186 Hz is to give the owner or manufacturer something to brag about—not to give better sound?

10. What accounts for the gradual increase in pitch of the sound as you fill a bottle with water? (Hint: Think about what's happening in the air left in the bottle)

11. Give some reasons why a boy's voice deepens as his mouth and nasal cavities get larger and as his vocal cords get heavier.

12. Imagine you live on Mars and have never seen an earthling. If you hear a recording identified as an earth man's voice and one identified as an earth woman's voice, what could you predict about the relative length and/or massiveness of the male and female vocal cords, and about the relative size of their nasal and mouth cavities?

The Ear

13. Your ear canal leading the sound to your ear drum is probably about 2.5 cm long. An air column this long at 37°C, closed at one end and open at the other, has a fundamental resonant frequency of about 3500 Hz. Do you think these statements have anything to do with your ears being most sensitive in the 3 to 4-kHz frequency range?

14. The basilar membrane in the cochlea of your inner ear is made like a harp. It has many tiny fibers that are longer and thicker as the distance from the oval window increases. The vibration of these fibers sets up impulses in nearby nerve endings, and these impulses are sent to the brain. Why do different frequencies excite vibrations in different parts of the basilar membrane? From which end of the membrane does your brain get impulses if the sound is of high frequency?

Objective and Subjective Sound Properties

15. What physical property of sound mostly determines its loudness? its pitch? its quality?

16. Suppose you call your dog with an average intensity level of 80 dB. How much more sound energy would you have to put into each call to increase that to 90 dB?

17. Table model radios seldom have speakers larger than about 15 cm in diameter. Even though such speakers probably cannot reproduce sound of pitch lower than about middle C (262 Hz), we certainly hear lower pitches from such radios. Why?

18. How is the sound of a saxophone physically different from that of a trombone playing the same note?

Music and Noise

19. Why does the pitch of wind instruments increase, and that of stringed instruments decrease, as an orchestra warms up?

20. A student told me that he places his autoharp on the toilet to play it because the sound produced is much richer that way. Compare this resonance phenomenon with the effect of the sounding board of a piano.

21. For a stringed instrument, the wavelengths of the sound produced are different from the wavelengths of the standing waves in the string. For a wind instrument, the wavelengths are the same for both the standing waves in the instrument and the sound waves produced. Explain.

22. Do you suspect a relationship between the sound whistling through the trees and the way in which sound waves are initiated in wind instruments?

23. Which would you take to be pinker, the noise of a jet airplane overhead or that of a bulldozer leveling ground for a new runway?

Special Sound Effects

24. Why is it so hard to understand what your opponent says to you on a racquetball court unless you're very close to her?

Problems

1. Middle C on a piano corresponds to a frequency of 262 Hz. Calculate the wavelength of the sound waves carrying this sound through air.

25. Many automatic-focusing cameras send out "chirps" of inaudible ultrasonic sound that reflect from the subject back to the camera. The time between emission and return of the sound pulse determines the distance for which the camera automatically focuses. Compare this with the way bats locate their prey.

26. Moths, a favorite epicurean delicacy of bats, have a furry covering on their bodies that absorbs sound waves. Their wings, however, are good reflectors. Why do you suppose moths instantly fold their wings and drop when they detect any ultrasonic sound?

27. Even though the speed of sound is much higher than the speed of air in wind currents, why does sound not travel nearly as far along the ground "against" the wind as "with" the wind?

28. When you're in your room and your friends are making lots of noise in the hall, does *partially* closing the door decrease the noise in the room an appreciable amount? Why not? Does it matter where you are in the room? (Assume soundproof walls and door.)

29. The pulse of ultrasound emitted by some bats varies in frequency during the pulse. How does this variability give the bat more information about the size of objects from which the waves are reflected? How does the use of the Doppler effect let it distinguish moving objects from the stationary background?

30. If you're on the side of the road and a police car passes with its siren sounding, why does the siren pitch drop as it passes?

31. A *Concorde* SST leaves Washington, D.C., and passes over Philadelphia and New York before heading out over the ocean. It accelerates through the speed of sound just as it passes over Philadelphia. True or false: The people in Philadelphia hear a sonic boom but the people in New York do not.

(The speed of sound in air at 20° C is 344 m/s.)
Answer: 1.31 m

2. A particular violin string has a fundamental frequency of 380 Hz. What is the frequency of the third overtone in the sound from this string?

    Answer: 1520 Hz

3. The standing waves in a 1-m-long organ pipe have a node at one end and an antinode at the other. What is the wavelength of the standing wave in this pipe? What is the wavelength of the sound traveling out from the pipe to a listener? (See Question 21.)

    Answer: 4 m; 4 m

4. The sound from a 1.0-m-long piano string has a frequency of 440 Hz. What is the wavelength of the waves *in the string*? If the tension in the string is doubled, what is the wavelength of the waves *in the string*? Why does it not change, even though the pitch of the sound certainly does?

    Answer: 2.0 m

5. Going up an *octave* on the musical scale corresponds to doubling the frequency of the sound. If an A note is 440 Hz, what is the frequency of a note two octaves higher?

    Answer: 1760 Hz

6. Musical tones that differ by an interval of a *fifth* have a frequency ratio of 3 to 2; those that differ by a musical *fourth* have a frequency ratio of 4 to 3. What are the frequencies of the tones a fifth higher and a fourth lower than A above middle C—440 Hz?

    Answer: 660 Hz; 330 Hz

7. The frequencies 352 Hz, 440 Hz, and 528 Hz form a musical *major triad*. The same ratio of frequencies holds for any major triad. What is this ratio?

    Answer: 4 to 5 to 6

8. If, after being yelled at by your physics instructor, you reduce the intensity level of your whispering from 20 dB to 10 dB as heard by your friend across the isle, how much less power are you putting into your whisper?

    Answer: 1/10

9. The volunteer firefighters in Smithville are called to a fire by a siren mounted on a tall pole in the center of town. If the intensity level of the sound 1.00 m from the siren is 120 dB, what is the intensity level 3.16 m away? Assuming no wind or other disturbances in the air, within what distance of the siren would the sound have an intensity level of at least 50 dB? (Comment: The intensity of the sound from a small nondirectional source *decreases* in proportion to the *square* of the distance from the source, provided there is nothing around for the sound to reflect from. That is, the intensity follows an inverse square law. Recall, $\sqrt{10} = 3.16$.)

    Answer: 110 dB; 3.16^7 m, or 3.1 km

10. A hoarse mockingbird sings loud enough that the intensity level is 20 dB 1.00 m away. What is the intensity level 3.16 m away? Notice the comment in Problem 9, and explain why the relative effect of moving by $\sqrt{10}$ is so much greater here than in that problem.

    Answer: 10 dB

11. The Apostle Paul reported that when he was about to be shipwrecked in the Adriatic Sea, the sailors "sounded and found it twenty fathoms" (The Acts of Apostles 27:28). At that depth, how long would it be between the emission and return of a sound reflected from the bottom? A fathom is 1.8 m (6 ft) and the speed of sound in water is 1450 m/s. From your answer, do you think it likely that their sounding was by way of sound reflection?

    Answer: 0.05 s

12. An automatic-focusing camera sends out a 60-kHz sound pulse that is reflected from the subject and returns to the camera in 12 ms (milliseconds). How far away is the subject? What is the wavelength of this ultrasonic sound in air?

    Answer: 2.0 m; 5.7 mm

13. If a bat is to detect the presence of a 5-mm-long insect by means of reflected ultrasonic sound, roughly what must the minimum frequency of the ultrasound be?

    Answer: 70 kHz

Home Experiments

1. Borrow a tuning fork from your instructor. Set it in vibration and notice how long it takes for the sound to die out. Set it in vibration again with about the same intensity and hold the base in contact with a tabletop. Notice the difference in the total sound output and in the time required for the vibration to stop. Explain the time difference in terms of the energy conservation principle.

2. Measure the speed of sound by timing the return of an echo. Find a flat field bounded on one side by a building. One timing method that is fairly accurate is to clap rhythmically, starting slowly and increasing the frequency until the echo returns from the wall on the "off beat." With your watch lying where you can observe it, count the number of claps per 10 s. Step off the distance to the wall. From your data, calculate the speed of sound and compare it with the accepted value.

3. Take a cardboard mailing tube, or the tube on which wrapping paper is wound, and hold it near your mouth. Sing a steady "aah" into the tube and vary the pitch slowly from low to high until you pass through resonance. You should not only hear a slight increase in intensity, but you can feel the difference in your throat as the standing waves are set up. To help convince yourself that you are passing through resonance, cover the other end and notice the absence of a change at the old resonant frequency. Find a new resonance with the end covered. It should be about an octave higher (that is, at twice the frequency). Try a tube of another length, and compare the pitch at resonance.

4. Stand in an enclosed shower stall and hum from a low pitch gradually up to a high one. Listen to the natural resonances as standing waves are set up in this three-dimensional enclosure.

5. Hold down the damper pedal of a piano and yell various kinds of sounds into the region containing the strings and sounding board. Yell out the various vowels, and notice that the particular frequencies in the sound are picked up and maintained for several seconds. Discuss the relationship of this to the frequency response mechanism of the ear (Question 14).

References for Further Reading

Crawford, F. S., Jr., *Waves* (McGraw-Hill Book Co., New York, 1968). Gives many interesting problems and home experiments that show the interaction of sound and other forms of waves with our daily lives. Several of the home experiments in this chapter are variations of Crawford's experiments.

Devey, G. B., and P. N. T. Wells, "Ultrasound in Medical Diagnosis," *Scientific American*, May 1978, p. 98.

Gelatt, R., *The Fabulous Phonograph* (J. B. Lippincott Co., Philadelphia, 1955). A history of the reproduction of sound.

Hall, D. E., *Musical Acoustics: An Introduction* (Wadsworth Publishing Company, Belmont, California, 1980).

Hansen, J., "Now You Hear It, Now You Don't," *Science 84*, May 1984, p. 74. Discusses experiments that use interference of sound waves to cancel unwanted sound.

Kryter, K. D., "Sonic Booms from Supersonic Transport," *Science*, Jan. 1969, p. 359. A study of the effects of sonic booms on people.

Neuweiler, G., "How Bats Detect Flying Insects," *Physics Today*, Aug. 1980, p. 34.

Readings from *Scientific American, The Physics of Music* (W. H. Freeman, San Francisco, 1978). This compilation of reprints contains material on the basic physics of music, on musical instruments, and on architectural acoustics.

Roederer, J. C., *Introduction to the Physics and Psychophysics of Music* (Springer-Verlag, New York, 1973). Gives not only the physics of sound, but considers specifically the sensation of hearing and how we react to various sounds.

Stanford, A. L., Jr. *Foundations of Biophysics* (Academic Press, New York, 1975). Includes details of the hearing mechanism.

The Way Things Work Book of Nature (Simon and Schuster, New York, 1980). Includes a section on sound and noise.

Walker, J. "Some Whispering Galleries Are Simply Sound Reflectors, but Others Are More Mysterious," The Amateur Scientist Column, *Scientific American*, Oct. 1978, p. 179.

PART FOUR

ELECTRICITY AND MAGNETISM

The experiments combine to prove that when a piece of metal (and the same may be true for all conducting matter) is passed either before a single pole, or between the opposite poles of a magnet,. . . electrical currents are produced across the metal transverse to the direction of motion.   M. FARADAY

Michael Faraday, born at Newington, Surrey, England, in 1791, was the son of a blacksmith. After becoming an apprentice to a London bookbinder at age 14, Faraday became fascinated with electricity after by chance reading an article on the subject in an encyclopedia brought in for rebinding. His scientific career began as a chemist, and he discovered benzene while working as an assistant to Sir Humphry Davy. However, his major contributions were in the field of electricity and magnetism. The law named in his honor expresses the relationship between changing magnetic fields and electric currents. Faraday is shown here with his wife at Christmas, 1821.

CHAPTER 14

ELECTRIC CHARGE

When the electricity goes off, life as we know it in this country comes to a standstill. A power failure following a storm might be fun for a while—classes and work might be called off, or you might enjoy eating dinner by candlelight if you get it cooked before the storm. But a blackout quickly loses its appeal. Certainly, you miss the obvious: lights, electric ranges, TV, electric heating and air conditioning, and the hundreds of electrical appliances that both simplify and complicate our lives. But you might be in for a surprise if you have a gas-fired furnace that doesn't heat because the electrically operated valve doesn't open to let the gas in, or if you have to walk up stairs rather than ride the elevator to your high-rise apartment, or if your escape to the next town becomes impossible because your tank is empty and the gasoline pump at the station won't operate, or if you can't purchase snacks at the grocery store because the cash register is dead without electricity.

Not only are our personal lives greatly disrupted, but most industry is completely shut down without electrical power. Sudden lack of electricty has even spawned massive social disorder, resulting in rioting and looting during blackouts in large cities.

Our involvement with electricty doesn't stop with gadgets that we plug into a wall receptacle. Ever increasing in numbers are the battery-operated tools and toys—calculators, radios, cassette players, intercoms. Even though the electric car is still relatively rare, the ordinary gasoline-fueled automobile depends on electricity to start the engine and to produce a spark thousands of times per minute in each cylinder of the engine. And where would we be without that electrical wonder, the telephone?

Even if we could live without all of these enjoyable electrical conveniences, we would still depend on electricity to control our own bodies.

The messages sent back and forth between our brains and all parts of our bodies travel electrically. Electrical signals control even our heartbeat and breathing.

14.1 WHAT IS ELECTRICITY?

Suppose you have an ordinary table lamp with a 60-watt bulb that you expect to use for reading. You plug the lamp into a wall receptacle and then turn the switch on. If everything is working right, light will emanate immediately from the bulb. It *appears* that *something* is waiting inside the receptacle to get into the bulb and create light as soon as a path is provided through the cord and a "gate" is opened by turning on the switch.

A similar thing happens when you use an electric shaver. When you plug it in and turn on the switch, you make a path for "something" waiting inside the receptacle to get out and into the shaver to make it go.

There are a few general requirements for the working of electrical lights and appliances that you already know about from your own experience: (1) This "something" needs a path—a cord—to get to the device, and this path has to be continuous—if a wire breaks or comes off one of the screws in the plug, your TV doesn't work. (2) It takes two paths going to the device, not just one—there are two wires in your hair drier cord. And (3) this "something" waiting in the receptacle to get into your electrical apparatus is choosy about what it travels through—it will go through the metal wire but not through the rubbery insulation outside the wire. You might have found, to your regret, that it sometimes chooses to travel through you.

Consider another familiar electrical device—a flashlight—that operates on batteries rather than from a power cord. A person who knows nothing about electricity still realizes that something in the battery gets into the bulb and makes it light when the switch is turned on. Again this "something" seems to be waiting, ready to get into the bulb as soon as it has a path that the switch somehow provides.

Let's look at how the flashlight works. Figure 14.1 is a diagram of the inside of a typical flashlight. When you push the switch on, the metal slide moves to the right, contacting the ring around the metal casing of the bulb. This contact provides a continuous *metal* path (dashed line) from the terminal labeled "+" on the right battery, through the filament of the bulb, through the metal slide on the switch, through the flashlight case, through the spring, and to the "−" at the end of the left battery. The "−" end of the right battery contacts the "+" end of the left one. Evidently, the batteries cause the "something" to flow through this continuous path, but they are unable to cause a flow when the switch is off (open) and the path is no longer continuous. This flow causes the bulb to produce light.

Metal slide   Switch   Insulator   Contacts

Reflector

Bulb

Filament

Metal ring

Spring

Metal

Figure 14.1 A flashlight bulb gives off light when there is a closed path from the positive terminal of the battery, through the filament, and back to the negative terminal (dashed line).

The desk lamp produces, or doesn't produce, light for the same reason. Figure 14.2 shows a simplified diagram of how the lamp works. When you move the switch so that there is a continuous path for our "something" to flow from the receptacle through the filament, the bulb gives off light.

To explain these effects we need to answer three questions.

1. What is this "something" that flows?

2. What makes it flow?

3. How does it do its jobs in the bulb, the motor, the heater, or whatever?

The answer to the first question is that what flows is a fundamental quantity called *electric charge*. To understand the properties of electric charge, and to answer the second question—"What make it flow?"—we need to start by looking at the structure of the atoms that make up all matter. We must work from there toward a definition of a quantity called electrical *potential difference*, or often called *voltage*. The third question is a broad one, and we will find the answers to it as we go through the next few chapters.

In this chapter, we'll look at the basic ideas and quantities that we need in order to describe electric charge and its properties. Most practical electrical devices make use of *moving* charges—that is, a *flow* of charges called *electric current*. Chapter 15 will build on the basic ideas concerning electric charge as it defines the terms needed to discuss current. Chapter 16 then discusses practical applications of electric currents. Chapters 17–19 treat various aspects of magnetism and its relationship to electricity. Chapter 20 takes an overall look at electrical energy: how it's produced from other sources and how it's used to do specific kinds of jobs.

A word of encouragement if this is your first study of electricity: Chapters 14 and 15 lay the groundwork for electricity by defining a number of new terms that may initially seem somewhat disconnected. By the time you get through Chapter 16, these seemingly separate concepts should all fit together.

Filament

Switch

Figure 14.2 An ordinary light bulb produces light after it's plugged in only if the path is complete from one prong of the plug to the other.

Chapter 14   Electric Charge

Figure 14.3 Image from an electron microscope showing individual atoms in a gold crystal.

14.2 ATOMIC STRUCTURE AND ELECTRIC CHARGE

All forms of matter—this book, the pencil with which you are supposed to be taking notes, the clothes you're wearing, as well as you and I—are made up of atoms. Your body consists of about a billion billion billion (10^{27}) atoms, which means that they are much too small to be seen with the eye or with the most sensitive light microscope. In recent years electron microscopes have been developed that are sensitive enough to display images of individual atoms. An example is shown in Figure 14.3.

No microscope, however, is capable as yet of showing the *internal* structure of an atom. For an understanding of the internal structure, we depend on indirect evidence that lets us develop models for the atom—representations that help us to visualize it in terms of more familiar concepts. A model can help to do that, even if it isn't correct in every respect.

We know that each atom has a nucleus made up of neutrons and protons at its center, and electrons distributed about this nucleus. A model you have certainly seen of the atom is shown in Figure 14.4. We might call this representation a *planetary* model because of its similarity to the planets and sun of our solar system. This simple model is not strictly correct, but it does help to give us a mental image of the atom. To describe the atom correctly, we use what is called *quantum mechanics*, which we will discuss in Chapter 24.

Figure 14.4 In the planetary model of the atom, electrons (*e*) are considered to orbit the nucleus (*N*), which is made up of neutrons and protons (*n* and *p*).

Aside from the need for quantum mechanics, there is another very misleading aspect of pictures like Figure 14.4, which people so often use to describe atoms. *The relative size of the nucleus is greatly exaggerated.* If we draw the atom to scale, with the nucleus the size shown in this figure, the electron orbits would have to be about one-tenth of a mile in diameter. The atom is almost all empty space! (Roughly the same proportion of empty space as in our solar system.) Nevertheless, at least 99.95% of the mass of the atom is contained in its tiny nucleus.

Atoms differ greatly in complexity. The simplest one is the hydrogen atom whose nucleus of one proton has one electron in orbit around it. But atoms that exist in nature range in complexity on up to the uranium atom with 92 electrons orbiting a nucleus of 92 protons and 146 neutrons.

Electric Charge

A quantity called **electric charge** exists on every electron and every proton. We can't really say what charge *is*; that is, we can't define it in terms of something more fundamental. We can say only that it exists— in the same sense that mass, time, and length exist—and we can describe its properties.

First, charge exists in two types that, following the terminology used by Benjamin Franklin, we call *positive* and *negative*. **Positive** charge is the kind held by the proton, and **negative** charge, the kind held by the electron. The *amount* of charge on the electron is exactly equal to the amount on the proton. The neutron has no net charge; we say it is electrically **neutral**.

The terms *positive* and *negative* are more than just identifying names. Just as positive and negative numbers balance each other, equal amounts of positive and negative charge balance each other. Since an atom in its normal state has the same number of protons and electrons, the total charge of such an atom is zero. As a *whole*, the atom is electrically neutral. A hydrogen atom, for example, with one proton and one electron, has no *net* charge. If a neutral atom gains or loses an electron, it becomes negatively or positively charged. We call such a charged atom an **ion**.

Ordinary-sized objects contain many, many atoms and therefore many, many electrons and protons. When the total number of electrons and protons in an object is equal, the object is electrically neutral. We say that such an object is **uncharged**, although what we mean is that it has the same amount of negative as positive charge. A **charged** object is one that has an *excess* of negative or positive charge. Because the charge on any object is made up of individual electron and proton charges, the total charge must be some whole number times the charge of the electron. In other words, charge is *quantized*—it exists in discrete bundles. These bundles are so small that, for ordinary purposes, the charge seems to be able to take any value. (If you have two truckloads of sand and are only interested in shovelfuls, you don't care that the sand is quantized in individual grains.)

We usually use the symbol q (or Q) for charge and measure it in a

Chapter 14   Electric Charge

unit called the **coulomb** (C), named in honor of Charles Augustin de Coulomb, an eighteenth-century pioneer experimenter on electric charges. The size of the coulomb is such that it takes 6.25×10^{18} electrons to give a charge of 1 C. Turning this around, the amount of charge on the electron (or proton) is 1.60×10^{-19} C. This amount of charge is very special in that it is the smallest quantity of charge known to exist on any object. This amount of charge has its own symbol—the letter e.

Physicists now accept the theory that particles such as protons and neutrons are made up of more elementary particles called *quarks*, which can have a charge of either $\pm\frac{1}{3}$ or $\pm\frac{2}{3}$ that of the electron. However, we believe that quarks cannot exist as free particles, but are always bound together to form particles with a net charge of either zero or a whole multiple of e. Section 25.5 discusses quarks in more detail.

Normally, your body is electrically neutral and contains about 10^{28} electrons and protons. If you were to lose one electron out of every billion, you would have a net charge of a few coulombs. A coulomb of charge is actually a *large* amount. Ordinary-sized objects can usually be charged to about a millionth of this amount.

Experiments show that the net amount of charge in any region where no charge enters or leaves remains constant. That is, no *net* charge can be created or destroyed; if a net positive charge appears in one place, a net negative charge must appear in another so that the total amount of charge does not change. This is a statement of the principle of **conservation of charge**, a law just as fundamental as the energy and momentum conservation principles.

Forces between Charges

Another important property of electric charge is that any two charges exert a force on each other. If both charges are positive or both negative, then the force between them is repulsive—they push on each other. If one charge is positive and the other negative, the force is attractive—they pull on each other. Consider, for example, the two protons of Figure 14.5(a). Each feels a force trying to push it away from the other. Similarly, two electrons (Fig. 14.5-b) feel a force trying to push them apart. However, an electron and a proton (Fig. 14.5-c) experience a force toward each other.

The force of attraction between the positive charge on the protons and the negative charge on the electrons holds the electrons in orbit around the nucleus of each atom. When atoms combine with each other to form molecules, the charges in the atoms rearrange themselves so that electrical forces hold the molecule together. The atoms or molecules that make up a solid are held in place by electrical forces. The charges arrange themselves so that when you try to pull the solid apart, the forces of attraction between unlike charges dominate, and the solid is held

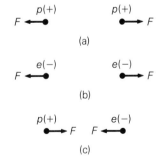

Figure 14.5 Like charges experience a force F repelling them from each other. Unlike charges experience a force of attraction.

Particles found inside the proton

STANFORD, CALIFORNIA (JANUARY 3, 1971)—In experiments performed at the Stanford Linear Accelerator Laboratory over the past few years, physicists have found strong evidence for the existence of tiny particles inside all protons—particles so small their size cannot be measured. Protons, along with neutrons, form the nuclei of all atoms. In this massive particle accelerator that tunnels under a California expressway, electrons get a two-mile-long running start to acquire the energy needed to penetrate inside the speck called a proton that is only about 40 millionths of a billionth of an inch across.

Physicists explain their evidence in this way: If you shot bullets at a glob of Jello, you'd expect the bullets to pass on through the Jello, with little departure from their original paths. But if a few other bullets were embedded in the Jello, you would expect an occasional fired bullet to bounce back at a large angle as it collided with one in the Jello. That's exactly what the scientists found with the high-speed electrons shot at protons: Some of the electrons scattered at large angles rather than passing on through the proton as they would if the proton were a uniform distribution of matter. Such evidence seems to confirm present theories that protons are made up of smaller particles.

By measuring the speed with which the scattered electrons come off at particular angles, physicists could calculate some of the properties of the particles doing the scattering. By comparing the results from the Stanford experiments with those of similar experiments done in Switzerland over the past year, scientists have shown that the electrical charges of the particles inside the proton are either $+\frac{2}{3}$ or $-\frac{1}{3}$ that of the proton itself. These are exactly the charges predicted for particles called "quarks," which since 1964, have been thought to be the constituents of protons as well as neutrons.

These constituent particles, presumably the predicted quarks, have one very strange behavior. If they really behaved like the bullets embedded in Jello, they would be knocked out of the proton by the high-speed electrons. But, though they seem to be free to roam about inside the proton, they apparently cannot get out even if struck an extremely forceful blow. One physicist put it this way, "The proton is like an ideal prison: the quarks have perfect freedom inside but are unable to escape."

together. When you try to compress the solid, the forces between like charges dominate and prevent you from compressing it. There is some "give" in each direction and, if the applied force is strong enough, it can overcome these electrical forces holding the solid together, and either break it or crush it.

The situation just described means that, *aside from the gravitational force attracting you toward the earth, every force that you experience directly is electrical in nature.* (Gravitational forces from other bodies act on you but are so weak that they're negligible.) Think about how sweeping this statement is and try to reconcile it with the forces you've seen in action today.

QUESTION   In ordinary table salt, sodium chloride, each chlorine atom takes one electron from a sodium atom. Does the sodium ion have a positive or negative charge? Does the chlorine ion have a positive or negative charge? What would you expect to be the nature of the force holding the two ions together in a sodium chloride molecule?

14.3 CONDUCTORS AND INSULATORS

In Section 14.1 we mentioned that charges sometimes *flow*. If charges exist on the parts of the atom, then how can they flow? Some materials have charges that are free to move about. We call such a material an electrical **conductor**. All materials conduct charge to some extent. Some are much better conductors—that is, the charge moves through them much more freely—than others. Those materials that are poor conductors are called **insulators**.

An extension cord, for example, consists of metal (usually copper) conductors surrounded by rubber or plastic insulator to prevent the flow of charge from one wire to the other or from a wire to your hand.

In general, metals are good conductors. Their atoms have a few (one, two, or three) electrons, called valence electrons, that are farther out from, and not as tightly bound to, their nuclei as the other electrons. In a solid metal these electrons are usually about the same distance from two or more nuclei. This makes them free to move throughout the metal as a sort of "electron gas." Because of the specific ways the atoms are arranged in different metals, some conduct more easily than others. Silver, copper, and gold are the best conductors at ordinary temperatures, with aluminum not far behind.

Another important class of materials called *semiconductors* ought to be discussed here. This would be logical from the physics involved. But to avoid introducing any more new ideas now, we'll postpone discussing semiconductors until Chapter 16.

Ionic Conductors

Pure water is an insulator. Yet we are often cautioned not to be in water or have wet hands when using an electrical device. Are these statements contradictory?

The catch is in the word *pure*. When certain salts are added to water, they form positive and negative ions that can move easily through the liquid. Ordinary tap water contains many impurities that form ions, which makes it a good conductor.

Figure 14.6 If your body is charged negatively, it induces positive charge onto the knob nearest your hand, which in turn induces more negative charge onto your fingertips. The closer your hand gets to the knob, the greater this effect.

Rubber

Fur

(a)

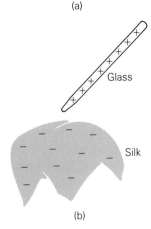

Glass

Silk

(b)

Figure 14.7 (a) When fur and rubber come into close contact by rubbing, the rubber atoms attract some electrons from the fur atoms, leaving the rubber charged negatively and the fur charged positively. (b) For glass and silk, the silk is the electron "thief."

There is one important difference in the way charge is conducted in an ionic solution. In a metal, only electrons—negative charge carriers—move. In an ionic conductor, both negative and positive ions are free to carry the charge.

In Chapters 15 and 16 we'll make extensive use of conductors as we consider the flow of charges—electric currents.

14.4 CONTACT ELECTRIFICATION

We can gain some evidence for the existence of two types of charge, and for the nature of the forces between charges, by considering some situations in which charge is transferred from one body to another through contact. We call this transfer of charge *contact electrification*. This effect is responsible for some "shocking experiences" that you would probably prefer to avoid.

For example, as you walk across a carpet, the soles of your shoes come in close contact with the carpet fibers, which are of course made of a different material than the shoe soles. It is usually true that the atoms of one substance will have a stronger attraction for electrons than those of another substance. If the two come into close contact, the substance with the stronger attraction steals some electrons from the other, giving the thief a net negative charge, and leaving the victim with a net positive charge. The close contact between shoe soles and carpet allows such a charge transfer. This net charge on the soles is conducted upward and spreads out over your body. As your hand, which is then charged, approaches an uncharged doorknob, the charge jumps across the small gap between hand and knob just before touching, and you feel an annoying but harmless shock.

To get much of a shock, you need to touch metal (or some conductor). The charge you pick up spreads over your body (which is a conductor) including some to your hand. Assume your body to be negatively charged. As your charged hand comes near the knob, it repels negative charge to the far knob, leaving a positive charge on the knob nearest your hand (see Fig. 14.6). This charge attracts more negative charge to your hand from the rest of your body, which in turn makes the nearest

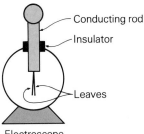

Conducting rod

Insulator

Leaves

Electroscope

(a)

Charged object

(b)

Figure 14.8 When you charge an electroscope, the leaves spread out because of mutual repulsion of the charges.

knob more positive. This redistribution is extremely fast, giving a larger concentration of charge, the nearer the hand is to contact. Just before actual contact is made, the concentration is large enough to cause charge to jump across the gap.

We often demonstrate contact electrification in the classroom by rubbing together two objects with different electron affinities. For example, if you rub a rubber rod with a piece of fur, the rubber attracts electrons from the fur, becoming negatively charged and leaving the fur positively charged (see Fig. 14.7-a). If you rub a glass rod with silk, electrons transfer to the silk from the glass, leaving the glass positive and the silk negative (Fig. 14.7-b).

We can show the presence of these charges by what is called an electroscope (Fig. 14.8-a), which consists of a metal rod with two very thin "leaves" of metal—usually gold—attached to the bottom. The end of the rod containing the leaves is housed inside a metal box with windows for viewing. The rod is insulated from the box.

If we touch a negatively charged object to the metal rod, some of the charge from the object distributes itself along the electroscope rod and leaves. The repulsion between the charges on the leaves causes them to spread apart (Fig. 14.8-b). When the charged object is removed from contact with the electroscope terminal, excess charge stays on the electroscope leaves, and they remain separated.

We get the same effect by touching the electroscope with a positively charged object. Then electrons escape the electroscope leaves, causing them to diverge due to the repulsion of the net positive charge.

QUESTION What would you expect to happen to the electroscope leaves if a charged object is brought near but does not touch the electroscope? What would happen if the object were then removed?

If the electroscope is charged negatively and a positively charged object brought near, the leaves will come together. You see the same effect if you bring a negative charge near a positively charged electroscope. For example, if you charge the electroscope with the glass rod, bringing the silk near the electroscope will cause the leaves to converge. Remember, the rubbing created no charge; it only transferred some.

Your instructor may invite you to volunteer to become a human electroscope by standing on an insulating stool and putting your hands on some kind of charging device (Fig. 14.9). Your hair becomes the electroscope leaves as each charged strand repels the others. As long as the charge comes into and out of your body slowly, you will not feel a shock.

Figure 14.9 Laurie's hair demonstrates repulsion between like charges as the charged strands repel each other. (*Photograph by Bob Veltri.*)

14.5 ATTRACTION BETWEEN A CHARGED AND AN UNCHARGED BODY—POLARIZATION

Let's now find out why a charged body is attracted to an uncharged one: why your comb will pick up little bits of paper after you comb your hair, why a balloon will stick to the wall after you rub it on your sweater or hair, why the dust sticks to the phonograph record, and why dust and pollutants can be filtered out with an electrostatic filter or pre-cipitator.

Consider again the charges induced on different parts of a conductor by a nearby charge. Suppose you wad a small piece of aluminum foil into a ball and hang it from a string. If you then run a comb through your hair a few times and bring it near the foil ball, the ball will be attracted to it (Fig. 14.10). The amount and the sign of the charge picked up by the comb will depend on the material of the comb. Assuming the comb has a negative charge, it induces positive charge on the side of the ball nearest the comb, leaving the other side negative. That is, the ball becomes **polarized**: There is a separation between the positive and neg-ative charge. The positive side of the foil is attracted to the comb, while the negative side is repelled. However, since the strength of the electric force decreases rapidly with distance between charges, the attractive force is stronger, and the ball swings to the right.

If you then remove the foil ball from the string and bring the charged comb near the string, you will find the string itself—an *insulator*—

Figure 14.10 The nega-tively charged comb attracts the uncharged conductor because the conductor's charge gets polarized: posi-tive charge nearer the nega-tively charged comb.

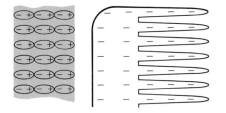

Figure 14.11 If the charged comb approaches an insulator, the comb's negative charge attracts the positive charge of each molecule and repels the negative. The individual molecules polarize, giving a net positive charge closer to the comb, a net negative charge on the far side. If the comb were positively charged, the insulator would polarize the other way, still giving a net attraction.

attracted to the comb. The reason is that the charge in each molecule of the string behaves as the charge on the conducting ball. That is, the charge within a molecule shifts so that the *molecule is polarized.* As Figure 14.11 shows, all the molecules near the comb get polarized and line up with their positive ends toward the negatively charged comb. The interior charges cancel each other, leaving a net positive charge on the right, a net negative charge on the left. Thus, the insulating string *as a whole* is polarized, with the positive side closer to the negative charge of the comb, giving a net attraction.

The *polarization* of insulating materials accounts for the attraction of uncharged bodies—insulators or conductors—to charged bodies. For some materials—water being a prominent example—the molecules are permanently polarized: The center of the positive charge is separated from the center of the negative charge. When this type of material is placed near a net charge, the molecules merely rotate into alignment, as in Figure 14.11. The overall result is the same whether the polarization of the molecules is permanent or induced. (See Questions 15 and 16.)

· ·

APPLICATION   **XEROGRAPHY**

In the mid-1930s Chester Carlson invented a process that many of us now take to be one of the necessities of life. This process, called *xerography,* is the dry-copying method used in the Xerox machine and in similar electrostatic copying machines.

In addition to electrostatic principles, such a machine uses one other process we have not yet talked about: the *photoelectric effect.* We will consider this effect in more detail in Chapter 23. For now, we'll just point out that when light strikes certain materials—for example, selenium—electrons are given off. Thus, those parts of a piece of negatively charged selenium exposed to light would lose their charge; neutral selenium exposed to light would become positively charged because of the loss of electrons.

In a typical electrostatic copying machine, static charge is applied uniformly to a selenium-coated drum. The image of the writing to be copied is focused onto the drum, lighting the entire drum *except* for the position of the image. The light removes the charge, leaving only the image on the drum charged. This image then passes near a black dust called toner, whose particles get polarized and attracted to the image. A precharged sheet of paper then contacts the drum and attracts the toner to the paper, where it is later heated briefly and fused to the paper. The mechanism for carrying out these few steps is quite complicated, as you can see by looking inside one of these machines.

Laser printers used with computers operate on a similar basis. A narrow light beam from a laser charges those portions of the drum that the beam shines on, by the ejection of electrons. The beam, under computer control, is turned on and off as well as scanned back and forth to form charged images of the letters. The remaining processes of picking up toner and depositing it on paper are similar to those in the electrostatic copier.

14.6 COULOMB'S LAW OF FORCE BETWEEN CHARGES

In the late eighteenth century Charles Coulomb, the scientist for whom the unit of charge was named, found that the law governing the force between charges has the same form as that governing the gravitational force between masses. Figure 14.12 illustrates how this force varies with the size, sign, and distance of separation between two charges. As Coulomb found, the strength of the force is proportional to the amount of each charge, but is inversely proportional to the square of their distance of separation. Mathematically,

$$F \propto \frac{q_1 q_2}{d^2}$$

where q_1 and q_2 are the amounts of each charge, d is their distance of separation, and "$\propto$" means "proportional to." Putting in a proportionality constant, we have

$$F = k \frac{q_1 q_2}{d^2}$$

where k has the value 9.0×10^9 N·m²/C² in the SI system of units. Thus, to use this equation in a calculation, we must express q in coulombs and d in meters, giving the force in newtons.

The force equation shows that doubling either of the charges doubles the force, but doubling their separation reduces the force to $\frac{1}{4}$ its original strength, as Figure 14.12 shows. Physicists call this law of force between charges *Coulomb's law.* You get a positive value for F if q_1 and q_2 are either both positive or both negative. If only one is negative, the force is negative. Thus, we interpret a negative force to mean attraction, a positive force, repulsion. Newton's action-reaction law tells us that the

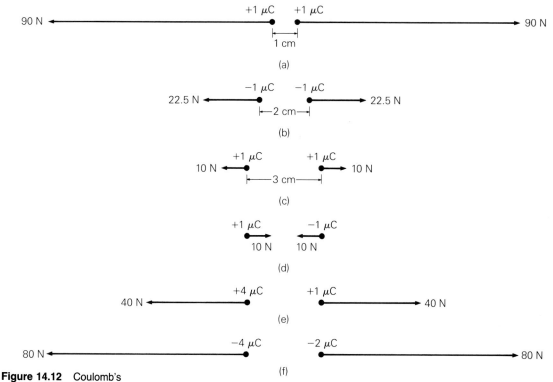

Figure 14.12 Coulomb's law describes the force between two charges. Each line (a through f) represents either a different pair of charges or a different separation. The length of the force arrows is proportional to the strength of the force (1 μC = 10^{-6} coulombs).

force on each charge must be the same strength, since these forces are an action-reaction pair. Coulomb's law is another of nature's "inverse-square" laws, in which a quantity varies as the inverse of the square of a distance.

· ·

EXAMPLE 14.1 Verify the force shown in Figure 14.12 when the separation between two positive 1-μC charges is 3 cm.
From Coulomb's law,

$$F = k \frac{q_1 q_2}{d^2}$$

$$= \frac{(9.0 \times 10^9 \ \text{N·m}^2/\text{C}^2)(1 \times 10^{-6}\text{C})(1.0 \times 10^{-6}\text{C})}{(0.03 \ \text{m})^2} = 10 \ \text{N}$$

· ·

EXERCISE If you were standing an arm's length (about 1 m) from a friend and 1% of your 10^{28} electrons suddenly jumped from you to your friend, what force then would act on each of you? What is its direction? Compare

(a)

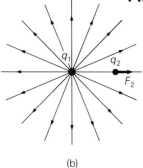

(b)

Figure 14.13 (a) The charge q_1 produces an electric field that surrounds it. (b) When the charge q_2 is in this field, a force acts on it.

(a)

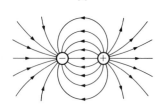

(b)

Figure 14.14 The combined field of (a) two positive charges; (b) a positive and a negative charge.

this force with the weight on earth of an extremely dense rock with mass equal to that of the moon (7.4×10^{22} kg).

Answer: 2.3×10^{24} N; attractive; electrical force 3 times as strong

Actually, the answer to this exercise is only a rough approximation. Because the charges spread out over the two bodies, not all of the negative charges are 1 m from the positive ones. Furthermore, the size of the force shows that this charge transfer is a *very* unlikely event.

14.7 ELECTRIC FIELDS

Look again at a pair of the charges in Fig. 14.5. How can we explain the interaction between them? They are not physically in contact the way your hand is in contact with a door when you push on it. Yet one charge pushes or pulls on the other. The interaction is sometimes called "action-at-a-distance," but we usually picture it differently.

Consider one of the pairs of positive charges and call the one on the left q_1, the one on the right q_2. There is certainly something at the position of q_2 that causes the force on it. That something we call an **electric field**. We picture the interaction as follows: (1) an electric field results from, and surrounds, charge q_1; (2) when in this field, charge q_2 experiences a force.

We represent the fields surrounding q_1 by lines outward from the charge, as in Figure 14.13(a). Figure 14.13(b) then shows the force on charge q_2 when it is placed in the field. The arrows on the field lines outward from the positive q_1 represent the direction of force on the positive charge q_2. If q_1 were negative, the force on q_2 would be toward q_1. That means the electric field is directed *toward* a negative charge and *away* from a positive charge. If q_1 were positive but q_2 negative, the force on q_2 would again be toward q_1. But since the field of positive q_1 is outward, we conclude that the force on a negative charge is in the opposite direction of the field.

We can understand the force on charge q_1 by similar reasoning. It results from the electric field surrounding charge q_2.

Summary

(1) The electric field is outward from a positive charge, inward toward a negative charge. (2) A positive charge in an electric field experiences a force in the same direction as the field, a negative charge experiences a force in the opposite direction to the field. The wider spacing of the field lines with increasing distance from the charge causing them symbolizes that the field gets weaker with increasing distance from the charge.

Figure 14.13(b) shows only the field of one of the charges. The overall field is a combination of that due to both charges. Figure 14.14(a) shows the combined field of two equal positive charges. If both were negative, the line shapes would be the same, but the arrows would head the other way. Figure 14.14(b) shows the combined field if the charges have opposite signs. Two equal but opposite charges separated by a small distance make up an electric *dipole*. When we study magnets, we'll define a magnetic dipole, and we'll want to look back at this figure then. A third charge in the vicinity of these would experience a force in the same or opposite direction as the combined field, depending on whether this third charge is positive or negative.

. . . .

Key Concepts

Electric charge exists in two types: **positive** charge on protons, and **negative** charge on electrons. If a particle or body has equal amounts of positive and negative charge, we say it is **uncharged**, or that it is electrically **neutral**. An atom with a net positive or negative charge is an **ion**.

Charge is measured in **coulombs** (C). The amount of charge on the electron or proton—the smallest quantity of free charge—is 1.60×10^{-19} C. The principle of **conservation of charge** states that a *net* charge cannot be created or destroyed—the total charge in an isolated system remains constant.

A **conductor** is a material with charges free to move. An **insulator** is a material that conducts poorly.

Contact electrification occurs during contact between two different substances and results from the different electron affinities of the two substances. An atom, molecule, or body is **polarized** when the centers of the positive and negative charges are separated from each other. Polarization causes the attraction between a charged and an uncharged body.

A *force* exists between any two charges. Like charges repel each other; unlike charges attract each other. The forces on the separate charges are an action-reaction pair. The force between two charges is given by *Coulomb's law*:

$$F = k\, \frac{q_1 q_2}{d^2}$$

Surrounding every charge is an **electric field**. A positive charge in an electric field experiences a force in the same direction as the field; a negative charge experiences a force in the opposite direction.

Questions

Atomic Structure and Charge

1. The distance from the center of a hydrogen atom to the most probable location of the electron is about 100,000 times the radius of its nucleus, which is a single proton. If you built a scale model of the hydrogen atom using a golf ball as the nucleus, how far away would you need to locate the electron?

2. If you and your chair are made of atoms, and atoms are mostly empty space, why don't you pass right through the chair when you sit on it?

3. What does it mean to say that an atom is uncharged? What does it mean to say that a baseball is uncharged?

4. A chemistry graduate student claims that he has an ionic solution in which each ion has a charge of 2.4×10^{-19} C. Is there a reason that you should doubt the accuracy of his claim?

5. Particle-physics theories postulate the existence of several kinds of quarks that have been given interesting names. These include: "up" quarks with charge $+\frac{2}{3}e$; "down" quarks with charge $-\frac{1}{3}e$; "strange" quarks with charge $-\frac{1}{3}e$; "charmed" quarks with charge $+\frac{2}{3}e$; and "bottom" quarks with charge $-\frac{1}{3}e$. (Here e is the amount of charge on the electron.) In these theories, protons and neutrons are thought to be made of three quarks. What combinations of three up quarks and down quarks give the right charge to each of these "ordinary" particles? (The other quarks are needed for particles not quite so ordinary.)

6. Justify the statement in the chapter that, aside from the gravitational force attracting you and the objects around you toward the earth, every force that you experience directly is electrical in nature.

Conductors and Insulators

7. True or false: Insulators conduct absolutely no charge when an electric field is present in them.

8. Name at least two purposes for insulation on an extension cord.

Contact Electrification

9. Dust particles in the air usually have an electric charge they have picked up by rubbing on other objects or from ions in the air. What does this have to do with the fact that dust will collect on vertical walls? When dust settles on a table, why do you have to wipe it off to get rid of it; why can't you just blow it off since it's so light?

10. Why is dust so hard to remove from phonograph records?

11. Blowing sand or snow on the plains sometimes charges wire fences enough to knock down a man or animal who touches such a fence. How does the fence get this charge?

12. You might have noticed sparks flying between your pajamas and the sheets when you turn over on a cold, dry night. How do you explain this event?

13. Smooth well-waxed floors are preferred in hospital operating rooms to help keep down the spread of infection. Smooth well-waxed floors are the kind you can easily pick up an electrostatic charge from. Do you see a problem here for hospitals and their patients?

Attraction between Charged and Uncharged Bodies—Polarization

14. Would you guess that electrostatic effects have anything to do with the "stickiness" of clear plastic food wrapping?

15. If you rub a balloon on your hair or sweater and touch the rubbed spot to a wall, the balloon stays there. Why?

16. If you run a comb through your dry hair on a dry day and then hold the comb near (not in) a thin stream of water from a faucet, the stream will "lean" toward the comb. Why? Why will the comb also pick up bits of paper?

17. When you make several copies on a Xerox machine, or other electrostatic copier, why are the pages sometimes hard to stack on top of each other immediately afterwards?

Coulomb's Law

18. Read the situation described in Problem 5 in this chapter. If the balls have the same mass, could you tell from the way they hang whether or not they have the same amount of charge?

If one has twice the charge of the other, how would the forces on the two balls compare?

Electric Fields

19. People sometimes incorrectly try to associate a force with the field of a single charge. How does this attempt compare with trying to be a participant in a kiss when you are the only person in the room?

20. Compare the effect of the electric field of one charge on another charge with the effect of the earth's gravitational field on the mass of a coin held in your hand. What happens if you release the coin? What would happen if the second of the two charges were released?

21. Arrows on the electric field lines tell you its direction. How can you tell from a diagram of the field what its relative strength is at different locations?

Problems

1. How many electrons would you have to remove from a sheet of paper to give it a net charge of 50 μC*?

    Answer: 3.1 × 10¹⁴

2. A charge of 54,000 C passes some point in the heating element of an electric heater in 1 h. How many electrons make up this charge?

    Answer: 3.3 × 10²³

3. What is the force of attraction between the electron and proton in a hydrogen atom when the electron is 53 pm* from the proton? Compare your result with the answer to Problem 4.

    Answer: 8.2 × 10⁻⁸ N

4. What is the force of repulsion from the electrical force between two protons 1.5 fm* apart in a nucleus? Compare your result with the answer to Problem 3. With the much stronger forces acting inside a nucleus than between the nucleus and electrons, do you get a hint as to why we can get much more energy from a kilogram of nuclear fuel than from a kilogram of coal?

    Answer: 100 N (23 lb)

5. A physics instructor sprays two small styrofoam balls with conducting paint and hangs them by 50-cm strings from the same hook. Using a charged glass rod, she gives a charge to the balls so that they hang 10 cm apart as in Figure 14.15. If the mass of the balls is 2 g, the

Metric prefixes are identified in Appendix III.

instructor can use the angle of the string to calculate that the force holding the balls apart must be 0.002 N. How much charge is on each ball if they have equal charge?

Answer: 47 nC

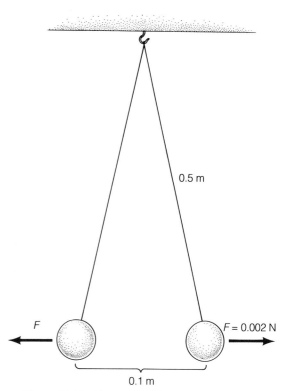

Figure 14.15 Two 2-g styrofoam balls charged with a glass rod might hang as shown.

6. The mass of the earth is 6.0×10^{24} kg; that of the moon is 7.4×10^{22} kg. How much charge would have to be added to both the earth and moon, in equal amounts, to overcome the force of the gravitational attraction? Assume there is one electron for every 4×10^{-27} kg of moon mass (slightly more than an average of one neutron and one proton per electron). What *fraction* of the moon's electrons would have to be removed to give this amount of charge?

Answer: 5.7×10^{13} C; 1.9×10^{-17}

Home Experiments

1. Try the three experiments mentioned in Questions 15 and 16. Try combs of different materials and test them under different conditions of humidity.

2. Salt and pepper unmixer: Sprinkle some salt and ground pepper on the table together. How can you get them unmixed? Rub a plastic spoon on your sweater or your hair, and hold it just above the mixture. Both salt and pepper will be attracted, but the pepper, being lighter, will jump up to the spoon.

3. Charge a plastic spoon as in Experiment 2 and hold it above a bowl of puffed rice. The rice will first be attracted to the spoon, and then will fly off in all directions. Why?

References for Further Reading

Garr, D., "Electronic Air Cleaners: What They Can Do For You," *Popular Science*, Sept. 1972, p. 58. The workings and advantages of electrostatic air filters.

Halliday, D., and R. Resnick, *Fundamentals of Physics*, Second Ed. (John Wiley & Sons, New York, 1981). If your mathematical abilities include calculus, you will find that this book contains an excellent introduction to electricity.

Moore, A. D., "Electrostatics," *Scientific American*, March 1972, p. 47. Some modern applications of electrostatics.

O'Dwyer, J., *College Physics*, Second Ed. (Wadsworth Publishing Company, Belmont, California, 1984). An example of the many good books that discuss basic electricity using techniques somewhat more mathematical than in this text, but nothing more advanced than algebra and trigonometry.

ELECTRIC CURRENT

If you have ever canoed through the falls of a river, you were certainly aware of the water current carrying you along with the flow. The word *current* that we use to describe the flow of water in the river is the same one we use to describe the flow of charge in a conductor.

In the preceding chapter we looked at the properties of electric charge and at the nature of the force between charges, independently of whether or not the charges were moving. Now we need to consider moving charges—what it takes to make them move, how we describe and measure their motion, what limits their motion, and how we quantify the energy interchanges associated with this motion.

As mentioned in Chapter 14, most practical uses of electricity depend on the motion of charges. Thus, we can now use some familiar practical devices to illustrate our points, and we can define the electrical terms that most of us have heard or read on appliances.

15.1 ELECTRIC CURRENT: CHARGES IN MOTION

If we have some way of producing and maintaining an electric field in a conductor that has charges free to move in this field, we can have a continuous flow of charge. One device that can maintain an electric field is a battery (Fig. 15.1). In Chapter 20 we'll go into the details of how a

Figure 15.1

Electron flow   Electron flow

+   −

Battery

Figure 15.2 The electric field set up by the charges on the battery terminals causes a flow of electrons through the wire and bulb from the negative to positive terminal.

battery uses chemical reactions to maintain one of its terminals—points for connection—more positively charged than the other. That charge difference causes an electric field outside the battery, which is generally directed from the "positive" to the "negative" terminal.

Suppose we connect a wire from the negative terminal of a battery to one side of a bulb filament and connect a wire from the positive terminal to the other side of the filament, as shown in Figure 15.2. The electric field acting on the free electrons in the wire causes them to move through the wire and filament in the direction shown.

This flow of charge we call an electric **current**. The term *current* means the same here as it does when used elsewhere, such as water currents in a river or air currents that we call wind. Sometimes we get careless and talk about a "flow of current." Since this description would literally mean a "flow of flow of charge," it's probably better to leave off the word *flow* when using the word *current*.

More specifically, electric current means *the rate of flow of charge*. In Figure 15.3, if charge is flowing in a section of the wire as shown by the arrows, the amount of charge passing through some surface (such as the plane shown) per unit of time is the current. We usually use the symbol I for current, Then,

$$\text{current} = \frac{\text{charge}}{\text{time}}$$

or

$$I = \frac{q}{t}$$

where q is the amount of charge passing some point in time t. Expressing the charge in coulombs, and the time in seconds, gives current in coulombs/second (C/s). We call this unit the **ampere**:

1 ampere (A) = 1 coulomb/second(C/s)

In conversation, we usually shorten the word *ampere* to "amp." Sometimes the current itself is called "amperage."

. . . .

EXAMPLE 15.1   If a charge of 20 C enters and leaves the bulb filament in Figure 15.2 in 1 min, what is the current?

Converting 1 min to seconds, we get

$$I = \frac{q}{t} = \frac{20\text{ C}}{60\text{ s}} = 0.33\text{ C/s} = 0.33\text{ A}$$

EXAMPLE 15.2   How many electrons pass a point in the filament of a flashlight bulb in 1 h if it carries a current of 0.20 A?

Since current is the charge flowing per unit of time—that is, the charge divided by the time—the total charge flowing is the current multiplied by the time:

$$q = I \times t = 0.20 \text{ A} \times 3600 \text{ s} = 720 \text{ C}$$

Since 6.25×10^{18} electron charges make a coulomb,

$$\text{no. electrons} = 720 \text{ C} \times 6.25 \times 10^{18} \text{ electrons/C}$$

$$= 4.5 \times 10^{21} \text{ electrons}$$

Figure 15.3 The current is the amount of charge passing through some area in each unit of time.

Section of a wire

q

q

Direction of Current

I'm now going to ask you to accept something that may at first seem quite illogical. After a while you'll get used to the idea. The flow of negative charge in one direction is completely equivalent in external effect to the flow of positive charge in the opposite direction. If the charges moving to the right in Figure 15.3 are electrons, this would have the same effect as *positively* charged electrons moving to the left. It is customary to refer to the direction of current as the direction of flow of *positive* charge. We call this *conventional current*. Then, if only electrons are flowing, as in a metal conductor, the conventional current is in the direction opposite to electron flow. In Figure 15.2, even though electrons are flowing from right to left through the bulb, we say that the *current is from *left* to *right*—from the positive to the negative battery terminals.

15.2 ELECTRICAL POTENTIAL ENERGY AND POTENTIAL DIFFERENCE

When we studied mechanics, we found that we could learn a great deal about the motion of a body (or group of bodies) by studying energy changes, without directly knowing anything about the forces on the bodies. We can do something similar in studying electricity. Knowing the exact electric field and the resulting forces on the charges in Figure 15.2, for example, is extremely difficult. For practical purposes, we don't need to know them anyway; we only need to know the energy changes involved.

Work done in moving
charge = 12 joules

q = 2 C
leaves

q = 2 C
enters

Lantern
battery

6 volts

Figure 15.4 The electric field gives up energy in doing the work of moving the charge.

Potential Energy

Suppose an amount of charge q leaves the positive terminal of the battery in Figure 15.4, and an equal amount of charge enters the negative terminal. (We'll talk in terms of movement of positive charge, even though negative charge actually moves in the opposite direction here.) The electric field does work in moving this amount of positive charge. This work must come from *some* form of energy. In this case, it comes from the potential energy the charge has in the field. *The change in electrical potential energy equals the work done in moving the charge.* Inside the battery, the same amount of charge moves from the negative to the positive terminal. The chemical energy of the battery does the work of moving the charge from negative to positive.

This flow of charge is similar in principle to a toy many young children enjoy (Fig. 15.5). The child acts as the "battery" that increases the gravitational potential energy of the marble, which loses its potential energy in rolling down the track to the floor (where its anxious admirer quickly replenishes its supply). In the electrical case, the charge is "raised" to the higher potential energy by the battery and then loses its potential energy in "falling" through the wire and bulb.

Potential Difference

The battery maintains its ability to move charge whether or not it is actually doing so. In rating this "ability," we need a measure that is independent of the amount of charge being moved. The quantity we use

Figure 15.5 Marshall is the "battery" that lifts the marble to the top of the zigzag track, where it tumbles to the floor as it loses potential energy.

is called *electrical potential difference*. We define the electrical **potential difference** as *the difference in potential energy per unit of charge moved.* That is,

$$\text{electrical potential difference} = \frac{\text{electrical potential energy difference}}{\text{charge moved}}$$

If, for example, charge q in Figure 15.4 is 2 C and the work done in moving it from "+" to "−" through the wire and bulb is 12 J, then the potential difference between battery terminals is:

$$\text{potential difference} = \frac{12 \text{ J}}{2 \text{ C}} = 6 \text{ J/C}$$

If q were 6 C, the work needed would be 36 J and the potential difference still would be 6 J/C. In other words, the potential difference is independent of how much, if any, charge actually moves.

The unit of potential difference is the volt (V). That is,

1 volt (V) = 1 joule/coulomb (J/C)

Because this unit is universally used, potential difference is often called **voltage**. The symbol we use for voltage is V. (Voltage is one of the few quantities for which the symbol used for that quantity and the symbol for its unit are the same letter.)

What is the significance of the word *difference* in the term *potential difference*? To answer, let's compare electrical and gravitational energy. You know that when you say the gravitational potential energy of an apple is 3 J, you mean relative to some arbitrarily chosen zero of potential energy. You are really talking about a gravitational potential energy *difference* between the point in question and the point you've called zero. The same holds true for electrical potential energy and electrical potential. You can talk about the actual value of the potential at a point only if you assign it a value of zero somewhere.

Consider two examples. A flashlight battery "has" a voltage of 1.5 V, meaning that there is a potential difference of 1.5 V between the positive and negative terminals. We can say that the positive terminal is *at* 1.5 V only if we agree to say that the negative one is at 0 V.

A second example is that of commercial electricity. Ordinary light bulbs, for example, operate on a voltage of 110 V, meaning that the bulb needs a potential difference of 110 V across the filament to operate normally. We often consider the earth (ground) to be at 0 V. Then there is 110 V of potential difference between a 110-V power line and the earth. The term **ground**, when used in electricity, means whatever region is being defined as having *zero potential*. This region is often, *but not always*, the earth.

When discussing the voltage between two points, we often label the point at the higher voltage "+" and the one at the lower voltage "−", even though they may not be positive or negative relative to ground. Then a positive charge, if free, would try to move from + to −, a negative charge from − to +.

People use several different terms in talking about potential difference. We might refer to the voltage "across" the power lines, the voltage "supplied to" the TV, the voltage that a battery "has," or the voltage that a calculator "operates on." We never say "voltage through"—it is the current that goes *through* the device. In every case, whatever the term used, when we talk about voltage, we mean the potential difference *between* one point and another. If it's used concerning some electrically operated device, we usually mean the potential difference needed between the two connections on the device to make it operate normally.

15.3 ELECTRICAL CIRCUITS

So far, we have been cutting out individual pieces of the electricity puzzle. Now let's try to put some of these pieces together and get an idea what the picture looks like—at least that part of the picture that shows currents.

Three basic ingredients are needed for the flow of electrical charge—a current—to do some practical job:

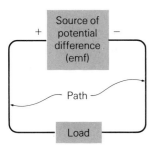

Figure 15.6 For a current to exist, we need a source of potential difference and a completely closed path.

1. A source of potential difference, or voltage.

2. A complete closed conducting path through which the charge can flow in response to this voltage.

3. A load—the device that does a job as the current passes through it.

Figure 15.6 shows these ingredients schematically. We'll look at each component separately.

Sources of Potential Difference—emf

Maintaining a steady current through a conductor requires some device for keeping a steady potential difference across the conductor. A potential difference produced by such a device historically has been called an "electromotive force." Since it isn't really a force at all, but a voltage, we now usually call it by its initials—**emf**. We've already talked about one source of emf, the battery. Generators and solar cells are among the other sources of emf that we'll consider in detail in later chapters.

Basically, a source of emf is a source of electrical energy. It converts some other form of energy—chemical energy in the case of a battery—into electrical energy. Being a particular potential difference, the emf of the source is, from the definition of potential difference, the work done on a unit of charge as it passes through the source:

$$emf = \frac{work}{charge}$$

Turning this around, we see that if some known charge moves through a particular emf, the work done on it is:

$$work = emf \times charge$$

. . . .

EXAMPLE 15.3

How much work is done by one of the 1.5-V flashlight batteries of Example 15.2 during 1 h of operation?

$$work = emf \times charge$$
$$= 1.5 \text{ V} \times 720 \text{ C} = 1100 \text{ J}$$

(a)

(b)

Figure 15.7 (a) Charge cannot flow when the conducting path has a gap. (b) Charge can flow when the circuit is complete.

This is about the amount of mechanical work a 60-kg (132-lb) person would perform in doing six push-ups.

The Complete Circuit

Suppose we have the situation shown in Figure 15.7(a)—a circuit containing a battery, a light bulb, and a conducting path. The conducting path contains a gap across which no charge can flow. Even though the battery, the conductor, and the bulb filament all contain free electric charges, no current can exist anywhere in the circuit because it cannot get across the gap where there is no free charge to move. Repulsion between like charges doesn't let charge build up at some point in the circuit; if it cannot flow in one part of the circuit, it cannot flow anywhere.

However, for the situation of Figure 15.7(b) there is no gap, and the charge can move freely through all parts of the circuit. Once the circuit is continuous and contains an emf for producing a potential difference in the circuit, an electric field propagates at nearly the speed of light throughout the circuit. This electric field acts on the free charge in the conductors, setting up a current almost instantaneously.

The situation is similar to the one shown in Figure 15.8. Here, we have a pump and a pipe to carry water to a waterwheel, and then from the other side of the waterwheel back to the pump. The pump continually produces a difference in pressure. Figure 15.8 shows the higher pressure by a postive sign to the left of the pump. This means that the pump is trying to push water around the circuit in a clockwise direction, but when the valve is closed, there is no path for the water to get through that part of the circuit. Water is still in the pipe on both sides of the valve, still in the paddlewheel, and still in the pump. The pump is still producing the pressure difference, but the water is unable to flow until

Figure 15.8 The flow of charge in a conducting circuit can be compared with the flow of water in a "water circuit."

the valve is opened. Once the valve is open, water flows throughout all parts of the circuit.

If we compare this arrangement with our electrical circuit, we see that the pump producing the pressure difference is analogous to the battery that produces the electrical potential difference. Both are produced whether or not the circuit is complete. The flow of water in a pipe is like the flow of charge—the current—in the electrical circuit. The charge (water) is waiting to flow under the action of the potential (pressure) difference once the circuit is complete. The valve in the water circuit and the gap in the electrical circuit perform the same functions. Charge (water) cannot get across the gap (closed valve) in the conductor. Turning an electrical switch on means closing a gap in the circuit so that charge can flow. The closed switch is like the open valve in the water circuit.

This analogy breaks down in one place. Water flows because the water molecules on the high pressure side of the pump collide with those in front of them, pushing them as a locomotive would push a group of freight cars. But the electric charge is not bumped along in this manner. Instead, the electric field that propagates around the circuit acts on each charge, causing it to move.

The Load

The **load** in a circuit can be a light bulb, a motor, a heater, or any device that operates on electrical energy supplied by a current. A circuit may have more than one load, as we'll see later in the chapter. Whatever the load, it uses the energy given to the charges set in motion by the electric field in the conductor.

One type of load is called a **resistance**. It acts much like a friction that impedes the flow of charge. Incandescent light bulbs and electric heaters are examples of loads that dissipate electrical energy by means of their resistance. We will consider resistance now, and other types of loads in later chapters.

15.4 RESISTANCE AND OHM'S LAW

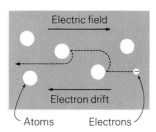

Figure 15.9 Electrons heat the conductor as they collide with its atoms, which get in the electrons' path.

We have said that charge moves freely through a conductor whenever there is a potential difference across it. But we haven't really said what we mean by moving "freely." How free is "free"? We have talked so far as if the electrons that carry the current are sitting at rest until a potential difference causes them to move one way or another. Actually, due to their thermal motion, the electrons are moving around randomly at extremely high speeds. When a potential difference is put across the conductor, an overall drift in one direction or the other is superimposed on this random motion. When talking about currents, we are talking about this overall drift, and we can usually ignore the random motion.

As electrons carrying a current move through a conductor, they collide with atoms that get in their paths (Fig. 15.9). When they do, they transfer some of the kinetic energy they had obtained from the electric field. This transferred energy increases the thermal energy of the atoms and the electrons collided with. Thus, the conductor is heated. (The process is somewhat akin to a burglar running from the police through a crowd of people. The burglar's motion is greatly impeded by the people and, every time he collides with one of them, he transfers some of his kinetic energy, which causes an increased jiggling of all the people in the vicinity.) Each time an electron loses kinetic energy by a collision, it is reaccelerated by the electric field.

We describe this interference with the electrical current by the term *resistance*, which means just that: The atoms of the conductor get in the way and *resist* the flow or charge. Since the atoms of each different substance are put together differently, each different substance would have a different resistance for the same-shaped conductor.

How would you expect the voltage needed for a certain current in a conductor to change with its resistance? Since the voltage is the energy needed to drive each unit of charge through the conductor, you probably would expect the voltage needed to be higher, the higher the resistance. That's like saying it takes more work to push a wheelbarrow through tall dense weeds than over smooth grass. The higher resistance of the tall weeds means that more work is needed to push each wheelbarrow.

In fact, the potential difference needed is directly proportional to the resistance for a given current. So, we can write

voltage = current × resistance

or

$$V = IR$$

where the symbol R stands for resistance. We don't need a proportionality constant because the unit of resistance is defined so that this relationship holds as it is. The unit of resistance is the **ohm**, usually symbolized by Ω (the Greek omega).

Georg Ohm found by experimentation in 1826 that the resistance for most ordinary materials is almost constant, whatever the current and voltage. This fact is known as Ohm's law, although many people refer to the relationship $V = IR$ as Ohm's law. This relationship, written as $R = V/I$, is no more than the defintion of resistance. The name ohm was given to the unit of resistance in honor of Mr. Ohm.

. . . .

EXAMPLE 15.4   What is the resistance of the flashlight bulb of Examples 15.2 and 15.3? We can arrange the equation

$$V = IR$$

to give

$$R = \frac{V}{I}$$

Since two batteries are used, the total voltage across the bulb is 3.0 V. For the current of 0.20 A,

$$R = \frac{3.0 \text{ V}}{0.20 \text{ A}} = 15 \ \Omega$$

When we talked about "good" conductors and "poor" conductors in Section 14.3, we were really talking about whether the resistance is low or high. A "good" conductor is one with low resistance. But even materials such as silver, copper, and gold, which are ordinarily the best possible conductors, have some resistance.

Any electrical energy lost by the current in a resistance is converted to heat. This heating effect can be useful. It is, for example, the way ordinary light bulbs produce their light. The current passing through the filament produces heat because of the resistance of the filament. As a result, the filament gets white hot. Almost all everyday electrical heating devices—ranges, toasters, hair driers, clothes driers, percolators, and so forth—generate their heat from this *resistive heating*. A device specifically intended to add resistance to a circuit is called a *resistor*.

Resistive heating can also be a nuisance. Since all conductors have some resistance at ordinary temperatures, there is always some loss of electrical energy to heat when current passes through conductors. In transmission lines, this represents a loss of usable energy. Even batteries and other sources of emf have resistance. This internal resistance of batteries usually increases with use.

APPLICATION   SUPERCONDUCTORS

A strange things happens to some metals when their temperature is lowered to near absolute zero: Their electrical resistance drops to *zero*—no resistance at all. This circumstance means that a current set up in such a material would continue indefinitely with no additional input.

The resistance of all metals decreases as their temperature drops, mostly because less thermal agitation of the atoms causes less interference with the flow of charge. Ordinary good conductors, such as copper, continue to decrease gradually in resistance down to low temperature, as shown by the graph in Figure 15.10. But for some metals—tin, for example—the resistance suddenly drops to zero as a certain temperature is reached. This transition temperature varies from metal to metal, and as the graph shows, it comes at 3.72 K for tin. We say that the material then becomes a *superconductor*.

You can easily imagine how a material with zero resistance could have many practical uses. The transfer of electrical energy over wires of zero resistance would eliminate the loss of electrical energy in transmission. However, the problem of keeping the conductor at a temperature within a few degrees of absolute zero itself requires a considerable expenditure of energy. Nevertheless, these temperatures can be maintained by keeping the conductors immersed in liquid helium. This technique has been used for a number of years in superconducting electomagnets that maintain large currents in a fairly small region of space.

For large-scale uses of superconductors, materials that become super-conducting at much higher temperatures—ideally around room temperature—are needed. For *seventy-five years* after the discovery of superconductivity in 1911, massive efforts had raised the superconducting temperature to only about 23 K, or −250°C. However, in early 1987 a dramatic breakthrough led to a material that is superconducting at just below 100 K.

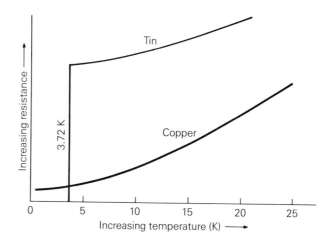

Figure 15.10 A graph of resistance versus temperature for two conductors. Tin is one of the metals that becomes a *superconductor*—its resistance drops to zero—at a few degrees above absolute zero of temperature.

It may seem no "big deal" that the temperature jumped from around 20 K to over 90 K—a blistering −175°C. The big deal is that this temperature is above the 77-K boiling point of liquid nitrogen. Thus, the superconducting temperature of such a material can be maintained by immersion in liquid nitrogen—a substance much easier and cheaper to obtain. Continued improvements in the temperature threshold, physical properties, and availability of superconducting materials will undoubtedly result in their more widespread use in technological devices.

Circuit Diagrams

To simplify drawing electrical circuits, people usually use a *schematic* diagram with standard symbols for electrical components. The conventional symbols used are as follows:

A battery or source of steady emf. The + and − signs are usually not written. You need to remember that the long line stands for the positive terminal.

A resistance.

An open switch.

A closed switch.

A conductor with zero resistance, or one whose resistance is small enough to be neglected.

Figure 15.11 A schematic diagram of the circuit of Fig. 15.7, containing a battery *B*, a switch *S*, and a resistor *R*.

Figure 15.11 is a schematic diagram, using these symbols, of the circuit of Figure 15.7(a). The schematic diagram for 15.7(b) would be the same, except that the switch would be closed (on). For most purposes, the resistance of the wires leading to a device is so much less than the resistance of the operating device itself that the wire resistance can be neglected. For example, the resistance of a 100-W light bulb is about 140 Ω; that of 3 m of ordinary extension cord is about 0.06 Ω.

15.5 ELECTRICAL ENERGY AND POWER

If you look at the specifications stamped or printed on an electrical device, you usually find the operating voltage and either the current *or* the power requirements listed. If you are a consumer and purchaser of electrical energy, the power a device uses (the rate of energy use) has a direct impact on your pocketbook, since it determines the energy used in a given period of time. Our next task is to find out how voltage, current, and power are related.

Remember, the potential difference, or voltage, between two points is the work done on each unit (coulomb) of charge in moving it from one of the points to the other. Therefore, the total work done, or electrical energy used, in moving some charge between these points is

work done = energy change = voltage × charge

The **power** is the rate of doing work:

$$\text{power} = \frac{\text{work}}{\text{time}} = \frac{\text{voltage} \times \text{charge}}{\text{time}}$$

But the charge flowing per unit of time is the current. Therefore,

power = voltage × current

or

$$P = VI$$

If we know the power used during a certain time t, we can obtain the total energy used during this time from

work done = energy used = power × time

$$= VIt$$

When the current moves through a battery, the potential difference is the emf of the battery. The power is then being supplied by the battery at the rate $VI = \text{emf} \times I$, since V in this case is the emf. A resistance or some device that depends on electrical energy, uses it at the rate VI, where V is the voltage across the device carrying current I. The *power* used by some devices—typically lamps, toasters, electric blankets—is marked on them in watts or kilowatts. On others—typically motors, refrigerators, radios—the *current* rating is listed. Whichever is given, you can find the other by a simple calculation.

· ·

EXAMPLE 15.5   What is the power delivered by the two flashlight batteries of Examples 15.2 through 15.4?

The voltage of the two batteries together is 3.0 V and the current delivered is 0.20 A. Then,

power = voltage × current

$$= VI = 3.0 \text{ V} \times 0.20 \text{ A} = 0.60 \text{ W}$$

EXAMPLE 15.6   How much current passes through a 100-W light bulb that operates on a potential difference of 110 V?

Rearranging the power equation by dividing both sides by V gives

$$I = \frac{P}{V} = \frac{100 \text{ W}}{110 \text{ V}} = 0.91 \text{ A}$$

A good rule of thumb for comparing different devices is to remember that a 100-watt bulb carries a current of *about* 1 amp.

QUESTION   Figure 15.11 is a circuit diagram for a flashlight. If R is the resistance of the bulb filament, how does the electrical power dissipated by the bulb filament compare with that delivered by the battery? How does it compare with the rate of use of chemical energy in the battery? How does it compare with the rate of heat plus light energy given off by the bulb?

EXERCISE   How much current passes through a 60-W bulb that operates on 110 V? At 9¢ per kilowatt-hour (kWh), how much does it cost to operate this bulb for 10 h?   *Answer*: 0.55 A; 5¢

15.6 A LOOK BACKWARD AND FORWARD

Figure 15.12 summarizes the relationship between the various quantities we have defined. In the vicinity of an electric charge there is an electric field. This field exerts a force on any other charge in the field. If this charge moves, work is done, and the potential energy changes. The potential energy change per unit of charge is called the potential difference, or voltage.

To maintain a continuous flow of charge—a current—a device is needed that keeps a potential difference between its terminals. This particular potential difference is called an emf. If a continuous path is connected across the terminals, an electric field will be set up through the conductor. This field acts on the charges, producing a current as the charges gain energy from the field. The moving charges then give up their energy to the load. If the load is a resistance, this energy transfer takes place as the moving charges collide with atoms in the resistor, transferring their energy to thermal energy (heat).

Thus, we now see how the basic quantities we have defined in this chapter and the last one fit together to produce a current in a closed

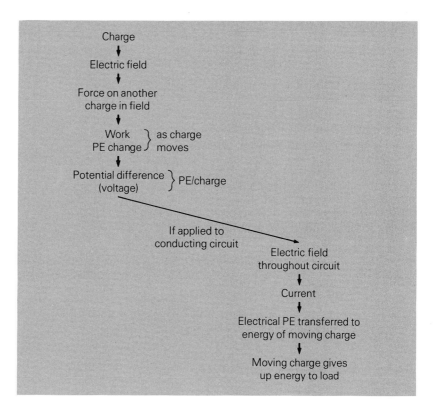

Figure 15.12 The cause-and-effect relationships between electrical quantities. *PE* stands for potential energy.

circuit. In the next section we will look at more complicated loads—namely, combinations of resistors. In the next chapter we'll see how these basic principles can be applied to doing practical jobs.

15.7 COMBINATIONS OF RESISTORS

You can plug several lamps into the same electrical outlet. Is there some special way these lamps have to be connected? What's inside that little adapter that lets you plug three lights into one outlet? Let's find out why loads have to be connected in a particular way in a circuit.

Resistance in Series

Compare the circuits of Figure 15.13(a) and Figure 15.13(b), where the only difference is an extra bulb in circuit (b). The two identical bulbs are in **series**—they are connected so that whatever current passes through one must pass through the other. How would you expect the current in circuit (b) to compare with that in circuit (a)? Since both resistances are impeding the same current, the resistance is twice as much in (b) as in

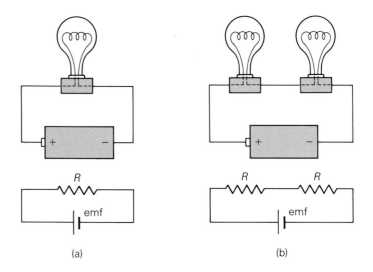

Figure 15.13 Actual and schematic diagrams of (a) a battery and one bulb; (b) a battery and two bulbs in series.

(a)   (b)

(a). Since the emf is the same for both circuits, we know from the relationship $V = IR$ that the current in (b) is $\frac{1}{2}$ that in (a).

If the resistances are not identical, you just add all of them, whatever they are, to get the total resistance.

• •

EXAMPLE 15.7   If each bulb in Figure 15.13 has a resistance of 200 Ω and the emf of the battery is 110 volts, what is the current in each circuit? What is the power used by each bulb?

First let's consider circuit (a). From $V = IR$, we have

$$I = \frac{V}{R} = \frac{110 \text{ V}}{200 \text{ Ω}} = 0.55 \text{ A}$$

From $P = VI$, the power is

$$P = 110 \text{ V} \times 0.55 \text{ A} = 60 \text{ W}$$

Now for circuit (b), the resistance is twice that in (a), or 400 Ω. The current is

$$I = \frac{110 \text{ V}}{400 \text{ Ω}} = 0.275 \text{ A}$$

The power to *both* bulbs is then

$$P = 110 \text{ V} \times 0.275 \text{ A} = 30 \text{ W}$$

or $\frac{1}{2}$ what it was to the one bulb of circuit (a). But since the bulbs are identical, the energy use is divided equally between them: 15 W to each bulb, or only $\frac{1}{4}$ the power to one bulb alone! Do you expect $\frac{1}{2}$ as much light from two bulbs as from one? Evidently, we don't connect multiple light bulbs in series when we want to maximize the light output.

A note of caution about light bulbs as resistors. If your instructor decides to demonstrate the correctness of Example 15.7 by hooking up and operating the circuits in Figure 15.13, it will probably look as if the light output of the bulbs is almost as predicted. But your eye is not a good judge of such matters. If a meter is hooked up to measure the current, don't be surprised if the current doesn't drop from (a) to (b) as much as expected. The reason is that the resistance of the bulb filament increases with increasing temperature. When the filament is at its normal operating temperature of roughly 2500°C, its resistance is about 10 times what it is at room temperature. Since the bulbs do not get as hot under the lower current of the series connection, the resistance of each bulb is less in the arrangement of circuit (b).

Furthermore, the power we are talking about is the *electrical* power, not the light output. Since the filaments do not get as hot, the fraction of the energy that goes into light is much less. The actual light output in Figure 15.13(b) is therefore much less than $\frac{1}{2}$ what it is in (a).

Resistances in Parallel

In Figure 15.14 we have added another bulb in **parallel** with the one of Figure 15.13(a). Now the current I passing through the battery has two alternative paths between point x and point y. (A parallel connection does not mean that the paths are necessarily geometrically parallel.) Do you now expect more current out of the battery, or less, than for only the one bulb of Figure 15.13(a)?

Consider an analogy. Suppose there are 50,000 people trying to get out of a stadium after a football game. If there is only one gate open, people will flow through the gate at a certain rate. If the stadium officials open another identical gate, people will pass through it at about the same rate. The total rate of flow of people from the stadium is about twice what it is with one gate open. We're assuming that people's urge to get out (their "emf") is about the same, whatever the number of gates open.

A similar effect occurs in electrical circuits. When we add path 2 in parallel, the same emf tries to push charge through the second path as through path 1. Since path 2 can carry the same current as path 1, the total current through the battery is twice as much as it would be with only one resistor.

Since more current passes through the battery with the two resistors in parallel than with one, the overall resistance of the combination is *less* than the resistance of one resistor by itself. If the two resistors have

Figure 15.14 Two identical resistors in parallel need twice the current from the battery as one resistor alone. Both the pictorial and the schematic diagrams are shown.

unequal resistance, the larger current flows through the lower resistance. (More people would leave a stadium through a larger, lower-resistance gate.) For more than two resistors, the current through each one is inversely proportional to the resistance, with the overall resistance of the combination always being lower than that of any one resistor.*

When resistances or other loads are connected in parallel across an emf, each one is independent of the others unless so much current is drawn from the source of emf that its voltage drops below normal. When we plug different devices into electrical outlets in a house, we are connecting them in parallel across the source of emf. That way, each one "sees" the full voltage supplied by the line and can operate as designed. The total current is shared by the various devices.

By comparison, for resistors in series the current is the same. The voltage, however, must be shared so that no one device experiences the full source emf.

EXERCISE   Draw a schematic diagram showing a light bulb, a toaster, a percolator, and a TV connected in parallel across a source of emf.

Water analogy.   To keep all these circuit ideas straight in your mind, it sometimes helps to compare electrical circuits with the flow of water in a water circuit. Figure 15.15 shows the water version of the three circuits in Figures 15.13 and 15.14. Recall that the current (the rate of flow of charge) is analogous to the rate of flow of water. The potential difference (voltage) of the battery is similar to the pressure difference produced across the pump that keeps the water flowing. The small wiggly pipes offer high resistance to water flow, whereas the large ones offer almost none. We'll assume all sections of wiggly pipe are the same, and therefore offer the same resistance to the water flow.

Let I represent the rate of flow of water in part (a). Then, in (b), the flow is $\frac{1}{2}I$ because the water has to go through two resistances. In (c), each resistance is connected independently across the pump, and each draws a flow of I. Then the total flow through the pump is $2I$.

15.8 ALTERNATING CURRENT

Batteries produce a steady emf in one direction. The current they cause when connected in a closed circuit is a steady current in one direction around the loop—what we call a **direct current**, or DC. This is the

*The effective resistance R of a group of parallel resistors is $\frac{1}{R} = \frac{1}{R_1} + \frac{1}{R_2} + \frac{1}{R_3} \ldots$

Figure 15.15 The current in an electrical circuit can be compared with the flow of water in a pipe. The small wiggly pipes offer resistance to water flow. All components are identical. If (a) I is the *water* current through the pump with one resistor, then (b) $\frac{1}{2}I$ is the current with two resistors in series, and (c) $2I$ is the current through the pump with two resistors in parallel.

situation we have been describing until now. But the emf produced by the generators at the power plant—and delivered to your room, your house, the place where you work, and your classroom—changes its direction 120 times per second. Therefore, the current it causes in your lights changes its direction 120 times per second—an example of what we call **alternating current**, or AC. We call such a current 60-hertz (Hz) AC, where 1 Hz is 1 cycle/s. There are 60 complete cycles each second, with current first in one direction and then in the other. Most countries other than the United States and Canada use 50-Hz AC. It is fairly common to use terms like *AC current* or *DC voltage*, even though their strict interpretation is not very logical.

These changes in direction are not abrupt but are smooth variations.

Figure 15.16 In 60-Hz alternating current, the emf and the current reverse their direction 120 times per second, varying smoothly with time as shown in this graph.

Figure 15.16 is a graph of emf versus time for AC. The part of the curve above the horizontal axis stands for emf in one direction, that below the axis, for emf in the opposite direction. We usually refer to this emf as positive when it is in one direction, and negative when it is in the other. The emf increases gradually to a maximum positive value, goes back to zero, then to a maximum negative value.

As the emf applied to a closed circuit varies according to the graph, the electric field throughout the circuit increases smoothly to a maximum first in one direction and then in the other. These changes in field propagate through the conductor at almost the speed of light. As a result, the current alternates smoothly at 60 cycle/s in proportion to the emf. A graph of current versus time would look just like Figure 15.16, where positive current is in one direction and negative in the other.

Niagara plant favors Westinghouse over Edison

BUFFALO, N.Y. (NOVEMBER 8, 1891)—The Cataract Construction Company has decided to use alternating-current, rather than direct-current, generators for the large hydroelectric plant they are to build at Niagara Falls to supply electricity to Buffalo. Even though great controversy has surrounded this decision, if it proves successful, it may set the pattern for future plants.

Thomas Edison has strongly opposed the use of AC rather than DC for such a purpose—a highly understandable attitude, since Edison Electric Light Company has invested vast sums of money in DC plants for many large cities. Edison's argument for DC is the following: Since electricity is used only for lighting, a need that exists just a few hours a day, the plant could store in batteries energy generated during the remainder of the day for use during peak-demand hours. No such storage device is available for AC-generated energy.

However, the Westinghouse Electric Company, under the leadership of George Westinghouse, has been building AC generating systems for about five years. Since any generator actually produces AC, Westinghouse's generators do not have the complicated devices called commutators that convert AC to DC at the generator output. Furthermore, Westinghouse has shown that with AC they can use a transformer to step up the voltage for transmission cross-country, and then step it down again on the user end. This voltage change, which is not possible with DC, substantially reduces electrical energy lost to heat along the lines.

Only time will tell if this decision was the correct one.

The statements we have made about circuits and about combining resistances apply equally well whether we are talking about AC or DC. The details of the design of some pieces of equipment—for example, motors that run hair driers and fans, and electronic devices such as TV sets—depend on whether the device in question is to operate on AC or DC. But light bulbs and electric heaters, which depend only on resistance, don't "care" whether they receive AC or DC.

When we specify the amount of alternating current or voltage, we are giving the *effective* value—the AC that has the same heating effect as an equal amount of DC. If Figure 15.16 shows the voltage variation for 110-V AC, the voltage actually fluctuates smoothly between +155 V and −155 V. These values illustrate the general situation that the effective value of voltage or current is 0.707 (actually $1/\sqrt{2}$) times the *peak* (highest) voltage or current during each cycle.

• •

Key Concepts

Electric **current** is the rate of flow of charge. It is measured in **amperes** (A). A steady current requires: (1) a source of potential difference—an **emf**; and (2) a continuous conducting path across the potential difference. There is usually a **load** in the circuit—a device for using the energy associated with the current.

The **resistance**, current, and voltage across a conductor are related by: voltage = current × resistance; or $V = IR$. Resistance is measured in **ohms** (Ω). All electrical energy dissipated in a resistance is converted to heat.

Electrical **power** is given by: power = voltage × current; or $P = VI$. This is the power delivered *by* a source with emf V, or the power delivered *to* a load across which there is potential difference V. The total energy delivered or used in time t is power × time or VIt.

When two or more circuit elements are connected in *series*, whatever current passes through one must pass through them all. The total resistance of resistors in series is the sum of the individual resistances.

When circuit elements are connected in *parallel*, the current divides so that the sum of the currents to and from the individual components equals the current leading to the parallel combination. In parallel resistors, the current to the various branches divides in inverse proportion to the resistances in the branches.

In 60-Hz *alternating current* (AC), the current and voltage smoothly reverse directions 120 times per second. The effective current and voltage are 0.707 times their peak values.

Questions

Current

1. Does the amount of current in a wire have anything to do with how fast the charges are moving? Hint: Is the amount of water passing per minute under a bridge any different for a bridge where the river is narrow and shallow, than for one where the same river is wide and deep?

2. If you connect a copper wire from the positive terminal of a battery to a bulb, and another wire from the bulb back to the negative battery terminal, what is the direction of electron flow in the bulb? What is the direction of the conventional current?

3. Which is a better analogy for electrons carrying current in a conductor: people standing on a moving conveyor belt or people dancing to fast music on a moving conveyor belt?

Potential Energy and Potential Difference

4. When a mass gets closer to the earth, its gravitational potential energy always decreases. When a small charge gets closer to a large one, does its electrical potential energy always decrease?

5. When an electron moves from the negative terminal of a battery, through a light bulb, and to the positive terminal, does its potential energy increase or decrease?

6. Is an emf a force? Why do you suppose it was called "electromotive force" historically?

7. The potential difference, or voltage, of a battery is the work/charge to move a charge between its terminals. Does this mean that the potential difference across the terminals of a 9-V battery is actually zero if the battery is disconnected so that it is not moving any charge?

Circuits

8. Is it necessary to have a load in order to have current through a circuit?

9. In what way does the analogy of opening a door to let people out of an auditorium fit the turning on of a light switch? In what way is this comparison bad?

10. In what way does a water faucet perform the same type of function as an electrical switch on the wall?

11. If a pump is used to pump water through a pipe that forms a closed circuit, the amount of water flowing is proportional to the pressure difference across the pump. In the analogous electrical circuit, which quantity—voltage or current—is analogous to pressure difference, and which to the rate of flow of water?

12. Why is it that on some sets of Christmas tree lights, if one bulb burns out, they all go out?

Resistance

13. Is there any similarity between the use of the word *resistance* in connection with electrical circuits and its use in other areas? (The suspect offered no "resistance" when arrested by the police.)

14. The resistance of a metal conductor comes partly from the fact that electrons moving through it collide occasionally with atoms that get in the way. Since these atoms vibrate farther from their equilibrium positions as the temperature of the metal increases, how would you expect the resistance of a metal to change with temperature?

15. A 200-W bulb that operates on 110 V needs twice the current that a 100-W bulb used on the same voltage. How do the resistances compare for the two bulbs?

16. What is the resistance of a light switch that is off (open)?

17. One method used to turn off automatic clothes driers when the clothes are dry is to sense the resistance to current through the clothes as they dry. The resistance increases with dryness. What minimum number of sensing points must be inside such a dryer?

18. A container of powdered carbon changes its electrical resistance as the pressure on the carbon changes. Can you speculate how this process is used in a telephone microphone to convert pressure variations in a sound wave to current variations in the telephone wire?

19. The internal resistance of a car battery increases with decreasing temperature, mostly because ions travel much slower through the battery solution when it's cold. What is one reason cars are much harder to start on a cold morning?

Energy and Power

20. True or false: A 5000-W water heater uses twice the energy that a 2500-W heater does.

21. In some European countries, the voltage on standard power lines is 220 V. What is the cur-

rent supplied to a 100-W bulb operated on this voltage if it takes 0.9 A to operate a bulb of the same power on 110 V?

22. A portable electrical heater has two settings: 1000 W and 1500 W. Is there any difference in the total amount of energy needed to keep a room's temperature near 20°C with this heater when you use its different settings?

23. Batteries are often rated in *ampere-hours*. Show that this rating, along with the rated voltage, tells you the total energy that can be supplied by the battery without recharging. Does it tell you anything about the number of months the battery will remain good?

Combinations of Circuit Elements

24. If you're hooking up two baseboard electric heaters and you make a mistake and wire them in series when they should be in parallel, how will the current through the heaters compare with what they need for rated heat output? Assume the resistance to be independent of temperature.

25. A 150-W bulb is connected to a 110-V power line.
    a. How would the current supplied by the power line change if two such bulbs were connected in parallel?
    b. How would the current through each bulb in parallel compare with that through the one bulb?
    c. How would the voltage across each bulb in the two-parallel-bulb case compare with that in the one-bulb case?
    d. If we had two such bulbs in series, how would the current supplied to each compare with the current with one bulb? How does the current supplied by the power line compare with that through each bulb? Assume the resistance of each bulb stays constant with changing current.
    e. Compare the total power supplied to one bulb, two bulbs in parallel, and two bulbs in series, assuming the resistance of each bulb is the same in each case.

26. When your car battery "dies" and you start the car by hooking jumper cables to a friend's battery, you should connect them as shown in Figure 15.17. Draw the schematic diagram for the circuit. Are the two batteries in parallel or in series? Why do you hook them "positive-to-positive," and what would happen if you hooked them "negative-to-positive"?

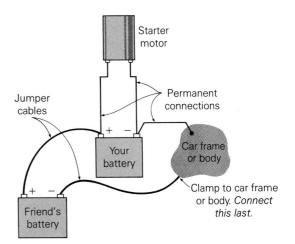

Figure 15.17 When hooking up jumper cables to your car battery, you hook positive to positive and negative to negative. (Connect the cable used for negative-to-negative *last* to some point on the body, frame, or engine so that any spark produced will not cause an explosion of the gases released in the battery.)

Alternating Current

27. Why are there 120 reversals of current direction per second in 60-Hz alternating current rather than just 60?

28. What is the average value of the voltage across a 110-V, 60-Hz power line? What is the average value of the current from this line to a 600-W hair drier? Why do we rate AC in "effective" values rather than average values? (The answer to the first two questions is *zero*.)

Problems

1. A 60-W light bulb carries a current of 0.55 A. How much charge leaves the filament in 8.0 h of operation?

   Answer: 16,000 C

2. A 100-W light bulb carries a current of about 0.91 A. How many electrons enter the filament per second?

   Answer: 5.7×10^{18}

3. If q in Figure 15.18 is 4.0 μC, and 0.60 J of work are needed to move it from where it is to point P, what is the potential difference between these points?

   Answer: 150 kV

Figure 15.18 Charge q is moved to point p, closer to charge Q.

$$Q \qquad P \qquad q$$
$$\bullet \qquad \times \qquad \bullet$$

4. What is the resistance of a 100-W bulb that operates on 110 V? What is it for one that operates on 220 V?

   Answer: 120 Ω; 480 Ω

5. A hot plate with a resistance of 24 Ω is designed for a current of 9.1 A. What voltage should it be operated on?

   Answer: 220 V

6. One wire of an ordinary lamp cord 25-m long has a resistance of about 0.51 Ω. What is the voltage drop in this wire if it carries a current of 3.0 A?

   Answer: 1.5 V

7. A current of 100 mA (milliamperes) through your body would probably stop your breathing. If you touched the two sides of a 100-V power line—one hand in contact with each wire—and the combined resistance of your body and the two points of contact were 1000 Ω, what would be the current through your body?

   Answer: 110 mA

8. A freezer is rated at 5.5 A on 120 V. What power is used by this freezer when operating on rated voltage?

   Answer: 660 W

9. If the freezer of Problem 8 is actually running 20% of the time, how much energy does it use in a year? How much does it cost to operate at 9¢/kWh?

   Answer: 1160 kWh; $104

10. How much electrical energy is used by a 100-W bulb operated for 24 h? How much water could you heat from 20°C to 100°C on this amount of energy?

    Answer: 2.4 kWh (2070 kcal); 26 liters

11. The resistance of the filament of an incandescent light bulb at room temperature is about $\frac{1}{10}$ what it is at operating temperature. How does the power consumption the instant after the bulb is turned on compare with that after it reaches operating temperature?

    Answer: 10 times as much initially

12. It takes about 0.02 s for an incandescent bulb to reach operating temperature. Assume the *average* power used during this time is five times what it is when operating normally (see Problem 11). For how long can you leave the light on and use less electrical energy than by turning it off and back on again at the end of the time?

    Answer: 0.08 s

13. A 12-V car battery is rated at 120 ampere-hours. How much energy in kilowatt-hours and in joules can be supplied by this battery before it is discharged?

    Answer: 1.44 kWh; 5.2 MJ

14. Resistances of 10 Ω, 20 Ω, and 30 Ω are connected in series across a 120-V emf. What is the current through the 20-Ω resistor?

    Answer: 2.0 A

15. Resistances of 10 Ω, 20 Ω, and 30 Ω are connected in parallel across a 120-V emf. What is the current through each resistance, and the

current through the battery? What is the effective resistance of the parallel combination? Hint: Knowing the battery current and emf, you can calculate the resistance it "sees."

Answer: 12 A; 6.0 A; 4.0 A; 22 A; 5.4 Ω

16. Check your answer to Problem 15 by using the equation $1/R = 1/R_1 + 1/R_2 + 1/R_3$ from the footnote in Section 15.7. Which method do you think is easier? Which method gives you a *greater understanding* of a parallel circuit?

17. Calculate the actual value of the current, voltage, or power as required by the various parts of Question 25. Calculate the overall resistance in each arrangement.

Answer: 1.36 A, 81 Ω; a. 2.72 A, 40 Ω; b. 1.36 Ω; c. 110 V; d. 0.68 A, 162 Ω; e. 150 W, 300 W, 75 W

18. What is the peak voltage in 220-V AC?

Answer: 311 V

Home Experiments

1. Take a flashlight apart and examine the parts. Trace out the circuit and see how it is completed and broken by turning the switch on and off. Flashlights often need a "whack" to make them come on. If you have one like that, try to find out why, and see if a judicious bending of some part will cure the problem.

2. Take the cover off and examine the switch mechanism, cord connections, and bulb contacts on a table lamp that uses an incandescent bulb. *Remember to unplug it first!* Follow the path of the current. If you have a lamp that doesn't work, try to locate its problem and fix it. If you can't find a broken wire or a loose connection between the wire and plug or switch, you may need a new switch and bulb socket from the hardware store. *Be certain that any part that carries a current is well insulated!*

References for Further Reading

Dresner, S., "Superconductors," *Popular Science*, Oct. 1974, p. 70. Discusses some of the fundamental aspects of superconductors.

Giancoli, D. C., *Physics Principles with Applications*, Second Ed. (Prentice-Hall, Englewood Cliffs, N.J., 1985). One of many good algebra-based college physics books that give a more detailed treatment of electric current.

Swartz, B. B., and S. Foner, "Large-Scale Applications of Superconductivity," *Physics Today*, July 1977, p. 34.

E V E R Y D A Y   A P P L I C A T I O N S   O F   E L E C T R I C I T Y

Everbody knows that electricity has many practical uses. *You* now know (after having read Chapters 14 and 15) the basic principles of electricity. But what do the practical and the basic have to do with each other? Now we want to see how basic electrical principles are applied to practical everyday uses.

Residential and automobile circuitry will provide down-to-earth examples of circuit principles. Resistive heating due to currents has many practical uses, but it can also be a nuisance or a hazard—we'll discuss both aspects. We will take a brief look at the electrical processes that go on in the building blocks of many electronic devices, including the omnipresent computer. We'll discuss an area important to everyone that uses electricity—shock, and how to avoid it.

We conclude with a discussion of atmospheric electricity, including lightning.

16.1 RESIDENTIAL AND AUTOMOBILE ELECTRICITY

Our first job is to see how the fundamentals of electrical circuits are applied in the distribution of electrical energy in the home and car. We will not learn any new physics in this section, but we will see how what we have already learned applies in some practical situations.

Residential Circuitry

Figure 16.1 is a schematic diagram that shows the overall distribution scheme in a house. It is not intended to look like the real thing, but rather to show its workings. We'll discuss the parts of this diagram separately.

Voltages. Home electrical devices are designed to operate on a potential difference of either 110 V or 220 V.* Lights and low-power appliances— TVs, shavers, sewing machines, typewriters—use 110 V; high-power appliances—ranges, clothes driers, water heaters, many air conditioners—use 220 V. There is nothing sacred about these values—electrical industries in the United States and Canada have simply agreed to use them. Other countries use other voltages.

Three wires (shown in the upper right-hand corner of the diagram) run to each house from the main lines of the power plant. One of these wires, called the *neutral* one, is electrically connected to the earth (the one with the ground symbol $\equiv$). This one's potential is defined as zero, or as electrical ground. Each of the other two wires is at 110 V relative to the neutral line. We often call these the "hot" lines, so named because they are at a voltage above that of ground. The two hot lines are at 220 V relative to each other. For AC this means that when one line is at its maximum positive voltage, the other is at its maximum negative voltage, giving an effective 220 V between them. (For DC this would mean that one line is at + 110 V and the other at − 110 V relative to the neutral line at 0 V.)

The energy meter. The incoming lines run to the energy meter, which is usually mounted on the side of the house (Fig. 16.2). The rotation of the dials on the meter is proportional to the energy used. What you pay for directly is the total energy (the number of kilowatt hours). As we will see in Chapter 20, the rate at which you use energy (the power) at certain times of the day may also influence your bill. (You read the meter from left to right—the reading of each dial being the smaller of the two numbers that the pointer is between. Verify that the meter shown reads 31182.)

Distribution of energy throughout the house. Figure 16.1 shows schematically how wires are connected for distribution of energy throughout the house. We'll mention a few pertinent points.

1. The input lines are connected to large conductors (bus bars) inside the distribution, or service, panel (usually called the *fuse box* or breaker box—see Fig. 16.3). Cables are connected to these conductors for running through the walls to various parts of the house.

*Sometimes you see these specified as 115 V and 230 V, 120 V and 240 V, or 125 V and 250 V. Light bulbs, for example, are usually labelled 125 V. The "110-V" line is usually at a voltage somewhere between 110 and 120 V.

Figure 16.1 A schematic diagram of the overall distribution of electrical energy to a residence. Notice that this diagram is not drawn to scale but is intended only to show the connections.

Figure 16.2 An electrical energy meter.

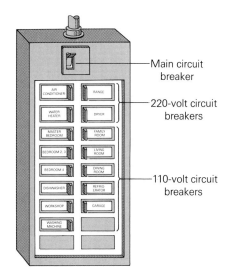

Figure 16.3 The service panel or circuit-breaker box. The large grey rectangle in Figure 16.1 schematically shows the connections inside this box.

2. Each cable consisting of three wires runs from the breaker box to a small group of lights and receptacles (electrical outlets). This is called a *circuit*. A large appliance, such as a water heater or range, usually has its own private circuit.

3. The two insulated wires in the cable actually form the current-carrying loop. For a 110-V circuit, the two wires are hooked, one each, to a hot line and the neutral incoming line via the bus bars. A 220-V circuit is hooked to the two hot lines.

4. It's customary in this country to follow the color codes indicated in the diagram. In 110-V circuits, the wire with the black insulation is used for the hot wire; the one with the white insulation, for the neutral wire. This code merely identifies which wire is which so that connections can be made properly.

5. The two current-carrying wires are connected to the terminals of a receptacle as shown in Figure 16.4, again following conventional color codes.

6. Wall switches to lights are connected in the hot line. Figure 16.5 shows how the actual switch closes and opens a gap in the closed conducting path.

7. When a wall switch is turned on (closed), a complete closed circuit is made from the power plant, through one of the supply lines, through the light, and back through the other line. This closed circuit lets current pass through the light bulb, "pushed" by the emf of the generator at the power plant. When other loads, such as the lamp in

Figure 16.4 The connections to a receptacle.

Figure 16.5 A wall switch opens and closes the circuit when turned off and on. When the switch is flipped on, the back part of the U-shaped metal moves up and closes the gap between the conductors attached to each terminal screw.

Ground terminal

Silver-colored terminal

Black wire

White wire

Wire nut

Ground wires

Ground terminal

Brass-colored terminals

Terminals

Switch lever

Switch lever

Switch lever

OFF

ON

Yoke

Spring

Off position

On position

the lower right of Figure 16.1, are plugged into a receptacle in the circuit, the connection is made in *parallel* with the light or other devices plugged into another receptacle.

8. Excess current in the wires causes them to overheat. This "over-load" can come from operating too many devices on the circuit or from faulty insulation, which allows the hot and neutral wires to come into contact (a "short circuit"). To guard against the dangers that may result, a fuse or circuit breaker is in series with each hot line. A fuse has a conductor through which the current in the circuit must pass. When the current exceeds that for which the fuse is rated, this conductor melts (the fuse blows), forming a gap that stops the current. The size of fuse is coordinated with the size of the wire in the circuit, which in turn is coordinated with the amount of current needed to operate the device or devices in the circuit.

Most houses now use circuit breakers rather than fuses. These

Silver-colored
terminal screw

Brass-colored
terminal screw

Green terminal
screw

Figure 16.6 The wire attached to the third (round) prong of the plug is connected to the housing of the appliance.

operate by means of a bimetallic strip that is heated by the current to the circuit. When the current exceeds the rated value, enough heat is produced to bend the strip until it opens the circuit. Rather than replacing the breaker as you would a fuse, you just reset it by throwing a switch *after finding and correcting the trouble that caused the excess current.*

Grounding. The bare wire in the cable for each circuit is connected inside the distribution panel to the neutral bus bar, which in turn is hooked to a water pipe or something else in good contact with the earth. This wire is then connected to the terminal on each receptacle (Fig. 16.4) into which the odd prong on a three-pronged plug fits (Fig. 16.6). A device such as a drill, with this third prong on its connecting plug, has its third wire connected to its housing. Should the insulation inside the drill fail and the hot wire contact the casing, the casing will still remain at zero volts because it's in contact with the ground. Since you, the operator, normally would be at zero potential relative to the ground, there would be no voltage across your body, and therefore no reason for a current to flow through your body. Instead, the ground wire would carry the excess current from the "short" in the drill to ground.

· ·

QUESTION   Analyze the diagram of Figure 16.1 carefully, and convince yourself that all devices plugged into receptacles and all lights turned on with wall switches are connected in parallel across the main supply lines.

Ground as Part of the Circuit

As we have said, electrical *ground* means whatever region we define to have *zero* potential. The symbol we use for electrical ground is $\perp$. In home electrical circuits, ground is *the* ground—earth.

The conductor that serves as ground is often used as part of the circuit carrying the current. For example, when certain rural areas of the country first got commercial electrical power, they operated temporarily with only one wire from the power plant to the homes. The earth served as the other half of the conducting loop. As another example, electronic devices such as TV sets use their metal chassis as ground and as part of the conducting paths for currents within the sets.

In an automobile, only one wire runs to each of the electrical parts and accessories. The frame and body form the other part of the conducting path. The starter motor, for example, could be supplied with energy by two wires from the battery, shown as the heavy solid line containing the switch and the dashed line of Figure 16.7(a). Instead, one terminal of the battery, usually the negative, and one terminal of each device are connected to the frame of the car. One wire containing the

Figure 16.7 An automobile uses the frame and body as part of the conducting circuit. The body is *ground* for the car. The pictorial representation in (a) is shown schematically in (b).

B = Battery   S = Starter motor   T = Taillight   ∿ = Fuse

(a)

(b)

fuse and switch is then run from the positive battery terminal to the device.

Figure 16.7(b) shows the schematic diagram for the battery, the starter, and the taillight of the car. Each connection to ground implies a connection to a common conductor carrying current back to the negative terminal of the battery.

The wiring to the electrical parts of a car is divided into circuits in much the same way as the wiring in a house. Each circuit has a separate fuse and supplies current to a limited number of devices.

QUESTION   What is ground on a flashlight? Does it carry the current? Draw a circuit diagram in the format of Figure 16.7(b) for a flashlight.

16.2 RESISTIVE HEATING

How many ways can you think of in which electrical energy is used to produce heat? We've already talked about the incandescent light bulb producing light as the filament gets white hot from current passing through it. Electric furnaces heat some of our houses, and most of our food is cooked on electric ranges. The small and large appliances that heat electrically are almost endless—toasters, electric blankets, water heaters, and coffee makers, to name a few.

However, there are also situations in which electrical heating is a nuisance. Since all conductors have *some* resistance at normal temperatures, heat is produced as electrical energy is transmitted over power lines and through the wires in your house. This heat represents wasted electrical energy, and it can be a hazard if too much is produced in a given area.

Rate of Heat Production

If we know the resistance of a conductor and the current passing through it, we can calculate the rate at which heat is produced. We saw in Chapter 15 that power (rate of delivering energy) from a source of emf or to a load is given by

$$power = voltage \times current$$

or

$$P = VI$$

To find an expression specifically for the power delivered to a resistor, we can use the relationship between its resistance, the current through it, and the voltage difference across it:

$$voltage = current \times resistance$$
$$V = IR$$

Using this for voltage in our power equation, we get

$$power = (current \times resistance) \times current$$
$$P_{heat} = (IR)I$$

or

$$P_{heat} = I^2R$$

This equation tells us the rate of production of heat in a conductor of resistance R carrying current I. The subscript "heat" on the P explicitly reminds you that this power all goes to heat production.

· ·

EXAMPLE 16.1   A typical 15-cm-diameter surface unit on an electric range has a resistance of 43 Ω. What is the rate at which heat is produced in this unit, when operated at 220 V?

From $V = IR$, the current is given by

$$I = \frac{V}{R} = \frac{220 \text{ V}}{43 \text{ } \Omega} = 5.1 \text{ A}$$

The heat production rate is then

$$P = I^2R = (5.1 \text{ A})^2 \times 43 \text{ } \Omega$$
$$= 1100 \text{ W}$$

It might be interesting to find out how long it would take to boil away 1 liter (about a quart) of water, if all this heat goes into the boiling water.

Energy to boil water = (mass of water) × (heat of vaporization)

One liter of water has a mass of 1 kg, so

$$\text{energy} = 1 \text{ kg} \times 540 \text{ kcal/kg} \times 1.16 \text{ Wh/kcal}$$
$$= 630 \text{ Wh}$$

Since energy = power × time,

$$\text{time} = \frac{\text{energy}}{\text{power}} = \frac{630 \text{ Wh}}{1100 \text{ W}}$$
$$= 0.57 \text{ h} = 34 \text{ min}$$

Most electrical cooking appliances produce their heat as current passes through a conducting heating element on the appliance, such as a surface unit on a range. An interesting exception is a "hot-dogger" (Fig. 16.8-b), named of course for what it cooks. As its schematic diagram (Fig. 16.8-a) shows, the current passes directly through the hot dog. The resistance of a hot dog turns out to be just right to give the amount of heat needed to cook it in about 1 min when connected across a 110-V supply. Since the hot dogs are connected in parallel across the emf source, each one is

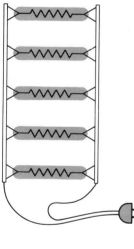

(a)

Figure 16.8 In a "hot dogger," the hot dog cooks from the heat produced as current passes directly through the hot dog. Each hot dog has a resistance in parallel with the others.

(b)

independent of the other, and cooking time is independent of the number of hot dogs. The total current through the cord is, of course, proportional to the number of dogs.

· ·

QUESTION   The wiring diagram of a popular portable electric room heater is shown in Figure 16.9. Why does it use only six resistance elements for a high-heat setting but all eight elements for low heat?

· ·

EXERCISE   If the portable heater of Figure 16.9 operates on 120 V and the resistance of each segment is 1.5 Ω, what is the heat output for each setting?

Answer: 1600 W; 1200 W

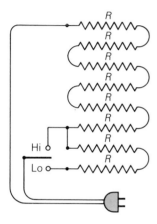

Figure 16.9 One type of portable electric heater gets its high-heat setting with six spans of resistance R in series. Adding two more in series gives the low-heat setting.

Figure 16.10 A circuit containing an air conditioner and a light bulb in parallel. Voltage drop along the wires due to current through their resistance causes lower-than-normal voltage across the devices being operated.

Wire sizes. Did you ever notice the lights dim briefly or the TV blink when a large appliance such as an air conditioner, or maybe a refrigerator, turned on? If so, you were seeing the effect of the resistance of the wires carrying the current. We usually ignore the resistance in copper wiring, since it is small compared with the resistance of the appliance. But in this case, the resistance in the wire, even though small, is the key to understanding what happens.

Suppose you have an air conditioner and an incandescent lamp connected in parallel in the same circuit (Fig. 16.10), shown operating on a battery for simplicity. When the air conditioner first turns on, it needs a large current to get started. Since the wires have some resistance, there is a voltage drop given by $V = IR$ along the wires between the battery and air conditioner. This situation means that the voltage across the bulb is less than it would be if no current was going to the air conditioner. The bulb gets dimmer because, with less voltage across it, less current goes through it.

If a wire has resistance R and carries current I, heat is produced in the wire at a rate I^2R. This heat represents a loss of electrical power that does not reach the devices being operated. At the least, this is wasteful, but in extreme cases enough heat can be produced to cause a fire hazard.

For a given current in the wire, the only way to reduce heat production is to reduce the resistance, which can be done by increasing the wire sizes. The larger the cross-sectional area of the wire, the less its resistance. Compare this situation to the flow of water in a pipe: The larger the pipe, the more water it can carry for a given pressure difference between the ends of the pipe. Also, as you might expect, the resistance of a wire is proportional to its length. Thus, minimizing the length minimizes heat production.

If you notice the sizes of the wires used in high-current circuits, such as a clothes-drier circuit, you see that they are much larger than those in lighting circuits.

QUESTION Why do refrigerators often come from the factory with a tag that warns you not to plug them in by way of a simple "dime-store" extension cord?

Loose connections. When you unplug your hair drier (or shaver or drill or back scratcher), do you pull on the plug or do you grab the cord and yank? If you do the latter, don't be surprised if after a few yanks the plug or cord starts to get hot when you use it. If your yanking loosens the connections to the prongs of the plug or breaks some of the strands in the wire, you greatly increase the resistance to current through that point. The current to operate the device, passing through the increased resistance, produces heat—in extreme cases, enough to start a fire. Aside from the potential danger and electrical energy loss to heat, the voltage drop in this high resistance along the current path means that you get a lower voltage at the device and therefore less efficient operation.

Another place where the same effect can cause trouble: If your car starter won't turn the engine and yet the battery is good, corrosion may have formed an insulating layer that causes a bad connection between the battery and the cable connected to it. The resistance can become so high there that not enough current can get through to run the starter. Usually, however, enough current exists so that the connector gets extremely hot—hot enough to burn if you touch it—while you're trying to start the car.

Any time the conducting path to or from an electrically operated device has too much resistance, either from bad connections at some point, or from too small or too long an extension cord or cable, energy will be lost to heat, and the voltage drop will reduce the efficiency of operation. A 10% drop in voltage lowers the output power of a motor by about 20%, and the light output of an incandescent bulb by about 30%.

QUESTION On portable electric heaters that have been in use for many years, the plug probably gets hot during operation. Why is this, and what should be done about it?

16.3 ELECTRONICS AND THE COMPUTER

Electronics is that special branch of practical electricity that deals with devices whose components control the flow of charge in specific ways in order to do a particular job. Electronic devices include TVs, radios,

Figure 16.11 Several forms of integrated circuit chips ("micro-chips") used in electronic circuits—shown with a penny for size comparison. The circuit itself is only a small fraction of the chip. Most of the size is needed for the external connections that you can see along the sides of the rectangular chips.

calculators, tape recorders, a vast range of research and industrial equipment, and that most versatile of all electronic machines—the computer.

Most modern electronic circuits depend almost exclusively on "solid state" components called *transistors* that are made of materials known as **semiconductors**. Transistors can be made so small and compact that hundreds of thousands of them can be combined into a single *integrated circuit*, or *IC*, that is a silicon "chip" only a few millimeters across (Fig. 16.11). ICs perform a wide variety of functions, depending on their particular design. For example, some serve as the memory in a computer, while others carry out the actual calculation or data manipulation.

A brief outline will illustrate how all this comes about. First, we need to consider the basic properties of semiconductors. We can then see how these materials function in transistors that are grouped together in ICs to form an electronic system.

Semiconductors

As their name implies, semiconductors typically have a resistance between that of conductors and insulators. One abundant semiconductor is silicon. Its atoms have four valence electrons: that is, four electrons in their outermost orbits. A pure piece of silicon is a crystal with atoms arranged in a regular array. (See Fig. 16.12. A real array is three-dimensional rather than two-dimensional as shown.) Each atom is bound tightly to its neighbors by sharing one valence electron with the four nearest atoms. Since this arrangement binds the electrons tightly, pure silicon is a very poor conductor.

Semiconductors become useful when small quantities of certain other substances are added to them. Suppose we add to the silicon a small amount of an element such as arsenic, whose atoms have 5 valence electrons. As Figure 16.13 shows, each arsenic atom has an extra electron. These loosely attached electrons are free to move about as negative charge carriers, making the impure silicon a good conductor.

We can also make silicon a conductor with *positive* charge carriers. Suppose that, instead of a valence-5 impurity, we add one whose atoms have 3 valence electrons, such as gallium. At the position of the gallium atom one electron is missing, leaving a "hole" in the electron structure (Fig. 16.14). Another electron can move in to fill this hole, leaving a hole where *it* was. Some other electron can fill this hole, leaving a hole in its old slot (Fig. 16.15). Thus the hole acts as a charge carrier as it moves around freely in the silicon. Since the hole represents an *absent* electron, it acts as a *positive* charge carrier.

Impurity semiconductors with excess electrons are called *n-type* (for negative) semiconductors; those with excess holes are called *p-type* (for positive) semiconductors. Their effectiveness as conductors depends on the relative amount of impurity added.

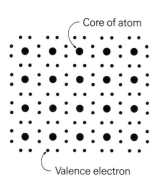

Figure 16.12 The atoms in pure silicon are arranged in a regular crystal lattice so that each atom shares one electron with each of its four nearest neighbors.

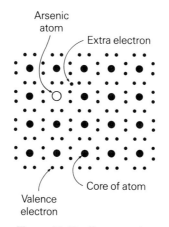

Figure 16.13 If an arsenic atom with five valence electrons replaces a silicon atom, the extra electron is free to roam about.

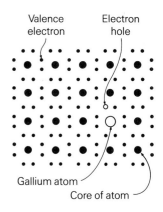

Figure 16.14 If a gallium atom with three valence electrons replaces a silicon atom, it leaves a hole in the electron structure that acts as a positive-charge carrier.

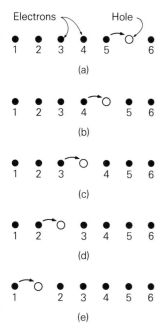

Figure 16.15 Conduction by a hole in a line of electrons. (a) The hole is between electrons 5 and 6. (b) As electron 5 fills the hole, the hole moves to the left. In each successive step—(c), (d), and (e)—the hole is conducted one step to the left.

Transistors and Integrated Circuits

Now let's see how semiconductors can be used to make a device for controlling the flow of charge—a *transistor*. We'll look at only one type—the *metal oxide semiconductor*, or *MOS*—that is very common in watches and hand-held calculators and in many computers. Other types vary in detail, but do the same basic job.

Consider the arrangement shown in Figure 16.16(a). The piece of semiconductor is mainly *p*-type but has two *n*-type regions known as the source and drain. A metal conductor, the gate, spans these three regions but is completely insulated from them by a thin layer of silicon oxide.

If a potential difference is placed across *S* and *D* in this situation, no charge can flow between these points. For example, if *D* were positive and *S* negative, electrons would be attracted toward *D* in the drain, and holes would be repelled in the neighboring *p* region. Thus, there could be no current across this boundary.

Now suppose, as in Figure 16.16(b), a positive voltage relative to *S* is placed on the gate. By electrostatic attraction, excess electrons will be attracted into the central region near the gate, converting this into an *n*-type region. Thus, we have a continuous *n*-type conducting region between *S* and *D*. A potential difference between *S* and *D* would then result in a current between these points.

The use of transistors can be divided into two classifications, *linear* and *digital*. In a typical linear application, the transistor is used as an amplifier. The conductivity, or ability to conduct, of the transistor varies

Metal (gate)

G

Silicon oxide
insulator

S

D

n-type

n-type

p-type

(a)

G

+5 V

S

D

n-type

n-type

p-type

(b)

Figure 16.16 The basic make-up of a metal oxide semiconductor (MOS) transistor. G = gate, S = source, D = drain.

in proportion to the voltage on the gate, since more electrons are attracted to the gate region for higher voltages. Thus, the current between S and D can be made to vary in proportion to the voltage on G. In this way a weak time-varying electrical signal can be converted to an identical but stronger one. This is the function of the amplifier in your sound system, for example.

In digital applications such as watches, calculators, and computers, the transistor is basically a switch. When a voltage of a few volts is put on G, the transistor becomes a very good conductor with nearly zero resistance. Thus, by changing the voltage on G from zero to a few volts (typically 5 V), we switch the transistor from *off* (high resistance) to *on* (low resistance). The result is identical, in principle, to switching a light switch from OFF to ON.

Whether for use in linear or digital applications, transistors are always used in combinations. Many uses require hundreds of thousands of transistors. Fortunately, one IC chip containing this many transistors can be made about as cheaply as a single transistor. Most functional circuits make use of not only transistors but also resistors and charge-storing devices called capacitors. These can be built directly into the chip. For example, the desired resistance between two points can be fixed by adjusting the amount of n- or p-type impurity in the intervening semiconductor. If you are interested, you can probably find some readable books in your library that discuss the details of the manufacture of ICs. Some suggestions are in the references at the end of this chapter.

APPLICATION

THE COMPUTER

A few (maybe a million or so!) simple transistor switches can be made into a computer that will do something useful. We cannot cover all of the many complicated details in the space of this application. What I want to do is outline how combinations of two transistor switches can be used to add *1* and *1*. Then I will trust you to take a quantum leap of faith and believe that, if these numbers can be added once, they can be added repeatedly to get any desired number. The computer components do not have to deal directly with any number higher than *1*. That's because the computer first converts any number into the binary system—a system that represents any number by a string of *0*s and *1*s. Thus, the computer can add any number by adding *0*s and *1*s.

But what about more complicated operations: subtraction, multiplication, and division, for example? Well, subtraction is just the inverse of addition. Multiplication is repeated addition—you can get 3 × 4 by adding 3 four times—and division is repeated subtraction. Thus, a computer does all its arithmetic by adding *0*s and *1*s. It's really quite dumb! And what about jobs such as word processing? Every alphabet letter is also represented by a string of *0*s and *1*s. Thus, the computer handles this kind of job in a similar way.

Now let's go back and see how our transistor switches can add *1* and *1*. Consider the diagram in Figure 16.17. The two transistors are the circles labeled "T". Recall that the transistor is a good conductor (switch is closed) with 5 V on the gate, and it is a poor conductor (switch is open) with 0 V on its gate. (These voltages are measured relative to the source terminal of the particular transistor.) Therefore, only if there is 5 V on *both* points, A and B, will both switches be closed and the circuit containing R be completed. In that condition, the circuit will be simply a 5-V battery connected to the resistor R, giving 5 V across R. If either transistor switch is open, there is no current through R, and the output voltage is zero.

We represent our two numbers to be added by A and B. The system is wired so that a value of *1* at either A or B is signified by 5 V on that terminal. Only if both A and B are *1*, do we get an output of 5 V. Similar circuits can determine if both A and B are *0*, or if one of these is *0* and the other *1*. By combining these three separate circuits, we get an *adder* that determines the sum of A and B and gives the answer in binary *1*s and *0*s as appropriate 5-V signals.

Now you have at least a hint at how a computer computes. Some of the end-of-chapter references describe how the various components fit together to make the entire system function.

Figure 16.17 Two transistor switches (T) arranged in a circuit to give an output of 5 V across *R* only when both *A* and *B* are *1*s—that is, when they have 5 V on their gate (G) inputs.

16.4 SHOCK

Everything your body does is controlled electrically. Therefore, it shouldn't be surprising that an electric current forced through your body by an outside potential difference would greatly disrupt your bodily functions. We'll look at some of the conditions for causing such a current, which we call an electrical *shock*.

First of all, notice that it's the *current* that gets you, *not* the voltage. The amount of voltage is of no *direct* consequence other than the fact that it takes some voltage across your body to drive the current through.

Let's illustrate by a specific example of how you can be shocked. Suppose your roommate has an electric toothbrush plugged into a receptacle in the bathroom and accidentally closes the medicine cabinet door on the cord. Suppose, in the process, a metal part of the door cuts through the insulation and makes contact with the wire at 110 V. You come along later to open the medicine cabinet. In so doing you touch its metal frame. As Figure 16.18(a) illustrates, as long as you are standing on the dry bathroom rug, or your body is insulated from ground in some way, nothing happens, even though your body goes to 110 V relative to its surroundings. It is not part of a complete closed circuit. However, if (as in Figure 16.18-b) while you are touching the medicine cabinet, you reach down to turn on the water, *zap!* Your body provides a path between the 110-V line touching the medicine cabinet and the grounded water pipe at 0 V. Current can then pass through you giving you a potentially serious shock.

Figure 16.18 (a) Your *whole* body is at 110 V relative to ground. There is no potential difference across it, and therefore no current through it—no shock. (b) There is 110 V across your body, giving a current that follows the path shown by the arrows.

(a)　　　　　　　　　　　　　(b)

Any time your body forms a path across a potential difference, it can carry a current that gives you a shock. In Figure 16.18(a), your body is at 110 V, but no harm is done because there is no potential difference across your body, and therefore no current through it. In Figure 16.18(b), the potential difference produces a current. Figure 16.19 shows *living* evidence that the voltage doesn't kill you if you provide no path *across* a potential difference.

Why do some people receive only a small jolt, but others get electrocuted by contact with a 110-V line? The main difference is the resistance between the body and the conductors at the points of contact. Figure 16.20 is a schematic diagram of the situation of Figure 16.18(b), but, for simplicity, the source of potential difference is shown as a battery. Here R_B stands for the resistance of the body; R_L and R_R are the resistances of the contact points at the left and right hands. Any current passing through your body has to pass through all three resistances. If any one is high, the current will be low. Since the body contains many ions, it is a pretty good conductor. The current, therefore, is determined mostly by the resistance at the contact points.

Fortunately, dry skin has a resistance high enough that the current from contact across 110 V is usually too small to be deadly to an adult, although certainly not comfortable! However, if the skin is wet from perspiration or water, contact is much better, and the resistance can be low enough that the current may be fatal. This is why shock hazards are worse in bathrooms and laundry rooms. A person in one of these rooms is often in contact with water that is, in turn, in contact with grounded water pipes. Water can easily seep into the inside of appliances picked up with wet hands, thereby providing a path for the current.

Figure 16.19 A resident of Irving, Texas, sits on a live 138,000-V power line, where he landed during a parasail accident. He received only burns on his hands and feet, probably from the surge of charge that initially brought his body up to 138 kV.

QUESTION If you are listening to the radio in the bathtub, and the radio falls into the tub with you while it's plugged in, what should you do? Remember the water is usually grounded by the connecting pipe, and to get a shock your body has to provide a path across a potential difference. What would happen to you if you just stayed in the tub with the plugged in radio? Discuss the consequences of the various possibilities for action.

For most people, it takes a current larger than about 1 mA (1 mA = 0.001 A) to be felt. At about 16 mA, the current stimulates the muscles enough that you have no control over them and cannot let go of the conductor. Between 25 and 100 mA breathing stops, and between 100 and 200 mA the heart stops pumping and goes into uncoordinated contractions (ventricular fibrillation). At higher currents the heart completely stops, irreversible damage may be done to the nervous system, and serious burns can result from the heating effect of the current.

If a hospital patient has wires connected to internal electrodes in the heart, currents as low as 0.01 to 0.1 mA in these wires can cause ventricular fibrillation. Leakage currents this large exist in many types of ordinary equipment. Even though extreme precautions are now usually taken to avoid even small currents around patients with electrodes implanted internally, many accidental electrocutions are alleged to occur each year among these patients.

16.5 ATMOSPHERIC ELECTRICITY

Figure 16.20 The schematic circuit diagram for Fig. 16.18. R_B is the resistance of the body; R_L and R_R are the resistances of the contacts made at the left and right hands. In Fig. 16.18(a), R_R is infinite, so there is no current. The ground is, in effect, a conductor shown by the dashed line.

Atmospheric electricity is an extremely complex and, in some ways, mysterious field. It is by no means thoroughly understood by anyone. We certainly cannot explore all the details here, but we will look at a few overall features. We will consider first some atmospheric effects that are independent of weather conditions and then consider nature's own spectacular fireworks—lightning. We'll see that atmospheric electricity is a combination of electrostatics and slowly varying currents.

Charges and Electric Fields in the Atmosphere

Ordinary air is a very poor conductor (good insulator) of charge, but it does conduct some. The reason is that about one of every few million billion air molecules is an ion—a molecule that has either lost or gained an electron and therefore has a net charge. These charges conduct a current if there is an electric field in the air.

Where do these ions come from, and why don't they merely exchange electrons and become neutral? They do this, but new ion pairs are continually being formed. Small amounts of naturally radioactive materials near the earth's surface give off electrons and gamma rays that collide with air molecules, knocking off some electrons and thereby forming ions. In addition, cosmic rays—high-energy particles from space—enter the atmosphere and knock electrons from molecules. Near the earth's surface about 10 million ion pairs are formed this way per second per cubic meter of air. The rate of formation increases with increasing altitude, since the cosmic ray density increases.

Experiments have shown that the surface of the earth carries a net negative charge that averages about $-\frac{1}{2}$ million coulombs over the entire earth's surface. The upper atmosphere carries an average net positive charge of the same amount, making the earth and atmosphere together electrically neutral (Fig. 16.21). Because of this charge separation, there is a downward electric field in the atmosphere. In clear weather, this field is about 130 volts/meter near the earth's surface. The overall potential difference between the earth and the upper atmosphere averages about 300 to 400 kV.

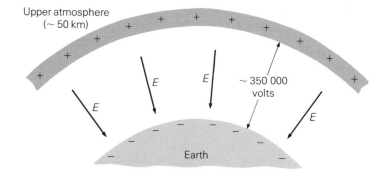

Figure 16.21 The earth's surface has a negative charge of about $-\frac{1}{2}$ million coulombs, the upper atmosphere, an equal positive charge. The potential difference between them is 300–400 kV.

Since you are probably close to 2 m tall, does this mean there is a 250-V potential difference between your head and your feet when you stand up? Why don't you get zapped? Your body, being a conductor, stays at the same potential as the earth with which you are in contact. This potential "gradient," or electric field, starts above your body or any other conductor in contact with the earth.

Since there is a voltage between the earth and upper atmosphere and the air in between is a conductor (admittedly poor), there is a continual current from the upper atmosphere to ground. That current averages about 1800 A through the entire earth's atmosphere. From the equation for electrical power ($P = VI$) the power dissipated in this current is around 600 megawatts, roughly half the output of a large modern power plant.

The variation of this current with time of day is interesting. When the current over the oceans is measured at various times during the day, and averaged over many days, we find that it has its maximum value all over the world at about 2:00 PM eastern standard time—11:00 AM Pacific standard time. No matter *where* you make the measurement, its largest value usually comes when it's 2:00 PM in Washington, D.C. This rather surprising occurrence is related to something else that has its peak at that time, as we will now see.

Charging the Earth

The current of 1800 A reaching the earth would discharge the $\frac{1}{2}$ million-coulomb negative charge on its surface in about five minutes. Yet the current continues, and the earth keeps its rather constant negative charge. How is this possible? Where is the "battery" that keeps this charge flowing?

Figure 16.22 is a picture of this "battery."* Lightning flashes all over the earth carry a negative charge to the earth at an average rate of about

Figure 16.22 Lightning is the "battery" that keeps the earth charged negatively.

The idea of lightning as a "battery" is borrowed from R. P. Feynman, R. B. Leighton, and M. Sands, The Feynman Lectures on Physics, Vol. II (Addison-Wesley Publishing Co., Reading, Mass., 1964). See Chapter 9.

1800 A. This keeps the charge separation between the earth and atmosphere relatively constant. As you might guess from the last section, the average rate of thunderstorm activity all over the earth has its maximum value when it's about 2:00 PM in Washington, D.C. Even if you're tempted to attribute this effect to the return from lunch of the United States Congress, this is also the time of day when the sun is over the tropical regions of main thunderstorm activity. Such storms are mainly generated by convection currents in air that is heated near the surface of the earth.

Just how thunderstorms come about is a complicated process discussed in the references at the end of the chapter. However, we will discuss the events that occur in a lightning flash.

· ·

APPLICATION

LIGHTNING

Cloud

Ground

Figure 16.23 The stepped leader of a lightning flash follows a zigzag path.

Thunderclouds are extremely complex systems that are usually polarized so that there is mostly positive charge near the top, mostly negative charge near the bottom. This large negative charge at the cloud bottom, usually at a height of about 5 km, induces a positive charge on the part of the earth just below the cloud in the same way your charged hand induces an opposite charge on the nearby doorknob as your hand approaches. A potential difference between 25 million and 300 million V exists between the cloud base and ground.

It takes about 300 kV per meter for a spark to develop through air. Therefore, continuous sparks from cloud to ground cannot form. Instead, electrons in the base of the cloud are accelerated by the tremendous electric field. They collide with air molecules, knocking out electrons and producing a conducting path. This path, filled with electrons, progresses about 50 to 100 m in less than a microsecond (10^{-6} s). After a pause of about 50 μs, this *stepped leader*, as it's called, advances another step of 50 to 100 m. This process continues in a zigzag pattern, often with several branches, until it reaches ground (Fig. 16.23).

When the leader touches ground, the negative charge at its bottom immediately leaks off. Then the charges from successively higher parts of the leader flow to ground, until all the negative charge in the leader has been deposited in the ground. This process, called the *return stroke*, can have currents that exceed 300,000 A but are typically about 25,000 A and deposit an average of 10 to 20 coulombs of charge. Most of the light is produced during the return stroke.

The return stroke is usually followed in about 0.04 s by another leader, called a *dart leader*, that darts smoothly along the original channel to ground. It's also followed by a return stroke that usually carries much less charge than the original one. A typical flash has three or four leaders and return strokes, but occasional flashes have twenty or more.

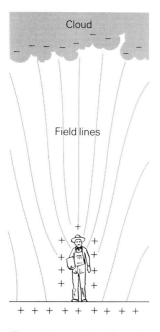

Figure 16.24 The effect of a person, or other conductor, on the electric field under a thundercloud.

The surge of current in the flash can heat the air in the channel to a temperature of 30,000°C. The gases expand quickly into the surrounding air, creating the sound waves we call thunder.

What does lightning strike? Suppose you're standing in an open field with a thundercloud above, as shown in Figure 16.24. (The person's height relative to the distance to the cloud is greatly exaggerated.) Positive charge will flow onto your body—actually, electrons will leave—concentrating the electric field at the highest part of your body. If a lightning strike is developing in the area, it will probably follow the path of the strongest electric field; namely, the path to your head. For the same reason, lightning is more likely to strike any object sticking up that has a conductivity greater than that of air—a lone tree, a house, a mountaintop, or particularly a metal tower.

What about *lightning rods*? They were invented by Benjamin Franklin over 200 years ago and are still used in essentially the same form. But do they *really* do any good aside from giving their owners a *feeling* of security?

The rod is a well-grounded conductor with a pointed end that sticks up higher than any part of the building it protects. It has two intended functions. The first is related to the fact that, because like charges repel each other, charge concentrates at the tip of a pointed object, increasing the surrounding electric field. On the lightning rod, the other end of this pointed conductor is connected to the ground. The strong electric field around the points leaks some of the earth's charge into the surrounding atmosphere. The charge leaked during a storm may amount to several coulombs. The idea is to neutralize some of the atmospheric charge, reduce the electric field, and thereby reduce the probability of a lightning strike.

The second function of the rod is to provide a good conducting path between a point higher than the building and ground. Should lightning then strike in spite of the charge leakage, it will strike the rod and be conducted to ground.

There is no doubt about the beneficial effect of the second function. It has been proven many times. There is also no doubt that substantial amounts of charge are leaked into the atmosphere by the pointed rods. Specialists in the study of lightning feel that this leakage may help in some circumstances; in others, however, it may intensify the electrification of the storm.

What about electrical shielding? We're sometimes advised that a safe place during a lightning storm is inside a metal automobile, airplane, or train. The metal body conducts the charge around the person, where it jumps the small distance to the ground. You may get the same benefit if a power line falls on your car. You are safe from electrical shock if you stay inside because the car body conducts the current around you. (However, you may not have this option if sparking ignites a flammable part of the car.)

Key Concepts

Practical electrical devices are connected in parallel across the power lines. The grounded body of an electrical device can safely form part of the electrical circuit.

The rate of **heat production** from electrical power is given by:

$$power = (current)^2 \times resistance$$

or

$$P_{heat} = I^2 R$$

Most electronic devices use *transistors* made of semiconductor materials. **Semiconductors** are intermediate in conductivity between conductors and insulators. Those used in transistors are either *n*-type, with valence-5 impurities providing negative-charge carriers, or *p*-type, with valence-3 impurities providing holes that act as positive-charge carriers.

Shock is caused by *current* passing *through* your body, not by the voltage of the body. This current exists when your body forms a path *across* a potential difference.

A lightning flash to earth starts with a *stepped leader* that is followed by a *return stroke*. There usually follows one or more *dart leaders*, each followed by a return stroke. Lightning flashes average 25,000 A and deposit an average of 10 to 20 C of charge. Any conducting object projecting above the earth increases the electric field directly above it and is more likely to be struck by lightning.

Questions

Residential and Automobile Electricity

1. Does the "electricity meter" on a house register energy or power?

2. In house wiring, why is the wire with the black insulation used for the "hot" wire?

3. What is the safety advantage of connecting a wall switch in the "hot" line?

4. Many appliances, including lamps, now come with polarized plugs that will plug in only in one orientation. From the schematic diagram of lamp connections in Figure 16.25, which prong of the plug should contact the "hot" side of the receptacle? Consider, for example, the safety of changing a bulb if the wrong side is electrically "hot" (that is, at high voltage).

5. What are the purposes of a fuse and a circuit breaker? Does one of these have an advantage over the other? What determines the "size" of fuse or circuit breaker that's needed?

6. When you plug a radio into one receptacle and a desk lamp into another receptacle in your

Figure 16.25 One side of the filament of a lamp makes contact with the circuit through the threaded connector.

room, are you connecting them in parallel, in series, or in some other fashion? Does it matter if these two receptacles happen to be on different house circuits?

7. What determines the size of wire needed for a circuit or appliance?

8. Suppose you have two 2200-W hot plates, one that operates on 110 V and one that operates on 220 V. If you planned to operate nothing else on the circuits used for each hot plate, how would the size fuse or circuit breaker needed for the 110-V plate compare with that needed for the 220-V one?

9. Why do large power-using devices (ranges, water heaters, clothes driers, etc.) usually operate on 220 V rather than 110 V?

10. Draw a circuit diagram for a DC power supply having three output wires arranged so that one wire is grounded, so that there is 110 V between two pairs of wires, and so that there is 220 V between the other pair.

11. A ground-fault interrupter (GFI) is a special circuit breaker that quickly turns off a circuit's current if the current in the circuit's hot line exceeds that in the neutral line. The GFI is especially recommended for circuits in high shock-risk areas, such as bathrooms. Why does the GFI provide added safety from serious shocks?

12. The circuit diagram for a three-way bulb and its accompanying switch is shown in Figure 16.26. The bulb has two filaments, and the switch has four positions that give 0 W, 50 W, 100 W, and 150 W of power used. The four switch positions correspond to: both switches

Figure 16.26 Circuit diagram for a three-way bulb and switch that gives three light levels corresponding to 50 W, 100 W, and 150 W.

open; S_A only closed; S_B only closed; and both S_A and S_B closed. Explain how this gives three light levels. Examine such a bulb and socket and you will see an extra contact point on the bulb base and in the socket, not found on ordinary bulbs and sockets. Why do you have to turn the switch two notches to get a change when you use an ordinary bulb in a three-way socket?

13. Figure 16.27 is a circuit diagram for a "three-way" switch that lets you control one light with two switches. What are the two switch-setting combinations that turn the light on? What are the two that turn it off?

14. Does the ground wire hooked to the casing of an electric drill prevent current from passing through the casing?

15. Why does the tail light of your car need only one wire running to it rather than two, as in a table lamp used in the house?

16. Why do the lights sometimes dim when a room air conditioner or other large appliance comes on?

Figure 16.27 Circuit diagram for a "three-way" switch.

Resistive Heating

17. Which has a thicker filament, a 100-W or 200-W incandescent bulb? (Assume the length of the filament to be the same in each.)

18. A portable electric heater for operation on 110 V has two heating elements: one 16-Ω and one 12-Ω element. Of the four alternatives—the 16-Ω element alone, the 12-Ω element alone, both in series, both in parallel—which way would you get the most heat? Which way would you get the least heat?

19. If you connect two identical hot plates in series, will you get a higher or lower rate of heat production from the two together than from one by itself? If the resistance of the heating elements does not change with temperature, how much higher or lower is the rate?

20. Why does the plug on an extension cord sometimes become hot if the connection between wire and prongs is loose?

21. Why does a hot dog cook when you pass a current through it?

Electronics

22. Suppose you were going to make a resistor out of n-type semiconductor. To get higher resistance, would you expect to have a higher or lower percentage of valence-5 impurity?

23. Explain how the movement of a hole in a p-type semiconductor is like the movement of an available auditorium seat when everybody on a row with one seat on the left side moves, one at a time, one seat to the left. Which way does the hole move?

24. MOS transistors can be made by using n-type for p-type and p-type for n-type semiconductors in the scheme shown in Figure 16.16. To make the transistor conducting, would you then put positive or negative voltage on the gate? (In computer literature you may see reference to CMOS electronics—*complimentary* metal oxide semiconductor. Such devices use combinations of transistors with n and p charge carriers.)

25. Design a circuit of the type shown in Figure 16.17 that has a 5-V output if *either A or B* is a 1—is at 5 V. Hint: In the circuit shown the transistors are in series. What would be the result if you put them in parallel and then put this parallel combination in series with the battery and resistor?

26. The circuit of Figure 16.17 is called an AND gate. That described in Question 25 is called an OR gate. Justify the use of these terms.

Shock

27. Does wearing thick rubber-soled shoes assure you of not being shocked?

28. If you grab a high-voltage power line, you will not necessarily be seriously shocked. Why? Why aren't birds killed when they land on a high-voltage line?

29. If you're in an automobile accident and a power line falls across the car, what are the safe and unsafe alternatives for action? In particular, what's wrong with *stepping* out of the car? How should you get out?

30. A cattle prod is a rod used by cattle handlers that gives a shock to cattle when the rod's end is touched to them—the purpose being to encourage them to move along. If the prodee but not the proder is to be shocked, at how many points must the prod make contact with the animal?

Atmospheric Electricity

31. What keeps the earth charged negatively relative to the upper atmosphere?

32. Discuss the relative probability of lightning striking the following objects during a thunderstorm: a mountaintop; a valley between mountains; a flat, open field; a single tree in a flat, open field; a person standing in a flat, open field; a person squatting in a flat, open field; a man and a boy walking across a field; a particular tree in a forest of many similar trees; a building on a flat terrain; a metal tower; a lightning rod.

33. Why is there a stronger electric field near the point of a lightning rod than anywhere else nearby?

34. Why is an automobile or airplane a safe place to be during a lightning storm?

35. Would a car with a fiberglass body be a protection during a lightning storm?

Problems

1. What must the minimum rating of a fuse be if it is used in a circuit for a 4800-W water heater operating on 240 V?

   Answer: 20 A

2. A 120-V bathroom circuit with a 20-A circuit breaker supplies a 1500-W heater, a 40-W fluorescent light, and a receptacle. Can you operate a 1200-W hair drier from the receptacle without tripping the circuit breaker when the heater is on?

   Answer: no, you need 23 A

3. Calculate the rate of heat production in a 40-Ω resistor when connected to a 220-V line.

   Answer: 1200 W

4. In Problem 3, calculate the total heat produced when two such resistors are connected in series. In parallel.

   Answer: 600 W; 2400 W

5. A resistance of 10 Ω is connected across a 100-V power supply. How much heat is produced in the resistor? How much heat is produced if the resistance is doubled?

   Answer: 1000 W; 500 W

6. For each of the alternatives of Question 18, what is the rate of heat production?

   Answer: 760 W; 1000 W; 430 W; 1800 W

7. An electric heater is rated 1200 W at 120 V. If the voltage is actually 115 V and the resistance does not change, what is the actual rate of heat output? The resistance of the heating element is probably less on 115 V than on 120 V, because it does not get quite as hot. Would the real heat output be slightly more or slightly less than

you calculate? Hint: Calculate I and R for the rated voltage, then I and P for the lower one.

   Answer: 1100 W; more

8. The tungsten filament of an incandescent light bulb has a resistance at operating temperature of about 10 times what it has at room temperature. How does the rate of heat production when the bulb is first turned on compare with that a small fraction of a second later, when it reaches operating temperature?

   Answer: 10 times as high initially

9. How does the *diameter* of the wire needed to supply current to a 5000-W range that operates on 220 V compare with the *diameter* of the wire needed for a 5000-W range that operates on 110 V? Hint: The cross-sectional area of the wire should be proportional to the current to keep the amount of current through each unit of area the same.

   Answer: $\sqrt{2}$ times as big for 110 V

10. If R_L and R_R in Figure 16.20 add up to about 10^5 Ω when you grab the two sides of a 110-V power line, what is the approximate current through your body? Is this current likely to hurt you badly? If the body resistance is 500 Ω, approximately how low would the contact resistances need to be to cause ventricular fibrillation?

    Answer: 1.1 mA; 600 Ω

11. If a lightning bolt flashes across a potential difference of 200 MV between cloud and ground, and carries down a charge of 20 C, how much energy is dissipated? How high could this much energy lift a 20,000-kg bulldozer if converted to mechanical work?

    Answer: 4.0 GJ; 20 km

12. It takes about 7800 kWh of energy to heat the average house in Atlanta for a year. How many lightning flashes of the intensity described in Problem 11 does it take to dissipate this amount of energy?

Answer: 7

13. From the average potential difference between the earth and the upper atmosphere (about 350 kV), and the average current through the atmosphere (1800 A), what is the resistance of the atmosphere as a whole?

Answer: 194 Ω

14. What is the heat generation in the atmosphere from the current and resistance described in Problem 13?

Answer: 630 MW

15. Based on the answer to Problem 11, how many such lightning flashes would need to occur over a 1-km^2 area to dissipate the amount of energy released in the atmosphere as 2 cm of rainfall condenses? (Heat of vaporization of water = 540 kcal/kg.)

Answer: 11,000

Home Experiments

1. If you live in an apartment or house, figure out which lights, receptacles, etc. are served by which circuit breakers or fuses in your service panel. (You can do this by turning off one circuit at a time.) Calculate the maximum current in your most heavily used circuit, and compare it with the fuse or breaker rating. If you don't have such a residence, perform this service for a friend who does.

2. Take two wires, paper clips, or other conductors in contact with the terminals of a 1.5-V flashlight battery, and touch them to your tongue. Notice the distinct taste. You won't get the taste without the wires contacting the battery terminals. Does this experiment give you a hint that taste is an electrical phenomenon? Try a 9-V calculator battery; it tingles a little, but the taste is still there. Why don't you feel that tingle by putting your finger across the terminals? (Some people test batteries in this way—by "tasting" the terminals.)

References for Further Reading

Blatt, F. J., "Nerve Impulses in Plants," *The Physics Teacher*, Nov. 1974, p. 455. Electricity plays a role in plants as well as animals.

Dalziel, C. F., "Electric Shock Hazard," *IEEE Spectrum*, Feb. 1972, p. 41. Discusses the effect of various current levels on the body, special precautions needed in hospitals, and equipment for reducing shock hazard.

Few, A. A., "Thunder," *Scientific American*, July 1975, p. 80. Describes how the sound of thunder is used to give information on the lightning flash.

Feynman, R. P., R. B. Leighton, and M. Sands, *The Feynman Lectures on Physics*, Vol. II (Addison-Wesley Publishing Co., Reading, Mass., 1964). See Chapter 9. An interesting and readable account of atmospheric electricity.

Fiess, M., "How Is Lightning Sparked?" *Science 83*, June 1983, p. 104. Summarizes current research and theories concerning the still-unanswered question as to why clouds polarize into positive and negative parts.

Jensen, K., "How to Troubleshoot Electrical-Accessory Failures," *Popular Science*, Aug. 1977, p. 124. Discusses wiring in automobiles and how to locate failures therein.

Malan, D. J., *Physics of Lightning* (The English Universities Press Ltd., London, 1963). A detailed description of lightning and atmospheric electricity. Many of the numerical data cited in this chapter were taken from Malan's book.

Mazda, F. F., *Integrated Circuits* (Cambridge University Press, Cambridge, England, 1978). Chapters 3 and

4 show how basic electronic components carry out various arithmetic functions.

Orville, R. E., "The Lightning Discharge," *The Physics Teacher*, Jan. 1976, p. 7. Franklin's study and the current understanding of the lightning flash, including some interesting lightning photographs.

Reid, T. R., "The Chip," *Science 85*, Feb. 1985, p. 31. The people and the story behind the development of the integrated circuit chip.

Sinclair, I. R., *Beginner's Guide to Integrated Circuits* (Butterworth & Co., London, 1977). A readable introduction to transistors, ICs, and their applications.

Time-Life Books, *How Things Work in Your Home (and What to Do When They Don't)* (Time-Life Books, New York, 1975). Contains a very good discussion of the layout and principles of home electrical wiring. Also gives a well-illustrated description of most of the ordinary electrical appliances that produce heat.

ELECTRO-
MAGNETISM

Almost everybody has played with magnets at some time. They're fun as toys and they're interesting, but are they useful? Magnets hold cabinet and refrigerator doors closed; magnetized screwdrivers are handy for putting screws in inaccessible places; compasses, really just small magnets, assist in navigation; magnets sometimes do specialized jobs, as in medicine (Fig. 17.1); but are magnets *widely* used? Do they actually influence your life very much? The answer is yes *only* if you do such unusual things as listen to a record, turn on a tape recorder, start your car, shave with an electric shaver, wash your clothes in a washing machine, beat your eggs with an electric mixer, ring a doorbell, listen to a radio or watch a TV, talk on a telephone, read under an electric light, or go into an air conditioned building.

As we learn the fundamentals of magnetism in this and the next two chapters, we will see what it has to do with ordinary everyday activities. And we will see that we are not really changing subjects from the last chapter, that magnetism is actually part of electricity.

We start this chapter by discussing permanent magnets and describing their effects in terms of magnetic fields. We'll see that these ideas apply not only to magnets of ordinary size, but to magnets ranging in size from that of the earth down to that of the atom. We'll then show that all magnetism results from electric currents, even that of permanent magnets. In Chapter 18 we turn our viewpoint around and examine how magnetic fields exert forces on moving charges. This will equip us to describe how motors and generators work. In Chapter 19 we'll find that

Swallowed objects removed with magnets

CHAPEL HILL, N.C. (AP)—A pot-holder magnet inserted in the end of a rubber tube is an ideal instrument for the removal of bobby pins, needles or nails from the stomach, according to two local radiologists.

Drs. George M. Himadi and Gary L. Fischer of North Carolina Memorial Hospital suggest that this easily assembled device can be used for the removal of elongated foreign bodies frequently swallowed by children.

And the technique often makes surgery unnecessary, the radiologists say.

The physicians recently used the magnet taken from a common kitchen potholder to remove a bobby pin from a child's stomach. The tube was passed through the child's mouth into the stomach where rapid contact was made with the foreign body under fluoroscopic control. The pin was removed easily within two minutes.

Figure 17.1 As this actual news article shows, people sometimes use magnets in surprising ways.

a changing magnetic field can produce an emf, a fact that is the basis for the functioning of transformers. We'll close that chapter by discussing electromagnetic waves, in particular those sent out by radio and TV stations.

17.1 PERMANENT MAGNETS

The most commonplace magnets are small bar- or horseshoe-shaped pieces of iron that attract nails and other objects made of iron or steel. These are called permanent magnets because they hold their magnetism indefinitely. If you stick one in a bucket of nails, the nails will cling only to the ends (Fig. 17.2). That is, the force attracting the nails is concentrated at the ends of the magnet. We call these regions of force concentration magnetic *poles*.

You probably already know that the two poles on each magnet are of different types that we call *north* and *south* poles. And, as Figure 17.3 shows, like poles—two north poles or two south poles—repel each other. Unlike poles—a north and a south pole—attract each other. Compare this effect with what we said about forces between like and unlike charges. There is one important difference: The individual charges can exist by themselves, but magnetic poles always come in pairs. On every magnet there is a north and a south pole. If you cut the bar magnet into two pieces, you always end up with two magnets, each of which has a north and a south pole. You cannot isolate the poles.

Figure 17.2 Only the poles of a magnet attract iron objects.

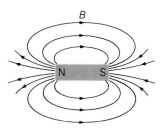

Figure 17.3 Like poles repel each other; unlike poles attract each other. N stands for north pole, S for south pole. F is the force on each pole due to the other one nearby.

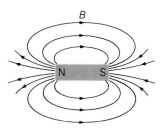

Figure 17.4 The field lines surrounding a bar magnet.

Figure 17.5 (a) A north pole experiences a force in the direction of the magnetic field. (b) A south pole experiences a force in the opposite direction.

Magnetic Fields

In our discussion of forces between charges, we described the interaction in terms of the electric field surrounding any charge. A second charge placed in the electric field of the first experiences a force because of the field. We can describe magnetic interactions in the same way. Surrounding any magnetic pole there is a **magnetic field**. If we place another magnetic pole in this field, there will be a force on the second pole. Just as we did for the electric field, we represent a magnetic field by lines that tell us two things about the field: (1) the direction of the line represents the direction of the field and (2) the stronger the field, the more closely the lines are spaced. (We use the symbol B to represent the magnetic field.)

Since magnetic poles always exist in pairs, we don't have straight field lines extending outward from the poles as we do for an isolated electric charge. Instead, the magnetic field lines curve around, as illustrated in Figure 17.4 for a straight bar magnet. For magnets of different shapes, the field lines take on different configurations, but in a general north-to-south direction.

The direction of the magnetic field is the direction of the force on a north pole placed in the field (Fig. 17.5-a). The north pole of the right-hand magnet feels a force to the left—is attracted to the center one—because the field there points to the left. The north pole of the left-hand one experiences a force to the left—is repelled by the center one—because the field *there* points to the left. Figure 17.5(b) illustrates that the force on a south pole is in the opposite direction of the field at that point. The (unshown) forces on the center magnet result from the fields of the outer two magnets.

The Earth as a Magnet

A compass needle aligns itself nearly in a geographic north-south direction. The same end of the needle that points north will point toward the *south* pole of a bar magnet. That's because the earth has a magnetic field similar to that of a bar magnet and oriented as shown in Figure 17.6. The axis of the magnetic field is at a slight angle to the earth's axis of rotation, which determines geographic north and south.

(a)   (b)

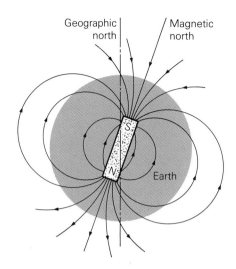

Geographic north

Magnetic north

S

N Earth

Figure 17.6 The earth's magnetic field is very much the same *as if* it were a large bar magnet whose axis is tilted slightly relative to the earth's axis of rotation.

The names "north" and "south" for magnetic poles came about because suspended magnets aim themselves approximately toward the poles of the earth. Since a north magnetic pole is attracted to a south magnetic pole, the earth's *north* pole must be a *magnetic south* pole.

· ·

QUESTION   Is the end of a compass needle that points north a magnetic north or a magnetic south pole?

N

↑↑↑↑↑
↑↑↑↑↑
↑↑↑↑↑
↑↑↑↑↑
↑↑↑↑↑
↑↑↑↑↑
↑↑↑↑↑
↑↑↑↑↑
↑↑↑↑↑

S

Figure 17.7 In a permanent magnet, the atomic magnets are aligned with their north poles (shown by arrowheads) predominantly in one direction.

Atomic Magnetism

We mentioned that if you break a magnet in half, you do not get two isolated poles, but two magnets, each of which has a north and a south pole—each is a *dipole* (two poles). This effect will continue, no matter how many times you break the magnet, until you get to the individual atoms. Each atom of a permanent magnet is a tiny magnetic dipole.

Figure 17.7 shows how you can get a bar magnet with only two poles from many small "bar" magnets. The arrows represent atomic magnets, with the arrowhead standing for the north pole of each small magnet. All the atomic magnets are aligned in the same direction. In the interior, north and south poles are adjacent and cancel the outside effect of each other. But on one end there is an exposed set of north poles, on the other end an exposed set of south poles. The overall result is a net north pole at one end and a south pole at the other. In an unmagnetized material, the atomic magnets are arranged randomly so that there are no overall magnetic poles.

Figure 17.8 In iron, all the atoms in a given domain have their magnetic dipoles pointing in the same direction. When the material is unmagnetized, the domains are oriented randomly.

Figure 17.9 (a) The nail is unmagnetized, and the domains are oriented randomly. (b) The magnetic field of the magnet causes a rotation of the dipoles, giving a net south pole at the end closest to the north pole of the magnet.

In iron, the magnetism of each atom is so strong that the atoms group themselves into clusters called *domains*, with all the atoms in a domain pointing in the same direction. In an unmagnetized piece of iron, the domain directions are oriented at random, giving no overall preferred orientation and no overall outside magnetic effect (Fig. 17.8). Even though each cluster contains many atoms, the domains are still extraordinarily tiny by ordinary standards, and we can think of them as small single dipoles.

If we put a piece of iron in a magnetic field, the atomic north poles experience a force in the direction of the field, the south poles a force opposite to the field. As a result, the dipoles shift their orientation more nearly in line with the field. The stronger the field, the better the alignment. In relatively pure iron, called "soft" iron, the domains regain most of their random orientation when removed from the outside magnetic field. But in steel and in certain other alloys, they cannot shift so easily, and the material stays magnetized even without an outside field. The amount of magnetism retained depends on the nature of the material.

Let's see why unmagnetized iron is attracted to *either* pole of a magnet. The nail of Figure 17.9(a) has its atomic dipoles oriented at random when there is no outside field. When the north pole of a magnet comes near, as in Figure 17.9(b), the magnetic field causes the dipoles to rotate so that a net south pole is at the end of the nail nearest the magnet; a north pole is farther away at the other end. The nail is then attracted to the magnet. If a south pole were brought near the nail, the dipoles would rotate so that the end of the nail toward the magnet becomes a north pole, again giving attraction. The alignment is usually not as complete as implied by the figure.

(a)

(b)

· ·

QUESTION Explain why the nail is attracted to the magnet if one of the poles is brought near the middle of the nail.

17.2 THE MAGNETIC FIELD OF A CURRENT

In 1820 Hans Christian Oersted reported the discovery of an effect that makes possible the widespread use of electricity that we enjoy today. He discovered the connection between electricity and magnetism. That con-

(a)

(b)

(c)

(d)

Figure 17.10 (a) If the wire is perpendicular to the compass needle, then (b) nothing happens when the switch is closed. (c) If the wire is first parallel to the needle then (d) it turns until it is perpendicular when the current comes on.

nection makes it possible to generate electricity in the huge quantities used by our society.

Even though he had been searching for a link between electricity and magnetism for some time, Oersted apparently made the actual discovery during one of his lectures. Having a battery and compass on his lecture table for other demonstrations, he noticed that when he ran a current through a circuit, the nearby compass needle deflected. On closer examination, he found that if he ran a current-carrying wire directly above the compass, with the wire perpendicular to the needle, as in Figure 17.10(a), nothing happened to the compass when he closed the switch (b). But if the wire were first parallel to the needle (c), the needle would swing around perpendicular to the wire when he turned the current on (d). This discovery illustrated that a *magnetic field exists in the region surrounding every electric current.*

To better understand this field, imagine a wire perpendicular to and through the page, as shown in Figure 17.11, with a current upward in the wire (out of the paper). If you place compasses around the wire, they will point in the directions shown: The north pole of each points counterclockwise around the wire, the south pole points clockwise. Apparently the field around the wire is in a direction tangent to a circle centered on the wire. Thus, the field lines are concentric circles around the wire (Fig. 17.12).

We can remember the direction of the field (whether clockwise or counterclockwise) by what is called a *right-hand rule:* Imagine that you grab the wire with your *right* hand so that your thumb, extended along the wire, points in the direction of the conventional current (Fig. 17.13). Your fingers then point in the direction of the field around the wire. Your hand, of course, has nothing to do with the field. This rule is merely a way to remember its direction.

The Field around Curved Conductors

Suppose that, instead of being straight, the wire carrying the current is bent into a circular loop, as shown in Figure 17.14. If you use the right-hand rule near any segment of the wire, you can find the direction of the magnetic field there. For the current direction shown, the field is upward everywhere *through* the loop; it is downward everywhere outside in the plane of the loop. The field lines curve around as shown in the figure.

EXERCISE Use the right-hand rule to explain the direction of the field lines near the wire of Figure 17.14.

Figure 17.11 Compasses around a straight current-carrying wire (current up through the paper) point in a direction tangent to a circle around the wire. The arrows represent the north poles of the compass needles.

Figure 17.12 The magnetic field lines around a straight current are concentric circles around the wire. The field is everywhere tangent to the circles and is in the direction of the arrows.

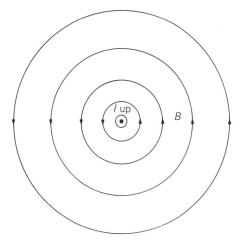

If we have several *loosely-wound* turns of wire, the field from the various turns combines to give an overall shape as in Figure 17.15. A straight coil of many *closely-spaced* turns, called a *solenoid*, gives field lines outside the solenoid that are similar to Figure 17.16(a). Notice that these field lines are very much like those for a bar magnet (Fig. 17.16-b). This similarity means that a solenoid carrying a current acts as a bar magnet. There are some useful differences. You can vary, or turn on and off, the magnetism of the solenoid merely by varying, or turning on and off, the current. Also, you can reverse the north and south poles of the solenoid by reversing the current's direction. Compare the magnetic field outside these magnetic dipoles with the electric field surrounding the electric dipole of Figure 14.14(b).

Figure 17.13 Right-hand rule: *Imagine* that you grab the wire with your right hand so that your thumb is in the direction of the current; your fingers then point in the direction of the magnetic field.

Figure 17.14 The magnetic field around a circular loop of current.

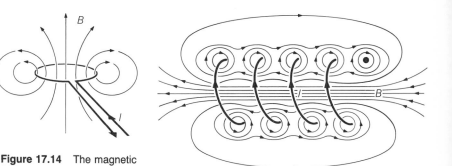

Figure 17.15 The field of several turns of current.

17.3 MAGNETISM IN THE ATOM

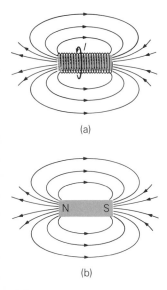

(a)

(b)

Figure 17.16 (a) The field outside a solenoid is identical to (b) the field outside a bar magnet.

We made the statement in the opening remarks of this chapter that all magnetism results from electric currents. Yet a permanent magnet needs no wires connecting it to a source of emf to make it work. There are two ways in which electric currents establish the magnetic fields of permanent magnets. Both of these are related to the electrons within the atoms of the magnet. The nuclei of the atoms also contribute a small amount, but this is negligible for most purposes.

As you know, an electron has a negative charge. In the planetary model of the atom (a useful approximation for many purposes), the electron is pictured as orbiting the nucleus much as a planet orbits the sun (Fig. 17.17-a). The movement of this negative charge in orbit makes up a conventional current along the path of the orbit (in the direction opposite to the movement of the electron). This current sets up a magnetic field in the direction shown just as the current in a loop of wire would.

The second and larger contribution to the atom's magnetism comes about because each electron acts *as if* it were a charged object spinning about an axis through its center. At this time we do not know anything about the internal structure of the electron, or if it even has one. And we cannot say whether or not it actually is spinning. But the magnetic field produced by the electron is what we would expect if that were the case. Imagine the electron to be a spinning charged sphere as shown in Figure 17.17(b). Then each little chunk of charge would move in a circle and form a current loop in the direction opposite to its direction of motion. Each of these current loops would contribute to a magnetic field in the direction shown.

These two effects of the atomic electrons give each atom its magnetism and cause it to act as a tiny bar magnet. The combined effect of all these atomic magnets gives the permanent magnet its magnetism. The reason iron can be so strongly magnetized is that several electrons in each iron atom have their spin axes pointing in the same direction, making each atom strongly magnetized. The description in terms of planetary motion and spinning electrons is only a model that assists us in visualizing how the field comes about. This model should not be taken too seriously as an actual picture of the atom or the electron.

17.4 ELECTROMAGNETS

When we put a material medium inside the windings of a solenoid, the magnetic field inside and outside is changed due to the action of the field on the atoms that themselves act as little magnets. If this material, called a core, is iron, it becomes strongly magnetized. As a result, the field outside the solenoid is much stronger than it is without the core. If the core is soft iron, it demagnetizes almost as soon as the current is turned off.

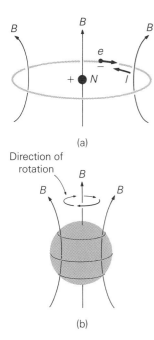

Figure 17.17 (a) In the planetary atomic model an electron e orbiting the nucleus N of an atom constitutes a current I that causes a magnetic field B. (b) If the electron were a spinning charged sphere, each bit of moving charge would be a current that would contribute to the overall magnetic field.

We call such an arrangement—coils of wire wound around soft iron—an *electromagnet*. Electromagnets have hundreds of practical applications. We'll mention a few.

Applications of Electromagnets

The reason electromagnets are so useful is that their fields can be varied easily by controlling the current to their coils. Industry uses many kinds of electromagnets, such as the large ones that pick up scrap iron and release it at will. But you and I use many electromagnets in our everyday routines.

Door chimes, for example, are a simple application of electromagnets. Figure 17.18 shows their basic components. When you push the button, it closes a circuit that passes current through the solenoid. This draws the core into the solenoid, making the nonmagnetic core extension hit the tone bar on the right (bing!). When you release the button, the magnetic field vanishes so that the spring pushes the core against the tone bar on the left (bong!). Most door chimes have more solenoids than this, but the basic principle is the same.

Relays are solenoids that operate switches in a second electrical circuit. A current through the solenoid of Figure 17.19 attracts the armature on the right, which swings over, closing the contacts. This allows current from the large battery, or other source of emf, to pass through the device being operated. Why not just put the switch S in the circuit on the right and avoid the solenoid? This arrangement is sometimes appropriate—for example, in a table lamp—but for many applications the switch used cannot carry enough current to operate the device.

Figure 17.18 (a) Door chimes are electromagnets that pull the core into the solenoid when you close the switch (push the button), thus (b) causing the core extension to hit the tone bar on the right. When you release the button, the spring pushes the core back to the left (c), hitting the tone bar on the left.

Figure 17.19 A relay: When the switch *S* in the low-current circuit is closed, the electromagnet closes the switch in the high-current circuit.

Figure 17.20 In a loudspeaker the coil vibrates in response to the force between the coil and the magnet. This force varies in proportion to the current through the coil.

For example, we discussed in Chapter 8 how thermostats turn heating systems on and off. It isn't practical to have the bimetallic strip of the thermostat carry the full current needed to operate a furnace. The thermostat is used as the switch *S* in Figure 17.19. This low-voltage circuit then operates the relay that closes the main circuit, which sends a much larger current to the furnace.

You can find solenoids and relays in all types of useful devices: They turn on air conditioners when the thermostat calls for more heat removal; they operate the mechanisms in tape recorders; they release the toast when it's done in a pop-up toaster; they control the filling and mixing of water in washing machines, to give the selected water temperature; they connect your phone to your friend's when you dial her number.

How would your life be different without loudspeakers? Everything from sound systems for listening to music to telephone receivers would vanish without them. Most speakers use electromagnets as their main component and are typically arranged as in Figure 17.20. The purpose of a loudspeaker is to convert variations in electrical current to vibrations in the air molecules. The current fed to the loudspeaker passes through the coil shown in the diagram. The force of attraction between coil and permanent magnet is proportional to the current in the coil. The coil, therefore, oscillates in proportion to the current. The resulting vibrations in the cone set up vibrations in the air that have the same variation with time as the current in the coil.

Superconducting Magnets

Large electromagnets use thousands of amperes of current in their windings. Even if the windings have very low resistance, you can easily estimate the huge amount of electrical energy lost to heat in such magnets. (Remember the rate of heat production is I^2R.) If the resistance of the windings can be made *zero*, this loss of electrical energy is eliminated.

As we discussed in Chapter 15, some materials do become superconducting—have zero resistance—at temperatures near absolute zero. Magnets with superconducting coils can carry very high currents and can routinely produce fields six or seven times those conveniently reached with conventional magnets. The big drawback is that the coils have to be kept at a temperature near absolute zero; no easy feat in itself. The usual way of doing this is to surround the coils with liquid helium at a temperature of −270°C (3 K). With the development of materials that become superconducting at much higher temperatures, superconducting magnet technology may change considerably for a number of years.

. . . .

Key Concepts

Permanent magnets always exist with two *poles*, a north pole and a south pole. Two north poles or two south poles repel each other; a north and a south pole attract each other. Surrounding a magnet is a **magnetic field** whose direction is the direction of the force on a north pole.

A magnetic field surrounds every current. Around a long, straight wire carrying a current, the field is directed tangent to a circle concentric with the wire. *Solenoids* have fields similar to those of bar magnets. A solenoid wound on an iron core forms an electromagnet.

The atoms of a magnetic substance act as tiny bar magnets because of the electron spin and orbital motion. In iron, the atoms group themselves into *domains* in which all the atomic magnets point in the same direction. In a magnetized piece of iron, all the domains are oriented predominantly in the same direction.

Questions

Permanent Magnets

1. What are the similarities and differences in the forces between magnetic poles and the forces between electric charges?

2. Suppose you were locked inside a room made of glued-together wood. You have a bar magnet and an identical-looking piece of iron, but nothing else. The guard will let you out *only* if you toss out the magnet and keep the iron bar. How can you be sure your choice will not condemn you for life?

3. Why can't you create a single isolated magnetic pole by breaking a bar magnet in half?

4. Why is a nail attracted to *either* pole of a magnet, but another magnet is attracted only to one of the poles?

Magnetic Fields of Currents

5. A magnetic field surrounds a current-carrying wire. Is there also an electric field around such a wire? Why not?

6. We represent magnetic fields by lines in the vicinity of the current producing the field. Does this practice imply that the field exists only at the position of the lines? Does the fact that we put arrows on these lines imply that there is something "going around" the current-carrying wire?

7. The magnetic field around a long, straight current decreases in strength in inverse proportion to the distance from the current. If we draw 12 lines/cm² to represent the field 3 cm from a straight current-carrying wire, how dense should the lines be 12 cm away?

Figure 17.21 Why does the bell ring continuously when the button is pushed? (The button is a switch that makes the wire continuous when it is pushed.)

To electric power supply

8. Could you find the location of the wires in the walls of your house using a compass? Could you do it if the current were DC?

Magnetism in the Atom

9. Sketch the magnetic field of an electron based on a model that assumes it to be a tiny rotating cylindrical shell (like the cardboard in a roll of paper towels) of negative charge. Is this model any better or worse than the one in Figure 17.17(b)?

10. Geologists believe that the earth's magnetic field direction has reversed at least 12 times in the last nine million years. This theory is based in part on the orientation of the grains of magnetic rock called magnetite that have been deposited in lake bottoms over the years. What can you say about the relative orientation of

the iron atoms in adjacent layers of this magnetite?

11. Justify the statement that all magnetism, even that of permanent magnets, results from electric currents.

Electromagnets

12. Would it matter if the battery in Figure 17.18(a) were reversed? How about the small one in Figure 17.19?

13. Design an addition to the door chime of Figure 17.18 that has another solenoid for the back door—one that "bings" but doesn't "bong."

14. Figure 17.21 is a diagram of a doorbell that rings continuously while the button is pushed. Explain its operation. Notice that as the clapper moves to the left, the electrical circuit is broken.

Home Experiments

1. Make a homemade compass. Magnetize a steel sewing needle by rubbing it on one pole of a magnet. Rub always in the same direction and move the needle away from the magnet on the return for the next stroke. Test your efforts and make the compass by floating the needle horizontally on a cork in a cup of water. It should align itself in a north-south direction.

2. Make a small electromagnet by winding a couple of hundred turns of small, thinly-insulated

wire around a nail. (Such wire is available inexpensively from local retail electronics stores.) Test your magnet on various iron and non-iron metal objects by holding the ends of the wire to the terminals of a flashlight battery. Minimize the time you maintain contact with the battery in order to minimize running down the battery. The longer and thinner the wire, the higher its resistance, and therefore the less current "drain" on the battery.

References for Further Reading

Bobeck, A. H., and H. E. D. Scovil, "Magnetic Bubbles," *Scientific American*, June 1971, p. 78. Discusses the way in which magnetic domains are formed into bubbles for use in computer memory.

Gould, J. L., J. L. Kirschvink, and K. S. Deffeyes, "Bees Have Magnetic Remanence," *Science*, Sept. 15, 1978, p. 1026. Magnetic materials within honey bees let them orient themselves to the earth's magnetic field.

"Is the Earth's Field Heading for a Flip?" *Science 84*, March 1984, p. 14.

Levi, B. G., "Magnetite in Bacteria Linked to Orientation," *Physics Today*, Nov. 1979, p. 22. Evidence that not only bacteria but honey bees and pigeons use magnetic material for navigation.

Mee, C. D., *The Physics of Magnetic Recording* (North-Holland Publishing Co., Amsterdam, 1968). A technical discussion of the techniques and components of magnetic tape-recording devices.

Schwartz, B. B., and S. Foner, "Large-scale Applications of Superconductivity," *Physics Today*, July 1977, p. 34. Applications of superconducting magnets.

The Way Things Work, an Illustrated Encyclopedia of Technology (Simon and Schuster, New York, Vol. 1, 1967, Vol. 2, 1971), or an equivalent publication, *How Things Work, The Universal Encyclopedia of Machines* (George Allen & Unwin Ltd., London, Vol. 1, 1967, Vol. 2, 1971). Discusses the physical principles and construction details of many electromagnetic devices.

C H A P T E R   1 8

ELECTROMAGNETIC FORCES

You can hardly spend more than a few minutes awake without seeing the effect of a magnetic field. The forces that magnetic fields exert on moving charges are responsible for the operation of electric motors that appear in every conceivable piece of electrical apparatus. They are responsible for the ability of generators to produce most of the electrical energy used in our society. And they are even responsible for the picture you see on your TV.

In this chapter we investigate the basic phenomena associated with the force on moving charges in a magnetic field. We consider these forces on moving free charges and on charges that make up the current in a wire. Once we understand the fundamental properties of these forces, we can see how they operate in the practical devices mentioned.

18.1 FORCES ON MOVING FREE CHARGES IN A MAGNETIC FIELD

We saw in Chapter 14 that a net charge, either stationary or moving, gives rise to an electric field. Furthermore, another charge put in such an electric field experiences a force on it caused by the field. In the last chapter we saw that currents—*moving* charges—give rise to *magnetic* fields. Comparing effects in electric and magnetic fields, you might ask whether or not a second charge moving in this or another external mag-

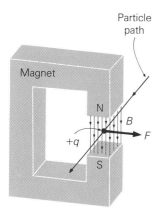

Particle path

Magnet

N

B

F

+q

S

Figure 18.1 A positive charge +q moving in the direction shown through a magnetic field B feels a force directed to the right that is perpendicular both to the charge's direction of motion and to B.

netic field would experience a force due to this field. The answer to that question is *yes*, provided its motion is in the right direction. Just as a stationary charge produces no magnetic field, a stationary charge in a magnetic field experiences no force.

Consider the situation shown in Figure 18.1, where a positively charged particle moves between the poles of a C-shaped magnet. The magnetic field of the magnet is down (north to south), and the particle moves in the direction shown by the arrows on its path. You might expect a force to be in the direction of the magnetic field B. Instead, the force is perpendicular to both the magnetic field and the direction of motion of the charge: to the right in this case.

The force has the following characteristics:

1. The force is proportional to the amount of charge.

2. The force is proportional to the speed of the charge. (This implies *no* force on a stationary charge.)

3. The force is proportional to the strength of the magnetic field.

4. For a given charge, speed, and magnetic field, the force is strongest when the particle motion is perpendicular to the field. It decreases in strength as the angle between its path and B decreases, becoming zero when the motion is parallel to the field.

5. The force is directed along a line perpendicular to both the particle direction and the magnetic field.

Notice that the field is *not* the force; the field *causes* the force on the *moving* charge.

Item 3 in this list concerns the *strength* of the magnetic field. The term usually used to express this strength has the unfortunate name *magnetic induction*. (The symbol B we've been using is the one normally used for magnetic induction.) The SI unit of magnetic induction is the tesla (T).* To give you an idea of the size of the tesla, here are a few examples. The field very near a magnetic cabinet latch is typically of the order of about $\frac{1}{100}$ to $\frac{1}{10}$ T. The earth's average magnetic field at its surface in the United States is about 0.000055 T. Large electromagnets usually have fields no stronger than about 2 T. The strongest fields that can now be produced are about 30 T.

You might see magnetic fields expressed in the unit gauss (G), where 1 G = 0.0001 T.

Item 5 doesn't tell us which way along the line the force acts. We find the direction by the right-hand rule shown in Figure 18.2. Imagine that you hold your right hand so that its thumb points in the direction of

Quantitatively, when the particle's velocity v is perpendicular to B, the force is given by $F = qvB$. This equation shows that the tesla is that magnetic induction that would cause a force of 1 N on a charge of 1 C moving perpendicular to the field with a speed of 1 m/s.

Figure 18.2 The right-hand rule for the force on a positive charge: If the thumb points in the direction of motion of the charge, and the fingers point in the direction of the magnetic field, the force is in the direction you would push with your palm.

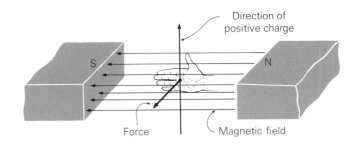

motion of a *positive* charge, and your extended fingers point in the direction of the magnetic field. Then the force is in the direction you would push with your palm. The force on a *negative* charge is in the *opposite* direction to this. Again, remember that your right hand has nothing to do with the force. It merely gives you a way to remember the direction.

You may be wondering *why* the force is in this particular direction. That's not a question we can answer. We can say only that that's the way nature is. Here, as always, the laws of physics describe *what* happens but not *why* it happens.

Our diagram, Figure 18.1, shows the particle moving in a straight line through the field. If this is a free particle moving through the region, the force will change its direction so that the path will bend to the right. The amount of bending depends on the force and on the mass and speed of the particle.

APPLICATION   THE TV PICTURE TUBE

By now you're probably saying, "Big deal! Who, besides some ivory-tower physicists, cares if there's a force on a charged particle moving through a magnetic field?" Well, *you* do if you enjoy watching TV.

Figure 18.3(a) shows the overall structure of a TV picture tube. Electrons emitted from the "electron gun" (Fig. 18.3-b) travel along the tube until they strike the inner surface of the screen that's coated with a phosphor. The phosphor gives off light (we'll see why in Chapter 23) at any point hit by electrons. If the electron beam could be swept quickly back and forth as well as up and down over the screen, the entire screen would give off light. That's where our force on particles moving in a magnetic field comes in.

Figure 18.3(c) shows the coils (solenoids) that are above and below, as well as on the left and right, of the tube neck. Current in the vertical coils gives rise to a vertical magnetic field that causes a horizontal force on the electron beam, bending its path to the right or left. Current in the horizontal coils gives a field that bends the beam up or down. For example, an upward vertical field bends the (negatively charged) electron beam to the left in the diagram (to the right facing the front of the screen). A field to the right bends the beam upward. Varying the current

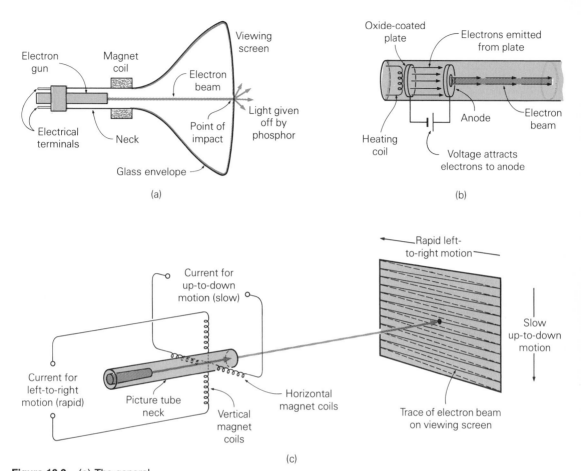

Figure 18.3 (a) The general layout of a TV picture tube, as if you take a slice through its middle. (b) The "electron gun." Electrons given off by the heated oxide coating are accelerated by the voltage between this surface and the anode. The anode has a hole in it that lets electrons pass through to the neck of the tube. (c) Varying the current through the magnet coils varies the magnetic field, causing it to sweep the electron beam back and forth, up and down along fixed lines on the screen.

in the magnet coils in just the right time sequence makes the beam trace out from left to right the top line on the simulated screen, return quickly along the dashed path, trace out the second line, and so on down the screen. In TV sets in North America, 525 such lines are traced out across the screen every $\frac{1}{30}$ second. That's so fast that your eyes see a uniformly bright screen.

But a solid white TV screen doesn't make a very interesting picture. Suppose you want a TV picture of a black circle on a white piece of paper. Figure 18.4 shows the lines traced out by the electron beam on the screen. (On a real screen these lines are much closer together.) By turning the beam off just as it gets to the edge of the circle in each sweep, and back on at the other edge, we get a dark circle on a light background. We can get pictures *even more* exciting than this by varying the intensity of the electron beam to give the right shade of grey at just the right spot on the screen. (We're assuming black and white TV.) To produce a moving picture, each scan of the screen depicts the proper distribution of light intensities at $\frac{1}{30}$-s intervals.

The detailed methods of getting the variations in beam intensity exactly synchronized with the beam sweep to produce a real picture are interesting engineering techniques discussed in some of the references at the end of this chapter. In Section 22.2 we'll talk about how to get color TV.

• •

QUESTION   Would you expect the picture of a TV to shift when you rotate the TV from facing north to facing east? If so, which way? If you expect a shift, but don't see it when you try, why do you think you don't see it?

18.2 FORCES ON CURRENT-CARRYING CONDUCTORS

Figure 18.4 We get a picture of a dark circle on a light background if the beam intensity in the TV picture goes to zero just as the beam gets to one edge of the circle and comes back to full intensity at the other edge. The horizontal lines represent the traces of the electron beam sweep. On a real screen, the lines are so close together that you don't see individual lines.

Figure 18.5 If the moving charge in a magnetic field is a current *I* confined in a wire, the entire wire feels a force *F* in the direction shown.

To get a force on a charged particle moving through a magnetic field, that particle does not have to be free; it can be inside a material medium. In fact, the moving charges can be those carrying a current through a conductor.

Figure 18.5 shows the magnet of Figure 18.1 with the moving free charge replaced by a current-carrying conductor. A current *I* in the direction of the arrows gives a force to the right on each moving charge. Since the charges cannot get out of the conductor, *the whole wire feels a force in the direction of F.*

From the factors that determine the force on a moving *free* charge, it is probably reasonable to you that the force on the wire is proportional to:

1. the current

2. the length of wire in the magnetic field

3. the strength of the magnetic field

As with the free charge, the force is a maximum when *I* and *B* are perpendicular, and it decreases to zero when they are parallel. The direction of *F* is perpendicular to both *I* and *B* and is given by the same right-hand rule as for the free charge. The direction of the current is the direction of motion of positive charge.*

As we will see in the next section, the forces on moving charges and on currents form the basis for both electric motors and generators.

When a length L of conductor perpendicular to a uniform magnetic field B carries a current I, the equation F = BIL gives the magnitude of the force F on the conductor.

18.3 MOTORS AND GENERATORS

Motors

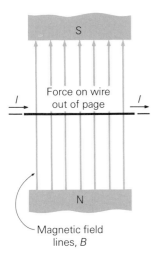

Figure 18.6 The force on the current-carrying wire is directed out of the page. The shaded area shows just the poles of a magnet, such as the one in Figs. 18.1 and 18.5.

Electric motors are the most widely used devices we have (aside from our hands) for doing mechanical work. They operate not only large industrial machines, but also home appliances such as refrigerators, washing machines, and alarm clocks, as well as hand-held devices such as drills, mixers, and shavers. If a device runs on electricity and hums or whines when it operates, it probably has an electric motor in it.

Suppose we have a wire carrying a current I to the right in the magnetic field of Figure 18.6. (To have a current in the wire, it of course has to be part of a complete circuit. We show only the part of the circuit in and near the magnetic field.) There is a force on the wire directed toward you out of the page. If the wire is free to move, it will move toward you because of this force.

Now suppose that, instead of a single wire, we have a rectangular loop in the magnetic field. Assume the plane of the loop to be in the plane of the page, with the current clockwise around the loop, as shown in Figure 18.7. The current in the upper wire is in the same direction as in the straight wire we talked about. It therefore feels a force directed *out* of the page. The ends of the loop, where the current is moving up or down, feel no force, since the current is parallel to the field. In the lower segment of the loop, where the current is moving to the left, the force is directed *into* the page. The two forces tend to rotate the loop about the dashed line through the center. We have here the basics of an electric motor: a device that has a rotatable shaft that can do mechanical work as a current passes through its internal coil. In this case, the coil is only one loop.

Figure 18.8 shows the other parts needed to have a continuously rotating coil. Each side of the loop connects to one side of a split circular piece of metal, called a *commutator*. Each half of the commutator slides against a *brush* connected to one terminal of a DC power supply. As the coil rotates, the commutator rotates with it, each side staying in contact with one of the stationary brushes. Figure 18.9 shows some end views of the loop and commutator at different orientations as they rotate. The direction of the force on the sides of the coil is shown at each position. When the coil is exactly horizontal (slightly beyond the position in Figure 18.9-c), the forces are directly opposing each other, giving no torque to the loop.* However, once the coil is moving, rotational kinetic energy carries the loop on past this orientation. Each time the coil passes through the horizontal position, the brushes lose contact with one side of the commutator and make contact with the other. This reverses the current

Figure 18.7 The force on the top of the loop is outward from the page; that on the bottom is into the page. These two forces tend to rotate the loop about the dashed line.

---

*A torque, defined in Chapter 6, results when a force is applied about an axis in such a way that it causes, or tends to cause, a change in rotation.

Figure 18.8 A motor that uses DC has a split-ring commutator that rotates with the coil, each side thereby maintaining contact with a brush connected to one side of a DC power supply. The magnet poles and one field line are shown in grey lines. The arrows *F* show the direction of the force on each side.

Figure 18.9 An edge view of the current-carrying loop at various orientations. The commutator sides interchange brushes when the coil is horizontal, thereby keeping the force to the right on whichever side of the coil is highest.

direction in the loop, keeping the force always to the right in the upper branch and to the left in the lower. The torque is always clockwise as viewed from the commutator end.

QUESTION

Except when the plane of the loop is parallel to the magnetic field, there is a force on each end of the loop. Why do these forces not affect the rotation of the loop?

For a motor that runs on AC, the current reversal is handled by the AC power supply. Each side of the loop connects to a continuous *slip ring*, as in Figure 18.10. The speed synchronizes with the frequency of the voltage reversal so that the loop passes the orientation perpendicular to the field just as the AC reverses direction.

Chapter 18   Electromagnetic Forces

Figure 18.10 A motor for AC has slip rings instead of a commutator. Each side of the loop is connected to a continuous ring that slides on one of the brushes, thereby maintaining contact between one side of the power supply and the same side of the loop. The AC power supply handles current reversal: Just as side 1 of the loop moves into the lower half of the field region, the current reverses in the external circuit, as indicated by the dashed arrows.

Figure 18.11 The components of a practical electric motor (or generator).

Practical electric motors have several rotating loops wound on an iron core—called the *armature*—in order to increase the torque and make it more steady. Also the magnet providing the field is usually an electromagnet fed by the current supplied to the motor (see Fig. 18.11).

An electric motor uses electrical energy to do mechanical work. This happens as the electric field set up by the power supply transfers energy to the charge to move it through the motor windings. This current in the armature experiences a force that moves the armature windings, thereby doing work on them as they rotate. In turn, the device hooked to the armature shaft does work on whatever it's designed to move or turn: a clipper through your hair, a fan blade that ventilates your room, a compressor in your air conditioner, or a saw that slices logs into boards for your house or desk.

Generators

Most electricity for home and industrial use is generated by devices made almost exactly like electric motors. To see how these *generators* work, let's go back and consider the single wire perpendicular to a magnetic field.

Suppose the wire of Figure 18.6 does *not* have a current passing through it. Instead, imagine that you grab the wire and quickly pull it toward you (straight out of the page). The charges in the wire would then experience a force perpendicular to both the motion and to the field—the positive charges, a force to the left, the negative charges, a force to the right. If this wire were then part of a complete conducting circuit, current would flow to the left in the wire. (Remember that negative charge moving to the right is a *conventional* current to the left.)

Figure 18.12 is like Figure 18.8, with the battery replaced by a light bulb. If you grab this coil and rotate it clockwise as viewed from the commutator end, the force on the free charges causes a current to the left in the top of the loop, and to the right in the bottom. The same amount of current passes through the bulb from top to bottom, as shown. Just as the plane of the loop is perpendicular to the field, the brushes switch contact to the other halves of the commutator, thus keeping the current in the bulb always in the same direction.

If slip rings such as those of Figure 18.10 were used instead of a commutator, the current through the bulb would be first in one direction and then in the other, depending on which part of the loop is at the top. We would then have an AC generator, sometimes called an *alternator*. Alternators are used for charging the batteries in most automobiles. Since a battery can be charged only with DC, the AC alternator output is converted to DC in devices called *diodes*, which allow current through them in one direction but not in the other.

The same device can act as both a motor and a generator, depending on its input. Figure 18.11, therefore, also shows the internal construction of a typical generator. In a motor, electric current is input causing the armature to rotate and do mechanical work. In a generator, the armature is mechanically rotated by an outside source of work, thereby generating

Figure 18.12 The basic DC generator. If the coil is rotated about its axis in the direction shown, positive charge will be pushed to the left in the top and to the right in the bottom of the loop, giving a current throughout the circuit in the direction shown. The commutator keeps the current always in the same direction in the outside circuit.

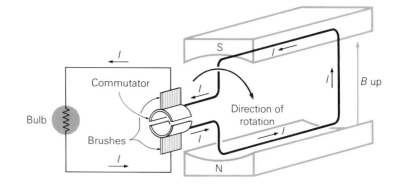

current through the attached circuits. The generator, then, uses mechanical work to generate electrical energy. In power plants, the mechanical work of turning the generator armature is usually done by a turbine operated by steam produced in a boiler or nuclear reactor. Hydroelectric plants use water turbines to turn the generator armature.

As in a motor, *electro*magnets usually supply the magnetic field of a generator. *Some* of the current supplied by the generator is fed back into these *field coils* to produce the necessary magnetic field.

. . . .

QUESTION   Does this last statement imply that the generator is running itself? Once a generator is operating, could you use some of the current from the generator to run a motor that, in turn, runs the generator, thereby eliminating the need for a turbine or other outside source of mechanical work? (The principle of conservation of energy will provide the answer.)

. . . .

Key Concepts

A charged particle moving through a magnetic field experiences a *force* whose direction is perpendicular to both the field and the direction of motion. The strength of the force is proportional to the charge, the particle speed, and the magnetic field strength; the force increases from zero when the field and particle direction are the same to a maximum when they are perpendicular. A current in a conductor—being a moving charge—experiences a force with the same characteristics.

The SI unit of magnetic field is the *tesla* (T).

An electric *motor* uses electrical energy to do mechanical work by means of the forces on the two sides of a current-carrying loop in a magnetic field. A *generator* uses mechanical work to produce electrical energy. Its basis is the force on the free charges inside a conducting loop as it rotates in a magnetic field.

Questions

Moving Charges and Currents in a Magnetic Field

1. If the charged particle in Figure 18.1 is an electron, how does the force compare with that on a proton moving at the same speed?

2. A charge of 0.5 C is at rest in a 2-T magnetic field. What is the force on the charge due to the magnetic field?

3. If there were a magnetic field parallel to this page and directed from the bottom to the top of the page, what would the direction of the force be on a negatively charged chlorine ion if it were moving toward the page from above, along a line perpendicular to the page? What if it were moving from top to bottom of the page? What if it were moving from the lower-left to the upper-right corner?

4. If a magnetic and an electric field are superimposed, describe the relative orientation of the fields and particle direction that would give no force on an alpha particle (a helium nucleus)

moving through the fields. Would it make a difference if the charge were negative?

5. If you have a Ping-Pong ball with a net positive charge on it, how could you swing a bar magnet in the vicinity of the ball and make it jump without actually hitting it? How would the ball move in relation to the magnet and its motion? Hint: Would an observer riding on the magnet say that the magnet is moving or that the charged ball is moving?

6. When a charged particle enters a uniform magnetic field perpendicularly, the force is always perpendicular to its direction of motion. What kind of path will the particle follow in the field? Will the particle be speeded up or slowed down by the field?

7. If a TV set were placed between the poles of a large magnet as shown in Figure 18.13, what would be the effects on the picture be?

8. Two parallel wires carrying a current in the same direction are attracted to each other. Why?

9. Do the two conductors in a lamp cord experience a force that tends to either compress the cord or pull it apart? If so, which? Does it matter whether the cord carries AC or DC?

10. One straight current-carrying wire lies across another so that they are perpendicular to each other. What is the nature of the forces on each wire?

Motors and Generators

11. If you run a large current through a horseshoe lying on a table, what would you expect to happen to it? (Remember the earth's magnetic field.)

12. The basis for electromagnetic meters that measure currents is similar to that of motors. Figure 18.8 would have to be changed a little to show their operation. In such meters instead of the brushes and commutator, each terminal is connected permanently to one side of the loop. The loop has a spring attached to it that opposes its rotation and lets it rotate only a limited amount.

The angle of rotation is proportional to the torque applied to the loop. Explain why the angle of rotation is proportional to the current through the loop.

13. How could you use the same device as both a motor and a generator?

14. True or false: If you take a ring off your finger and spin it on the tabletop, there will be a current in it. (Remember the earth's magnetic field.)

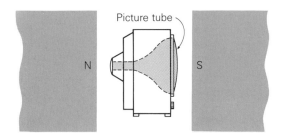

Figure 18.13 What happens to the picture on this TV?

15. Po-Dunk Power and Light Company supplies electrical power to the city of Po-Dunk Holler. How does the mechanical work used to turn Po-Dunk Power's generator in any one hour compare with the combined work done (or energy expended) by all the electric lights, tooth brushes, can openers, sawmills, aircraft assembly plants, and the like in Po-Dunk Holler during that hour?

16. When you turn on an electric light, it's a little harder to turn the armature of the generator supplying your power. Explain this fact on the basis of the energy conservation principle.

17. Imagine that the generator of Figure 18.12 connects to the bulb through a set of slip rings (as in Fig. 18.10). Now, draw a graph of bulb current versus time. What is the orientation of the loop relative to the magnetic field when the current is a maximum and when it is zero?

Problems

*1. A baseball with a net charge of 3.0×10^{-5} C is thrown with a speed of 40 m/s into the 13-T field of a superconducting magnet. If the ball moves perpendicular to the field, what is the force on the ball? If its mass is 0.15 kg, what is the acceleration of the ball due to this force?

Answer: 0.0156 N; 0.10 m/s² (0.01*g*)

*2. An electron in a TV picture tube is moving with a speed of 3.0×10^7 m/s in a direction perpendicular to the earth's magnetic field of strength 5.5×10^{-5} T. What is the magnetic force on this electron? What is its acceleration due to this force? (The electron mass is 9.1×10^{-31} kg.)

Answer: 2.6×10^{-16} N; 2.9×10^{14} m/s² ($3 \times 10^{13}g$)

*3. A straight wire carrying a current of 5.0 A passes through 1.5 m of a 2.0-T magnetic field. If the current direction is perpendicular to the magnetic field, what is the magnetic force on the wire?

Answer: 15 N

*4. Suppose the loop of Figure 18.8 is a square 20 cm on a side, and it carries a current of 3.0 A. If the magnetic field strength is 0.90 T, what is the torque on the loop when its plane is parallel to the magnetic field?

Answer: 0.11 N · m

5. A DC motor is 90% efficient and operates on 110 V. It has an output work rate of 250 W ($\frac{1}{3}$ horsepower). What is the current to the motor when operating at full power?

Answer: 2.5 A

6. A generator is 98.0% efficent and has an output voltage of 6.00 kV. What must the power provided by the turbine turning the generator be if the load draws 20.0 A of current?

Answer: 122 kW

Home Experiments

1. Hold a strong magnet near the front of a working *black and white* TV and notice what happens. Do *not* do this to a color set; you may magnetize the metal mask in back of the screen!

2. Swing a finger back and forth quickly in front of a lighted TV screen. You see the finger only every $\frac{1}{60}$ s as the beam sweeps down the screen. (The beam sweeps across every other line each $\frac{1}{60}$ s, so that it covers all lines every $\frac{1}{30}$ s.) Estimate how fast you're moving your finger by the spacing between the points where you see it.

3. Make a list of all the devices in your room, apartment, or house that contain an electric motor.

References for Further Reading

Aldridge, B. G., and G. S. Waldman, *Automobile Ignition Systems* (A Physics of Technology Module, McGraw-Hill Book Co., New York, 1975). Discusses electromagnetism by focusing on the automobile ignition system.

Fink, D. K., and D. M. Lutyens, *The Physics of Television* (Anchor Books, Doubleday & Co., Garden City, N.Y., 1960). Very good discussion of the physics of television broadcasting and receiving. Written at the level of this text, and includes discussion of the modulation of the carrier wave.

Maugh, T. H., "The Magnetic Ship," *Science 81*, May 1981, p. 82. An unusual application of forces on currents that may have practical importance in the future.

*Solutions for Problems 1, 2, 3, and 4 depend on the equation in the footnotes of Sections 18.1 and 18.2.

Rubin, L. G., and P. A. Wolff, "High Magnetic Fields for Physics," *Physics Today*, Aug. 1984, p. 24. A discussion of how high magnetic fields are obtained and used.

Ruby, D., "GE's Superconducting Generator," *Popular Science*, Jan. 1984, p. 54. Progress toward a super-efficient generator.

Schwartz, B. B., and S. Foner, "Large-scale Applications of Superconductivity," *Physics Today*, July 1977, p. 34. Applications of superconducting magnets.

Time-Life Books, *How Things Work in Your Home (and What to Do When They Don't)* (Time-Life Books, New York, 1975). Well-illustrated discussion of the construction and principles of electric motors and how they are used in appliances.

The Way Things Work, an Illustrated Encyclopedia of Technology (Simon and Schuster, New York, Vol. 1, 1967, Vol. 2, 1971), or an equivalent publication, *How Things Work, The Universal Encyclopedia of Machines* (George Allen & Unwin Ltd., London, Vol. 1, 1967, Vol. 2, 1971). Discusses the physical principles and construction details of many electromagnetic devices.

C H A N G I N G

M A G N E T I C   F I E L D S

Those properties of magnetic fields that we have discussed so far have related to steady magnetic fields, or at least have been independent of whether or not the fields have been changing. Some different effects come into play when the fields change, and some of these effects turn out to be very useful.

When the magnetic field changes in the vicinity of some conductor, or conducting circuit, an emf is induced that can lead to an electric current. We can describe the principles of the electric generator in terms of this changing magnetic field rather than in terms of moving charges as we did in the previous chapter. This approach is somewhat simpler because then we don't need to consider the rather complicated electromagnetic force.

The emf induced by a changing magnetic field forms the basis for the operation of transformers that increase or decrease AC voltages to any desired level. In the last section of this chapter we see that electromagnetic waves—the stuff of which light, x rays, radio waves, and thermal radiation are made—are oscillating electric and magnetic fields.

19.1 ELECTROMAGNETIC INDUCTION

Suppose you have the items shown in Figure 19.1(a): a permanent bar magnet and a solenoid with the ends of the wires connected to a sensitive current-measuring meter. If the magnet and solenoid are not moving

(a)

(b)

(c)

(d)

Figure 19.1 (a) and (c) When the magnet is stationary relative to the coil, there is no current. (b) When the magnet is being inserted into the solenoid, there is a current in one direction. (d) Removing the magnet induces a current in the other direction.

relative to each other, there is no current in the coil—the meter reads zero.

Now suppose you quickly shove the magnet into the coil as in part (b). During the movement you will find the meter registering a current I in the direction shown. Once the magnet is at rest inside the coil (c), the current stops. If you pull the magnet out (d), a current will register in the opposite direction, but only during the motion. While the magnet and coil are moving relative to each other, a current exists. When there is no *relative* motion, there is no current, no matter where the magnet is. Furthermore, the faster the relative motion, the higher the current.

The current is induced during the motion because the magnetic field through the coil is changing. Any time the magnetic field in *any* circuit changes, an emf is induced. If it's a closed conducting circuit, a current results from this emf. Michael Faraday discovered this effect in 1831. The law that now bears his name states that *the emf induced in a circuit is proportional to the rate of change of the magnetic field through the circuit.* This process we call **electromagnetic induction**. If the changing field is within a coil of many turns, the same emf is induced in each turn. Therefore, *the total emf is proportional to the number of turns of the coil through which the field is changing.*

The change in magnetic field *through* a circuit can result not only from a change in the field itself, but also from a movement or rotation of the circuit so that it intercepts more or less of the field. For example, when the armature loop of a generator is in the orientation of Figure 19.2(a), it intercepts a maximum number of field lines. When it's oriented as in (b), it intercepts *no* lines. As the loop rotates, the amount of field intercepted continuously changes. Instead of the way we described the generator's action in the last chapter, we could explain it in terms of changing magnetic fields. You get the same result from either analysis.

To make the loop of Figure 19.2 into an AC generator, or alternator, you need a set of slip rings for contact with the external circuit (see Fig. 19.3). When the loop is in the orientation shown in Figure 19.3, the field through the loop is changing so as to give a current I in the external circuit in the direction shown by the arrows. As the loop *approaches* an orientation perpendicular to the field (that of Fig. 19.2-a), the amount of field through the loop is *increasing*. Just *after* passing this orientation, the field through the loop is *decreasing*. Thus, the emf and therefore the current reverse in direction just as the loop passes through the perpendicular orientation. The result is an emf that repeatedly, but smoothly, reverses its direction as in Figure 15.16, and a current that follows in proportion.

For a DC generator, you need the commutator arrangement of Figure 18.12 instead of the slip rings. The current in the external circuit is then always in the same direction even though it reverses within the loop itself.

Figure 19.4 shows another way a changing magnetic field can induce a current. When the switch is closed in the left-hand circuit, the current

(a)

(b)

Figure 19.2 The number of magnetic field lines that pass through a generator loop changes as the loop rotates.

sets up a magnetic field inside its coil. Since this magnetic field is confined mostly to the iron ring, almost the same magnetic field passes through the coil on the right. Immediately after the switch is closed, the field is changing, and a current exists in the meter. Once the current is steady, the field is not changing, and there is no current on the right. When the switch is opened, the current exists in the opposite direction while the field is decreasing. Faraday used an arrangement such as that shown in Figure 19.4 in his original discovery of electromagnetic induction.

An interesting application of induced currents may be seen in magnetic levitation for use in high-speed trains (Fig. 19.5). The idea is that as the train moves along, the strong magnetic field of a *superconducting* magnet on board the train gives a *changing* field in the track below. This induces a circulating current in the track. The magnetic field of this circulating current repels the magnet inside the train, lifting it off the track. The only friction is then due to air resistance on the train, which rides along, suspended on the magnetic field. The train is propelled by the interaction of the field of the on-board magnet with that due to an alternating current fed successively into short sections of the track. As the current moves from one section of the track to the next, the train moves along at the same speed—much as a surfer is propelled along by a wave on the ocean.

19.2 THE TRANSFORMER

The main reason alternating current is so widely used instead of direct current is that AC can be run through a *transformer*, and the voltage can be stepped up or down easily. Let's see how this is done and why it's useful.

Figure 19.3 Schematic diagram of an AC generator.

Figure 19.4 A current exists in the meter immediately after the switch is closed or opened but not while the current in the left-hand circuit has a steady value.

Figure 19.5 A Japanese experimental train that uses magnetic levitation produced by superconducting magnets on board the train.

Figure 19.6 is a slight variation of Figure 19.4. An alternating emf replaces the battery and switch; a resistor replaces the meter; and the iron core has a different shape, but most of the field is still confined inside it. An alternating emf and therefore an alternating current exist in the left-hand coil—called the *primary* winding. Since the current is continually changing, the magnetic field is continually changing. This change induces a sustained alternating emf in the right-hand coil—the *secondary* winding.

If the primary and secondary coils have the same number of turns, the emf at the secondary equals that input to the primary. Since the emf induced is proportional to the number of turns, the voltage output at the secondary winding can be changed by changing the number of secondary turns. The relationship is:

$$\frac{\text{voltage at secondary}}{\text{voltage at primary}} = \frac{\text{number of turns in secondary}}{\text{number of turns in primary}}$$

If N stands for the number of turns and V, for the voltage, this proportionality can be written in symbols as

$$\frac{V_s}{V_p} = \frac{N_s}{N_p}$$

(The subscripts indicate "secondary" and "primary.")

For example, if the input voltage to a transformer is 110 V AC, the primary winding has 1000 turns, and the secondary has 4000 turns, what is the output voltage? Multiplying both sides of the above equation by V_p, we have

$$V_s = \frac{N_s}{N_p} \times V_p$$

Therefore,

$$V_s = \frac{4000}{1000} \times 110 \text{ V} = 440 \text{ V}$$

Figure 19.6 A transformer consists of two coils, a primary and a secondary, wound on the same iron core.

Primary Secondary

Alternating emf

B

The heating element for the electron emitter in the electron gun of a TV picture tube operates on 6.3 V. Design a transformer to step down normal line voltage (110 V) to this value.

Answer: For example, N_p = 10,000 turns; N_s = 570 turns

With this ability to step up the voltage in a transformer by merely having more secondary than primary turns, are we getting something for nothing? The principle of conservation of energy tells us *no*. From this principle, we know that the total energy input to a transformer per unit of time—that is, the power input—must equal the total power output. Some electrical energy is converted to heat by the currents within the transformer, but this fraction is small in a well-designed transformer. Therefore, to a good approximation we can assume the electrical power output equals the electrical power input. Thus,

power from secondary = power to primary

or

$$V_s I_s = V_p I_p$$

where we have used the fact that the electrical power is the product of the voltage V and current I. Dividing both sides by V_s and I_p, and remembering that V is proportional to N, we get

$$\frac{I_s}{I_p} = \frac{V_p}{V_s} = \frac{N_p}{N_s}$$

The current is *inversely* proportional to the voltage and, therefore, inversely proportional to the number of turns. *When the voltage is stepped up, the current is stepped down proportionally.*

The biggest advantage to using AC, and the reason for its widespread use, is the ability to step the voltage up to high values for transmission over long distances and to step it down to whatever voltage is needed. For the same power transmission, the current is stepped down and then up proportionally. Since the power dissipated as heat in a conductor is proportional to the current squared (I^2), this transformation greatly reduces the loss of electrical energy to heat along the transmission line. Some typical voltages between generator and consumer are shown in Figure 19.7.

EXAMPLE 19.1

Suppose 550 W of power are to be transmitted a distance of 100 km over 1.2-cm-diameter copper wire that has a total resistance for this length of 16 Ω. Compare the electrical power lost to heat if the transmission voltage is 110 V with that lost at 110,000 V.

Recall that

$$power = voltage \times current$$
$$P = VI$$

and that the rate of heat generation is the current squared multiplied by the resistance, or

$$P_{heat} = I^2R$$

At 110 V:

$$I = \frac{P}{V} = \frac{550 \text{ W}}{110 \text{ V}} = 5.0 \text{ A}$$

$$P_{heat} = I^2R = (5.0 \text{ A})^2(16 \text{ } \Omega) = 400 \text{ W}$$

At 110 kV:

$$I = \frac{P}{V} = \frac{550 \text{ W}}{110,000 \text{ V}} = 0.0050 \text{ A}$$

$$P_{heat} = I^2R = (0.0050 \text{ A})^2(16 \text{ } \Omega) = 0.00040 \text{ W}$$

Figure 19.7 Typical voltages at various stages in the distribution of power from power station generator to user.

765,000 V

138,000 V

12,000 V

(Long-distance transmission)

Substation transformer

(To various secondary substations)

Turbine — Generator

4160 V

220 V

Secondary substation transformer

User

Transformer

(Throughout city or countryside)

For transmission at 110 V, 400/550 or 73% of the power is lost along the way as heat. At 110 kV, the heat loss is only 0.000073%.

Notice also that the voltage drop, given by $V = IR$, in the transmission line for 110 V at the source would be

$$V = (5\text{ A})(16\ \Omega) = 80\text{ V}$$

There would be only $110 - 80$ V or 30 V potential difference across the line at the user's end. You show that at the higher voltage, the voltage drop is negligible.

. . . .

EXERCISE Show that the transmission of as much as 1100 W at 110 V would be impossible over the line described in Example 19.1, but that 1100 kW could be transmitted at 110 kV with a loss of only 1600 W or 0.14%.

. . . .

QUESTION Why are heavy power users in the home (electric ranges, water heaters, clothes driers, etc.) usually designed to operate on 220 V rather than on 110 V?

The ignition coil of an automobile is just a transformer with typically about 200 turns in its primary winding and about 20,000 turns in its secondary. The primary is connected in series with the battery and a switch—the breaker points—located inside the distributor. Just before each spark is needed, a rotating cam separates the breaker points, cutting off the current to the primary. The rapid change in field induces a high voltage (about 20,000 V, or more) in the secondary winding that is connected by cables to the appropriate spark plug to initiate the spark.

Transformers do lots of jobs around the house. For example, they step down the 110-V line voltage for operating door bells, toy electric trains and racing cars, calculator battery chargers, and furnace thermostat circuits, to name a few.

19.3 ELECTROMAGNETIC WAVES

In Chapter 10 we described the transfer of thermal energy from one body to another by electromagnetic waves of the type called *infrared* radiation. Many other forms of electromagnetic radiation are very important to our lives. These include radio waves, light, x rays, and gamma rays. All of these are the same stuff; they differ only in wavelength (distance from crest to crest) and in the way they are produced. We'll examine the

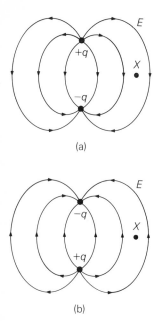

(a)

(b)

Figure 19.8 The electric field lines E around an electric dipole when (a) the upper charge is positive and (b) when it is negative.

production of radio waves—electromagnetic waves that carry broadcast signals from radio and TV stations to our homes.

Consider the electric *dipole*—a positive and an equal negative charge separated by some distance—shown in Figure 19.8, along with its accompanying electric field. In part (a), the upper charge is positive and the lower, negative. In part (b), the charges are reversed. The electric field at point X is therefore downward in Figure 19.8(a) and upward in (b). If the charges in the dipole could somehow be made to change continuously from positive to negative, the electric field at point X would continuously alternate between up and down.

One way to get such an *oscillating dipole* is to connect two metal rods (see Fig. 19.9) to an alternating emf. As the emf changes sign, the ends of the rod are alternately charged positively and negatively. The field at point X then alternately changes from up (as shown) to down. The change is gradual from zero to maximum up, back to zero, to maximum down, and so on.

The field extends from the charges out to large distances. But the *change* in the field takes a finite time to propagate outward. In vacuum, it propagates at a speed that is the same anywhere in the universe: *the speed of light.* Thus, at an instant when the field at point X is up, the field a long distance away at point Y may be down. At both points the field is continuously changing from up to down and back.

We get a *wave* in the electric field that moves along as shown in Figure 19.10(a). This graph shows the electric field at some instant at various distances from the oscillating dipole: the *antenna*. This up and down variation in electric field moves outward from the antenna at the speed of light (3.0×10^8 m/s, or 186,000 miles/s).

Look again at Figure 19.9 and notice that when charge is flowing in the antenna so that the top is becoming negative and the bottom positive, a net downward current exists in the antenna. This situation requires a *magnetic* field at nearby point X directed *outward* from the page, at a right angle to the electric field. As the charge oscillates up and down, the antenna current is first up and then down, meaning that the magnetic field at X alternates between in and out. This magnetic field change also propagates outward at the speed of light, so that when the field is out (indicated by the ·) at X, it may be in (indicated by the ×) some distance away at Y. Far from the antenna the variations in electric and magnetic field are in phase—they both have their maximum values at the same time at any point—but are directed at right angles to each other. This overall **electromagnetic** *wave* at any instant is shown in Figure 19.10(b).

Figure 19.9 When an alternating emf is applied to two metal rods as shown, they form an antenna that radiates field variations outward. If, at some instant, the electric field is upward at X, it could be downward a long distance away at Y. The magnetic field would be outward (·) at X and inward (×) at Y.

Chapter 19   Changing Magnetic Fields

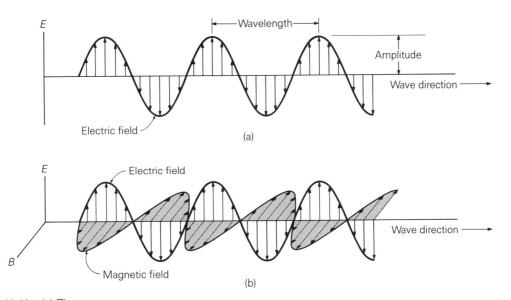

Figure 19.10 (a) The wave in the electric field that is produced by an oscillating dipole. (b) In an electromagnetic wave, the electric and magnetic field variations are in phase, but directed at right angles to each other.

The electromagnetic wave propagates out from the transmitting antenna. When it impinges on a receiving antenna, the variations in electric and magnetic field induce current variations in direct proportion to the field variations, which in turn are in direct proportion to current variations in the transmitting antenna.

We use the term *frequency* to describe the number of oscillation cycles per second, and we measure it in the unit hertz (Hz), where 1 Hz = 1 cycle/s (recall Section 12.1). AM radio stations broadcast in the frequency range 550 to 1700 kHz, while FM stations use the frequency range 88 to 108 MHz. TV stations broadcast between 59 and 890 MHz. You may have noticed that you can get the sound from Channel 6 on your FM radio, since its frequency is within the FM band.

The broadcast frequencies we have mentioned are called the *carrier* frequencies. The sound or picture information is "carried" along with this wave by superimposing on it a variation in amplitude or frequency. For example, in AM (amplitude modulated) waves, the carrier wave of Figure 19.11(a) is varied in amplitude according to the waveform of Figure 19.11(b), which contains the sound information. The resulting wave takes the form of Figure 19.11(c). In FM (frequency modulated) waves, the frequency rather than the amplitude is varied (see Fig. 19.12). In TV broadcasts, the sound is transmitted by FM and the picture by AM, both on the same carrier wave.

The Electromagnetic Spectrum

Any form of electromagnetic radiation is a wave of electric and magnetic fields that propagates at the speed of light. The various forms differ in

(a)

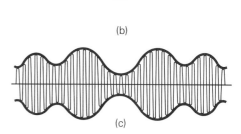

(b)

(c)

Figure 19.11 (a) The carrier wave, modulated in amplitude by (b) a lower-frequency wave, gives (c) an overall AM wave.

frequency and in wavelength—the distance from crest to crest, as indicated in Figure 19.10.

Figure 19.13 shows the approximate frequency and wavelength range for each type radiation in the entire **electromagnetic spectrum**. The boundaries between types are not well-defined, and there is overlap between the various regions. The figure also shows the specific regions in which particular devices operate. There are no upper and lower boundaries to the spectrum. In principle, it continues indefinitely to longer as well as shorter wavelengths.

All the radiation from microwaves up to the longest wavelengths (lowest frequencies) come under the general category usually called *radio* waves, which are all produced electrically in some way. Radiations from x rays through infrared are all produced within individual atoms or molecules, as we will see in Chapter 23. Gamma rays are emitted from the nuclei of atoms as the neutrons and protons inside rearrange themselves. (This is a form of radioactivity that we'll consider in Chapter 25).

As you look at the spectrum in Figure 19.13, be careful to notice the vast range of wavelengths shown. Between any two vertical lines, the wavelengths increase (or decrease) by a multiplication factor of 100. The ratio of the longest to the shortest wavelength shown is about the same as the ratio of the size of the earth's orbit around the sun to the size of a single proton in the nucleus of an atom. Viewed in this perspective, you see that visible light is just a "speck" in the vast electromagnetic spectrum.

Figure 19.12 In FM (frequency modulation), a wave form (b) is transmitted by varying the frequency of the carrier wave (a) in proportion to the amplitude of the wave of interest. The result is shown in (c).

(a)

(b)

(c)

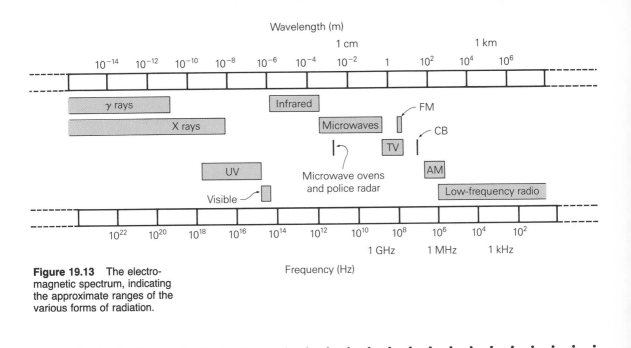

Figure 19.13 The electromagnetic spectrum, indicating the approximate ranges of the various forms of radiation.

. . . .

Key Concepts

A changing magnetic field through a circuit induces an emf proportional to the rate of change of magnetic field. This effect is called *electromagnetic induction*. In a transformer,

$$\frac{\text{output voltage}}{\text{input voltage}} = \frac{\text{input current}}{\text{output current}}$$

$$= \frac{\text{number of secondary windings}}{\text{number of primary windings}}$$

Electromagnetic waves are oscillating electric and magnetic fields that propagate out from their source at the speed of light. They include x rays, light, microwaves, and radio waves.

Questions

Electromagnetic Induction

1. If the bar magnet in Figure 19.1 were dropped to fall straight through the loop, describe the deflection of the meter at various points during the fall.

2. As the armature coil of an electric motor turns during operation, the changing magnetic field through the coil induces an emf just as in a generator. The induced emf opposes the current that causes the armature to rotate. Explain why there is a surge of current to a motor when it is first turned on that decreases as the motor attains operating speed.

3. Explain why the principle of energy conservation would be violated if the second statement in Question 2 were not true.

4. When the refrigerator turns on in some houses, the lights dim momentarily and then come back to near full brightness. Will the same thing happen if an electric heater is turned on? (See Question 2.)

Transformers

5. Why isn't commercial electricity DC rather than the more complicated AC?

6. If the current meter in Figure 19.4 were replaced with a light bulb of the proper voltage rating, would the bulb glow continuously while the switch is closed? What could you substitute for the battery that would make this happen?

7. There is 220 V effective potential difference between the two hot lines supplying power to a typical residence. Why isn't a transformer needed to get 110 V for the ordinary lights and appliances? (Recall Section 16.1.)

8. Why don't power companies provide only 110-V lines to residences, and then have customers use a small transformer to step up the voltage for ranges, clothes dryers, and other appliances that need 220 V?

9. Battery chargers use "rectifiers" to convert AC to DC for charging the battery, and they use transformers to change the voltage to the appropriate value. Which of these two devices would you expect to appear first (closest to the AC plug) in the circuit?

Electromagnetic Waves

10. Oscillating electric charges produce electromagnetic fields. Does this mean that the electric field lines extending outward from an oscillating charge have wiggles in them rather than being straight as from a point charge? (That is, does "shaking" a charge do the same thing to a field line as shaking the end of a rope does to a long, straight rope?)

11. Light passes through water. Microwaves do not pass through water; they're absorbed. (That's why you can heat water in a microwave oven.) Does that mean that light and microwaves are *not* the same stuff?

12. What is the difference between AM and FM radio waves?

13. Why can you get the sound from a Channel 6 TV station on an FM radio?

14. Visible light has wavelengths in the range from about 400 nm to 700 nm. Why is it a meaningless question to ask what fraction this is of the entire electromagnetic spectrum?

Problems

1. If the transformer that steps the voltage down from 4160 V to 220 V in Figure 19.7 has 20,000 turns in its primary winding, how many turns are in its secondary?

   Answer: 1060 turns

2. Suppose 100 A of current at 220 V are supplied to the user in Figure 19.7. Neglecting losses, what current is needed in each part of the transmission line to give this current?

   Answer: from user end, 5.3 A; 0.16 A; 0.029 A; 1.8 A

3. Design a transformer (determine an appropriate number of turns for each winding) to operate on 120 V and provide the 12,000-V power needed for a neon sign.

   Answer: $N_s/N_p = 100$

4. An automobile ignition coil typically has about 200 turns in its primary winding and 20,000 turns in its secondary. If you hooked the primary of such a coil to 110-V AC, what would the output voltage of the secondary be?

   Answer: 11 kV

5. The average wavelength of visible light is about 500 nm ($n = 10^{-9}$). When you see the police car ahead, about how many light waves span the 3-cm wavelength of the radar waves being used to detect your speed?

   Answer: 54,000

References for Further Reading

Gannon, R., "Electromagnetic Pollution—Are They Zapping You?" *Popular Science*, Dec. 1983, p. 98. Considers whether or not there are health hazards associated with radiation in the microwave range.

"*Ole Amigo,*" *Time,* Aug. 29, 1977, p. 32. An interesting example of the use of electromagnetic radiation in controlling ants, mice, gophers, and other pests.

Powell, E., "Induction Cooktops," *Popular Science*, Feb. 1981, p. 49. An application of electromagnetic induction that may show up in your kitchen.

"Railroad Tracks Are Good Conductors," *Science 83*, Dec. 1983, p. 8. An unexpected hazard from the changing magnetic fields of overhead power lines.

E L E C T R I C A L
E N E R G Y

Electrical energy is energy in its most convenient form for everyday use. Some of us use natural gas, oil, coal, or wood for jobs that are always done in the same place—heating our houses and water, or cooking our food. But try to think of a design for a portable hair drier, clothes iron, coffee pot, or automatic food mixer that doesn't use electricity.

In this chapter we'll look at the two ends of the electrical energy spectrum: its sources and its uses. Concerning sources, we'll consider the overall scheme for electrical energy production in a power plant, and we'll discuss the mechanism that converts chemical to electrical energy in a battery. We'll also discuss two sources of electrical energy that have not, as yet, found large-scale use—fuel cells and solar cells.

In examining electrical energy uses, we'll compare the energy requirements for different kinds of jobs. This section will tie together electrical and other forms of energy that we've discussed throughout the book, and it will provide some practical insights into conserving usable energy. Finally, we'll investigate the interaction between energy and power as they influence the cost of supplying electrical energy.

20.1 SOURCES OF ELECTRICAL ENERGY

We've mentioned electrical energy sources at various places in this book. We'll now discuss them in more detail.

Chemical energy

(Coal or other fuel)

Thermal energy (steam)

Boiler (87%)

(43%)

Turbine

Pumps

Water

Mechanical work (rotating shaft)

Generator (98%)

Electrical energy

Figure 20.1 A simplified block diagram of a fossil-fuel power plant. Efficiencies of each stage are given in parentheses. The energy to run circulating pumps and other accessories is included as part of the energy "loss" in the turbine.

Electrical Power Plants

In the previous two chapters we described the basic physical laws that govern using mechanical work to produce electrical energy in a generator. We'll now look at what else we need besides the generator to get electrical energy from a power plant.

First, notice that the amount of work done in rotating the generator armature coils must be at least as much as the energy supplied to the devices using the generator's current. For example, if ten 50-kW motors operate on the current from the generator, work must be done at a rate of *at least* 500 kW in turning the generator. What supplies this work?

Most power plants use steam turbines to turn their generators. A simplified block diagram of the overall processes in a steam plant that burns fossil fuel—coal, oil, or natural gas—is shown in Figure 20.1. We say "simplified" diagram because a real power plant has many auxiliary components along with each of the basic ones shown here. The result of the process is the conversion of chemical energy in the fuel to electrical energy, with the intermediate forms, thermal energy and mechanical work, along the way.

The diagram shows typical efficiencies of the devices performing the various processes. For each 100 kWh of energy entering the boiler as chemical energy in the fuel, 87% (or 87 kWh) are passed on to the steam. Of these, 43% (or 37 kWh) do useful work in turning the generator. The generator is quite efficient, converting 98% of this work (or 36 kWh) to electrical energy. A typical overall efficiency is thus 36%. The most efficient modern fossil-fuel plants have an overall efficiency of about 39%. This efficiency does not include the loss, typically about 8%, of electrical energy to heat in the transmission line between power plant and consumer.

For *nuclear* power plants the block diagram looks the same. The boiler is then a nuclear reactor that uses uranium or plutonium as a fuel. The nuclear energy released in the reaction produces heat that generates steam. Because of the need to guard against leakage of radioactive material, boilers in nuclear plants are run at lower temperatures to increase the structural strength of the materials, and they are run at lower pressures

Figure 20.2 The basic principle of a hydroelectric power plant: The conversion of gravitational potential energy to electrical energy.

to decrease the chance of rupture. This lower steam temperature and pressure reduce the turbine efficiency so that the overall efficiency is typically about 33%.

All the energy released from the fuel but not converted to electrical energy is converted to heat in the environment. Many power plants are located on large bodies of water so that they can dissipate most of this heat into the water. Other plants have large cooling towers that release the heat into the atmosphere. This *thermal pollution* can have an adverse effect on local conditions and must be controlled carefully.

In a hydroelectric plant, the process is somewhat different than in a steam plant (Fig. 20.2). There, the gravitational potential energy of the water at the top of the dam gives kinetic energy to water falling from the dam. This kinetic energy turns water turbine blades, providing the work to turn the generator. The overall conversion is from gravitational potential energy to electrical energy.

If you want to compare the fuel you consume when you use electrical energy with that consumed when you use energy in some other form, you have to consider the power plant and transmission efficiencies. For example, electrical baseboard heaters are nearly 100% efficient: They convert all the electrical energy supplied to them to heat. However, the typical oil-fired home furnace is probably no more than about 50% efficient. If your power plant burns fossil fuels, by which of these two methods of heating your house will you consume *less* of the world's fuel supplies? As we have said, a typical power plant is about 36% efficient, and about 8% of the electrical energy is lost in transmission. Therefore, in operating the 100%-efficient baseboards, you would use about $1\frac{1}{2}$ times as much fuel. Unfortunately, the choice of which of these methods gives the biggest overall benefit is not quite so clear-cut. For example, the power plant might use a much more plentiful fuel than you would use at home, thereby reducing the advantage of less consumption. The answer to which method is most advantageous overall may involve some complicated technical, economic, and social issues that are discussed in some of the end-of-chapter references.

You might question whether or not you really save the power plant any fuel when you turn your lights off, since the generator is turning anyway. From an overall energy viewpoint, if less electrical energy is used, less must be generated, and therefore less fuel is burned.

To understand why it's harder to turn the generator when more electrical devices are used, consider the generator and motor diagrams of Figures 18.12 and 18.8, respectively. The armature loop of the generator (Fig. 18.12) is turned at a constant speed. Only if there is an external load will there be current in the loop. But if there is current in the loop, the situation is exactly that of a motor: a current loop in a magnetic field. For the direction of rotation shown, the current is opposite to that shown in the motor diagram, Figure 18.8. Thus, there is a force in the opposite direction that tends to oppose the rotation. The higher the current, the stronger the force opposing rotation. Thus, the turbine turning the generator must do more work as the current through it increases. If this *counter torque* did not exist, the work of turning the generator would be independent of output electrical energy—a situation that would violate the conservation-of-energy principle.

· ·

QUESTION Even though you probably can't feel the difference, explain why you pedal a little easier on your bicycle if the bulb on your generator-operated light burns out. Why do you use *slightly* less fuel in your car if you don't use the electrically operated fan on your heater? (Again, you won't notice the difference.)

Batteries

A battery converts chemical energy directly to electrical energy. It is made up of one or more *cells*, each with the parts shown in Figure 20.3—two rods called *electrodes* of different metals in an acid solution called an *electrolyte*. (Strictly speaking, a battery means more than one cell; the word is used nowadays to mean one or more.) In 1800 the Italian physicist Alessandro Volta used copper and zinc electrodes in a saltwater solution for the first human-made battery.

The metals dissolve slightly in the solution, each dissolving atom leaving one or more electrons on the electrode and going into solution as a positive ion. Being of *different* metals, one rod dissolves more than the other. If the metal of rod B dissolves more, it leaves behind more electrons, becoming more negatively charged than rod A. This process means that rod B is at a lower potential than rod A.

The parts of the electrodes extending outside the battery are called *terminals*. If you connect a wire from the positive terminal through a light bulb (or other load) and back to the negative terminal, electrons will flow through this wire and bulb from the negative to the positive terminal. This flow corresponds to a conventional current from the positive to the negative terminal. The continued dissolving of electrode B provides a continuous supply of electrons. Electrons arriving at electrode

Figure 20.3 The components of a single-cell battery.

+

Zinc

Carbon

−

Ammonium
chloride
paste

Figure 20.4 The main components of a flashlight battery: a "dry" cell. The zinc case is the negative electrode.

A from the wire are absorbed by positive ions in the solution. Eventually, electrode *B* completely dissolves, and the battery quits operating. Some types of batteries can be *recharged* by forcing current in the opposite direction through the battery. This reverses the chemical reaction, restoring the electrodes and electrolyte. Car batteries are, of course, of this type.

The output voltage (emf) of the cell depends on the relative extent to which the electrodes dissolve in the electrolyte. This naturally depends on what substances are used for electrodes and electrolyte. In a conventional car battery, the positive electrode is lead peroxide, the negative terminal spongy lead, and the electrolyte dilute sulfuric acid. The emf of such a cell is about 2 V. Six cells connected in series give a 12-V battery.

As the battery discharges, the chemical reaction uses up sulfuric acid and produces water. Eventually, the solution is too dilute to support the reaction—the battery is "dead." Since sulfuric acid is denser than water, the density of the electrolyte increases with the concentration of sulfuric acid. When service station operators check the "charge" of your battery, they use a *hydrometer* that measures the electrolyte's density. This indirectly measures the extent of discharge of the battery.

A flashlight battery is called a *dry* cell because the electrolyte is in a paste that is thick enough not to run out when turned upside down. As Figure 20.4 shows, the container itself is zinc, which forms the negative electrode. A carbon rod down the center is the positive electrode. The electrolyte is ammonium chloride. This arrangement gives an emf of 1.5 V.

Many car batteries on the market today have no plugs for adding water, and many that do have them seldom need filling. You may wonder why the old type has to be refilled occasionally, but the new type does not.

When a battery is being recharged, small amounts of hydrogen and oxygen are given off at the electrodes and vented to the outside through the top of the battery. These gasses come from a decomposition of water during the charging process. Water needs to be added periodically to replenish that used up.

Manufacturers use various techniques that basically involve two modifications to minimize the need for adding water. First, the lead plates that form the electrodes have a small amount of "hardener" added to them to make them stronger. That hardener has traditionally been antimony. Many of the newer batteries use calcium as a hardener. This changes the chemical processes so that they produce less gas and therefore use less water. The second technique that some manufacturers use is to redesign the structure of the battery so that the plates rest on, or near the bottom of, the case. The electrolyte level can then be higher above the plates so that the battery can lose more water without affecting its operation.

Inside a battery, charge moves *against* the potential difference: Negative charge moves toward the negative electrode, positive charge, toward the positive electrode. The chemical reactions provide the energy to do the work of moving this charge. Outside the battery in the external circuit, the charges flow freely in the direction of decreasing potential energy. The work done per unit of charge moved though the battery equals its emf.

Other Electrical Energy Sources

Even though generators and batteries are the main sources of everyday electrical energy, other sources either are used for special purposes or are being developed for large-scale power production in the future. We'll briefly comment on a few of these.

Fuel cells. *Fuel cells* are first cousins to batteries, in that they, too, produce electrical energy from chemical energy. However, the fuel for the chemical reaction in a fuel cell is continuously supplied from outside the cell rather than being contained within the cell.

A fuel cell combines hydrogen and oxygen to form water, with the excess energy given up in the reaction going to electrical energy. If hydrogen and oxygen are merely combined, they interact violently. In a fuel cell, this reaction is controlled. One type of cell has porous carbon electrodes separated by an electrolyte (Fig. 20.5). Hydrogen is fed to one electrode and oxygen to the other. Chemical reactions at the electrodes liberate electrons at one and absorb electrons at the other, thereby giving positive and negative terminals, as in a battery.

Figure 20.5 A hydrogen-oxygen fuel cell.

Because fuel cells are compact and portable, they are well-suited for applications that have limited power needs, such as manned space flights. As the techniques for building and operating these cells improve, they may assume a more important role in the production of electrical energy for everyday use.

Solar cells. Tremendous quantities of energy reach the earth every day in the form of radiation from the sun. People use various schemes to capture some of this radiant energy to produce heat. But we can also use this radiation to produce electrical energy.

It is possible to let the heat from the radiation produce steam to run a turbine to turn a generator and generate electrical energy in the conventional way. It is also possible to convert radiant energy directly to electrical energy in a device called a *solar cell*. Solar cells are made of semiconductors—a class of materials we discussed in Section 16.3. Recall from that section that, by adding the appropriate impurity to the semiconductor, we can get a conductor with either negative or positive charge carriers. The *n*-type semiconductor conducts by the flow of negative charges—electrons. The *p*-type conducts by the motion of holes that act as positive charge carriers.

Figure 20.6 The heart of a solar cell is a *p-n junction* made from layers of *p*-type and *n*-type semiconductors.

The heart of a solar cell is a *p-n junction*—a layer of *p*-type and a layer of *n*-type semiconductors in contact (see Fig. 20.6)—with a wire connected to each side. When the two layers first make contact, some electrons flow from the *n* side to the *p* side, filling some holes on the *p* side and creating some holes on the *n* side. This causes a net negative charge on the *p* side, and a net positive charge on the *n* side. The charge separation produces an electric field across the boundary from the *n* to the *p* side, and this field eventually stops the flow of charge.

When light strikes the vicinity of the junction, it can knock electrons loose from the lattice, leaving *electron-hole pairs*. If this happens on the *p* side, the electron is accelerated across the boundary by the electric field there. If it happens on the *n* side, the hole moves across the boundary because of the field. Consequently, an excess of holes builds up on the *p* side; an excess of elecrons builds up on the *n* side. Thus, when the junction is exposed to light, the *p* side acts as the positive terminal of a battery, the *n* side, as the negative terminal. The emf of such a cell is about $\frac{1}{2}$ V when in sunlight.

As the technique for making solar cells improves and the cost of their production comes down in the future, solar cells may become a significant source of electrical energy. Keep in mind, however, that *the output power from a solar cell is limited by the rate of energy hitting it from the sun*. The sun beams an average of about 200 watts per square meter over the United States. If we assume a 20% conversion efficiency in solar cells (about the best that can be hoped for in the near future), this method would require an area of about 25 square kilometers covered with solar cells to equal the power output of a 1000-MW power plant.

Chapter 20   Electrical Energy

20.2 USING ELECTRICAL ENERGY

Since electrical energy does so many kinds of jobs, it helps to know just how much energy different kinds of devices use in doing their particular tasks.

You've learned several laws about the conversion of energy from one form to another. Now let's use some of these laws to see how ordinary quantities of energy in different forms compare in size. In this section, I'd like you to gain some understanding in at least two areas.

First, I'd like you to get an intuition—a feel—for the sizes of the numbers describing ordinary quantities of different forms of energy and how they compare in the different forms. Here are some analogies: If you have to buy a new suit, you probably have some feeling for what it will cost. If you have to mow the lawn, you probably have some intuition about how much effort is involved. If you have to make a 500-mile trip by car, you have some idea how tired you're going to be when you get there. This section is intended to give you the same kind of intuition about various forms of energy and their interdependence—how much energy it takes to operate a light bulb, use an electric blanket overnight, heat water for a shower, or air condition your room. We'll develop some typical relative numbers that will assist you in comparing the energy needed for various jobs.

The second thing I'd like you to get from this section is the know-how to figure out the energy needed for a specific job—maybe a job you do regularly, or a job you're thinking about doing and are counting the cost in advance.

Remember, power is the rate of doing work: power = work/time. Thus: work done = energy used = power × time. Since we usually give electrical power in watts (W) or kilowatts (kW), it's convenient to express energy or work in kilowatt-hours (kWh), which we get by multiplying power in kilowatts by the time in hours. For comparison, $1 \text{ kWh} = 3.60 \times 10^6 \text{ J} = 860 \text{ kcal}$. For "in-your-head" estimates, a handy conversion factor is that 1 Wh is roughly the same as 1 kcal.

Examples

Now, let's look at some general ways of using electrical energy. Some of the jobs we'll consider could be done with other forms of energy, but the same ideas apply anyway. Our calculations will be rough approximations intended to give a feel for the amounts of energy involved.

Light production. You're certainly familiar with electric light bulbs, and you probably have a fair intuition about how much light you get from a 100-W incandescent bulb. The electrical energy used by such a bulb in 1 h of operation is

$$\text{energy} = \text{power} \times \text{time} = 100 \text{ W} \times 1 \text{ h} = 100 \text{ Wh} = 0.1 \text{ kWh}$$

If you buy energy at 10¢ per kWh, that 1 h of operation will cost 1¢. We'll compare the energy needed for other jobs with this familiar one.

EXERCISE Since most of us identify better with how hard our pocketbook gets hit than with kilowatt-hours, you will benefit more from the illustrations in this section if you find out the average price of electrical energy in your area. Each time we calculate an energy use, translate that into cost. One way to estimate the price is to look at a monthly electric bill. Divide the charges by the number of kilowatt-hours used as stated on the bill. (This price probably won't be exact, because the rate usually varies with the total energy used.) You can find out the exact rate by contacting your power company.

Figure 20.7 The bulb package tells the light output in lumens.

Notice that 100 W (or whatever the power rating of a light bulb) is the electrical power, *not* the rate of *light output*. Different "sized" bulbs operate at different filament temperatures and therefore differ in light production efficiency. You can find out the relative efficiencies at the light-bulb counter of the grocery store. The package on each bulb tells you not only the power in watts, but the average number of *lumens*, a measure of the rate of visible light output (Fig. 20.7). From the information on the packages, you can see that it takes two 60-W (not 50-W) bulbs to give about the same light output as one 100-W bulb. Worse still, four 40-W bulbs give just about the same amount of light.

EXERCISE If you replace two 100-W bulbs with enough 40-W bulbs to give the same light, how much *more* energy will you use in 150 h of operation? How much *more* will it cost at 10¢ per kWh?

Answer: 8 bulbs; 18 kWh; $1.80

Mechanical work. To get a feel for the energy needed to do mechanical jobs, we'll first pick a job that you can do both by hand *and* with a power tool, so that we can compare the energy needed. Suppose you want to drill a 2-cm-diameter hole ($\frac{3}{4}$ in) in a 4-cm-thick piece of wood (for example, a "two-by-four"), and you have available both a brace and bit (Fig. 20.8) for hand drilling, and a portable electric drill rated at 3 A on 115 V: 345 W. Doing this job with the brace and bit might take 30 turns of the handle, which moves in a circle of radius 13 cm. We'll estimate the average force on the handle to be about 50 N (11 lb). The work done in turning the brace—force × distance per turn × number of turns—is

$$50 \text{ N} \times (2\pi \times 0.13 \text{ m}) \times 30 = 1220 \text{ J}$$

To convert to kilowatt-hours, divide this quantity by 3.6×10^6 J/kWh to get 0.00034 kWh.

With the electric drill, this job might take about 10 s. Operating the drill at full power (certainly an overestimate) for 10 s gives an energy use of 0.345 kW × 10/3600 h, or 0.00096 kWh.

The answer we get for drilling by hand is about $\frac{1}{3}$ that for the electric drill—not a bad comparison considering the estimates made, the difference in the nature of the drill bits, and the somewhat less than full power needed to drill the hole with the power drill. The important point for this discussion is the general amount—"order of magnitude"—of energy used, which is about 1/100th the energy used in our 100-W bulb in an hour.

Consider another mechanical example. How much work is needed by the electrically operated elevator to lift the mayor of New York City to the top of the 380-m-high Empire State Building. We'll assume the mayor's mass to be 70 kg: a weight of 700 N (157 lb).

$$\text{Work} = \text{force} \times \text{distance} = \text{weight} \times \text{distance}$$
$$= 700 \text{ N} \times 380 \text{ m} = 266,000 \text{ J}$$

At 3.6×10^6 J/kWh, this is 0.074 kWh, roughly the energy used by our 100-W bulb in $\frac{3}{4}$ hour. (We haven't included the energy needed to lift the elevator.)

Estimate the work done in other mechanical jobs that you're familiar with, and see how it compares with the ones done here.

Water heating. Most of us would have a hard time getting along without hot water. In electric water heaters, current in resistive heaters produces heat that raises the water temperature. Typically, there are two heating elements, each rated about 4 kW.

Recall from our chapter on heat energy that the heat needed to raise the temperature of a substance is

$$\text{heat} = (\text{specific heat}) \times \text{mass} \times (\text{temperature change})$$

For a specific quantity—say, 1 liter—of water, let's find the energy (heat) needed to raise its temperature an amount typical of a household water heater. Since that "typical amount" will vary with climate and temperature setting, we'll estimate some reasonable values. I measured the water temperature entering my house on a winter day and found 7°C (45°F). In a moderate summer this temperature might average 20°C

Figure 20.8 A brace and bit for hand-drilling holes in wood.

(68°F). Let's choose the *average* incoming water temperature to be 13°C and assume it is heated to 60°C (140°F). One liter of water has a mass of 1 kg, and the specific heat is 1 kcal/kg per C deg, which is 0.00116 kWh/kg per C deg. Raising the temperature of 1.0 liter of water by 47 C deg requires:

heat = specific heat × mass × temperature change

= 0.00116 kWh/kg·C deg × 1.0 kg × 47 C deg

= 0.054 kWh for each liter

In other words, it takes *roughly* $\frac{1}{20}$ kWh for each liter of water heated. If you think better in gallons, that's about $\frac{1}{5}$ kWh per gal, since 1 gallon is about 4 liters.

That may not seem like much, but try estimating how much hot water a family uses in a month. Suppose, for example, a family takes four baths a day at 60 liters per bath, washes dishes in a dishwasher once a day at 50 liters per wash, and washes six loads of clothes a week at 30 liters of hot water per wash. Aside from other uses, that comes to about 9400 liters in a month, needing about 470 kWh for heating. (Calculate the cost of this at your local rates.)

· ·

EXERCISE   Compare the energy needed in shaving for 5 min using a 12-W electric shaver with that needed to heat 6 liters of water for use in assisting a safety razor shave.   *Answer*: 300 times as much to heat water

Cooking.   Most Americans cook with electrical energy. Surface cooking units on electric ranges are typically rated between 1400 W and 2600 W for full-heat settings. Conventional (not microwave) oven heating elements typically use around 3000 W. Both surface and oven units are resistive heaters that maintain their lower than full-heat settings by cycling on and off for various times according to the setting of the controls.

· ·

EXERCISE   How much energy is used in cooking a turkey for 3 h in a 3.0-kW oven if the temperature control is set so that the heating element is on half of the time?
          Answer: 4.5 kWh

Space heating.   The energy needed for space heating—the heating of living, working, and playing space—depends not so much on the energy needed to change the temperature of the air and other contents of the space, as on the energy needed to replace the heat that leaks out through the walls, roof, and floor of the building. We considered the factors influencing that heat loss in our chapter on heat transfer. The needs vary widely with climate, house size, and the house's insulation. For example, in Philadelphia with about 2500 C degree-days* per year, a typical well-insulated house needs about 15,000 kWh of heat. One possible source of that heat is the conversion of electrical energy to thermal energy in resistance heaters. Essentially all the electrical energy delivered to such heaters is converted to thermal energy. However, with an electrically operated heat pump under moderate temperatures, you can get more heat delivered than electrical energy used (Chapter 11). That house in Philadelphia could get its 15,000 kWh of heat by supplying only about 8000 kWh to a heat pump.

Cooling. Most of us take for granted that refrigerators run on electrical energy and are a necessity of life. Air conditioners are almost as commonplace in many areas of the United States. The energy needed to pump heat from the inside of a refrigerator depends greatly on the refrigerator and on how it is used. A typical refrigerator uses about 500 W of power when the compressor is running, which is about one-third of the time.

An operating air conditioner typically removes about $2\frac{1}{2}$ times as much heat from the room as the electrical energy supplied to it. The amount that needs to be removed depends, of course, on the climate and on the room or house. A central air conditioner for a house might use about 5 kW, a one-room air conditioner, about 1 kW. One way that air conditioning engineers describe the extent of air conditioning needed in a given climate is by the number of "cooling hours" per year needed to maintain comfort in a properly air conditioned home. In Houston that's about 1900 h; in Seattle it's about 100 h.

Summary

The illlustrations we have included should give you a feeling for the relative energy needs of different kinds of everyday jobs. As you can see, mechanical jobs of the type we do regularly need comparatively small amounts of energy. When we think of conserving electrical energy, many of us think first in terms of turning off unused lights. Of course, every little bit helps, but as our examples have shown, lights are not the worst culprits. Aside from space heating and cooling, water heating is the biggest domestic consumer. This is not a plea for going around dirty, but conservation in the areas of hot water use, as well as in space heating and cooling, will help where it counts the most.

See Problem 7, Chapter 10 for a definition of degree-days.

Table 20.1 Average power ratings and annual energy consumption of various appliances

Appliance	Average Power (W)	Estimated Annual Consumption (kWh)
air conditioner (room)	860	860[a]
blanket	150	150
blender	300	0.9
broiler	1140	85
carving knife	95	0.8
clock	2.5	22
clothes drier	4856	993
clothes washer	512	103
coffee maker	600	138
dehumidifier	257	377
dishwasher	1201	363
fan		
attic	370	291
window	200	170
food mixer	127	2
frying pan	1200	100
furnace (electric)	—	19,000[b]
hair drier	600	25
heat pump	—	11,000[c]
heater (portable)	1322	176
hot plate	1200	90
humidifier	177	163
iron	1100	60
oven (microwave only)	1450	190
radio	71	86
range (with oven)	12,200	730
refrigerators/freezers		
standard (12.5 ft³)	—	1500
frost-free (17.5 ft³)	—	2250
sewing machine	75	11
shaver	15	0.5
television (solid state)		
black-and-white	45	100
color	145	320
toaster	1100	39
toothbrush	1.1	10
vacuum cleaner	630	46
water heater	4474	4811

[a]Based on 100 hours of operation.

[b]Assuming 3100 C degree-days (see Prob. 7, Ch. 10 for definition).

[c]Heating only, assuming 3100 C degree-days.

Table 20.1 gives some average power ratings of various devices, and their average total electrical energy consumption per year in the United States. The average time of use and percentage of "on" time when in use are built into the values given for energy consumption. Compare the entries in Table 20.1 with the illustrations discussed here and compare them with your estimates of energy use by other devices that are of interest to you.

· ·

EXERCISE Compare the electrical energy consumption of small appliances intended for specific jobs with that of general-purpose appliances intended for many jobs. Examples: a 1000-W toaster rather than a 3200-W broiler; a 1150-W electric skillet rather than a 2000-W range surface unit.

20.3 DO WE PAY FOR POWER OR FOR ENERGY?

Almost everything we have said in this book related to consumption of expendable energy has implied that the important consideration is the total energy used to do a job, not the power or rate of using this energy. Some examples: when you pay your electricity bill, you pay for the number of kilowatt-hours—the energy—used during the month; when you are concerned about the operating cost of some appliance, it is the total energy it uses in some time period that counts and not its operating power; the value of fuel is based on its energy content, not the rate at which it can be used. The implication of all this is that power is not important, and that the answer to our section title should be "Energy."

However, that isn't the whole story. Suppose, for example, that in the middle of a hot summer afternoon, most people in Metropolis turn on their air conditioners, later, their ranges for cooking dinner, and maybe their color TV sets. The *power* for operating these, and everything else electrical in town, must be supplied by the power plant. The mechanical power from turbine to generator must *exceed* the combined electrical power used by the consumers. (Why exceed?) In the wee hours of the morning when most of these devices are off, that high-power generator is used at a small fraction of its capacity. The power company needs large enough generators to meet the "peak" demands, but these generators are either unused or underused at off-peak hours.

You and I pay indirectly for this peak-power capacity. We pay financially because the power company charges enough per kilowatt-hour to let them buy the needed generating equipment. We pay in world fuel supplies because energy is needed to manufacture this equipment. This is why power companies encourage, either financially or morally, the scheduling of electrical energy consumption for off-peak hours.

The answer to the question in this section title is therefore "Both!"

Key Concepts

The main sources of electrical energy in present use are electromagnetic **generators** and **batteries**. In a steam power plant, generators carry out one step in a process that converts *chemical* or *nuclear* energy to *thermal* energy, then to *mechanical* energy, and then to *electrical* energy. Batteries convert *chemical* energy directly to *electrical* energy. Other sources of electrical energy include *fuel cells* and *solar cells*.

Solar cells use **p-n junctions**, formed by joining layers of **n-type** and **p-type semiconductors**.

Electrical energy consumption by ordinary devices generally ranks in the following order from lowest- to highest-energy users: appliances that do mechanical jobs, light sources, cooking and other heat-producing appliances, refrigerators/freezers, water heaters, space cooling and heating systems.

Questions

Electrical Energy Sources

1. Draw a block diagram like Figure 20.1 for a hydroelectric plant.

2. Given the efficiencies for each step in the power plant diagram (Fig. 20.1), how do you get the overall efficiency to be 36%?

3. Suppose Ann and Abby retire in identical side-by-side houses in Concord, Mass., and that each house needs 25,000 kWh of heat in a particular winter. If Ann heats her house with a 70% efficient natural gas furnace and Abby heats hers with a 92% efficient electrical furnace, who uses the most fuel if the power plant also uses natural gas?

4. A fuel cell, like a battery, converts chemical energy directly to electrical energy. In what way is it more like a power plant than a battery?

5. Why do battery electrodes have to be made of two different metals?

6. The emf of a flashlight battery is 1.5 V. Does that mean flashlight bulbs normally operate on 1.5 V? Why not?

7. When a dry cell operates, hydrogen and other gases build up around the carbon electrode. What would you expect this to do to the emf of the cell?

8. John Solarwatt claims to have invented a solar cell so efficient that an array of his cells 10

meters square can provide an average of 1 MW output. Is there some reason you should be skeptical about this claim?

9. What is the direction of the electric field across a *p-n* junction?

Electrical Energy Uses

10. A police officer stopped a driver on the Pennsylvania Turnpike for driving at night without lights. The driver's explanation was that she was saving gasoline. Was she?

11. Charlie Warmears keeps his bedroom at such a temperature that he will be warm under a regular blanket and not "waste" electrical energy on an electric blanket. Jane Coldfoot uses an electric blanket and sleeps in a cold bedroom. Discuss the economics—both energy and financial—of the two approaches.

12. When you turn on a hot water faucet after it's been off awhile, you get cold water at first, and then the water gets hot. The farther the water heater is from the faucet, the more cold water you get. At bill-paying time, does that cold water *count* as cold water or hot water?

13. Mrs. Brown insists that the house be kept at a temperature of 15°C, but Mr. Brown refuses

to take a bath unless the bathroom temperature is at least 20°C. If the Browns have a gas-fired central heating system so that they must heat the whole house in order to heat the bathroom, might they come out ahead if they installed an electric heater in the bathroom?

Problems

1. If a power plant's overall efficiencfy is 38%, and 7% of the electrical energy is lost to heat in the transmission line between plant and user, what is the efficiency for the entire process of generating and delivering electrical energy?

   Answer: 35%

2. How much chemical energy in the fuel do Ann and Abby use in the situation of Question 3? Assume the local power plant is 38% efficient, including transmission to the house.

   Answer: Ann—36,000 kWh; Abby—71,000 kWh

3. If the heat of combustion of natural gas is 0.40 kWh/ft³ and that of coal is 8.2 kWh/kg, how much fuel do Ann and Abby consume in the situation of Problem 2 if the power plant burns coal rather than natural gas?

   Answer: Ann—90,000 ft³; Abby—8600 kg (9.5 tons)

4. The Peachtree Power Company normally operates at an efficiency of 39.0% and supplies an average of 1000 MW to its customers at a price of 8¢/kWh. What would the financial loss to Peachtree's stockholders be in a year if the plant's efficiency dropped by 0.1%?

   Answer: $1,800,000

5. The Jones family uses an average of 1200 kWh of electrical energy per month. They decide to switch to solar cells to get this energy, and they use batteries to store the sunny-day energy for nights and rainy days. Assuming they have 20% efficient solar cells, and the incident solar energy density averages 200 W/m² for an average of 8 h per day, how much area would their solar cells have to cover?

   Answer: 125 m²

6. The U.S. consumes about 3 × 10¹² kWh of electrical energy per year. How many 60-kg people would this amount of energy lift to the top of the 440-m-tall Sears Tower?

   Answer: 41 trillion

7. Compare the energy used in toasting two pieces of bread in a 1050-W pop-up toaster that takes 2.0 min with that used to toast them in a 3200-W broiler that takes 3.5 min, including warmup time.

   Answer: 35 Wh versus 190 Wh

8. If your water heater is 10 m from your bathroom, and the hot water pipe is 1.27 cm ($\frac{1}{2}$ in) in diameter, roughly how much energy do you waste in running the cold water out of the pipe?

   Answer: 0.07 kWh

9. Assume the Browns, of Question 13 fame, have a house with hot-water radiators: a total of 10 radiators averaging 20 liters of water per radiator. How does the energy needed to heat the water in all 10 radiators from 15 to 60°C compare with that needed to operate a 1.5-kW electric heater for $\frac{1}{2}$ h?

   Answer: 10 kWh versus 0.75 kWh

10. Refer to Problem 9. If the Browns' heating system is 50% efficient and their power plant is 35% efficient, compare the fuel energy used in heating their bathroom by heating their whole house with that used by the bathroom electric heater.

    Answer: 20 kWh versus 2.1 kWh

11. A friend proposes saving electricity costs by switching to a small gasoline-powered generator at night when there are only a few appliances to keep running, rather than using energy from the power company. If the gasoline-engine generator set is 15% efficient, how

much electrical energy do you get per liter of gasoline? The heat of combustion of gasoline is 7820 kcal/liter. At your local gasoline prices, how much would this cost per kilowatt-hour? *Answer*: 1.36 kWh/liter; 19¢/kWh @ $1/gal

12. Show that 1 kWh is equivalent to 3.6 MJ.

Home Experiments

1. Stick a piece of copper and a piece of zinc into a lemon or a tomato. A copper wire and a galvanized (zinc coated) nail or a paper clip will do. Bring the other ends of each piece close together and touch your tongue. You feel a tingle and a distinct taste similar to that from a battery (see Home Experiment 2, Chapter 16). If you use well-cleaned strips of copper and zinc about $1\frac{1}{2}$ cm wide, and connect several such cells in series, you might be able to light a small bulb.

2. Run your cold water until the temperature stabilizes and then measure its temperature with a thermometer. Then measure the hot water temperature. (Be sure you have a thermometer with high enough temperature range.) If you have access to the water heater that heats your water, turn off the power to it, remove the cover plates, and check the thermostat settings. Compare with your measured temperature. Try to estimate whether any difference is from heat loss between heater and faucet or from faulty calibration on either the thermometer or thermostat. Calculate the energy needed per liter to heat *your* water and compare with the estimates in the chapter.

3. Read your electrical meter for two consecutive months on about the same day as it's read by the power company. Subtract the difference to get energy use. Calculate your bill based on a rate schedule available from the power company. Compare it with the bill they send you.

References for Further Reading

Derven, R., "Heat Pumps," *Popular Science*, Sept. 1976, p. 92. Includes information about the heating and cooling needs in various parts of the United States.

Edison Electric Institute (90 Park Ave., New York, 10016), "Annual Energy Requirements of Electrical Household Appliances," EEI-Pub #75-61. Information on average energy consumption of appliances.

Hassoun, V. S., and F. E. Hunt, "Electric Energy Usage in the Home: A Predictive Model," *Home Economics Research Journal*, March 1980, p. 252.

Hodges, L., "Resource Letter SE–2: Solar Energy," *American Journal of Physics*, Oct. 1982, p. 876. A vast array of references on solar energy, covering the whole range of discussion level.

Johnston, W. D., Jr., "The Prospects for Photovoltaic Conversion," *American Scientist*, Nov.–Dec, 1977, p. 729. Can solar cells contribute significantly to a solar energy economy?

Marion, J. B., and M. L. Roush, *Energy in Perspective*, Second. Ed. (Academic Press, New York, 1982). Discusses a broad range of energy considerations.

Pansini, A. J., *Basic Electrical Power Transmission* (Hayden Book Co., Inc., Rochelle Park, N.J., 1975). The problems and techniques of transporting electrical energy.

Powell, E., "How to Keep Your Appliances on an Energy Budget," *Popular Science*, Nov. 1975, p. 106. Gives suggestions for efficient use of appliances.

Ruby, D., "Cheaper to Leave Fluorescents On? Forget That Myth and Switch 'Em OFF," *Popular Science*, Jan. 1981, p. 133.

Solomon, B., "Will Solar Sell?" *Science 82*, April 1982, p. 71. The economics of solar energy.

L I G H T   W A V E S

The theory of light and colors, though it did not occupy a large portion of my time, I conceived to be of more importance than all that I have ever done, or ever shall do besides.

THOMAS YOUNG

Thomas Young, born at Milverton, Somerset, England, in 1773, mastered a number of Middle Eastern languages as well as Newton's *Principia* and *Optics* by the age of 17. A substantial inheritance enabled Young to become a gentleman scholar of broad interests. A major linguistic accomplishment of his was the translation of Egyptian hieroglyphics. Following a brief tenure as professor at the Royal Institute in London, he turned to medicine. Young conducted many experimental studies, especially on the eye, often using his own eyes for his investigations. He developed an understanding of the process by which the eye can focus on objects at different distances, and he was the first to demonstrate that all colors can be obtained by an appropriate combination of three primary ones. Young made a major contribution to physics by demonstrating the wave nature of light, making use of the phenomenon called interference.

L I G H T   R A Y S
A N D   O P T I C A L
D E V I C E S

For hundreds of years people have known that light, whatever it is, travels in a straight line in any one uniform medium. Early physicists developed rules that described what happened when light, traveling in one transparent medium, reached the boundary of a different one—for example, what happened when light traveling through air hit a piece of glass. Those rules gave them the know-how to make eyeglasses and other optical instruments. This understanding came before anyone knew that light was a wave.

In this chapter we follow the example of the pioneers and treat light as *rays* that travel in straight lines in any one medium. We usually represent these rays by lines with arrows on them to show the direction of light travel. We call this ray description of light *geometric optics*. In the next chapter, we'll look at light's wave nature.

For now we will be concerned mainly with two general effects: the *reflection* of light by smooth surfaces such as mirrors, and the *refraction*, or bending, of light rays as they go from one medium to another. We'll see how lenses, that depend on refraction effects, focus light in optical instruments.

Figure 21.1 One of nature's own mirrors.

21.1 REFLECTION

The reflections of light that you see probably range from those created for you by nature (Fig. 21.1) to the delightful sight you see in your bathroom mirror each day as you prepare to face the world. Light reflections follow one basic **law of reflection** illustrated in Figure 21.2: *The angle of reflection is equal to the angle of incidence.* Even though we aren't supposed to consider waves in this chapter, notice that this is just the general law of reflection for any kind of wave, including light. By tradition, we usually measure angles relative to the **normal**—a line perpendicular to the surface—rather than to the surface itself.

Most surfaces are rough, as far as light reflection is concerned. That is, the surface irregularities are large compared with the wavelength of light. The law of reflection is obeyed at any one spot, but a parallel beam of light is reflected in all directions from such a surface, as Figure 21.3(a) shows. That's the way you *see* the object—by *diffuse* reflection.

However, if the surface is smooth enough that surface irregularities are small compared with the wavelength of light, *regular* reflection occurs, as illustrated in Figure 21.3(b). There, the entire beam is reflected so that the angle of reflection equals the angle of incidence. Many surfaces—for example, a shiny red apple—give both diffuse and regular reflection. Our remaining discussion will be concerned with regular reflection, the kind you get from a mirror.

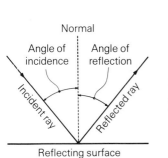

Figure 21.2 When light reflects from a surface, the angle of reflection equals the angle of incidence. The *normal* is perpendicular to the surface.

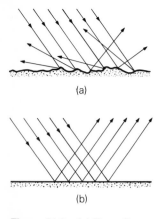

Figure 21.3 (a) The reflection from most surfaces is *diffuse*. (b) When the surface is smooth, the reflection is *regular*.

Reflection from Plane Mirrors

When you look at yourself in a mirror, the face you see is an *image* that appears exactly the same distance behind the mirror as your face—the *object*—is in front of the mirror. To see that this statement is generally true for a **plane** (flat) **mirror**, imagine that you're looking at the reflection of a flower, as in Figure 21.4. Concentrate for now on the light you see coming *only* from the center. The light that gets to either eye is reflected so that the angle of reflection equals the angle of incidence. Drawing the rays so that this situation is true, we see that the rays to each eye are at slightly different angles. As the dashed lines show, the rays appear to be coming from a point behind the mirror a distance equal to the distance from the mirror to the center of the flower. We call this point the **image**. If you draw rays similarly for other points on the flower, you can see that the image of the entire flower is as shown. The point from which the rays originate *before* hitting the mirror, we call the **object**.

Notice that the missing petal is on the *left* side of the *object* but is on the *right* side of the *image*. The "mirror image" is reversed relative to the object. If you draw ray diagrams for each side, you can see why this reversal happens. That's why you might think your photographs don't look quite like you. The left side of your face may be slightly different than the right, or maybe you part your hair on the side. In the mirror you see things reversed from the way others see you, and from the way you appear in a photograph.

Writing is reversed when viewed through a mirror. The writing on the front of some vehicles is backwards (Fig. 21.5-a) so that when viewed through a rear-view mirror (Fig. 21.5-b), it appears in the right order.

We can use the law of reflection to find out what size a mirror needs to be for proper viewing. Let's find the length of the shortest mirror in which Sandy Allan, the tallest living woman, at 232 cm, can see herself completely. Suppose the mirror is mounted vertically on her bedroom wall in Shelbyville, Indiana, and she stands so that the tips of her toes are directly under her eyes. Assume her eyes are 12 cm below the top of her head. Figure 21.6 shows the situation.

Figure 21.4 The image of an object—the point from which the rays appear to come—viewed in a plane mirror is the same distance behind the mirror as the object is in front. The mirror reverses the sides: Notice the missing petal.

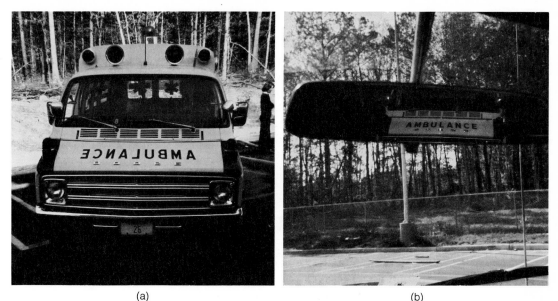

(a)   (b)

Figure 21.5 (a) The writing on the ambulance is reversed. (b) The writing appears in the right order when looked at through a rear-view mirror. (Photographs by Loren Long.)

Light from any point that gets to her eyes travels so that the angle of reflection equals the angle of incidence. Light from her toes hits the mirror at a level halfway between the floor and her eyes. Light from the top of her head hits halfway between head level and eye level. Light from other points hits between these two extremes. The minimum length needed is then 110 cm for viewing from the eyes down, plus 6 cm for viewing from the eyes up, or a total of 116 cm: $\frac{1}{2}$ her height. The mirror needs mounting with the bottom 110 cm from the floor: halfway between floor and eye level. Using a "floor-length" mirror merely lets her see more of the floor and other surroundings—to give added perspective—but doesn't show any more of her body.

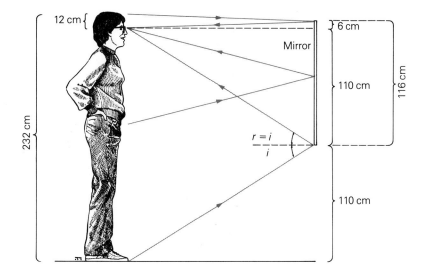

Figure 21.6 If placed vertically at the right height, a mirror needs to be only $\frac{1}{2}$ the height of the person for a full view. For a given ray, the angle of reflection r equals the angle of incidence i.

(1) If Sandy's size $19\frac{1}{2}$ EEE shoes extend forward of a point directly below her eyes, how will this change the minimum length the mirror must be to show her entire shoes?

(2) Does the rear-view mirror in Figure 21.5(b) need to be $\frac{1}{2}$ as large as the front of the ambulance for the driver to see the entire ambulance front? (Hint: The ambulance is farther from the mirror than the viewer's eyes. Draw ray diagrams to verify the answer you know from experience.)

Reflection from Curved Mirrors

Even though plane mirrors are more common, we do use curved mirrors as magnifiers (Fig. 21.7), as reflectors for various kinds of lights, as rear-view mirrors for cars and trucks (Fig. 21.9-a), and in many other ways.

First we'll find out how a curved mirror can act as a magnifier. Suppose we put our flower near the concave side of a mirror with a slight spherical curvature. Figure 21.8 shows two rays from the center of the flower, reflected so that the angle of reflection at the point of reflection equals

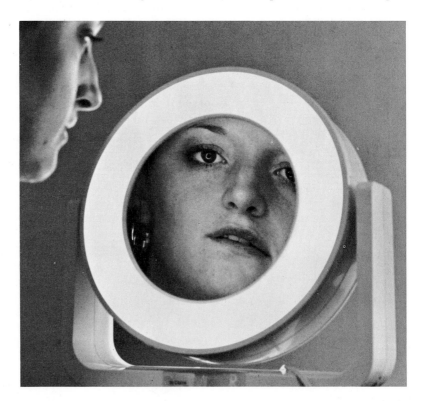

Figure 21.7 Dressing-table magnifiers are curved mirrors.

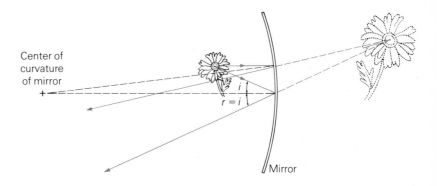

Figure 21.8 A ray from any point on the object reflects according to the law of reflection. The apparent intersection of two such reflected rays locates the image of that point. Doing the same thing for other points on the object shows that the overall image is magnified.

the angle of incidence. After reflection, these rays diverge as if they were coming from the center of the dashed image. *Any* ray from the center of the flower that makes a fairly small angle of incidence on the mirror will reflect so that it appears to come from the same point on the image. (Notice that the normal to any point on the mirror passes through the center of curvature.) Drawing this diagram as described locates the image of the center of the flower. Doing the same thing for other points on the flower locates the entire image. We see that the curved mirror causes the image to be *magnified:* The image appears larger than the object. This is the principle of the magnifying mirror (Fig. 21.7), and it works whether you're viewing a flower or your own face.

If you draw lots of ray diagrams such as the ones in this figure, you can demonstrate that for any object that is small compared with the radius of curvature of a spherical mirror, and that is located anywhere between the mirror and $\frac{1}{2}$ the distance to the center of curvature, you get a magnified image.* This fact can be shown mathematically, using a little geometry and trigonometry, but we'll settle for the ray-drawing approach here.

· ·

EXERCISE   Draw diagrams such as Figure 21.8 for an object near the mirror and for an object *almost* $\frac{1}{2}$ the distance to the center of curvature. Which way gives more magnification? If you have a magnifying mirror, check your answer by trying it.

Sometimes it's useful to have a curved mirror that reflects from the convex side. Figure 21.9 illustrates one such application.

*To do this, you can measure with a protractor the angle between the incident ray and the normal where the ray hits the mirror. The normal is a straight line from that point on the mirror through the center of curvature. Then draw the reflected ray at the same angle on the other side of the normal. The seeming intersection of any two such rays locates the image.

Figure 21.9 A rear-view mirror used by trucks. (a) The plane mirror reflects light toward the driver's eye from only one direction, showing the driver what is behind as seen directly beside the trailer body. The small spherical mirror at the bottom shows a greatly expanded, though distorted, view of what is nearby. (b) The actual location of the car that shows up in the spherical, but not the plane, mirror. With this mirror the driver can see cars, or rather the images of cars, that are almost beside the cab. (Photographs by Loren Long.)

(a)

(b)

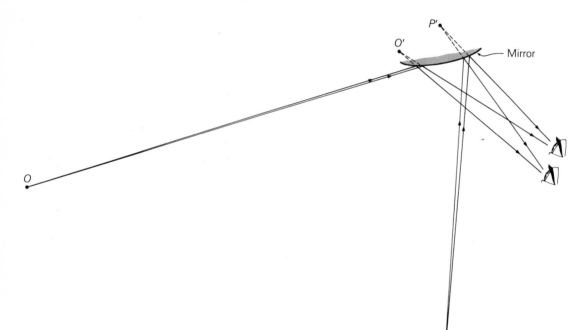

Figure 21.10 A convex spherical mirror reflects light from a broad range of angles into the direction of the viewer. The image of point O is at point O'; that of point P is at P'.

Figure 21.10 shows why the convex mirror gives a greatly expanded field of view. Light from an object at point O appears after reflection to be coming from point O'. That from an object at point P appears to come from point P'. Each ray obeys the law of reflection. The image is reduced in size and badly distorted, but you can tell that the object is there.

(a)   (b)

Figure 21.11 Because of the refraction of light rays, (a) the *one* fish in this photograph looks like *two*; (b) straws seem to do strange things.

21.2 REFRACTION

When light traveling in one medium enters a new medium in which its speed differs, some rather strange things can happen. One example is shown in Figure 21.11(a). This photograph is of *one* fish, but because of the shape of the boundaries of the different media transferring the light, it can look like *two* fish, one larger than the other. You may have noticed some other strange-looking effects: the apparent bending or separation of a straw at the surface of a transparent drink (Fig. 21.11-b); the short, squatty appearance of a person standing underwater; and the shallower look of the far end of a swimming pool, no matter which end is deeper. Some of the things that happen when light enters a different medium are fortunate: for example, one of them gives you the ability to see, as we'll soon describe.

To see why light should bend as it enters a medium in which its speed is different, consider an analogy. Suppose you're coasting on roller skates down a smooth sidewalk, and you go off the sidewalk into the grass at an angle so that one skate goes into the grass before the other. The skates meet more resistance, and therefore move slower, in the grass. Since one skate slows down before the other, you will gradually change direction until the other skate gets into the grass. You then move in a fixed *new* direction. The analogy isn't perfect, because a given skate doesn't instantaneously change speed at the grass boundary and then move at constant speed thereafter. But you can imagine what would happen if it did.

The bending of a light beam on entering a new medium is similar. We can get an indication of why this bending occurs if we again jump ahead and remember that light is a *wave*. Figure 21.12 shows parallel light rays incident on glass from air. The lines perpendicular to the rays are *wavefronts*, representing crests in the wave and are one wavelength apart. As the light enters the glass, where its speed is about $\frac{2}{3}$ that in air,

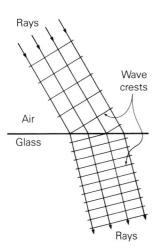

Figure 21.12 Light rays slow down as they enter the glass, so that the part that enters first is retarded relative to the other side of the beam. This retardation causes the beam to bend, or *refract*.

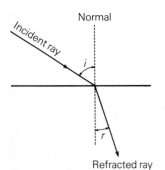

Normal

Incident ray

i

r

Refracted ray

Figure 21.13 We usually measure angles relative to the normal. The angle *i* is the angle of incidence; *r* is the angle of refraction.

the light travels less distance in one period; thus, the wavelength is shorter, and the wavefronts are closer together. To keep the crests in phase for the different rays, the rays must bend (remember the roller-skate analogy).

We call this bending due to speed difference **refraction**. It can occur in any kind of wave. In discussing refraction of sound we considered the gradual bending caused by a continuous variation in speed with position, rather than an abrupt change at a boundary.

As we did in considering reflection, we usually specify the angle a ray makes with the *normal* to the surface, as shown in Figure 21.13. When the ray hits perpendicular to the surface, there is no bending. The greater the angle with the normal, the more the ray is refracted. For a given angle, the amount of bending increases with increase in the ratio of initial to final speeds.

We've been discussing the case in which the speed in the second medium is lower than in the first. In that case the ray bends *toward* the normal: The angle of refraction is *less* than the angle of incidence. If the speed in the second medium is higher, the ray bends *away* from the normal.

As an example, we'll see how refraction causes a pool to seem shallower at the far end, no matter which end is deeper (Fig. 21.14). For light from the far end to get to the observer, it has to bend so that it appears to come from just below the water surface.

• •

EXERCISE   By drawing a diagram like Figure 21.14, show why a person standing underwater in a pool looks much shorter to someone on the side than he or she actually is. That is, show that light from the feet seems to come from a point higher than the feet.

Total Internal Reflection: Fiber Optics

Imagine you are under water with a flashlight, and you shine the beam upward. You shine it at various angles between straight-up and about 60°, as shown in Figure 21.15. At each angle some light is reflected and

Figure 21.14 The far end of a pool appears shallower than the near end, no matter what its depth.

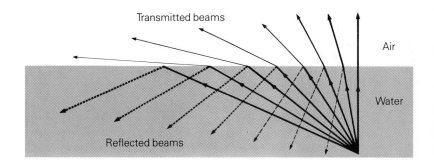

Figure 21.15 Light entering a medium with higher speed of travel bends *away* from the normal. When the incident angle between ray and normal exceeds that for which the refracted ray travels along the surface (about 48° for water), the light is totally internally reflected.

some transmitted. As the incident angle of the beam increases, the fraction of the beam that is transmitted decreases, and it bends farther from the normal as it emerges. Eventually (at about 48° in water) the emerging ray is bent right along the surface. For larger angles, the ray cannot get out and is totally internally reflected at an angle equal to the angle of incidence. We call the angle of incidence beyond which total internal reflection occurs, the **critical angle**.

Total internal reflection has many practical applications. In a prism, it can reflect light back parallel to its original direction (Fig. 21.16). This process comes in handy in instruments such as binoculars and periscopes.

A more recent development that is having far more impact on our lives is the use of total internal reflection in a field called **fiber optics**. Suppose you have a thin glass fiber bent into a curved path as shown in Figure 21.17. Light entering one end is totally reflected at various points inside the fiber and eventually emerges from the other end. Bundles of these fibers can be made flexible enough to curve around in fairly sharp bends.

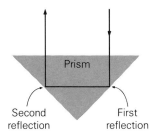

Figure 21.16 Total internal reflection can be used in a prism to reverse the direction of a light beam.

One important use of fibers is in medicine (Fig. 21.18). Tiny but intense light beams from lasers have replaced the scalpel for many types of surgery. Rather than having to aim the laser at the spot needing the surgery, the light beam is directed into tiny transparent fibers. The other end of the fibers can be bent easily and aimed carefully by the surgeon at the desired point.

Fiber optics has made it possible to diagnose certain medical problems with a minimum of disturbance and pain to the patient. Fiber-optic endoscopes—essentially small bundles of transparent fibers—can be inserted into parts of the body for viewing the inside of an organ in much the same way as you might shine a flashlight into a dark room to see what's inside. There's one difference: the light showing the condition inside the organ would have to come back outside by way of optical fibers. Physicians routinely examine the interior of functioning human hearts, for example, by inserting such an instrument into a blood vessel in the arm and threading it through the vessel into the interior of the heart.

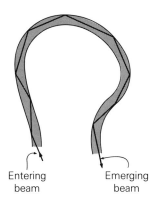

Figure 21.17 By total internal reflection in a small glass fiber, a light beam can be bent in any direction.

Another major use of fiber optics is in communications fields such as television and telephone. It is likely that, within a few years, optical fibers will take over a large part of the job now done in communication by metal wires. Rather than sending a signal down a wire by means of an electrical current, we can send it down a fiber by means of a pulsed light beam—the message being coded in the pulses. A receiver then decodes the signal back into a picture or sound or whatever is being transmitted. Because of the extremely high frequency of light waves, the beam can be pulsed at a very high rate. Systems under test carry about 45 million pulses per second, a rate sufficient to carry 672 one-way voice signals. Therefore, a cable of 24 fibers, separately pulsed, can carry about 8000 two-way telephone conversations—vastly more than can be carried in the same sized copper wire conductors.

There are still a few problems to be solved before fibers take over a major portion of the communications jobs. One is long-distance transmission without too much loss in light intensity. Another is the problem of connectors: It's easy to plug a telephone into a jack and have it work reliably after many pluggings and unpluggings; a comparably reliable and sturdy optical connector is much harder to make.

Figure 21.18 Fiber optics is being widely used in microsurgery.

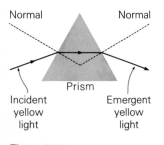

Figure 21.19 Light incident on a transparent prism refracts downward at both surfaces.

Refraction by a Prism

Many optical instruments make use of *prisms*—triangular-shaped pieces of glass through which light can pass. Let's consider what happens when a beam of light strikes a prism at an angle to the surface such as that shown in Figure 21.19. For reasons that will become clear in the next chapter, that beam needs to be of a particular color—for example, yellow. The incident beam bends toward the normal in entering the glass, which means it bends downward in the figure. When the ray reaches the other side, the slope is such that, in bending away from the normal, the ray again bends downward. The result is that the emerging beam deviates substantially from the original direction. The amount of deviation depends on the incident angle and on the angles between the sides of the prism.

• •

QUESTION   What happens to a beam of yellow light that hits at an oblique angle a piece of glass with *parallel* sides?

21.3 LENSES

Lenses abound in our lives. They form the heart of magnifying glasses, eyeglasses, cameras, projectors, microscopes, telescopes, and most important of all, our own eyes. Lenses depend on the *refraction* of light. We'll now see how they work, and how they are used in various optical devices.

The Focusing of Parallel Rays

Suppose we put two prisms base-to-base (as in Fig. 21.20) and bring in a beam of parallel light. The rays will refract at each surface, and bend toward the line formed by the bases. We'll call this line the *axis*. Any two rays coming in at the same distance on each side of the axis will cross at the same point on the axis.

If we could arrange to have *all* the rays cross at the *same* point on the axis, after passing through the refracting device, we would have a **lens**. Look at the way the rays bend, and you will see that to reach this goal, our prisms need to slope *less* near the bases, *more* near the points. That way, the rays near the axis will bend less, those away from the axis, more, so that all will cross at an intermediate point.

It turns out that if we make the refracting surfaces spherical in shape and keep the rays near the axis, they will all cross at the same point (as shown in Fig. 21.21). We call this point at which the rays *focus* the **focal point**.* We call the distance from the lens center to the focal point the **focal length**.

The path of a given ray is the same, no matter which direction along the path the ray is traveling. As Figure 21.22 shows, if we have a small light source placed at the focal point, those rays passing through the lens will move off parallel to the axis.

Figure 21.20 Base-to-base prisms cause parallel rays *equal distances from the axis* to cross the axis at the same place.

Figure 21.21 A lens with spherical surfaces brings all parallel rays near the axis to a focus at the same point *F* called the *focal point*.

*When the distance from the ray to the axis is large enough that the angles of bending are more than about 10°, a spherical surface no longer brings the rays to a sharp focus.

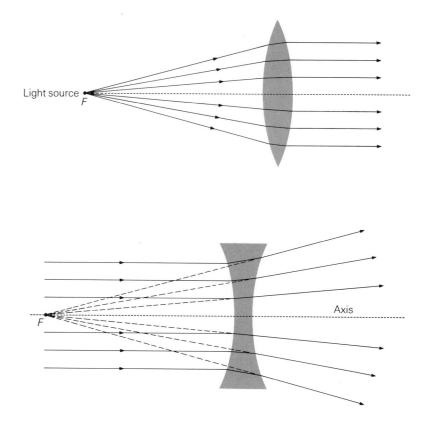

Figure 21.22 Light rays coming from the focal point travel parallel to the axis after traversing the lens.

Figure 21.23 If a lens with spherical surfaces is thinner in the middle than at the edges, parallel rays will appear to diverge from the focal point on the incoming side of the lens.

Next, notice what happens if we bring a parallel beam of light rays onto a lens with spherical surfaces arranged so that the lens is thinnest at the axis (Fig. 21.23). Because of the way the surfaces are curved, the rays bend outward and *diverge* away from the axis. They appear to be coming from a point on the incoming side of the lens, as shown by the dashed lines. The focal point is, therefore, on the *opposite* side of this lens from that for the previous lens.

We call a lens that causes parallel rays to converge, a **converging** lens. If parallel rays diverge after passing through the lens, it is a **diverging** lens. Converging lenses are always thicker in the middle than on the edges, but diverging lenses are always thinner in the middle. Both surfaces may curve the same way (as for most eyeglasses) but, depending on which surface has the greater curvature, the lens may be diverging or converging.

Notice that if you turn one of our lenses completely around, it would have the same effect. This statement means that there is a focal point on both sides of the lens, equal distances from the center of the lens. Since *both* surfaces determine the distance to *each* focal point, both focal lengths are equal, even if the curvature of the two surfaces is different.

Focusing Nonparallel Rays

Next look at a situation such as shown in Figure 21.22, but with the light source farther away than the focal length from the lens. In Figure 21.24(a) a small light source S is located on the axis of a converging lens. Compare this case with that in which the source is at the focal point. Because a given ray hits the lens at a smaller angle than if it had come from the focal point, it bends beyond the "parallel direction." The rays all converge to a point on the axis. In Figure 21.24(b), the source is located off-axis. The rays then come to focus at a point off-axis. Notice that, as you should expect from our discussion so far, the particular ray that comes in parallel to the axis goes off through the focal point. Also, the one that comes in through the other focal point goes off parallel to the axis. We'll make special use of these two particular rays.

Image Formation by Lenses

In talking about mirrors, we described the *image* formed by the reflected light (the image being the point from which the light appears to come after reflection). A similar situation exists for lenses. Light from an **object** forms an **image** on passing through a lens. The transmitted light *appears* to come from this image.

To simplify the drawing, we'll use an upright tower-shaped form to represent the object, as we've done in Figure 21.25. This tower sym-

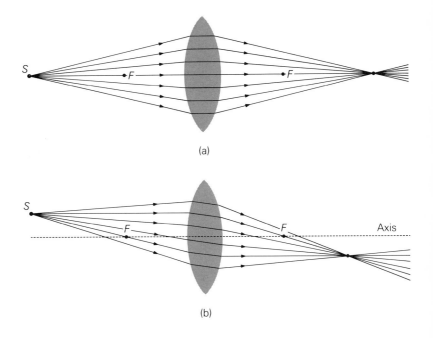

(a)

Figure 21.24 (a) Rays from a small light source on the axis in front of the focal point converge to a focus beyond the other focal point. (b) Rays from an off-axis source come to focus off-axis.

(b)

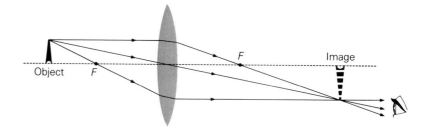

bolically represents anything that sends light through the lens, which is what we mean here by the term *object*.

In this figure, the object is shown *in front of* the focal point of a converging lens. We have located the image by drawing three rays, all of which can be drawn easily without measuring any angles. Here's why:

First, notice that the rays drawn locate the image of only the tip of the tower. Once it's located, the rest can be drawn in. But you should locate a few other points on the image the first time around, just to convince yourself.

Second, let's see what is special about these three rays. *One travels parallel to the axis, so that it then passes through the focal point. Another passes first through a focal point, so it then must move off parallel to the axis. The third passes through the center of the lens and is almost undeviated*, since the lens is thin and the sides are nearly parallel there. Since *all* rays (within the small-angle limit) must pass through the same point, we don't need to draw any more. Actually, the crossing point of *any two* of these rays is enough to locate the image. We call the three rays we've described **principal rays**.

If you look from any position to the right of the image at the light from the object, the object will *appear* to be at the image position. It will appear upside down, and the size you see will be that of the image, not that of the object.

· ·

EXERCISE   Draw two diagrams such as Figure 21.25, one with the object near but in front of the focal point, one with it far from the lens. Demonstrate that the nearer the object is to the focal point, the farther away the image is.

By comparing the triangle formed by the object, the axis, and the straight-through ray, you can see that the ratio of *image size* to *object size* is the same as the ratio of *image distance* to *object distance*:

$$\frac{\text{image size}}{\text{object size}} = \frac{\text{distance from lens to image}}{\text{distance from lens to object}}$$

For example, if the image is three times as far from the lens as the object, it will be three times as big as the object.

In drawing ray diagrams, you can usually ignore the thickness of the lens and draw the bending as if it all occurred at the center. We will continue to show the bending at each *surface*, since that's where it actually occurs. If you know the focal length of the lens and how far the object is from the lens, you can find out exactly where the image falls by drawing the ray diagram to scale. We'll do that in the next section when we discuss specific instruments.

The magnifying glass.   Look at the case in which an object is put *between* a converging lens and its focal point. Figure 21.26 shows the image location. The rays never cross after passing through the lens, but appear to cross at a point in front of the lens, on the left side of the focal point. Since the image is farther from the lens than the object, it appears magnified.

This is the principle of a simple magnifying glass, which is merely a converging lens placed so that the object you're magnifying is *inside* the focal length.

Diverging lenses.   We find the image of a diverging lens the same way, but it's a little trickier (see Fig. 21.27). A ray that comes in parallel diverges so that it appears to come from the focal point on the left. A ray headed for the focal point on the *right* is refracted by the lens so that it moves off parallel to the axis. The ray through the center is undeflected, as before. In other words, the role of the two focal points is reversed from that for the converging lens.

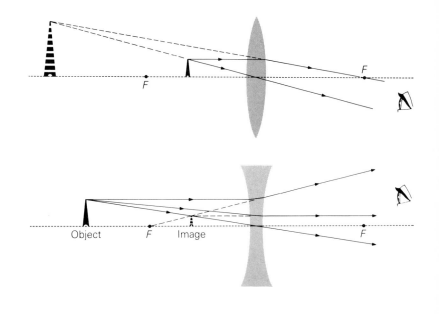

Figure 21.26   A simple magnifying glass is a converging lens with the object placed *inside* the focal length.

Figure 21.27   Rays diverge from a diverging lens so that the image is located on the same side of the lens as the object.

Again, any two of the three rays locate the image. No matter where you place the object to the left of the diverging lens, the rays will *always* diverge and appear to come from an image on the left side, reduced in size compared with the object.

· ·

EXERCISE   Draw ray diagrams for diverging lenses with several different object locations. Show that you always get the image on the same side as the object, upright, and reduced in size.

Notice that for the magnifying glass (converging lens with object within the focal length) and for all diverging lenses, the focused rays do not actually pass *through* the image. They only *appear* to come from the image. We call such an image a **virtual** image, as opposed to a **real** image, through which the focused rays actually pass. If you have a real image, you can put a screen there and see the image formed on the screen. You cannot do this for a virtual image, since the rays do not actually pass through the point of focus.

The Lens Equation

It's sometimes convenient to have a mathematical relationship between the image distance, object distance, and focal length. This relationship, called the *lens equation*, is

$$\frac{1}{f} = \frac{1}{o} + \frac{1}{i}$$

where o and i are the object and image distances, respectively, and f is the focal length.

In using the lens equation, remember that the focal length of a diverging lens is negative. If you find that an image distance comes out to be negative, this means that the image is on the same side of the lens as the object. For a single lens, you can always take the object distance to be positive.

· ·

EXAMPLE 21.1   A converging lens of focal length 18 cm is held 12 cm above an old coin. As viewed from above the lens, where does the image of the coin appear? If the numbers in the coin's date are 1 mm tall, how tall are they in the image?

From the lens equation,

$$\frac{1}{i} = \frac{1}{f} - \frac{1}{o} = \frac{1}{18 \text{ cm}} - \frac{1}{12 \text{ cm}}$$

$$= \frac{2}{36 \text{ cm}} - \frac{3}{36 \text{ cm}} = -\frac{1}{36 \text{ cm}}$$

Therefore,

$$i = -36 \text{ cm}$$

Thus, the image is 36 cm *below* the lens. To get the size of the image, remember that

$$\frac{\text{image size}}{\text{object size}} = \frac{i}{o}$$

Therefore,

$$\frac{\text{image size}}{1 \text{ mm}} = \frac{36 \text{ cm}}{12 \text{ cm}}$$

which gives an image height for the numbers of 3 mm—3 times the height of the numbers on the coin.

The lens equation also works for spherical *mirrors* (Section 21.1), and in that context we call it the *mirror* equation. The focal length for a spherical mirror is $\frac{1}{2}$ the radius of curvature. A convex mirror has a negative focal length. Images behind the mirror correspond to negative image distances.

21.4 OPTICAL INSTRUMENTS

In discussing optical devices that use lenses, we'll start with the camera. The camera directly uses the image-formation ideas we've been discussing, and it offers a good background for discussing the human eye.

The Camera

A camera operates by bringing the rays from an object into focus—that is, by forming an image of the object—on the film (Fig. 21.28). Suppose you're photographing a tree. When you open the shutter, you want the light from *one spot* on a leaf on the tree to hit only *one spot* on the *film*.

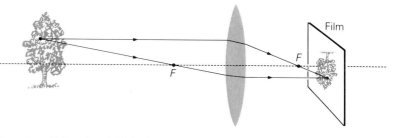

Figure 21.28 A camera uses a converging lens to bring the image into focus on the film.

No other light should hit that spot. In other words, the image of the tree should fall exactly on the film. Light in the image causes chemical changes in the film that, after processing, reproduce the colors in the original light.

How then, can we be sure the image is always on the film? The focal length of the lens is fixed, yet you can take pictures of the tree from various distances. The tree can be at various *object distances*. From the previous section, you know that changing the object distance for a lens changes the image distance. When you focus the camera for a particular object distance, you are changing the distance between the lens and the film, making this distance equal the image distance. Usually, the object distance is large compared with the lens focal length, making the image fall near the focal point for all object distances. As a result, only slight lens movement is needed in focusing.

. . . .

EXERCISE

If a camera has a 10-cm focal length, how far should the film be from the lens when taking a picture of a distant mountain? Hint: Rays hitting the camera from a point on a distant object are almost parallel. For nearby objects, would the lens be closer to, or farther from, the film?

Answer: 10 cm; farther

Camera and projector lenses are rated not only by their focal length but by an "f-number," for example f/2. (Don't confuse this "f" with the "*f*" we used earlier for focal length.) The f-*number is the ratio of focal length to lens diameter*. If our f/2 lens has a focal length of 48 mm, the effective lens diameter is 24 mm. Reducing the diaphragm opening to let less light enter the camera gives a higher f-number setting, and vice versa.

The Eye

The eye is essentially a high-speed "instant" camera that processes the picture quickly and sends the result to the brain rather than printing it out on paper. But Polaroid and Kodak have a long way to go to catch up with the technology in the eye!

Figure 21.29 is a diagram of the human eye. Light from an object

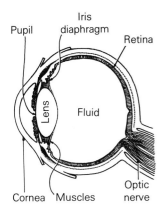

Figure 21.29 The human eye.

enters the eye through the pupil and is brought into focus on the *retina*. The pupil, the opening in the iris, varies in size, depending on the brightness of the object. The retina, which plays the analogous role of the film in a camera, contains the photosensitive cells that translate the visual image into the electrical signals sent to the brain by way of the optic nerve. A clear message is sent only when the image is focused clearly on the retina.

The focusing is done not only by the lens, but by the curved surface of the cornea, the outer protective layer of the eye. With varying object distances, something has to change if the image is always to fall on the retina. The camera accomplishes this feat through movement of the lens relative to the film. But the distance from the lens to retina in the eye is fixed. The eye adjusts to varying object distances by adjusting the focal length of the lens, letting the image distance stay constant. When a normal eye looks at a distant object, the muscles controlling the lens relax, and the lens takes on the shape needed to focus the image on the retina. When the eye looks at a closer object, the muscles pull the lens into a fatter shape so that the focal length is shorter and the image still falls on the retina.

The process of adjusting the focal length is called **accommodation**. Typically, at about the age of 40, a person's lens starts to stiffen, and can no longer accommodate as effectively. For the eyes to focus, the person then has to hold things farther away than normal. By the time a person's arm gets too short to read the newspaper, he usually gives up and visits the ophthalmologist. The accommodation ability continues to decrease with age, and usually vanishes by about age 75.

Eye defects. A normal eye brings images into focus on the retina (Fig. 21.30-a). But there are certain ordinary eye defects that can be corrected by eyeglasses, such as *nearsightedness* (myopia) and *farsightedness* (hyperopia). In a nearsighted person (Fig. 21.30-b), a distant object is focused in front of the retina. Glasses with diverging lenses cancel some of the converging effect of the eye and make the image fall on the retina. In a farsighted person (Fig. 21.30-c), nearby objects focus behind the retina. A converging lens corrects the problem.

Loss of accommodation may cause a form of farsightedness for close-up objects. Thus, an older person's "bifocals" may have a diverging region for seeing distant objects and a converging region for reading. You might call the use of bifocals "artificial accommodation."

Another eye defect that can be corrected with glasses is *astigmatism*. The cornea and/or lens of an astigmatic eye has more curvature, and therefore a shorter focal length, in one direction than in another—for example, more curvature in the vertical plane than in the horizontal. The remedy is glasses whose lenses are more curved in the plane that the eye is least curved in. Figure 21.31 shows the focusing effect of lenses that are purely converging compared with that of lenses for correction of astigmatism.

(a) Normal

(b) Nearsightedness

Uncorrected   Corrected

(c) Farsightedness

Figure 21.30 (a) Normal vision. (b) Nearsightedness is corrected by a diverging lens. (c) Farsightedness is corrected by a converging lens.

Uncorrected   Corrected

Visual acuity. Most people know that 20/20 vision is considered good and that some other ratio of numbers is not as good. But few seem to know just what these numbers mean. This ratio, called *visual acuity*, deals with your ability to view certain test letters or other symbols. It is defined as:

$$\text{visual acuity} = \frac{\text{distance at which test was made}}{\text{distance for which the letter was designed}}$$

A letter is designed (rather arbitrarily) to be read at a distance for which it subtends an angle of $\frac{1}{12}$ degree. This "design" agrees with what a person with good eyesight can see. For example, if a certain letter is designed to be read from a distance of 9 m (30 ft), and a nearsighted person cannot read it farther away than 6 m (20 ft), her visual acuity is 6/9 or 20/30.

Figure 21.31 The shadow of eyeglasses in a parallel beam of light. The upper lenses, mine, are purely converging with a focal length of 1.3 m. The lower lenses, my wife's, are slightly diverging and are astigmatic in addition. As you can tell, both the axis of the astigmatism and the amount of astigmatism are different for each of my wife's eyes.

A CLOSER LOOK AT EYEGLASSES

We can use the ray drawing techniques (or the lens equation) we learned earlier to find the specific focal lengths needed for particular nearsightedness and farsightedness problems. For these problems you can figure out for yourself approximately what your eyeglass prescription needs to be.

First, consider nearsightedness. A normal eye can focus on distant objects clearly. But a nearsighted person cannot see clearly beyond a certain distance. This distance is called the *far point*.

Let's find the focal length of lenses needed by a woman whose far point is 100 cm. Remember that the rays intercepted by the eye from a *far distant point* are almost parallel. These rays are not "in focus" for the woman, but we can put them into focus by having the woman use a lens that will make that far-distant object appear to be only 100 cm away. In Figure 21.32 you see that parallel rays would diverge from a point 100 cm away from the woman if a diverging lens of 100-cm focal length is used.

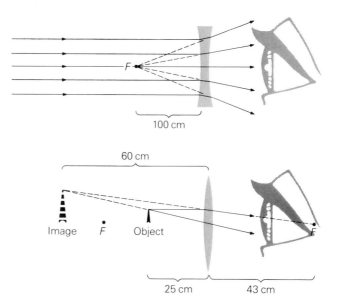

Figure 21.32 For a person with a far point of 100 cm, a diverging lens of focal length 100 cm makes a distant object have an image 100 cm away.

Figure 21.33 For a person with a near point of 60 cm, a converging lens with a focal length of 43 cm makes an object 25 cm away have an image of 60 cm.

Now, consider farsightedness. A normal eye can focus on objects *up to* a distance of about 25 cm. Thus, 25 cm is the normal *near point*. A farsighted person has a near point longer than this.

What lens is needed to correct the farsightedness of a man with a 60-cm near point? This man needs to fool his eyes into thinking an object at 25 cm is at 60 cm. In other words, an object 25 cm aways needs to have an image 60 cm away. We get the object and image located appropriately (as in Fig. 21.33) only if the lens is converging with a focal length of 43 cm. (The fact that one focal point of the lens is behind the observer's eye is not important.)

When you get a prescription for glasses, the magnifying power for each lens is specified by the ophthalmologist in units of **diopters**. The *power* in *diopters* is the *reciprocal* of the *focal length in meters*. Thus, if your prescription calls for a "+0.75" lens, that means a 0.75-diopter lens: one with a focal length of 1/0.75 or 1.33 m. The *positive* "+" sign means a *converging* lens. A *diverging* lens is specified as having a *negative focal length* and therefore a *negative power*.

Complex Instruments

Optical instruments such as microscopes, telescopes, and binoculars use two or more lenses to do their magnifying jobs. The light transmitted through one lens forms an image, and this image serves as the object for the lens that follows. In many instruments, "the lens" is not a *single* lens but several lenses grouped together as a unit with an appropriate overall focal length.

Key Concepts

Geometric optics treats light as rays that travel in straight lines in a given medium. In general, the rays change direction at the boundary between two media.

The **law of reflection**: The angle of reflection is equal to the angle of incidence.

The **image** formed by a *plane* mirror is as far behind the mirror as the **object** is in front of the mirror. A **concave** spherical mirror acts as a *magnifier* if the object is within a distance of $\frac{1}{2}$ the radius of curvature of the mirror. A **convex** spherical mirror increases the field of view of the observer.

When light travels from one medium into one in which its speed is different, it **refracts**. The angle the rays make with the *normal* is lower in the medium in which the speed is lower.

Light incident on a medium in which its speed increases undergoes **total internal reflection** if the angle of incidence exceeds the **critical angle**: the angle at which the refracted ray would travel parallel to the boundary.

Light rays parallel to the axis of a lens refract so that they pass through the **focal point** of a **converging** lens; they appear to have come from the focal point of a **diverging** lens. The **image** of an **object** that sends light through a lens can be located by the point of intersection of any two of the three **principal rays**.

A **magnifying glass** is a *converging lens* located so that the object is *within* the focal length. A camera is focused by changing the image distance, and the human eye is focused by changing the lens's focal length. This process in the eye is called **accommodation**.

Nearsightedness is corrected by a diverging lens, *farsightedness* by a converging lens, and *astigmatism* by a lens that focuses more strongly in one plane that in another. Lens prescriptions are written in units of **diopters**, the inverse of the focal length expressed in meters.

Telescopes, binoculars, and microscopes use two or more lenses (or groups of lenses that act as a unit). The image of the first lens serves as the object for the second.

Questions

Reflection

1. Show, by drawing diagrams that obey the laws of reflection, that the distance you stand from a *vertical* mirror has no influence on how much of yourself you can see in the mirror. Why is this not true if the mirror is over a dresser, and the dresser top shows in the mirror? Does the answer to either of these questions change if the mirror is not vertical?

2. I once rented a "U-Haul" trailer that had MAXIMUM SPEED 45 MPH painted across the *front* of the trailer. Was that because the sign painter had something stronger than milk for breakfast the day he painted the letters?

3. Figure 21.34 shows a top view of a clothing-store mirror used for seeing yourself on all sides. Trace the approximate path of a ray from your back to your eyes. Since you stand sideways as shown to see your back, does the center section of the mirror contribute anything to that view?

4. Bike riders sometimes use small rear-view mirrors attached to the frames of their eyeglasses. If the mirror is 2 cm in diameter, does the rider see only a 2-cm-diameter circle of the car following behind? Why more?

5. A short person driving a car moves the seat forward to better reach the steering wheel and pedals, while a tall person moves the seat far-

Figure 21.34 Top view of clothing store mirror that lets you see all sides of yourself.

Figure 21.35 Convex spherical mirrors are often used in retail stores for monitoring.

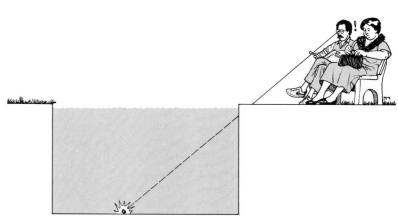

Figure 21.36 A straight line from the diamond to your eyes goes through the concrete pool wall.

ther back. Which person will see a wider field of view in a fixed rear-view mirror?

6. After taking a night physics class at Flat Mountain Community College, the manager of the local clothing store installed a convex spherical mirror in the corner of his store so he could keep an eye out for shoplifters (Fig. 21.35). Why did he use that type of mirror rather than a plane one?

Refraction

7. As you and your rich aunt converse beside her pool (Fig. 21.36), you spot her lost diamond ring on the pool bottom. Why would you not have been able to see it in that spot from where you're sitting if there had been no water in the pool?

8. The tube of liquid in a liquid-in-a-glass thermometer looks *much* bigger than it really is. Why? (In a clinical thermometer, the mercury column looks almost as large as the glass tube itself.)

9. A hot highway far ahead looks "glassy" or wet. You're really seeing light from the sky refracted (bent) into your eyes because the air near the road is hotter than that a little higher, giving light a slightly higher speed near the surface. Explain.

10. Highly reflecting road signs and bumper stickers consist of many tiny glass spheres embedded on a reflecting surface. Figure 21.37 shows one such sphere. The refractive properties of the glass bend a given ray of light so that it hits the back surface on a line through the center of the sphere and parallel to the incoming ray. Explain how such materials reflect any incident light back parallel to its incoming direction.

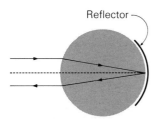

Figure 21.37 Reflecting tape and road signs have tiny spheres of glass embedded in a reflector. The reflected light is directed back parallel to the incoming beam.

11. Cats' eyes shine brightly at night, but only if you beam a light at them, and then most strongly right back along the path of the incoming light. Read the background part of Question 10 and then formulate a theory of shining cat eyes.

12. Optical cables for communication are made up of many tiny glass or plastic fibers in a bundle, rather than one larger fiber. This construction makes them more flexible. Why does this also make them work better optically? (Hint: Draw what happens to a ray in a sharply bent small fiber and in a large fiber bent to the same radius.)

13. By drawing a side view of the fishbowl of Figure 21.11(a), explain why you see two fish when there is only one. (Hint: Draw two rays from the fish to your eye—one through the top surface of the water and one through the curved side of the water and bowl.) By drawing a top view, explain why the "bottom fish" looks larger than the fish really is.

Lenses

14. If the sun is shining, how can you easily measure the focal length of a converging lens?

15. By drawing ray diagrams for converging and diverging lenses with different object distances, show that a real image is always inverted, and that a virtual image is always upright.

16. Many houses and apartments have little "peepholes" in their front doors with a device inside that lets you see someone standing *anywhere near* the front of the door, even though you look through a tiny hole. What kind of lens is in this peephole—converging or diverging? If you have one, check your answer by shining a flashlight through it from inside after dark, and watching what happens to the beam outside.

17. Draw a *converging* lens for which one surface has a radius of curvature of 20 cm, and one, 40 cm. Draw a *diverging* lens for which one surface has a radius of curvature of 20 cm, and one of 40 cm.

18. If you have a magnifying glass with focal length 15 cm, where do you hold the lens relative to an object to get maximum magnification?

Optical Instruments

19. Why does a photograph come out blurred if the camera is not properly focused when the picture is taken?

20. Why do you have to put slides upside down in a slide projector to have the image upright on the screen?

21. If your optometrist prescribes glasses with −0.50-diopter magnifying power, what will

the focal length of the glasses be? Would that indicate you are nearsighted or farsighted?

22. When your not-so-young second-grade teacher looked through her converging-lens glasses at a distant tree, it appeared blurry. Why? Hint: Where is the image if an object is beyond the focal length? Does that mean that she could see it clearly if she really had eyes in the back of her head, as she always told the class?

23. During an eye examination, an ophthalmologist usually adds drops that dilate your eyes. In addition to opening your pupils so that she can see inside your eye, the fluid temporarily paralyzes the muscles that control accommodation. Why would she want to eliminate your accommodation ability in order to accurately evaluate your eyesight? Why would this paralyzation be more important for younger than for older people?

24. Light passing near the edges of a lens crosses the axis slightly closer to the lens than light passing near the center. The blurring effect that results is called *spherical aberration*. Does this explain why you might get clearer pictures with an inexpensive camera if you use it in bright light so that the diaphragm opening can be set smaller, rather than in dim light that needs a large diaphragm opening?

25. Would you expect spherical aberration, defined in Question 24, to have anything to do with why you can read more clearly in bright light than in dim? (Remember what happens to the pupil when the light level changes.)

26. When people first get new glasses, why do they sometimes step too soon or too late when they get to a staircase. (Hint: The eye sees the image formed by the glasses, not the object.)

Problems

1. For a mirror mounted vertically on a wall, what must its minimum height be in order for you to see yourself completely? How high must the bottom be from the floor in order to get by with this minimum height? Would the same mirror at the same place work for a shorter person? Why not?

   Answer: $\frac{1}{2}$ your height; $\frac{1}{2}$ distance from floor to eyes; no.

2. If, for bike riding, you clip a 2-cm-diameter flat mirror onto your sunglasses so that it is 4 cm from your eye, what's the tallest object you can completely see 10 m behing you?

   Answer: 5 m

3. Suppose you look at a dime under a magnifying glass of focal length 5 cm. If the dime is 3 cm from the magnifier, how far away will it appear to be? How big will it appear?

   Answer: 7.5 cm; 2.5 times its actual size

4. A personalized slide projector uses a screen 60 cm from the projection lens. If the lens focal length is 10 cm, how far should the slide be from the lens?

   Answer: 12 cm

5. If you're 10 m from the queen and your camera has a focal length of 50 mm, how far should your film be from the center of the lens to have her in focus? (Because of the big difference between object distance and focal length, the ray-drawing method won't be very accurate. You'll need to use the lens equation.)

   Answer: 50.2 mm

6. If you discover at the ripe old age of 43 that you can no longer read the newspaper, and your ophthalmologist prescribes lenses of +1.0 diopter, what will your near point be, assuming you are fitted correctly?

   Answer: 33 cm

7. A nearsighted person cannot clearly see anything farther away than 3 m. What magnifying power should this person's glasses have to correct the problem?

   Answer: −0.33 diopter

8. A 35-mm color slide has film dimensions of 24 mm by 36 mm. You want to show the picture on a 1.0-m by 1.5-m screen. If the slide is 120 mm from the projection lens when the image is focused, how far away should the screen be?
Answer: 5.0 m

Home Experiments

1. Verify that you would need a mirror only $\frac{1}{2}$ your height to see your entire body if there are no obstructions such as a dresser top. Show that if the mirror is vertical, your distance from it makes no difference in how much you see.

2. Get in a swimming pool or pond with no one else in it to ripple the water. Wearing a skin-diver's viewing mask, swim to the bottom and look up. Why do things looks as they do?

3. If you have a converging lens available (for example, the projection lens of many slide projectors can be taken out along with the tube that contains it), hold it near a nonfrosted light bulb such as a high-intensity reading lamp. It should be slightly farther away than the focal length of the lens. Hold a paper out from the lens until the image of the filament comes into focus on the paper. You can easily see the filament coils. Vary the distance to the lens (the object distance) and see what happens to the image distance. If you know the lens focal length, measure the distances involved and draw ray diagrams to check the correctness of the ideas in the chapter. If you don't know the focal length, you can measure it this way.

4. If you wear eyeglasses, find out from your optometrist or optician what your prescription is. See if you can roughly verify this prescription by the way your glasses refract the *parallel* rays of the sun.

5. In driving some cars at about 55 mph, you may notice a curious effect. If you look through the rear-view mirror at the lug bolts of a truck behind you, the bolts may appear to stand still or turn backwards, as the spokes of a wheel do in a movie. As the truck passes and you look at the bolts directly, all you see is a blur. (You're more likely to observe this from a compact car.) Look for this effect and try to explain why it happens.

References for Further Reading

Boyle, W. S., "Light-Wave Communications," *Scientific American*, Aug. 1977, p. 40. Discusses the techniques in the first test of telephone communication via light waves.

Free, J., "Photonics," *Popular Science*, May 1982. Discusses the switches and other components that allow telephone communications with fiber optics.

Gregory, R. L., "Visual Illusions," *Scientific American*, Nov. 1968, p. 66. Discusses many of the interesting optical illusions that we see.

Kaufman, L., and I. Rock, "The Moon Illusion," *Scientific American*, July 1962, p. 120. An explanation of why the moon looks larger when it first "rises."

Lynch, D. K., "Atmospheric Halos," *Scientific American*, April 1978, p. 144. Describes how the misty ring we sometimes see around the sun and moon is caused by reflection and refraction of light by ice crystals in the air.

Physics Today, May 1976. This special issue covers the various aspects of lightwave communications.

The Way Things Work, an Illustrated Encyclopedia of Technology (Simon and Schuster, New York, 1967). An equivalent publication: *How Things Work, The Universal Encyclopedia of Machines* (George Allen and Unwin Ltd., London, 1967). Details of many optical instruments including cameras, telescopes, and microscopes.

L I G H T   A S   A   W A V E

And God said, "Let there be light"; and there was light. And God saw that the light was good; and God separated the light from the darkness. God called the light Day, and the darkness he called Night. (Genesis 1:3–5)

What *is* light? Is it merely the absence of darkness, as the sign on one dentist's darkroom implies (Fig. 22.1)? Where does it come from, and how does it behave? In the last chapter we treated light as rays that travel in straight lines in any one medium. But we made no attempt to find out what light really is. In this chapter, we begin to investigate more carefully the fundamental nature of light.

In particular, we consider the fact that light is a *wave*. We will build on what we know from Chapter 12 about waves in general. To start, we look at the particular characteristics of light waves, and then at what determines the color of light. We'll then find out what gives material objects the colors they have. Finally, we will see how two general types of wave behavior—polarization and interference—influence the light we see.

However, let me caution you not to get too attached to the idea that light is *only* a wave. In the next chapter you may be surprised to find out that sometimes light is not a wave at all!

22.1 THE WAVE NATURE OF LIGHT

Light is a **transverse electromagnetic wave**. Saying that the wave is *transverse* means that whatever oscillates to form the wave, oscillates in a direction perpendicular to the direction of wave travel. In contrast, a

Figure 22.1 The sign on this dentist's darkroom tells the story of one theory of the nature of light.

sound wave is *longitudinal*; that is, the oscillating molecules move parallel to the direction of wave travel.

When we say that the wave is *electromagnetic*, we mean that it consists of oscillating electric and magnetic fields. In Section 19.3 we described how to produce an electromagnetic wave in the radio frequency range. Figure 22.2 illustrates such a wave. Rather than the movement of particles, as on a string, the "displacement" consists of two parts. First, an electric field varies smoothly from zero to a maximum in one direction, then smoothly to zero again, and then to a maximum in the opposite direction. Second, in a plane perpendicular to that of the electric field, a magnetic field varies exactly *in phase* with the electric field: It has its crests at the same points as the electric field does, but in a perpendicular direction. This electromagnetic field variation propagates along just as waves in the ocean or on a string do.

An electromagnetic wave, unlike all other known kinds of waves, needs no medium to travel through. As a matter of fact, it travels at its fastest through vacuum—*nothing*. Expecting that light, like all other waves, needed a medium for propagation, nineteenth-century physicists went on many "wild goose chases" in search for the medium. You might find these searches interesting to read about in a science history book.

The Characteristics of Light Waves

The *speed of light in vacuum* is 3.00×10^8 m/s or 186,000 mi/s. This is roughly 30,000 times the highest speed reached by a rocket on its way to the moon. The speed of light in air is almost the same, but in glass is about $\frac{2}{3}$ this value.

The frequency range of *visible light*—light to which the human eye is sensitive—is between about 4×10^{14} Hz and about 7×10^{14} Hz. We call this the **visible region** of the *electromagnetic spectrum*. As we discussed in Chapter 19, other forms of electromagnetic radiation range in frequency from above 10^{22} Hz (x rays and gamma rays) to below 100 Hz (long radio waves). Even though we will be concerned mostly with visible light in this chapter, much of what we say will also apply to electromagnetic radiation of other frequencies.

From the speed and frequency range, we can calculate the wavelength range of visible light in vacuum or air to be between about 7×10^{-7} m

Figure 22.2 Light is an electromagnetic wave: It consists of electric and magnetic fields that oscillate at right angles to each other.

Figure 22.3 White light such as sunlight breaks up into its component colors—wavelengths—as it passes through a transparent prism.

and 4×10^{-7} m—700 nm to 400 nm.* As with any type of wave, the frequency stays the same as light goes from one medium to another. If the speed changes in the new medium, the wavelength must change also. When we talk about the wavelength of some particular light, we will mean its wavelength in vacuum, unless we specifically state otherwise.

Wavelength and Color

We showed in Figure 21.19 that when a light beam of a specific color passes through a prism, it deviates from its original direction because of the refraction at the two surfaces. If, however, we send a narrow beam of sunlight or light from an incandescent bulb through the prism, something different happens. As Figure 22.3 shows, the beam still deviates, but it also breaks up into a *spectrum* consisting of bands of all the colors of the rainbow. Letting this light fall on a screen, we see that the bands are not of the same width, and that one color gradually merges into another with no distinct boundaries.

The explanation for this spectrum of colors lies in the fact that light from a very hot object has wavelengths over the entire visible range. Furthermore, the speed of light in glass decreases as the wavelength decreases. Consequently, the amount of bending at the glass boundaries increases as the wavelength decreases. This variation in bending spreads the outcoming light according to its wavelengths. The different colors, therefore, correspond to different bands of wavelengths—red being the longest wavelength in the spectrum and violet being the shortest.

Light such as sunlight that contains this entire range of wavelengths produces the sensation of white when it enters our eyes. We therefore call it *white light*, even though it contains all the visible colors.

When you look at a rose petal and admire the deep red color, that means physically that light in the wavelength range from about 650 to 700 nm has entered your eye. The sensation of redness is your brain's interpretation of your eyes' response to that particular wavelength of light. The sensation of green, from looking at the leaves of the rosebush,

Many books use the unit angstrom (abbreviated Å) in describing light wavelengths. Since 1 Å = 10^{-10} m = 0.1 nm, moving the decimal point one place to the left converts angstroms to nanometers.

comes from your eye-brain response to light in the 500 to 550 nm range. When light containing all the colors of the visible spectrum—wavelengths between 400 and 700 nm—falls on your eye, the combined effect is the sensation of white.

· ·

QUESTION   From the fact that the visible region of the electromagnetic spectrum runs from 400 to 700 nm, and assuming the same wavelength interval for the six colors of the rainbow, what is the approximate wavelength range for each of the colors?

The rainbow.   An effect similar to light dispersion in a prism causes the colors you see in a rainbow. A rainbow appears when the sun shines on drops of water in the atmosphere (Fig. 22.4). A ray of white light enters a water drop at the point shown and is refracted at the surface. Each color is refracted a slightly different amount. The rays partly reflect at the back of the drop, and refract as they leave. The violet light leaves at about 40° to the incoming beam, the red at about 42°. Other colors are in between. Your eyes would intercept only a small portion of one color from any given drop. By looking in a slightly different direction, you see the other colors from other drops. The drops that reflect red into your eyes lie on a circular path that gives you the correct 42° angle with the incident sunlight.

22.2 HOW WE SEE THINGS AND WHAT COLOR THEY ARE

There are two ways that you see things. If you're looking at a light source, you see the light emitted by the source. If you're looking at something that is not a light source—for example, this page—you see it by light reflected from the object into your eyes.

Figure 22.4   A rainbow is formed by the refraction, reflection, and refraction again of sunlight in many, many water drops. The various colors are dispersed in different directions because they refract different amounts.

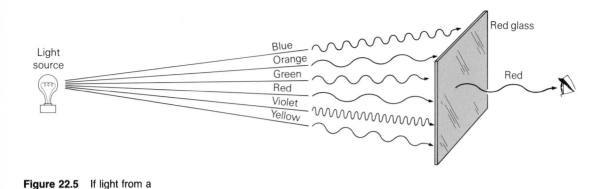

Figure 22.5 If light from a white light source passes through red glass, light of all wavelengths except red is absorbed. Only the red is transmitted. (The incoming light is usually not separated into colors, but is all combined as white light.)

The Color of Light Sources and Transmitted Light

If the light from a source enters your eyes directly, the color you see is the color of the light emitted. If, as is usual, the source emits a combination of wavelengths, the color you perceive is the combined effect of the various wavelengths.

Suppose you have a white light source, and you put a piece of red glass (or plastic or cellophane) between it and your eyes, as in Figure 22.5. The light that comes through is red because light of all other wavelengths has been absorbed by the glass. Only the red is transmitted. Any time you look at light coming through a colored "transparent" material, you are seeing the only color for which the material is transparent; the rest of the light is absorbed. The process is called *selective transmission* because the material selects only a specific color to transmit.

The Color of Everything Else

Everything you see that is not a light source, you see because light from a source *somewhere* shines on the object and then reflects into your eyes. But why are things the color they are? Why is grass green, the sky blue, red rose petals red, and the White House white?

To answer that question, consider the sketch of the daisy in Figure 22.6. The various parts of the daisy appear the color they are because of *selective reflection*. The leaves and stem absorb light of all wavelengths except green; they reflect the green, some of which gets to your eyes. The center of the flower reflects the yellow light and absorbs the rest. The petals of the flower reflect all colors of visible light, and therefore appear white. In general, the colors *not* absorbed by an object give it its particular color.

A black object is black because it absorbs *all* the light hitting it. Actually, it's impossible to get a truly black object that reflects no light, but that is the definition of black. You see black writing on a white page by the light reflected from the white background and the *absence* of light reflected from the writing.

Figure 22.6 Each part of the daisy has its own particular color because all other colors are absorbed. Light of that particular color is reflected.

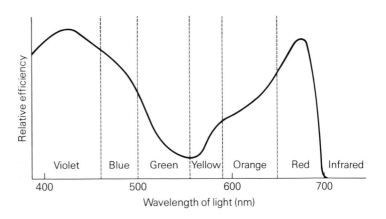

Figure 22.7 A graph of the relative efficiency of each wavelength of light in causing photosynthesis in plants.

· ·

QUESTION   From the facts that photosynthesis occurs in plant leaves, and that leaves are usually green, what colors of light would you expect to be needed for photosynthesis? After you've reached a conclusion, compare it with the graph of Figure 22.7 that shows the relative amount of radiation of each wavelength used in photosynthesis.

Why Is the Sky Blue and the Sunset Red?

Have you ever wondered why the daytime sky is blue and the sunset colors are generally reddish? The reason is actually the same for both. As light passes through air, the electrons of the air molecules absorb and reradiate light. This process *scatters* some of the light in all directions. But the probability of a scattering taking place increases with increasing frequency: It actually increases in proportion to the fourth power of the frequency, f^4. This fact means that blue light is about 6 times more likely to be scattered than red. As Figure 22.8 illustrates, the light you see from the daytime sky is scattered light and is therefore dominated by blue.

When you see the blue sky, you're actually *seeing* the atmosphere in the same way you *see* a leaf: by the scattering of light from it. Since scattering from an air molecule is a relatively rare event, it takes the entire thickness of the earth's atmosphere to scatter enough light to give the faint illumination you see from the sky.

But wait; violet light has a shorter wavelength than blue! Why isn't the sky violet? The answer lies in the sensitivity of your eyes to the various wavelengths of light. Violet is on the extreme upper-frequency end of the visible spectrum. Thus, the sensitivity to violet is considerably less than to blue. The result—you see the sky as being blue.

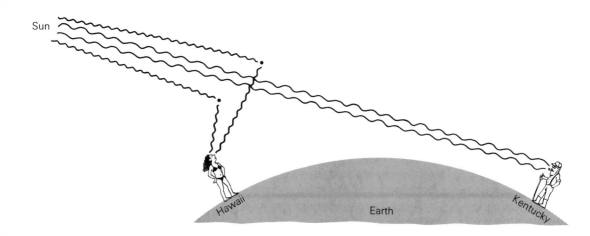

Figure 22.8 Short-wavelength light is more likely to be scattered by air particles than long-wavelength light. The Hawaiian sees the daytime light scattered in the sky, which is predominantly blue. This leaves mostly long-wavelength light (red to yellow) for a Kentucky sunset.

The same process explains why sunsets contain predominantly red-to-yellow colors. At sunset, the sun is at such an angle that the light getting to you must travel near the earth and through much more of the earth's atmosphere (Fig. 22.8). If mostly short-wavelength light is scattered out, what's left is mostly in the long-wavelength region—red to yellow. Even with such a long distance of travel through the atmosphere at sunset, pure air would not give enough scattering of short wavelength light to account for a colorful sunset. There must be dust or other foreign particles in the air to increase the probability of scattering. The nature and amount of such foreign material determines the color of the sunset at any particular time.

Combinations of Colors

Color is all in your head! The light emerging from a beautiful autumn forest has no color in it; it's merely electromagnetic radiation of various wavelengths. The leaves themselves do not *contain* the color; they merely reflect light of particular wavelengths. The color comes about when your eyes send different messages to your brain in response to different frequencies of light received. Color perception is your brain's interpretation of the receipt by the eye of certain frequencies. We usually go ahead and talk about the color *of an object*, or the color *of light*. For practical purposes it amounts to that. But keep in mind that, in reality, color comes about only in your head.

A single wavelength of light—monochromatic light—entering your eye gives the sensation of one color. But a combination of wavelengths gives some color sensation intermediate between the responses to the separate wavelengths. White light, as we've said before, is a roughly even mixture of all visible wavelengths.

We can get essentially any color by an appropriate mixture of red, green, and blue—usually referred to as the *additive primary* colors. Notice that the wavelengths for these colors come from the long-, middle-, and short-wavelength regions of the visible spectrum. Combining these colors in appropriate relative intensities can produce the sensation of any color we want.

A color television set produces its colors on this basis. There are three electron beams, rather than one as described in Chapter 18 for the black-and-white TV. The screen is covered with three sets of tiny phosphor dots that glow separately in red, blue, and green. Each beam is aimed to hit only dots of one color. Because the relative intensity of each beam hitting any one spot may be varied, the right color is produced at that point. The three dots of different color are close enough together to appear to the eye to be overlapped and producing only one combination color.

· ·

QUESTION   Plastic food-storage bags of one brand have a yellow side and a blue side that zip together to make green. In terms of the relative wavelengths of these colors, why is this effect reasonable?

Having an object appear the right color depends on having that color present in the illuminating light. A red object absorbs the short-wavelength (blue-violet) light. If you illuminate it with blue light, it will reflect no light and appear black. If certain wavelengths are accentuated in the illuminating light, they will be accentuated in the light reflected from the object. That's why you may have to take that coat outside the store into daylight to really find out what color it is before you buy it. Fluorescent and other artificial lights are usually not truly white light. Therefore, certain colors are emphasized more in fabrics viewed under them than if viewed under sunlight.

22.3 POLARIZATION

You've probably seen TV commercials attempting to convince you that if you wear "polarized" sun glasses, you can see underwater or see people in front of your car that are otherwise obscured by glare. Let's examine the process called **polarization** and see if these claims have some factual basis.

Consider again, as we did in Chapter 12, waves on a rope. If you tie one end of a rope to a post and shake the other end, you can produce waves. If you shake it back and forth in all directions perpendicular to the rope, you will get wave displacements in all directions perpendicular to the rope. But suppose (as shown in Fig. 22.9), you let the midpoint of the rope pass between two vertical boards with a gap just large enough for the rope. The only waves getting past the boards will be those with

Shake this end.

vertical oscillations. We say that the boards **polarize** the wave: that is, they transmit waves with vibration only in a single plane. We call the boards **polarizers**.

Polarization of Light by Absorption

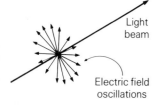

Light beam

Electric field oscillations

Figure 22.10 As viewed along the line of travel of an ordinary light beam, the electric field oscillates in all directions perpendicular to the beam.

There is a way to do the same thing to light. Imagine that there's a thin light beam coming directly toward you. From an ordinary light source, the electric field oscillations would be in all directions perpendicular to the beam, just as for the rope vibrations in front of the polarizer of Figure 22.9. If you could see the electric field as you look toward the beam, it would be oscillating up and down, back and forth in all directions as shown by the arrows of Figure 22.10.

In 1928 Edwin Land, while a freshman in college, invented a material he called "Polaroid" that does for light what the vertical boards do for the rope wave. (If you've taken pictures with a Polaroid Land camera, you can thank the same inventor.) When light passes through Polaroid, the light with electric field vibrations in one direction is transmitted; other light is absorbed. Figure 22.11 illustrates this point. Light with its electric field vibrations at an oblique angle to the transmission direction is only partially absorbed, leaving a fraction of that light transmitted. That fraction varies from 0 to 100% as the angle with the transmission direction varies from 90° to 0°. This fact means that, since the electric field direction of the incoming light is equally distributed in all directions, on the average, half the light is absorbed. The other half is transmitted with its electric field vibrations all in one direction. If the light then falls on another sheet of Polaroid with its transmission direction perpendicular to the first, no light emerges from the second sheet (Fig. 22.12).

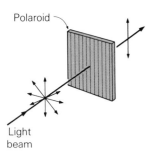

Polaroid

Light beam

Figure 22.11 Polaroid transmits light with electric field variations in one plane, but it absorbs other light. The vertical lines on the Polaroid indicate the transmission direction: The direction along which the electric field variations are transmitted.

"Polarized" sunglasses are made of Polaroid. They therefore cut out half the light from an ordinary unpolarized light source. Try looking through two pairs of polarized sunglasses held perpendicular to each other—you get no light through (Fig. 22.13).

An electromagnetic wave, such as light, consists of both electric *and* magnetic field vibrations that are perpendicular to each other. In dis-

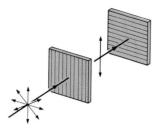

Figure 22.12 Two crossed Polaroids transmit no light.

cussing polarization, we haven't said anything about the magnetic field variations. These, of course, exist also. A material that transmits electric field vibrations in one direction, transmits magnetic field vibrations in a perpendicular direction. Otherwise, no light would get through.

Polarization by Scattering and Reflection

Remember that the scattering of light, for example by atmospheric particles, occurs when electrons are set into vibration by the light and then re-emit light in another direction. The electrons vibrate only perpendicular to the incident light, they reradiate most strongly perpendicular to their vibration, and the electric field oscillations of the scattered light are in the same plane as the electron vibration. Therefore, the scattered light is polarized. You can check this for yourself by looking through your

Figure 22.13 When the transmission directions of polarized sunglasses are parallel (top), you get almost no light *loss* in the second pair. With the directions perpendicular (bottom) you get *no light* through the second Polaroid.

polarized sunglasses at the blue sky. Rotate the glasses and notice that at certain orientations the sky looks almost black.

· ·

EXERCISE   With polarized sunglasses, check the direction of polarization of blue light from the sky. (The polarizing direction of the glasses is vertical as you normally wear them.)

Figure 22.14 The glare in (a) is reduced significantly by polarizing sunglasses (b).

The eyes of bees can detect the polarization of the light they receive. They use the direction of polarization of light from the sky to navigate, even when they are not in direct sunlight.

The reflection of light from a surface is basically the same process as scattering, except that it occurs from much more densely packed scattering centers. Therefore, light reflected from a smooth surface is partially polarized. The direction of polarization is *parallel to the surface*. The extent of polarization depends on the material and on the angle of reflection.

The glare from the surface of roads or lakes, for example, is light from the sky reflected by the surface. It obscures what you may be trying to see *on* or *under* the surface. Since this glare is partially polarized parallel to the surface, you can largely reduce it by viewing through material that polarizes vertically. That's why polarized sunglasses have their polarizing direction vertical, and why they are more effective than nonpolarizing glasses in reducing glare (Fig. 22.14).

22.4 INTERFERENCE IN LIGHT

Turn back to Section 12.3 and refresh your memory as to what we mean by *interference* in waves. Interference describes what happens when we have two or more waves at the same spot at the same time. At any point the net displacement is the sum of the displacements for each of the individual waves. Figure 12.10 illustrates the two extremes of *constructive* interference—two identical waves exactly in phase with each other—and *destructive* interference—two identical waves exactly out of phase with each other. For constructive interference, the amplitude of the overall wave is twice that of the individual waves. For destructive, the displacement is *zero* continuously: *there is no wave!*

It was by means of a demonstration of interference that Thomas Young first showed, in 1801, that light has a wave nature.

Interference from Thin Films

One of the most often observed interference effects results when light reflects from a thin transparent film of some material. This effect is

(Sees light)

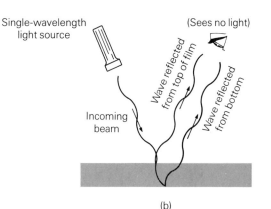

(a)

(b)

Figure 22.15 (a) If the film thickness is such that the light reflected from both surfaces is in phase, an observer sees reflected light. (b) If exactly out of phase, no reflected light is seen. (The film thickness and light wavelength are greatly exaggerated in size relative to the eye and light source.)

responsible for the spectrum of colors you see in soap bubbles or when small amounts of gasoline or oil are spilled on water.

Look at what happens when single-wavelength light hits a thin film of material, such as a soap bubble. Figure 22.15 shows two thicknesses of film, greatly exaggerated in thickness. Even though the film is transparent, some light always reflects from both the top and the bottom surfaces. When the thickness of the film is such that the light reflected from both surfaces is *in phase*, constructive interference occurs and you see reflected light (Fig. 22.15-a). When the thickness is right for the reflected waves to be exactly out of phase (as in Fig. 22.15-b), the waves interfere destructively, and you see no light. The film thickness for which constructive interference occurs depends also on the viewing angle, since the distance traveled through the glass by the wave reflected from the bottom increases as the angle of the light beam relative to the surface decreases.

Now suppose that you have white rather than monochromatic light incident on the film. When the film varies slightly in thickness from point to point, constructive interference occurs at different points for different wavelengths, and you see light of different colors reflected from different places on the film. Even when the film thickness is the same everywhere, you can see a spectrum of colors because of the slight variation in the viewing angle from point to point.

This effect is responsible for the different colors you see in soap bubbles, "blowing" bubbles, oily films on water, and sometimes on dishes that aren't thoroughly rinsed of detergent. For this to happen, the film thickness must be no more than a few wavelengths of light.

Destructive interference gives us the nonreflecting glass that is often used, for example, in picture frames. A picture covered with ordinary glass gives a glare at some angles that inhibits seeing the picture. Nonreflecting glass is covered with a coating that has just the right thickness to give destructive interference for yellow light—the center of the visible spectrum. This interference cuts down on reflection for the part of the visible light to which the eye is most sensitive. Since other wavelengths are still reflected somewhat, the glass has a bluish hue.

Key Concepts

Light is a **transverse electromagnetic wave** with wavelengths in air between about 400 and 700 nm. Its propagation speed in air is almost the same as in vacuum: 3.00×10^8 m/s. The color of light depends on its frequency and is the brain's interpretation of the eye's response to a particular frequency.

White light can be dispersed into its various colors by passing it through a prism, since the amount of bending depends on the wavelength.

The **color** of a light source is determined by the wavelengths emitted. The color of illuminated objects is determined by the wavelengths of the light reflected, scattered, or transmitted, but not absorbed.

Light is **polarized** if the electric field oscillations are all in the same direction. Light can be polarized by absorption, as in Polaroid, and by reflection and scattering.

Light waves can **interfere** *constructively and destructively*, depending on whether the waves are in phase or out of phase, respectively.

Questions

Light Waves

1. The first known attempt to measure the speed of light was made by Galileo. He and a partner located themselves on hilltops about a mile apart. Each had a lantern. One would uncover his lantern. When the second person saw the light he would uncover *his* lantern. The idea was that the time between the first person uncovering his lantern and seeing the light from the other lantern would be twice the time for the light to travel from one person to the other. Why do you think this attempt failed?

2. Rank the following forms of electromagnetic radiation in order of increasing wave frequency: red light, x rays, FM radio, AM radio, ultraviolet radiation, infrared radiation, green light. How would the order change if you ranked them in order of increasing wavelength?

3. Since different colors of light refract different amounts, all colors do not have the same focal point for a lens. Explain why this effect, called *chromatic aberration*, can give you rainbow-colored regions around white spots on a slide projected onto a screen by a cheap projector. (That's one reason "the lens" of a device is often two or more lenses of different materials that compensate for this aberration.)

Seeing and Color

4. We can speak of *single-wavelength* red, green, or blue light, for example. Can we legitimately speak of *single-wavelength* white light? Why?

5. If sunlight is white, why don't you see *white* when you look through a region of sunshine? What does it mean to "see" something?

6. What color is a red shirt illuminated in red light? in blue light?

7. Do you see a black cat because the light coming from it is black? What does it mean to say that a shirt is white?

8. Why do some of your pants appear to have a strange color under city street lights?

9. The technical crew for a stage performance can change the color of the performers' clothes in the middle of a performance. How do they do this? What is the physics of this change?

10. Does a red rose petal absorb long-wavelength or short-wavelength visible light more effectively?

11. Considering that we see objects that aren't light sources only by reflected light, explain why during the day it's hard to see the inside of a room through a window from the outside,

even if the window is open. Remember, light can only get to the room through the window, and is likely to have many reflections inside the room.

12. Explain why a black horse gets hotter in sunlight than a white one. What is the relative hotness of a brown horse?

13. Think about the cause and effect stated in Question 12. Should we really say that an object is dark-colored because it gets hotter, rather than that it gets hotter because it's a dark color?

14. If the earth had no atmosphere, what color would the sky be? Does the moon have a blue sky?

15. Dust or other foreign particles must be in the air to produce a colorful sunset. Are particles other than air molecules needed for a blue sky? Hint: What color is the daytime sky on a clear pollution-free day?

16. The make-up of some clouds is such that they absorb very little light. The light may scatter around several times inside, but eventually almost all the light that goes in comes back out. What color are such clouds? What must happen to much of the light that enters the clouds that we see as being "black"?

17. Why does the earth look mostly blue and white when photographed from the moon?

18. Water is a very good absorber of electromagnetic radiation over a broad range of wavelengths from x rays to microwaves, with a very important exception: Right at the visible region its absorbing ability drops to a very low value. If this rather amazing coincidence were not true, would water be transparent to visible light? How would that situation affect you and the limitless forms of underwater life?

19. What does the situation described in Question 18 have to do with your ability to cook with microwaves? What does the fact that the absorption drops off for longer wavelengths have to do with your ability to listen to the radio on a rainy day?

20. Deep, clear ocean water gets its color mostly

from incident light scattered by water molecules. As with scattering by air, short-wavelength light scatters more effectively. What color do you expect (and observe) clear ocean water to be?

21. My wife, a member of a relatively rare species called *native* Floridian, likes to impress me when we go to the beach by estimating the depth of the water and the location of sandbars by the color of the water. At least three factors are different where the water is shallower: the bottom is nearer the water surface and hence likely to scatter more incoming light; more solid particles are stirred up in the water in these regions; and the nature of the waves and ripples on the surface varies from that in the deeper regions. Explain how all three of these can influence the water's color.

Polarization

22. If you have two pairs of unlabeled sunglasses, how can you tell if they're both polarizing?

23. When you wear polarizing sunglasses, why can you often see the following more clearly:
    a. A person underwater when you look from above?
    b. Cars or obstacles on a highway when the sun is up but somewhat in front of you?
    c. The true colors of water at the beach?
    d. The occupants of a car you're following?

24. If you have papers on the dashboard of your car, their reflection in the windshield sometimes makes it hard to see what's ahead of the car. Why does wearing polarizing sunglasses reduce this interference? Why is the dashboard of most cars a dark color?

25. Viking sailors of the ninth century navigated their ships during the day by the position of the sun. By using "magical sun stones," they reportedly could locate the sun in any weather (see *Time*, July 14, 1967, p. 58). A natural material found in Scandinavia called cordierite has polarizing properties very similar to those of Polaroid. From the fact that sunlight scattered from the atmosphere is polarized, can you for-

mulate a theory of what these "sun stones" were and how they were used?

Interference

26. Some people mount their color slides between two pieces of glass for protection of the slides. When these are projected on a screen, why do you sometimes see curvy lines with all the colors of the rainbow superimposed in places on the picture?

27. Oily spots on a wet road often show all the colors of the visible spectrum. Why?

Problems

1. The *Apollo* astronauts placed a special reflector on the moon. A laser beam was aimed at this reflector and reflected back to earth. The time between the sending and return of a light pulse gave an accurate measure of the distance to the moon. If, at that instant, the reflector on the moon were 380,000 km away, what was the time interval between sending and receiving the light pulse? (See Question 1.)

   Answer: 2.5 s

2. What is the frequency of the 640-nm wavelength component of the red light of a neon sign?

   Answer: 4.69×10^{14} Hz

3. Based on Figure 22.7, approximately what two light frequencies are most effective in causing photosynthesis? What frequency in the visible region is least effective?

   Answer: 7.1×10^{14} Hz and 4.4×10^{14} Hz; 5.4×10^{14} Hz

4. The probability of light being scattered by atmospheric particles is proportional to the fourth power of frequency, f^4. If the frequency is doubled, how much more likely is the light to scatter?

   Answer: 16 times

5. If the speed of light in water is $\frac{3}{4}$ that in air, and that in window glass is $\frac{2}{3}$ that in air, what are the wavelengths of light of frequency 5.0×10^{14} Hz in air, water, and glass?

   Answer: 600 nm; 450 nm; 400 nm

6. Suppose a sheet of glass has a coating on the front in which light has a wavelength of 0.70 of that in air. How thick should the coating be to minimize the reflection of 550-nm light— that to which the eye is most sensitive?

   Answer: 96 nm

Home Experiments

1. View the spectrum of a white light source with a water prism. Put a dressing table or shaving mirror at an angle underwater in your wash basin as shown in Figure 22.16. Back away from the mirror and hold a flashlight near your head so that you can see its reflection transmitted through the water as shown. The situation is like half of the prism of Figure 22.3, with a mirror in the middle so that light is transmitted back through the front as it would be through the second half of the prism. One side of the reflected flashlight beam will appear bluish, the other, reddish. Put your hand over the flashlight with a small *horizontal* crack between your fingers. You'll then see the entire spectrum of the white light that emerges through the crack. Why does the crack need to be horizontal? The farther back you are, the more the spectrum spreads out. You might see a spectrum from other light sources or reflectors that are parallel to the water surface: for example, the reflection of the ceiling light from a shower-curtain rod.

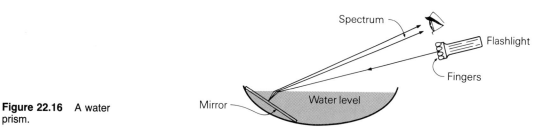

Figure 22.16 A water prism.

2. If you have a strong magnifying glass, look closely at the screen of an operating color TV. You can see that the screen is covered with small dots that are red, blue, and green. Any color is formed by different intensities of these dots. If you can get a stationary picture, such as a test pattern, you should be able to predict which dots are most intense for particular colors, and then test your prediction.

3. Look through polarized sunglasses at the glare from various horizontal smooth surfaces. Notice the difference without the sunglasses, and when they are rotated through 90° from the normal orientation. Compare the effect with sunglasses that are merely darkened glass.

4. Add a couple of drops of milk to a clear glass of water. Shine a flashlight beam through the glass. Looking from above at 90° to the beam, you see the beam by the light scattered from the milk particles. Look at the scattered light through polarized sunglasses. Then rotate the glasses 90° and notice the change in the amount of light you see. Put the sunglasses between the flashlight and milky water, rotate through 90°, and observe the effect on the amount of light scattered. Reconcile what you observe with the polarization expected from light scattering. Notice the bluish tint of the scattered light. Relate your results to the color of the sky.

5. Mix up a very soapy solution with dishwashing detergent, so that bubbles form on top. Look at the bubbles under white light. They first appear clear, then with all the colors of the visible spectrum, then clear again within a fraction of a minute. These effects happen because the film over the bubbles starts relatively thick, and then gets thinner with time. About how thick is it when you see the rainbow colors?

References for Further Reading

Küppers, H., *Colors* (Van Nostrand Reinhold Ltd., London, 1972). Describes qualitatively the various aspects of color, beautifully illustrated with color plates throughout.

"The Light Fantastic," *Science 84*, May 1984, p. 50. Shows how the beautiful colors of butterflies, peacocks, and many of nature's other colors result from thin-film interference.

Minnaert, M., *The Nature of Light and Color in the Open Air* (Dover Publications, Inc., New York, 1954). The physics of hundreds of everyday observations on light.

Pease, P. L., "Resource Letter CCV–1: Color and Color Vision," *American Journal of Physics*, Nov. 1980, p. 907. A survey of the literature on color vision.

Walker, J., "The Bright Colors in a Soap Film Are a Lesson in Wave Interference," The Amateur Scientist Column, *Scientific American*, Sept. 1978, p. 232.

Watterlond, M., "Stopping by Woods," *Science 81*, Oct. 1981, p. 92. Discusses the origins of the beautiful colors of autumn.

Wehner, R., "Polarized-light Navigation by Insects," *Scientific American*, July 1976, p. 106. An interesting application of the polarization of light.

THE ATOM

Humanity certainly needs practical men, who get the most out of their work, and, without forgetting the general good, safeguard their own interests. But humanity also needs dreamers, for whom the disinterested development of an enterprise is so captivating that it becomes impossible for them to devote their care to their own material profit.

MARIE CURIE

Marja Sklodowska, the youngest child of a physics teacher in Warsaw, was born in 1867. After her marriage to Pierre Curie in 1895, the two shared not only a happy marriage but an extremely tedious and intensive search that culminated in the discovery of the elements radium and polonium. Though saddened by her husband's death in 1906, she continued not only her detailed studies of radioactivity, but pioneered a mobile x-ray unit and founded a radiological school for nurses. She rejected the money, comfort, and many advantages offered by her great fame, crowned by Nobel Prizes in both physics and chemistry. She died of leukemia in 1934, apparently from prolonged exposure to radiation during her scientific investigations. (She is shown here with her daughters Eve and Irene.)

CHAPTER 23

LIGHT AS PARTICLES

When you were much younger, you may have identified with the feelings expressed in Figure 23.1. Maybe you even imagined the dark to be particles of some sort that plugged up the openings in your eyes. Well, in this chapter we are going to take just the opposite view of such matters: We're going to look at the particle-like nature of *light*.

In the last chapter, I did my best to convince you that light is a wave. It would seem that something that travels as a wave certainly could not travel as particles. Traveling particles would be like droplets of paint shot from a paint sprayer, although possibly much smaller in size. The two ideas—waves and particles—are inconsistent with each other. Yet we find that light is a rather strange entity; it has both a wave and a particle nature.

In this chapter, we're continuing to look for the answer to the question asked at the beginning of the last one: What *is* light. To find the other half—the particle half—of the answer, we need to look inside the atoms that make up all matter. Thus, we're beginning a new part of the book, a part that deals with the atom and its pieces.

The topics in this and the remainder of the chapters come under the category of what is usually called *modern physics*. That does not mean that what we have studied so far is out of date. It merely means that our knowledge of these upcoming topics has developed in relatively modern times—from about 1890 on.

Our plan for this chapter is the following: First, we investigate light sources—everything from ordinary light bulbs to lasers. We'll find that all light originates during energy changes within atoms. Furthermore, these sources emit light in discrete quantities called *photons*. We'll briefly consider x-ray emission and see that the basic ideas are the same as for light. Next, in considering the photoelectric effect, we will find that not only is light emitted in photons, but it can also interact with matter as individual photons. Finally, we will comment on this *dual*—wave and particle—nature of light.

23.1 LIGHT SOURCES

Aside from the sun, there are many types of artificial light sources that we use or see each day. Probably the most common type is the *incandescent* lamp, or what we usually call a "light bulb." The light source here is an extremely hot metal filament. However, before considering this very ordinary source, let's first examine a type of source in which we can more easily understand the origin of the light—gas light sources, such as neon signs.

In Section 14.2 we briefly discussed atomic structure, pointing out that all atoms consist of nuclei—made up of neutrons and protons—with electrons orbiting about them. Then we were content with an elementary satellite model in which the electrons are considered to orbit about the nucleus in much the same way as satellites orbit the earth. Even though that is not a very accurate picture, we'll not try to refine it in this chapter. Instead, we will use this rather intuitive model to gain an understanding of the light emission process. In the next chapter we'll touch up our understanding of the atom itself.

Light Emission from Gas Atoms

Imagine a marble sitting on a step of a staircase (Fig. 23.2-a). At that point the marble has a certain gravitational potential energy. As long as it stays on that step, its energy doesn't change. If it drops to a lower step, it loses some of its potential energy. Since *total energy* is conserved, we know that the loss in gravitational potential energy means an increase in another form (or other forms) of energy by the same amount. In this case, the potential energy goes to kinetic energy of the marble before hitting the next step, and then to heat as it hits. Each step the marble goes down results in the production of an amount of heat equal to the change in the marble's potential energy. The heat production takes place in discrete, fixed-sized chunks of energy. When the marble gets to the bottom of the stairs, it's in the lowest energy state and can't go any lower as far as that staircase is concerned. It then cannot give up any more potential energy to heat.

An electron in an atom has several characteristics in common with the marble on the stairs. First, it can exist stably only in certain *distinct energy levels* separated by *discrete energy intervals*. These energy levels are

"Don't put out the lights, Mommy. The dark gets in my eyes and I can't see."

Figure 23.1 *The Family Circle*, by Bil Keane, reprinted courtesy of the Register and Tribune Syndicate, Inc.

(a)

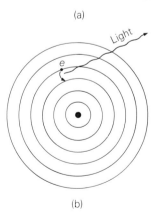

(b)

Figure 23.2 (a) Energy levels of a marble on a staircase. (b) Energy levels of an electron in an atom.

shown in Figure 23.2(b) as circles symbolizing electron orbits in the atom. The electron energy is made up of kinetic energy and electrical potential energy.

Second, the electron can *change* its energy only in *discrete steps* given by the difference in energy between the two levels. Third, when the electron goes to a lower energy level, energy is given off. This energy is in the form of *light*. *The energy carried away by the light exactly equals the energy change of the electron.* When the electron is in the lowest energy level for the atom—called the **ground state**—the atom can give off no light. The levels of higher energy than the ground state are called **excited states**.

When you see that bright orange-red neon sign advertising your favorite soft drink, you are seeing light given off as the electrons in the atoms of neon gas fall to lower energy states. The same process is happening in the mercury-vapor lamps and sodium-vapor lamps that are used so widely for city streetlights nowadays.

The light from the neon sign is reddish, but that from the mercury-vapor lamp is bluish white. From another gas, we get another color of light: Those green "neon" signs aren't really filled with neon, but with a gas that emits light that appears green.

To understand this light emission process, we need to look in some detail at the energy levels of the atoms of a particular gas. The simplest of all atoms, the hydrogen atom, can illustrate the principles most clearly. We'll show its energy-level diagram (Fig. 23.3) the way physicists usually draw them: A series of horizontal lines representing the levels, and separated by distances proportional to the energy intervals between levels. The figure shows the lowest five levels, with the energy of each level above the ground state—the **excitation energy**—written on the right. The energy unit used, the electron volt (1 eV = 1.60×10^{-19} J) is convenient for atomic energy levels because it's about the right size.*

Figure 23.3 An energy-level diagram for the five lowest energy states of the hydrogen atom. Vertical lines show possible transitions between these levels. The wiggly lines symbolize photons given off during transitions that have wavelengths λ as shown.

One electron volt is defined as the energy an electron acquires if it falls through an electrical potential difference of 1 volt.

Chapter 23   Light as Particles

The vertical arrows show possible downward changes in electron energy, with the numbers on each arrow giving the energy change. The chunk of energy given off as an electron makes a jump between any two levels we call a **photon**, or a **quantum** of energy. We say the light is **quantized**, meaning that the energy is emitted in discrete quanta. (We talked in an earlier chapter about the fact that electric charge is quantized in units of electron charge.)

Experiments show that the frequency of a particular photon's light is proportional to the energy it carries:

frequency ∝ energy

or

$$f \propto E$$

where ∝ means "proportional to."

The proportionality constant is a fundamental constant called **Planck's constant**, for which we use the symbol h. It has the value 6.62×10^{-34} joule-second (J·s). We can then write the equation as

$$E = hf$$

Since for any wave, frequency equals velocity divided by wavelength, we can calculate the wavelength of the photon's light if we know its energy. This energy is just the *difference* in energy between the two atomic levels involved in the energy change.

For example, let's find the wavelength of the light given off as a hydrogen atom goes from level E_4 at 12.75 eV to level E_2 at 10.20 eV. The photon energy is the energy change, 2.55 eV. First convert this energy to units of joules so that all quantities will be in the SI system:

$$E = 2.55 \text{ eV} \times 1.60 \times 10^{-19} \text{ J/eV}$$
$$= 4.08 \times 10^{-19} \text{ J}$$

The frequency is

$$f = \frac{E}{h}$$
$$= \frac{4.08 \times 10^{-19} \text{ J}}{6.62 \times 10^{-34} \text{ J·s}} = 6.16 \times 10^{14} \text{ Hz}$$

Then the wavelength is

$$\lambda = \frac{c}{f}$$
$$= \frac{3.00 \times 10^8 \text{ m/s}}{6.16 \times 10^{14} \text{ Hz}} = 4.87 \times 10^{-7} \text{ m} = 487 \text{ nm}$$

where c is the speed of light. The diagram shows the wavelength for this transition as well as for several others.

. . . .

EXERCISE Using the energies given in Figure 23.3 for the E_3 and E_1 levels, verify that the wavelength shown for this transition is correct.

Only the transitions *to* the second energy level E_2 give wavelengths in the visible region. The others are in the ultraviolet (beyond-violet in frequency) or infrared (below-red in frequency) regions of the electro-magnetic spectrum.

The light from hydrogen has these particular wavelengths. If a beam of it were passed through a prism, the light would break up into several distinct beams of different colors. The colors shown in Figure 23.4 would emerge, rather than the continuous range of colors we would get if the incoming beam were white light.

For other gases, a different set of distinct colors would be emitted. This light *spectrum* of an element can serve as its fingerprint for identifying the element—no two elements have the exact same spectra. The overall color appearance of the light from an element results from the combined effect of the wavelengths emitted.

We've left out one necessary ingredient, if the atoms of a gas are to emit light. Some of the atoms have to be in excited states in order to "fall" to lower states and give up energy. And for continuous light emission, it's necessary to continuously excite atoms to higher states. Otherwise, there would be no more light once all the atoms were de-excited—a fairly rapid process, since most excited states live only about 10^{-8} s.

One way to excite the atoms is by collisions from other atoms or electrons. If the gas is placed between two metal conductors across which there is a voltage, electrons or ions traveling between these conductors can collide with and excite the atoms of the gas (Fig. 23.5). This process is usually used in lighting applications such as neon signs.

Atoms can also be excited by *absorption* of a photon, *if the photon has exactly the right amount of energy to raise the atom to one of its possible levels*

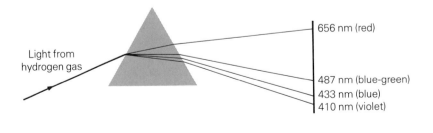

Figure 23.4 Light from hydrogen has only certain distinct wavelengths, as you can see by passing it through a prism.

Light from hydrogen gas

656 nm (red)

487 nm (blue-green)
433 nm (blue)
410 nm (violet)

(a)

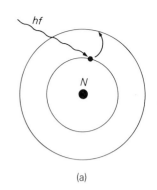

(a)

(b)

$E_2 - E_1 = $ energy lost by electron

(b)

$E_2 - E_1 = hf$

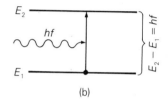

Figure 23.5 The excitation of an atom by a collision from another electron: (a) orbital representation; (b) energy-level representation.

Figure 23.6 The excitation of an atom by a photon having energy *hf* exactly equal to the difference in atomic energy levels.

(Fig. 23.6). If the photon energy is *either* higher than or less than the energy difference between levels, no excitation will occur. This process, which is just the reverse of photon emission, is a form of **resonance** similar in principle to the mechanical resonance we discussed in Section 12.5. We sometimes call this process *resonance absorption*.

When white light passes through a gas, the atoms of the gas absorb light of those wavelengths corresponding to upward energy transitions in the atoms. If, for example, you look at sunlight through a *spectroscope* (a device that measures wavelengths of light), you will find certain wavelengths missing from the continuous spectrum. The missing wavelengths are those absorbed by the gases of the sun's atmosphere as the light passes through. This is the way helium was first discovered.

QUESTION   Most city and campus street lights are either mercury vapor or sodium-vapor lights that emit only a few discrete colors of light, rather than white light. Why can you not tell what color something really is when viewed under such lights?

Figure 23.7 In a fluorescent bulb, electrons carrying current between cathodes collide with mercury atoms, thereby raising them to excited states. They de-excite by emitting ultraviolet photons that excite atoms and molecules in the phosphor coating of the tube. The phosphor atoms each emit several visible photons in going back to their ground states.

Fluorescence

There's a fair chance that you are reading this book under the light of a fluorescent lamp. A diagram of a fluorescent tube is shown in Figure 23.7. This tube contains mercury vapor mixed with an inert gas and heated electrodes at each end. Some electrons in the electrodes gain enough thermal energy to leave the metal—a process something like evaporation. These electrons get accelerated back and forth by the alternating voltage between electrodes. Some collide with mercury atoms, exciting them enough to emit radiation, much of it in the ultraviolet region. This radiation is absorbed by atoms and molecules of the *phosphor* coating inside the tube. The phosphor atoms then decay in smaller steps, emitting photons of visible radiation. Figure 23.8 shows the energy-level diagrams for these processes.

Fluorescence is the process by which a material absorbs ultraviolet radiation and de-excites in smaller steps by emitting visible radiation. This is the process just described in the phosphor coating on the inside of the tube of a fluorescent lamp. Sometimes one or more excited states live for extended periods of time. They may then glow for hours after exposure to ultraviolet radiation. This process is called *phosphorescence*, but there is not a clear dividing line between it and fluorescence.

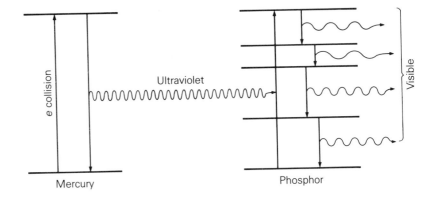

Figure 23.8 An energy-level diagram for the processes shown in Figure 23.7 for a fluorescent bulb.

I promised in Chapter 18 to discuss why a TV screen glows when the electron beam hits it. There, the phosphors are excited by the electron bombardment and emit visible light as they de-excite.

A few years back, the person doing the family laundry usually added a little "blueing" to the rinse water. That gave the white clothes a slight bluish tint and made them look whiter. Modern detergents with "whiteners" do the same job by leaving small amounts of a phosphor in the clothes that fluoresces in the blue region under sunlight.

QUESTION So-called "black lights" emit ultraviolet radiation. Explain how they can cause certain substances to glow without any apparent illumination when the "black light" is shone on them.

Incandescent Light Sources

In discussing light emission from individual atoms, we mentioned several times that the substance must be a gas. But we didn't say why. If the substance is a solid, liquid, or even a gas under high pressure, the atoms are close enough together that they interact strongly with each other, and a given electron is not confined to a single atom. As a result, the energy values that an electron can have are not quantized, but are spread over a wide continuous range.

An electron of such a substance can give up any fraction of its energy. This means that photon energies, and therefore wavelengths, do not take just discrete values, but are spread over a *continuous* range. Figure 23.9 shows how the number of photons given off varies with wavelength.

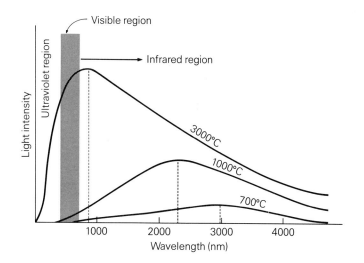

Figure 23.9 The relative amount of light radiated at various wavelengths for objects at three different temperatures. The vertical dashed lines show the wavelength for the most dominant radiation at each temperature.

At moderate temperatures, most of the radiation emitted is in the infrared region. As the temperature of the body is raised by adding heat, more of the electrons can gain, and then radiate, higher energy. Therefore, the frequency for the most dominant radiation shifts up, and the wavelength shifts down, with increase in temperature. At 700°C, there is enough radiation in the visible region that the body appears dark red. At 3000°C—a little above the temperature of the filament of an incandescent bulb—a large fraction of the radiation is in the visible region, and the object appears white. At about 5000°C, the peak is near the center of the visible region.

The relative intensity of the radiation at each wavelength depends only on the temperature and not on what the material is. Tungsten is used in incandescent light bulbs because of its high melting point. Electrical current passing through the filament heats it and converts the electrical energy to radiant energy.

Efficiencies of Light Sources

Having discussed the two types of light sources most often used for everyday lighting, let's now compare the efficiencies of these sources at producing light. Figure 23.9 shows that most of the energy radiated by a hot object is *not* in the visible region. This means, for example, that an incandescent bulb converts only a small fraction of the electrical energy supplied to it to light. In fact, a 40-W incandescent bulb converts only 1.7% of the energy to light, a 100-W bulb, only 2.5%: Their *absolute efficiencies* are 1.7% and 2.5%, respectively. Most of the rest goes to infrared radiation that serves only to heat the bulb's surroundings. In comparison, a 40-W fluorescent bulb is 9.4% efficient.

The rate of light output of a bulb is usually measured in *lumens*. The lumen is defined in the SI system in terms of the light output of a particular standard source. Rather than giving that definition, we'll point out that at 550 nm, the wavelength at which the eye is most sensitive, 1 lumen is equivalent to 0.00147 W of light energy. You can probably get a better feel for the "size" of the lumen from the fact that a 100-W standard incandescent light bulb has a light output of about 1750 lumens.

Table 23.1 Efficiencies of light sources

	Source (watt)	Light Output (lumens)	Luminous Efficiency (lumens/watt)
Incandescent bulbs (standard)	40	455	11.4
	60	870	14.5
	100	1750	17.5
Fluorescent bulbs	20	1000	50.0
	40	2560	64.0

To get the most for your money from a light source, you want the ratio of light output to energy input to be highest. This quantity, the *luminous efficiency*, is usually measured in lumens per watt of *electrical* power. A light bulb you buy at the grocery store has the light output in lumens, as well as the power needed, marked on the package. Table 23.1 gives the luminous efficiency of a few ordinary bulbs.

You might have wondered why your town or city uses the kind of streetlights that make everything look the wrong color. Mercury and sodium vapor lamps are much more efficient than incandescent bulbs. Considering the vast amount of electrical energy that must go into outdoor lighting, most taxpayers would rather accept the lack of color than the much higher bills that would result from using white light sources.

Flame

A burning flame has been the source of artificial light during most of human history. Even though the practical importance of this light source has diminished, the intrigue of an open fire or burning candle is not ever likely to escape the person who takes the time to leisurely stare at it.

As you might guess, the processes involved in the production of light in a flame are complicated chemical reactions and are, in some respects, not thoroughly understood. We can, however, describe the general nature of the light emission processes.

There are basically two types of effects that produce the light of a flame. First, the chemical reactions taking place in the flame leave many of the molecules in excited states. These molecules emit visible light in going to lower energy states. This mechanism is responsible for the blue light you see near the base of a candle flame, for example.

The second way that light is produced in a flame is somewhat different. Carbon in the fuel forms solid carbon particles that are heated to incandescence by the hot gases in the flame. These hot particles emit a continuous spectrum of light in the same way the filament of an incandescent bulb radiates. The temperature of the particles is such that the light usually appears yellowish. This mechanism is responsible for most of the yellow light of a burning candle or an open wood fire.

Lasers

The **laser** is a light source that is growing by leaps and bounds in its applications to technology (Fig. 23.10). Laser light is unique in that its output, which can be extremely intense, is in a narrow beam of *single-wavelength, coherent* light that broadens very little as it leaves the source. Coherent light is light for which the waves are all *in phase* with each other. Figure 23.11 illustrates this point. Part (a) of this figure shows *incoherent* waves in which the phases of the different waves vary randomly from each other. Ordinary light sources are not only incoherent, but

Figure 23.10 A laser.

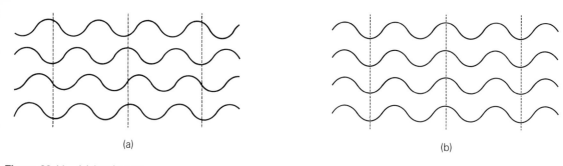

(a)   (b)

Figure 23.11 (a) Incoherent waves. (b) Coherent waves.

they produce a broad range of wavelengths. In part (b) of the figure the waves all have the same wavelength and are exactly in phase with each other: Their peaks and valleys occur at the same time and reinforce each other; they are *coherent*.

The name "laser" is an acronym for **l**ight **a**mplification by **s**timulated **e**mission of **r**adiation. *Stimulated emission* is closely related to our earlier discussion of emission and absorption of photons by atoms. Figure 23.12 compares the three atomic processes: *spontaneous emission, resonance absorption,* and *stimulated emission* of light.

Spontaneous emission (Fig. 23.12-a) is the process whereby the atom, of its own accord, goes to a lower energy state by spontaneously emitting a photon whose energy hf equals the change in energy levels of the atom. Resonance absorption (Fig. 23.12-b) occurs when the atom in the lower state absorbs a photon of exactly the right energy to raise it to a higher state. However, suppose the atom is already in the higher state, and a photon impinges on it that has energy *equal* to the energy difference down to the lower state. That incident photon can *stimulate* the atom to emit another identical photon while at the same time going to the lower energy state. We call this process **stimulated emission** (Fig. 23.12-c). To conserve energy, there must be *two* photons coming out, whereas only one went in. In other words, the amount of light is *amplified* by *stimulated emission*: precisely the process implied by the name laser.

Furthermore, the stimulated photon is exactly in phase with the incident photon, and it moves off in the exact same direction. Thus, the light carried by the photons is *coherent*.

However, two photons do not make a laser beam. Even a low-power laser emits about 10^{15} photons per second. For a working laser, we need many, many atoms in an appropriate excited state. The stimulated emission process can then cascade and amplify the light to high intensities.

This need presents a problem: In most substances, almost all the atoms are in their ground state. Any atoms that *do* get to excited states decay spontaneously almost immediately. Even if a photon *should* pass through, there aren't any atoms to stimulate to lower energy states. For a laser to work, we need more atoms in an excited state than in the ground state: We need a **population inversion**. This can occur only for a state with a relatively long lifetime—of the order of a few tenths of a second. Electrons then need to be *pumped* into this state until it is more highly populated than the one to which we want to stimulate a transition.

In the most common type of laser, the *helium-neon* laser, this pumping is a two-step process. Electrons traveling between electrodes in a helium-neon gas mixture collide with helium atoms, exciting them as shown schematically in the left side of Figure 23.13. It turns out that neon atoms have a long-lifetime excited state with the same energy as one excited by electron collisions in helium. Some of the excited helium atoms collide with and transfer energy to neon atoms, exciting them to the long-lifetime state. This state in neon can then be stimulated to emit photons of wavelength 633 nm in going to a lower-energy state. The light is reflected back and forth in the laser tube, stimulating the emission of many photons until the beam is highly intense. The emerging red light is a narrow coherent beam that broadens little with distance from the source.

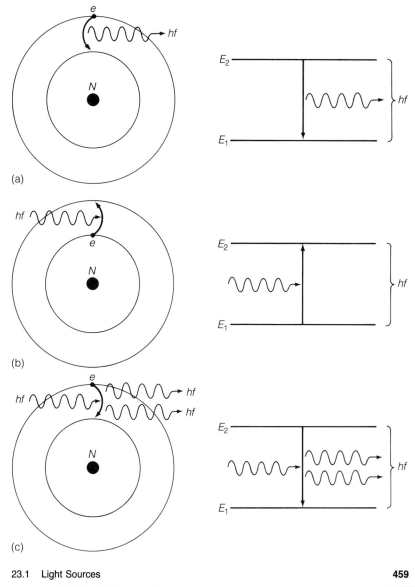

Figure 23.12 The orbital representation and energy-level diagrams for (a) spontaneous emission, (b) resonance absorption, and (c) stimulated emission.

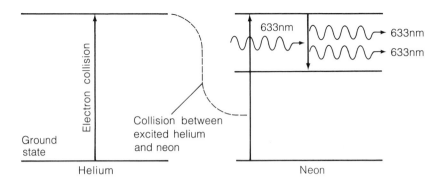

Figure 23.13 In a helium-neon laser, electron collisions excite the upper level shown in helium. Collisions between the excited helium atoms and neon atoms transfer the excitation energy to the neon state with the same energy. Stimulated emission occurs from this relatively long-lived state.

Laser Applications

We'll mention a few of the hundreds of important modern applications of laser beams. These include: reading the Universal Product Code (bar code) at the checkout counter of many supermarkets (Fig. 23.14); communicating over long distances by pulsed or intensity-modulated beams; detecting pollutants in the atmosphere by determining what wavelength of laser light is absorbed by the air; measuring the distance to the moon by determining the time for a laser-beam pulse to travel to and return from a special reflector left on the moon by the *Apollo* astronauts; drilling of tiny holes in hard substances; welding of special materials that are difficult to weld otherwise, including the "welding" of detached retinas in the human eye; inducing fusion in small hydrogen pellets as a way of producing and controlling thermonuclear reactions (Chapter 26); routing of airport luggage; and performing surgery. In this last category, the laser has replaced the scalpel in many types of operations. The laser beam can be focused to concentrate its energy into a tiny spot to vaporize a tumor or make an incision. The intensity and color can be matched to the specific effect needed for the particular body tissue.

· ·

APPLICATION   **HOLOGRAMS**

One fascinating application is in the making of *holograms*: recordings on photographic film that, when illuminated, produce a three-dimensional image. An ordinary photograph is two dimensional, in that you see the same thing from all vantage points. A hologram is a three-dimensional image: As you look at it from a different angle, you see a different part of the image, just as you would when you move around a real object and view it from different sides.

Several popular credit cards display holograms, which provide more than mere decoration. They substantially enhance security. Since part of the account number spans the hologram, the number cannot be altered without noticeable damage to the hologram. As you view such a hol-

Figure 23.14 The Universal Product Code that appears on most grocery store items can be read by laser beam at the check-out counter. The supermarket's computer then automatically adds the cost to the customer's bill and reduces the store's inventory by one jar of peanut butter, in this case.

ogram from different angles in bright light, some parts of the image seem to protrude and others seem to recede. Tilting the card shows the image from a different perspective.

Perhaps you've wondered how the laser scanners at automated supermarket checkout counters read the bars of the Universal Product Code for odd-shaped packages or when the package is passed over the scanner at random angles. Early scanners frequently missed reading the code; however, holography has greatly increased their effectiveness. The laser beam passes through an array of holograms, each designed to deflect the beam at different angles to cover the bottom and sides of the package. One such portion of the beam usually reflects from the bar code back into a light detector that sends an electrical signal to the computer.

Several of the references at the end of this chapter do an excellent job of describing techniques for making holograms and numerous practical applications of them.*

LEDs and LCDs

In view of the abundance of digital readout devices such as calculators and watches, we'll briefly discuss the light sources usually used in the displays of these devices. Most early hand-held calculators and digital watches used *light emitting diodes* (**LEDs**) for readout. Because of much lower power requirements and readability in ordinary light, *liquid crystal displays* (**LCDs**) have taken over as readouts for most portable devices. LEDs abound in clocks and other plug-in appliances.

In Section 16.3 we discussed semiconductors, and in Section 20.1 we showed how semiconductor diodes are used in solar cells for converting light energy to electrical energy. Light production in an LED is just the reverse of that process. If you have forgotten or never learned about semiconductors, you might want to read the part of Section 16.3 dealing with the basic properties of *p*- and *n*-type semiconductors.

I especially commend to you the March 1984 issue of National Geographic magazine. The cover contains an impressive hologram of an eagle, and its several articles on lasers and their applications are beautifully illustrated with color photographs and drawings.

Figure 23.15 A light emitting diode (LED) gives off photons with energy equal to that released when an electron and a hole combine near the *p-n* junction.

An *n*-type semiconductor has a small excess of electrons in an otherwise tightly bound atomic lattice. These electrons can act as negative charge carriers. A *p*-type semiconductor has electron vacancies, or *holes*, that act as positive charge carriers.

A semiconductor diode is a *p-n junction* formed by putting together a layer of *n*-type and one of *p*-type semiconductor (Fig. 23.15). When a voltage is applied across the junction in the direction shown, electrons move from the *n* side to the *p* side and combine with the holes. (We could equivalently say that holes move in the other direction and combine with the electrons.) This annihilation of the electron-hole pair results in a lower energy state. As a result, energy is released. If the energy released in one electron-hole combination corresponds to that of a light photon, and if the material is transparent to those photons, light is emitted: We have an LED. If these two conditions are not met, the released energy goes to heat in the diode.

LEDs can be shaped to form various segments of a digit or letter. Which segments are lighted determines what letter or digit is displayed. Typical red LEDs are made of gallium phosphide or gallium arsenide phosphide. They operate on about 2 V, and a single digit uses a few milliwatts of power.

APPLICATION   **LCDs**

Liquid crystal displays are not actually light sources. They merely reflect or transmit light that is incident on them. However, since they *look like* light sources when used in digital displays, and to contrast them with LEDs, we consider them here. Because LCDs only need to reflect light, rather than produce it, they typically use much less power than light-producing displays.

A liquid crystal is *liquid* in the sense that it takes on the shape of its container; it is *crystal* in the sense that there is a definite order to the molecules. Those used in displays are made up of long thread-like organic molecules.

With no voltage across such a crystal, the molecules are regularly arranged, and the crystal is transparent. When a voltage is applied, the molecules become randomly oriented and scatter any light incident on them. Figure 23.16 shows how this effect can be used in a display. The part of the liquid crystal with no voltage applied transmits the light to the retroreflective foil—a specially made foil containing tiny spherical reflectors that reflect any incident light directly back along the path of incidence.* This light is not seen, and the crystal appears dark. Where voltage is applied, the randomly oriented molecules scatter light in all directions, making that part of the crystal appear light.

*See Question 10, Chapter 21.

Figure 23.16 In a liquid crystal display (LCD), the molecules become randomly oriented and scatter the light only in those regions where there is an applied voltage across the crystal.

A given display digit is usually divided into segments. Any digit between 0 and 9 can be made visible by applying voltage to appropriate sections of the display.

The type of LCD we've discussed here, called the *dynamic scattering display,* typically uses less than a milliwatt of power per square centimeter of display. Another type of LCD depends on what are called *twisted nematic* liquid crystals, and it uses less than a microwatt per square centimeter— 1/1000 that needed by an LED. Because of its low power requirement, this is the display used in wrist watches and other instruments needing to operate on batteries for long periods. Arrays of these crystals form the displays for small flat-screen TVs and portable computers.

Twisted nematic displays make effective use of the polarization of light. As light passes through the crystal, the molecular structure of the crystal causes the direction of light polarization to twist through 90° when there is no voltage across the crystal. A display has a sheet of crystal sandwiched between two sheets of polarizing material, oriented with their polarizing directions 90° to each other (Fig. 23.17). A mirror is behind the second polarizer. Incident light is polarized and then has its polarization twisted so that it goes through the second polarizer, is reflected, and travels back out the front. The crystal would then appear light under external illumination.

Figure 23.17 The main components of a twisted nematic liquid crystal display are shown enlarged and separated from each other. Arrows on the light beam show the polarization directions. When *no* voltage is applied across the crystal, it twists the light polarization through 90°.

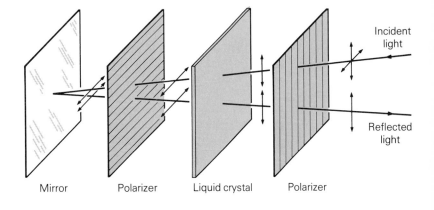

In those regions of the crystal in which a voltage is applied, the polarization is not twisted, and the light is absorbed by the second polarizer. These regions appear dark in contrast with the lighter surroundings. Look through Polaroid sunglasses at an LCD watch; then rotate it 90°. Any surprises?

23.2 X RAYS

Figure 23.18 An x-ray photograph of a foot.

You've probably had some part of your body—your teeth, your chest, your broken foot—x-rayed at some time in the past. The radiologist beams these highly penetrating rays at the part to be examined—let's say your foot. The film beyond the foot is sensitive to the radiation coming through. A crack or other abnormality absorbs a different amount of radiation and thus shows up on the film (Fig. 23.18). Various industries use x rays in the same way, to detect cracks that cannot be seen visually in pipes or structural materials.

Aside from medical and industrial uses, x rays are a very useful tool for understanding the structure and energy levels of atoms. Remember, atoms give off photons of electromagnetic radiation when they de-excite from a higher to a lower energy level (Fig. 23.19). If the energy change, and thus the photon energy, is 2 or 3 eV, the photon has a wavelength in the visible region. At 100 eV, the photon is in the ultraviolet region. If the energy change exceeds a *few hundred electron volts*, the photon is an *x ray*.

Figure 23.20 shows how an x-ray tube produces x rays. Electrons emitted from a heated filament are accelerated by a voltage of typically 40 to 140 kV. This high-energy electron beam hitting a heavy-metal target knocks electrons out of the target atoms, or raises the atoms to excited states. As these atoms go back to lower energy states, they give off x rays. The x-ray beam emerges through a narrow opening—a collimator—that confines the beam to hit only the part of the body that needs x-raying.

23.3 THE PARTICLE NATURE OF LIGHT

Remember that when light is emitted from a substance, *one* photon of light is given off as *one* electron goes to a lower energy state in an atom or a group of atoms. Furthermore, the energy carried away by this photon is *hf*—Planck's constant multiplied by the frequency—and it is exactly the energy given up by the atom or atoms during the electron transition.

If the energy is *given up* in bundles as photons, you might expect that the energy *carried* by light exists in discrete photons, or *quanta*. This

Figure 23.19 An atomic energy-level diagram showing energy changes in the visible, ultraviolet, and x-ray regions. Wavelengths in nanometers for each transition are in parentheses. Energy changes are shown in electron volts. The diagram is not to scale; if it were, you wouldn't be able to see the separation between the three highest levels.

Figure 23.20 The basic principle of an x-ray tube.

would mean that light, in some ways, would act as if it were made up of particles. In fact, certain experiments bear out this idea. One of the first processes to substantiate the *particle nature* of light was the **photoelectric effect**, which we'll now discuss.

The Photoelectric Effect

Figure 23.21 In the photoelectric effect, light incident on a solid surface ejects electrons (*e*).

When light hits the surface of certain solids, electrons are given off, as illustrated in Figure 23.21. This phenomenon is called the *photoelectric effect*. The number of electrons emitted from a particular substance depends on the intensity and wavelength of light, but the exact way these properties influence the number is quite surprising if you're thinking of light as a wave.

If electrons are to be knocked out by absorbing energy from a light wave, you'd probably expect to knock out an *occasional* electron, regardless of the light wavelength, even for low intensities. You'd probably also expect higher-energy electrons for higher-intensity light.

But what is observed in experiments is somewhat different. First, if the wavelength is larger than some particular value—that is, the frequency is less than some particular value—no electrons are ejected, no matter how intense the light. The particular wavelength depends on what the substance is. For copper, for example, the longest wavelength is 272 nm (frequency 1.1×10^{15} Hz). Second, if the wavelength is less than that particular amount, you immediately get *some* electrons even in extremely dim light. Furthermore the *number* of electrons increases with increasing intensity, but the *energy* of the electrons *does not*. (Notice that we're using the word *light* to include radiation outside the visible region.)

In 1905 Albert Einstein was able to explain these effects, which seemed rather odd at that time. It was for this discovery that he won the Nobel Prize in 1921. Based on Max Planck's earlier discovery that light is *emitted* in discrete quantities, Einstein interpreted the photoelectric effect to mean that light also *propagates* as discrete quanta, which we now call *photons*, each with energy *hf.* He pointed out that *each photon acts independently*, and he applied the principle of conservation of energy to each photon as follows:

energy of photon = (energy to remove electron) +

(kinetic energy of emitted electron)

or, in symbols,

$$hf = \phi + KE$$

In other words, a given photon carries energy *hf.* When it strikes the surface, a certain amount of energy goes to remove the electron from the substance. The *minimum* energy needed to remove an electron depends on the substance and is called the *work function* ϕ. The energy left over goes to kinetic energy *KE* of the electron. When the photon frequency *f* is so low that its energy is less than the work function, no electrons are emitted regardless of the intensity of the light. If the energy is high enough to eject electrons, increasing the intensity (increasing the number of incoming photons) gives more emitted electrons, but *not* more kinetic energy for any one electron.

There are lots of practical applications of the photoelectric effect. One widely known application is the so-called "electric eye" (Fig. 23.22). Light incident on the photosensitive surface of a *phototube*, or *photocell*, ejects electrons. As long as the light shines on the surface, electric current can flow in the circuit containing the battery, the phototube, and the electromagnetic relay. When the light stops, no current can get across the phototube, and thus the electromagnet is not activated. The relay can either close or open a switch that controls some other device.

A common application is in automatic door controls for supermarkets and other stores. A light beam shines across the entrance in front of the door and hits the phototube. When a person walks through the beam,

Figure 23.22 A circuit diagram showing how a phototube—an "electric eye"—can open and close a relay controlling some other device.

current no longer flows in the phototube, and the relay closes the switch that opens the door. In a similar way, some burglar alarms use phototubes sensitive to nonvisible ultraviolet light. When the burglar passes through the light beam, the alarm sounds. Photocells also form the heart of photographic exposure meters, and they turn the street lights on when the sun goes down. In one type of smoke detector, photocells detect smoke particles that scatter light; and they serve as the eyes of a television camera by converting the optical image focused onto many tiny photocells into an electrical signal for transmission to your home. Solar cells make use of the photoelectric effect to create the electron-hole pairs in semiconductors (Section 20.1) needed in the conversion of light to electrical energy.

23.4 THE DUAL NATURE OF LIGHT

What goes on here? The previous chapter went to great lengths to try to convince you that light is a wave. It has all the characteristics that waves have; it does all the things that waves do. Yet in this chapter we discussed light being emitted from sources in discrete quanta and propagating as individual photons. In the last section we even found that light sometimes interacts with matter as if it were made up of particles. The question arises: Just what is light? Is it a wave, or is it made up of particles? Surely it has to be one *or* the other. The properties of waves are not compatible with those of particles.

The debate among physicists over the true nature of light raged from Isaac Newton's time until the beginning of the twentieth century. Newton believed light to be made up of particles. But in the nineteenth century people accepted the wave nature because light beams were found to exhibit constructive and destructive interference, and only waves do that. In the early twentieth century we found light to be doing things—the photoelectric effect for example—that waves cannot do but particles can. The nature of light seemed to depend on the type of experiment being done—a rather disconcerting situation.

It is natural to want to categorize entities into neat packages that we think we understand. Unfortunately, the real world doesn't always lend itself to such convenient categorization. Physicists finally had to accept the idea that light has a **dual** nature: both **wave and particle**. In those situations in which light interacts with individual atoms, it acts as a *particle;* when light interacts with macroscopic quantities of matter, it acts as a *wave*. No experiment can show both the wave and the particle nature at the same time. These particle and wave ideas are merely *models* that represent certain aspects of light but do not alone tell the whole story of its nature.

· ·

Key Concepts

Light is emitted in discrete wavelengths from gas atoms as they give up energy. One **photon** is emitted as one atom de-excites. The *energy* of the photon is Planck's constant h multiplied by its frequency f, and it exactly equals the energy lost by the atom. A photon is sometimes called a **quantum** of light; the light energy is **quantized** in units of hf. An atom cannot emit light if it is in its **ground state**— the lowest energy state. Atoms can be raised to **excited** states by collisions with electrons or other atoms and by absorption of photons of energy exactly equal to the change in the atom's energy.

Fluorescence is the process by which an atom absorbs radiation of one wavelength and emits radiation of longer wavelength (lower-energy photons).

In solids and liquids electrons can exist in a *continuous range* of energies. *Incandescence* is the emission of light from such a substance. Since an electron in a solid or liquid can lose any fraction of its energy, the light emitted has a continuous range of wavelengths.

The *luminous efficiency* of a light source is the ratio of light output rate in *lumens* to power input in watts.

A **laser** is a source of *single-wavelength, coherent* light produced by *stimulated emission*.

X rays are emitted from atoms when their energy changes in steps of hundreds of electron volts or more.

Light has a *particle nature*, in that its energy is carried in photons of energy hf. This nature is demonstrated by the *photoelectric effect*.

Even though light has a **dual** nature, both wave and particle, in any one event it can act *either* as a wave *or* as a particle, but not as both.

Questions

Light Sources

1. One of the end-of-chapter references in the previous chapter says that "all light is the radiation of surplus energy by some substance." Do you agree with this statement?

2. How does the energy needed to excite an atom from state E_1 to state E_2 compare with energy of a photon given off when the atom de-excites from E_2 to E_1?

3. The visible light from hydrogen has only the four wavelengths shown in Figure 23.4. Why does the light of the longest wavelength result from the smallest energy change, and vice versa?

4. The transition that gives the 410-nm violet light shown in Figure 23.4 is not shown on the energy-level diagram—Figure 23.3. If the final

state in the transition is E_2, where must the initial state be on the diagram?

5. Why do the heating elements of an electric range get red when they get hot? Does the light contain only one wavelength?

6. Suppose your electrical power company imposes a "brown-out"—a reduction in operating voltage to reduce power use during peak-load periods. What would that do to the color of the light from your incandescent bulbs?

7. When you strike certain kinds of stones together you get sparks. The sparks are particles of stone knocked off that are heated by the collision. Do you expect the light from the sparks to have a continuous range of wavelengths, or to consist of discrete wavelengths?

8. Is the light from the flame of a gas range due mostly to atomic and molecular de-excitation or to incandescence, as in a candle flame? (Hint: What colors dominate most incandescent objects?)

9. As advertisements, businesses sometimes give away little yellow strips (containing their ad) that you can tape to your telephone. The strip glows in the dark to show you where the phone is after the lights are off. How does it work? (Some clock dials glow in the dark for the same reason.)

10. A laser is a device for "light amplification." It runs on electrical energy. Does this mean you get more light energy out than electrical energy put in?

11. If you were performing laser surgery in which you wanted the light energy to be absorbed by red blood cells but no other tissue, would you use an argon laser that emits blue-green light or a helium-neon laser that emits red light?

Hint: Recall the color of light that is *least* absorbed by green leaves (Section 22.2).

12. If a particular type of photographic film has a light sensitivity that makes it give correct colors when the pictures are taken in daylight, why are the colors in the picture not correct if taken under ordinary incandescent or fluorescent lights?

13. Suppose that when you look at a red traffic light and then at a green traffic light, the same amount of light *energy* entered your eye per second from each. Did your eye receive more photons per second from one light than the other? If so, which?

14. At some beaches, clear water appears very green, rather than blue as you might expect from scattering of white light by water molecules. This coloring is thought to be due to fluorescence in certain organic substances concentrated in the water. Explain how this might occur in the same way that the coating on a fluorescent bulb gives off light.

Photoelectric Effect

15. Many cameras have built-in exposure meters that set the diaphragm opening to let the proper amount of light reach the film. Figure 23.23 is a diagram that shows how a photocell based on the photoelectric effect can be used to indicate the intensity of incoming light. Explain why there is a current through the meter that is proportional to the light intensity.

16. Draw a schematic wiring diagram for a circuit that could use a photocell (such as shown in Fig. 23.22) to turn on an outdoor light after sunset.

Figure 23.23 Diagram of a light exposure meter.

17. The sound track of a movie film is a band on the side of the film that varies in darkness in proportion to the pressure variations of the sound wave it represents. When a bright lamp shines through the track, the transmitted light varies in intensity in the same proportion. What phenomenon that we've studied in this chapter could be used to monitor these intensity variations and convert them to electric current variations, which can be amplified and used to operate a speaker system?

18. Draw a schematic diagram for the steps described in Question 17.

19. Sunburn occurs when photons of radiation cause chemical changes that produce cell damage in the skin. Which *one* of the three causes sunburn: infrared, visible, or ultraviolet radiation? Justify your choice in terms of which photons have enough energy to cause the change.

20. If an electric eye is used to open a supermarket door when a person walks through a beam of white light, could you make the door open faster if you used a beam of blue light?

Problems

1. From the energies of the states in Figure 23.3, verify the wavelengths shown for light emitted in E_3-to-E_2 and E_5-to-E_2 transitions.

2. Calculate the excitation energy of the level E_6 that gives the light described in Question 4.
   Answer: 13.23 eV

3. How many photons per second are sent out by an FM radio station transmitting 50,000 W at a frequency of 100 MHz?
   Answer: 7.5×10^{29}

4. How much kinetic energy must be given up by an electron to excite a hydrogen atom in the ground state so that it can emit 656-nm light?
   Answer: 12.09 eV

5. If you burn table salt, the sodium will give off yellow light of wavelength 589 nm. What is the energy difference between the energy levels in sodium that give rise to this light?
   Answer: 2.11 eV

6. From Figure 23.9 estimate the energy of the photons most often emitted from an incandescent light bulb whose filament operates at 3000°C. What is the energy of the most abundant photons from a "red hot" fireplace poker at 700°C?
   Answer: 1.4 eV; 0.4 eV

7. The *absolute* efficiency of a 100-W incandescent bulb is 2.5%. How much light energy is pro-

duced in 8 h of operation of such a bulb?
   Answer: 20 Wh

8. Based on the absolute efficiency given in Problem 7, how many visible photons are emitted per second from a 100-W bulb? Take the average wavelength of the visible radiation to be 560 nm.
   Answer: 7.0×10^{18}

9. If you convert your 100-W kitchen incandescent light to a 40-W fluorescent, how much more light will you get? How much electrical energy will you save per month if you use the light for an average of 6 h per day?
   Answer: 46%; 11 kWh

10. A *helium-neon* laser emits light of wavelength 633 nm. What is the energy of the photons emitted from such a laser?
    Answer: 1.96 eV

11. About 4.8 eV of energy are needed to carry out a photosynthesis reaction in a plant. If eight 700-nm photons are needed to supply the energy for the reaction, what is the efficiency of light use in this process?
    Answer: 34%

12. If you were designing a light exposure meter to be sensitive to 550-nm light, what is the largest work function the photosensitive material in the photocell could have?
    Answer: 2.2 eV

Home Experiments

1. Your instructor can probably give you a small piece of inexpensive *diffraction grating*—a sheet of material that uses interference to display the individual colors that make up a light beam. If not, perhaps he or she can lend you a grating or prism. Use it to examine various kinds of nighttime light sources, and determine their color components. This will be much more fun than it sounds—even enough to justify being laughed at by your friends for doing it! Your instructor will explain how to hold the grating or prism *near your eye* for viewing.

2. Use a prism or grating (see Experiment 1) to determine the kinds of street lights used on your campus or town. The spectra of the most common types of lights are as follows: Incandescent—continuous; mercury vapor—red, yellow, green, blue, and a faint violet; high-pressure sodium—same as mercury plus a continuous band of reds; low-pressure sodium—single 589-nm yellow. (The mercury is probably the most interesting to look at; you will see the usual outline of the lamp, but in four distinct, separated colors. From which of these types of lights do you expect the least ability to distinguish colors of illuminated objects?)

References for Further Reading

Boraiko, A., "Lasers—A Splendid Light," *National Geographic*, March 1984, p. 335. This and a companion article on holography give vast insight into the usefulness of the laser.

Edelson, E., "Holography—Out of the Lab at Last," *Popular Science*, March 1984, p. 90. Practical benefits of holography, such as speeding our progress through the supermarket check-out.

Free, J., "The Laser-Disc Revolution," *Popular Science*, May 1985, p. 66. How lasers give us high-fidelity music on durable recordings.

Hewlett-Packard Optoelectronics Division Staff, *Optoelectronics Applications Manual* (McGraw-Hill Book Co., New York, 1977). Discusses light-emitting diodes and other optoelectronic devices, starting with a readable introduction and progressing to considerable detail.

Kallard, T., *Exploring Laser Light* (Optosonic Press, New York, 1979). Many aspects of the operation and use of lasers.

Kock, W. E., *Radar, Sonar, and Holography: An Introduction* (Academic Press, New York, 1973). Parts of the book discuss the making and properties of holograms.

Meier, G., E. Sackmann, and J. G. Grabmaier, *Applications of Liquid Crystals* (Springer-Verlag, New York 1975). Properties and applications of liquid crystals.

Physics Today, May 1982. A special issue on liquid crystals, including fundamentals and applications.

Walker, J., "Physics and Chemistry Underlying the Infinite Charm of a Candle Flame." The Amateur Scientist Column, *Scientific American*, April 1978, p. 154. An interesting description of the various processes going on in a candle flame, as well as experiments to support this description.

CHAPTER 24

THE ATOM:
A CLOSER LOOK

Our modern idea that all matter consists of tiny particles called *atoms* is actually not so modern. The Greek philosophers Leucippus and Democritus introduced the first atomic theory in the fifth century B.C. After considerable debate among the Greek philosophers, atomism became associated with atheism and fell into disrepute from Aristotle's time until reintroduced by the English chemist John Dalton in the early nineteenth century.

The word *atom* meant "uncuttable" or "indivisible" in Greek. As we study the properties of the atom and its nucleus in this and the next few chapters, we'll see that the word's origin doesn't quite fit the way the atom behaves.

In this chapter, we discuss the structure of the atom as we now know it. We will find that we need to abandon any ideas about the atom being made up of little particles that, if we look closely enough, are found to move in specified orbits. Instead, we'll need to describe these particles from the viewpoint of their wave nature. This means introducing *quantum mechanics*, the basic mechanics for describing atomic and subatomic systems. Under such a description we'll find that we can no longer specify the exact state of a particle with absolute certainty as we could, in principle, under the laws of "classical" mechanics. Rather, we will have to be content with probabilities for finding the particle within a certain region.

After using quantum mechanics to understand the basic structure of the simplest atom, we then can see how this understanding leads to a systematic description of all atoms—that its, how it explains the periodic table of elements.

24.1 THE RUTHERFORD-BOHR MODEL OF THE ATOM

In earlier chapters when we needed to talk about the atoms that make up all matter, we used a "planetary" model for the atom. In this model, the atom is pictured as consisting of electrons orbiting the nucleus in much the same way as planets orbit the sun. Each time we did this, we noted that this crude picture is useful for discussing *some* atomic behavior, but it has serious drawbacks. We'll now look at the origin of this model, its drawbacks, and what it takes to give a better description of the atom.

Rutherford and the Nuclear Atom

How would you try to learn about the composition of something so small that it takes about 10^{24} of them to make up this book—something about 10^{-10} m (0.1 nm) across? You certainly couldn't look at it with light, no matter *what* the magnification of your microscope. Something only 1/10,000 the wavelength of light is not about to scatter any light into the microscope. The atom is so small that, in a sense, you can't get "close" enough to see inside it.

Well, what do you do when you're standing beside a pond in a mountain stream, but you can't get close enough, or in it, to find out how deep it is or what the bottom is like? You throw rocks at the pond and watch what happens to them. If you can see what happens to the rocks, you can tell something about the pond. That's exactly how we can study the atom: We throw "rocks" at it and watch what happens to them. These "rocks" have to be a special kind—particles of atomic or smaller sizes—and we have to "watch" with equipment that can tell what these particles do.

Around 1910 the prevailing view was that the atom had a sort of "plum pudding" arrangement, with electrons spread out like plums in a sphere of positively charged "pudding." Between 1909 and 1911 the English physicist Ernest Rutherford and his associates found out differently. The "rocks" they used were *alpha* (α) particles—clumps of two neutrons and two protons—emitted from certain *radioactive* substances. (More about radioactivity later.) They aimed these α particles at a gold foil in an arrangement such as that shown in Figure 24.1. If the foil were made of "plum pudding" type atoms, these high-speed particles would move through almost undeflected from a straight line. Much to the experimenters' surprise, many of the α particles were scattered at large angles, some even straight backwards. In Rutherford's words, "It was

Figure 24.1 The apparatus that Rutherford and his associates used to study the scattering of alpha particles by atoms of a gold foil. An α particle striking the scintillator caused it to give off a small amount of light that could be seen in the microscope. The scintillator and microscope could be moved through 180° to find the number scattered at each angle.

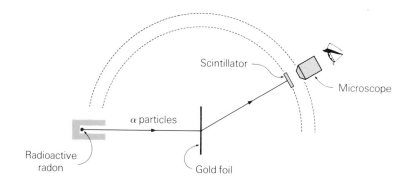

almost as incredible as if you fired a 15-inch shell at a piece of tissue paper and it came back and hit you."

Rutherford correctly interpreted the results to mean that *most* of the atom's mass, and *all* of its *positive* charge, are concentrated within an extremely tiny region—the **nucleus**. You can visualize the reason for the difference in scattering pattern by the following thought experiment. Imagine many, many billiard balls to be thrown in the general vicinity of a bowling ball hanging from a rope. Most would miss the bowling ball, but some would be scattered at large angles. Now imagine the billiard balls to be thrown at a *cloud* having the same total mass as the bowling ball. Most would hit the cloud, but you'd be very surprised to see any billiard balls bouncing backward. That was Rutherford's line of reasoning, but he was also able to calculate mathematically the correct number of α particles scattered at each angle. His insights brought him knighthood and a Nobel Prize.

The diameters of the nuclei of atoms range from about 10^{-15} to 10^{-14} m, 1/100,000 to 1/10,000 the size of the atom as a whole. Let's look at the atom on a much larger scale: If a typical atom were expanded, with everything keeping the same proportions, until the nucleus became the size of our sun, the outer electrons would be 500 to 1000 times as far away as the earth is from the sun.

Based on this idea of a *nuclear atom*—a positively charged nucleus with negatively charged electrons distributed in a much larger region around it—we can break our study of the atom into two areas of inquiry. First, how are these electrons distributed in the atom as a whole? Second, what are the structure and behavior of the nucleus? In this chapter we focus on the first of these questions and leave the second one for subsequent chapters.

The Bohr Model

In discussing light emission from gas atoms (Section 23.1), I rather casually stated, without justification, that the discrete wavelengths of light emission from gases comes about as electrons in the atoms go to one of

Level   Energy

$n = 5$ 13.1 eV
$n = 4$ 12.8 eV
$n = 3$ 12.1 eV

$n = 2$ 10.2 eV

$n = 1$ 0

Figure 24.2 An energy-level diagram for the hydrogen atom. Taking the *ground state* (level 1) to have *zero* energy, the energies of the lowest four *excited* states are given in electron volts (1 eV = 1.60 × 10⁻¹⁹ J).

the lower *quantized* energy levels in the atom. That discussion overlooked the years of painstaking effort on the part of many physicists in coming to that understanding. In fact, it was primarily through the study of light emitted from atoms that our present concept of atomic energy levels came about.

The energy-level diagram for the hydrogen atom is repeated in Figure 24.2. The levels are identified by the number n, starting with $n = 1$ for the ground state. The various levels correspond to different states of the electron in the atom. Recall that as the atom goes from a higher to a lower energy state, the energy of the photon given off, E_{photon}, is *exactly* equal to the *change* in energy of the atom. Also, the photon's energy is related to its wavelength, λ, by $E_{\text{photon}} = hc/\lambda$. Therefore, measuring the wavelengths of light given off by a particular atom tells you the energy levels of that atom.

The next question is, can we get from a knowledge of the energy levels to a knowledge of the actual structure of the atom? Let's start to answer that question by considering a planetary model, with the electron held in orbit by the attractive electrical force between it and the nucleus. If we were to calculate the total energy—the sum of the kinetic and potential energies—of the electron in the orbit, we would find that the only orbits that would give the correct energy levels for the hydrogen atom are the ones whose orbital radii are shown in Figure 24.3. Notice that you can get the radius of any orbit by multiplying the smallest (*ground state*) radius by n^2.

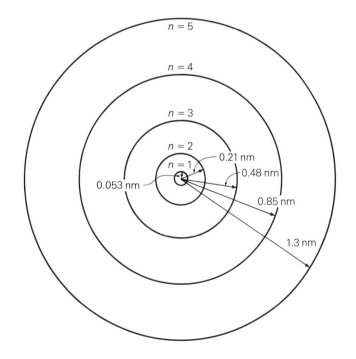

Figure 24.3 Electron orbits in the Bohr model for the lowest five states of the hydrogen atom. The radii of the orbits are given in nanometers (1 nm = 10⁻⁹ m).

We have two conditions here: (1) a planetary model with electrons in orbit around a massive nucleus and (2) discrete energy levels with their accompanying requirement of only certain orbits being allowed. Neither of these requires the other to exist, although we know from the light emission spectra that the second condition must be true.

In 1913 Niels Bohr tied these two ideas together by *postulating* an additional condition for the electron orbits. He postulated that the angular momentum of an electron in orbit—its linear momentum multiplied by the orbit radius—could not take on any arbitrary value. Rather, the angular momentum could have values of *only* some **integer** (1, 2, 3, 4, . . .) multiplied by Planck's constant h divided by 2π. That is,

$$\text{angular momentum} = n \times \frac{h}{2\pi}$$

where n is an integer. Remember from Chapter 23 that Planck's constant is a universal constant with the value of 6.62×10^{-34} joule-second. The particular orbits that satisfy this postulate give the exact energy levels for the hydrogen atom—those of Figure 24.2. We call this planetary atomic model, with *quantized angular momentum* values, the **Bohr model** of the atom.

The Bohr model quite accurately predicts the energy levels of the hydrogen atom: The simplest atom we have. But for more complicated atoms, this model doesn't quite give the right answers. We'll see in the next section what is needed to do that. The model is useful, however, because it gives us an intuitive picture of the atom, and it helps us to visualize many atomic properties.

From *our* historical perspective, it might seem that it should not have taken much insight to come up with the ideas that Rutherford and Bohr gave us. They seem perfectly obvious. But, for the state of atomic knowledge in 1910, their contributions were immense.

What Is a Model?

Before going further to describe a more up-to-date model of the atom, let's consider what we mean by this word *model* that we've been using. If you have ever built, or just looked at, a model airplane, you understand part of what we mean by a model. A model airplane represents the real thing, looks somewhat like it, and has many of its characteristics. But no one would claim it to *be* the real thing. It is as close as we can get to a representation of the real airplane using familiar, available materials. But it certainly doesn't tell all there is to know about the airplane.

A model of an atom or other entity is an attempt to describe a complex entity, perhaps one that's impossible to visualize, in terms of familiar concepts or systems. It gives us insight into the nature of the real thing, and it may correctly explain some of its features, but it does not tell us all there is to know about the actual system. For example, the wave picture of light is a *model* that correctly explains part of light's behavior, but not all; the particle model explains another part of what is observed. The better the model, the more closely it duplicates reality. Thus, when we consider a model of an atom, we attempt to describe its features in terms of familiar, though not necessarily 100% correct, concepts.

The fact that a model is not completely correct does not make it useless. The Bohr model gives us an intuitive picture that has been very useful in describing those aspects of the physical world that we have considered so far. But it falls short of explaining everything we can observe about the atom.

24.2 THE QUANTUM THEORY OF THE ATOM

The mid-1920s brought the introduction of **quantum theory** that we now use to describe the atom and its behavior. As background for discussing this theory, we'll summarize the status in about 1920 of the understanding of the atom and its parts.

1. From the fact that light is emitted in *discrete wavelengths* from individual atoms, the atom must have *discrete energy levels*. The energy of the photon emitted equals the *change* in energy of the atom emitting it.

2. The Bohr planetary model with quantized orbits could explain the energy levels of hydrogen. But it did this by a somewhat arbitrary assumption about angular momentum, and it didn't work for more complicated atoms. Physicists, including Bohr, knew that something new and different was needed.

3. Light was known to have a *dual* nature: Sometimes it acted as a wave; sometimes it acted as a particle.

The Wave Nature of Particles

In 1923 a young French graduate student named Louis de Broglie, in searching for a Ph.D. thesis topic, began thinking about the dual nature of light and about the symmetry that usually exists in nature. De Broglie's line of reasoning was something like this: Light, which is usually thought of as a wave, also has a particle nature. Since things in nature are normally not lopsided, it is reasonable that electrons, which we usually think of as particles, would also have a wave nature. In his thesis, de Broglie proposed just that. Furthermore, he was able to predict that the wavelength associated with the electron was given by Planck's constant divided by the electron momentum:

$$\lambda = \frac{h}{mv}$$

If we combine the de Broglie wavelength hypothesis with the Bohr model, a surprising coincidence appears. You find that for those orbits allowed by the Bohr model, the orbit circumference is *exactly* a whole number of electron eavelengths. (Fig. 24.4-a). This implies a sort of standing wave similar to those of a plucked guitar string. For other orbits, the wave does not close on itself (as Fig. 24.4-b illustrates).

What this standing-wave effect says is that there must be *something to* the idea that the electron has a wave nature. It still doesn't give us the picture of the atom that we need, but it points us in the right direction.

Three years *after* de Broglie made his hypothesis, Clinton Davisson and Lester Germer at Bell Telephone Laboratories showed that electrons undergo *diffraction*—something that only waves can do. Within a few years other experimenters showed that the other entities we usually consider to be particles sometimes also behave as waves.

We see, then, that Mother Nature has pulled another trick on us. In the last chapter we saw that light, which we usually think of as a wave, has a particle nature that is of equal importance. Now we discover that those chunks of matter we usually think of as being particles—electrons, protons, neutrons, for example—also have a wave nature. Furthermore, as we will see in the next section, it is the wave nature of these particles that lets us understand the environment in which they usually exist, namely inside the atom. Thus, the *wave-particle duality* permeates our microscopic view of the world.

· ·

APPLICATION   **THE ELECTRON MICROSCOPE**

The electron microscope represents a direct application of the wave nature of electrons. Figure 24.5 shows a picture of an electron microscope and an example of the magnification it can achieve. Just why is an electron microscope capable of so much greater magnification than an ordinary light microscope? Let's investigate.

A person might guess that it would be possible to see smaller and smaller objects with an optical microscope merely by increasing the magnifying power of the lenses used. But remember, we see illuminated objects because the light reflects or scatters from them into our eyes. To see an object under a microscope, light must scatter from the object before it passes through the microscope into our eyes. Recall that, if a wave is to scatter from an object rather than to diffract around it, the wavelength must be shorter than approximately the object size. Thus,

Bohr orbit

Superimposed wave

(a)

(b)

Figure 24.4 (a) For the orbits allowed by the Bohr model, the orbit circumference is a whole number of electron wavelengths. (b) For other orbits, the waves overlap randomly and cancel each other's effect.

we can never hope to see anything with visible light if it is smaller than the wavelength of the light—about 400 nm.

But what if we illuminate the thing we want to see with radiation of a shorter wavelength? It turns out that electrons qualify. We noted earlier that the *de Broglie wavelength*, as the wavelength associated with particles is called, is given by

$$\lambda = \frac{h}{mv}$$

Electrons accelerated through 10,000 V in a sophisticated version of an electron gun (Section 18.1) have a velocity of about 6×10^7 m/s. Therefore, the de Broglie wavelength for such electrons is

$$\lambda = \frac{6.6 \times 10^{-34} \text{ J·s}}{(9.1 \times 10^{-31} \text{ kg})(6 \times 10^7 \text{ m/s})} \simeq 0.01 \times 10^{-9} \text{ m} = 0.01 \text{ nm}$$

We see that, just from the standpoint of wavelengths, electron microscopes can let you see objects 1/10,000 the size of those seen with light. In practice, the electron's accelerating voltage can be 10 times as much, but practicalities such as focusing the electron beam have thus far limited the ability to see objects smaller than about 0.2 nm.

To use these ideas in an electron microscope, we can't just replace the light beam with an electron beam. Among other things, our eyes do not respond very favorably to being zapped with high-energy electrons. In the scanning electron microscope (Fig. 24.5), magnetic fields focus the beam of high-speed electrons and sweep them back and forth across the specimen in much the same way as the electron beam is swept across the picture tube of a TV (Section 18.1). The electrons are scattered by the specimen—the pattern of scattering depending on the shape of the specimen's surface. The scattered electrons strike a detector that senses the position and number of electrons. This information is then converted electronically to an image of the specimen shown on a TV monitor. This image can then be photographed, as was done to get Figures 14.3 and 24.5(b).

Quantum Mechanics à la Schrödinger

In 1926 the Austrian physicist Erwin Schrödinger invented a model that serves as the basis of our present understanding of the atom. Using de Broglie's idea of the wave nature of particles, he developed a **wave equation** that is the starting point in deriving a model for the atom. This model is a *mathematical* one and, unfortunately, is not something that can be visualized easily. We will describe in qualitative terms some

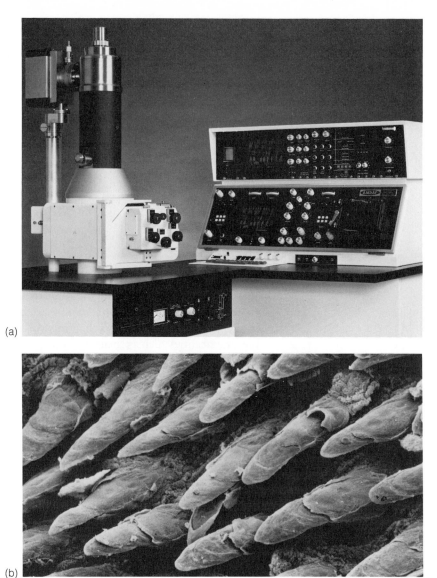

Figure 24.5 (a) Photo of a scanning electron microscope. (b) Photograph of an image from an electron microscope. The image is of the surface of a mouse's tongue, magnified 420 times. (Courtesy of Dr. R. L. Grayson.)

(a)

(b)

of the main features of the picture this equation gives of the atom and of atomic particles in general.

In case you'd like to see what the Schrödinger wave equation looks like, I'll write it down:

$$-\frac{\hbar^2}{2m}\nabla^2\Psi + V\Psi = i\hbar\frac{\partial\Psi}{\partial t}$$

kinetic energy + potential energy = total energy

(a)

(b)

Figure 24.6 The "cloud" density is proportional to the probability of finding the electron at that location. These "probability clouds" are for the (a) ground state and (b) one of the excited states of the hydrogen atom.

I won't bother you with the mathematical details of what those strange symbols mean. But, as implied by the line below the equation, it is in principle just an expression of the conservation of energy. In a mechanical system without friction, the total energy stays constant and is the sum of the kinetic and potential energies. Each term of the Schrödinger equation, as shown, represents one of these forms of energy for atomic particles.

When you apply the wave equation to the electrons of a particular kind of atom, what you get out is something called a **wave function**, which we usually represent by the Greek letter psi (Ψ). This wave function is a mathematical expression that has in it all the information that it is possible to know about the electrons.

The wave function cannot tell you precisely *where* an electron *is* in the atom. Instead, the square of the wave function, $|\Psi|^2$, gives the *probability*, or relative chance, of finding the electron in a certain region of the atom. Instead of having a precise electron orbit, we have a *probability distribution* for the electron that tells the probability of finding it at various points in space. The closest we can come to a visual picture of this is to draw a cloudlike picture in which the density of the cloud is proportional to the probability of finding the electron there. Figure 24.6 shows this "probability cloud" for the ground state and one of the excited states of the hydrogen atom.

This probaility distribution business in not nearly as satisfying as being able to say that the electron is at "such-and-such" a place in "thus-and-so" orbit. However, the creator of the atom has not allowed us the satisfaction of having such precise information. So far as physicists can tell at this time, it is not *in principle* possible to precisely determine the status of any given electron, but only to know the probability of finding it at a certain place with a certain velocity. But, within this limitation, we can calculate the structure of all the atoms.

Our knowledge of atomic electrons is something like the knowledge statisticians have of how many people will be killed on the highways over the Fourth of July weekend. They can tell you in advance with considerable accuracy how many people will be killed. But they can predict *nothing* about what will happen to Mary Elizabeth Johnson–Whippendorfer or you or any other individual. The *quantum mechanics* of the Schrödinger equation can tell you *how many* electrons are *where* in a given group of atoms, but it cannot tell you what a particular electron is doing.

A quantum mechanical calculation using the Schrödinger wave equation predicts the *highest probability* for the electron of the hydrogen ground state to be at the position of the first Bohr orbit. But the electron can also be, with high probability, at other locations. And using quantum mechanics, we can calculate the positions of the energy levels of *any* atom. In the ground state of a many–electron atom, electrons occupy the lowest possible energy states.

The Heisenberg Uncertainty Principle

If you're like many people, you probably visualize the electron as some tiny round speck. But as we've seen, the electron on an atomic scale acts more like a wave than a particle. When you *really, really* get to where the electron "is," how do you locate it?

Early in the days of quantum mechanics, Werner Heisenberg showed that you cannot simultaneously know a particle's position and momentum with infinite precision. If the momentum *mv* is *uncertain* by an amount $\Delta(mv)$, and the position *x* is *uncertain* by Δx, then $\Delta(mv)$ multiplied by Δx must be equal to or greater than Planck's constant divided by 2π. In symbols,

$$\Delta(mv) \cdot \Delta x \geq \frac{h}{2\pi}$$

The Δ here means uncertainty, and $\geq$ means greater than or equal to. We call this statement the **Heisenberg uncertainty principle**. It says that we can measure either position *or* momentum precisely but not both at the same time. Or, for example, suppose you know the electron is within an interval of 0.1 nm along some line. (That's about the diameter of a hydrogen atom.) The closest you can then determine its momentum is

$$\Delta(mv) \geq \frac{h}{2\pi \cdot \Delta x} = \frac{6.62 \times 10^{-34} \text{ J·s}}{2\pi(0.1 \times 10^{-9}\text{m})} = 1 \times 10^{-24} \text{ kg·m/s}$$

Because the electron mass is so small, 9.1×10^{-31} kg, the speed is uncertain by about 1000 km/s.

. . . .

QUESTION   Look at the size of the numbers involved, and explain why the Heisenberg uncertainty principle did not limit Babe Ruth's ability to locate and hit Charlie Root's curve ball into the center-field bleachers at Wrigley Field.

The Heisenberg uncertainty principle is closely related to the idea that, in order to measure anything, you *must* disturb it. In fact, the two ideas are inseparable. For example, suppose you wanted to measure the size of this book. If you lay a ruler across it, light must strike the book and reflect into your eyes in order for you to see where its edges are. (You'd have a hard time measuring it in the dark!) In the process of reflecting light, the atoms of the book absorb some energy from the light photons and then re-emit that energy in the form of reflected light.

For an ordinary-sized object, such as a book, the energy imparted to it by the light is much too low to disturb its position or size. But suppose the object you were trying to measure is an electron. To locate it, you would need to scatter at least one photon from it. Your best approximation of its location would be within one wavelength of the photon. Remember, the wavelength λ of a photon is related to its energy, E, by $E = hc/\lambda$: To get a short wavelength for the scattered photon, it needs to have a high energy. This, in turn, will increase the disturbance of the electron and possibly remove it from the state in which you want to observe it.

Suppose, for example, you wanted to locate an electon to within a distance of 0.1 nm—about the diameter of a hydrogen atom in its ground state. The photon would then need to have a wavelength of about 1.0 nm or less. The energy of such a photon would then be

$$E = \frac{hc}{\lambda} = \frac{(6.62 \times 10^{-34} \text{ J·s})(3.0 \times 10^8 \text{ m/s})}{0.1 \times 10^{-9} \text{ m}} = 2.0 \times 10^{-15} \text{ J}$$

$$= (2.0 \times 10^{-15} \text{ J})/(1.6 \times 10^{-19} \text{ J/eV}) = 12{,}400 \text{ eV}$$

This amount of energy is about 1000 times that needed to remove the electron from a hydrogen atom. It would certainly destroy the system you were trying to measure.

There is also a form of the uncertainty principle that relates to energy and time. It states that

$$\Delta E \cdot \Delta t \geq \frac{h}{2\pi}$$

In other words, the product of the uncertainty in energy of a system and the uncertainty in time during which it has this energy must be at least equal to $h/2\pi$.

The uncertainty principle still holds but is meaningless when applied to ordinary-sized objects. For example, the roughness of the sides of the smoothest possible object is still too large to make the uncertainty in its velocity measureable. We can measure its position or velocity to any precision allowed by the measuring device. But when dealing with atomic and subatomic particles, no matter how good your measuring devices, you *in principle* cannot *simultaneously* measure any two entities covered by the same statement of the uncertainty principle.

24.3 QUANTUM MECHANICS: THE FOUNDATION OF CHEMISTRY

You may be wondering what all this quantum mechanics has to do with anything relevant to you. The truth is that a person going about his daily routine does not *directly* see or experience anything quantum mechanical.

That's one reason quantum mechanical ideas are hard to grasp. They are not part of our ordinary experiences that give intuition about things. But *indirectly* the quantum mechanical behavior of atoms influences *everything* we do or think. Everything is composed of atoms, and these atoms interact by the rules of quantum mechanics.

Quantum mechanics does more than just satisfy our intellectual curiosity about the nature of the atom. For example, there is the rather large area of *applied physics* called **chemistry**. Quantum mechanics lets the chemist in on the laws of how atoms interact with each other and how to combine them to form substances that are very useful to you and me.

I would like to give a hint at how quantum mechanics, using the wave nature of particles, can provide information on the energy levels of atoms. We have already seen from the light emission spectra of atoms that atomic energy levels are quantized. We'll use a hypothetical example to show how a consideration of the wave nature of particles leads directly to quantization of energy. From that example I hope it will be reasonable that the same ideas carry over to real atoms.

Particle-in-a-Box

Consider a particle that can only move back and forth in a straight line and that, furthermore, is confined to a short region of that straight line. Even though we will be interested only in atom-sized distances, a large-scale model of the situation would be a Ping-Pong ball bouncing back and forth inside a pipe with closed ends, while experiencing *no* friction. In other words, we have a particle in a *one-dimensional box*. Figure 24.7(a) illustrates the point.

As discussed earlier in this chapter, the square of the wave function $|\Psi|^2$ for a particle tells the probability of finding the particle at a certain point. If the particle cannot get outside the box, then the wave function must go to zero at the ends of the box. Assuming we can represent the wave function by the usual form of a wave, the possibilities for wave functions that go to zero at the edges of the box must look something like the waves shown in Figure 24.7, parts (b), (c), and (d).

As this figure shows, the possible wavelengths are 2L divided by an *integer*. These wavelengths are evidently the possible de Broglie wavelengths for the particle. You will recall that this wavelength is given by $\lambda = h/mv$ and that kinetic energy is given by $(1/2)mv^2$. Since these wavelengths are quantized—they can take only discrete values—the possible kinetic energies must be quantized. For ordinary-sized particles, such as the Ping-Pong ball in the pipe, the allowable energies are so close together that they can, in practice, take any value. However, if the size of the box were about the size of a typical atom and the mass of the particle were about that of an electron, the magnitudes of the allowed energies would be of the order of a few electron-volts, and these energies would be *separated from each other* by a few electron-volts.

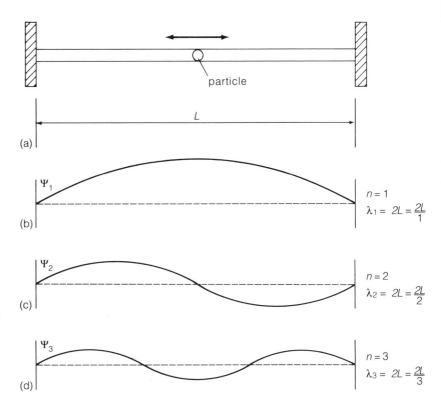

Figure 24.7 (a) The particle can move back and forth in a straight line of length L between two fixed ends. (b)–(d) Three of the possible wave functions that go to zero at the ends of the box.

Our hypothetical particle-in-a-box illustrates a general quantum mechanical principle: *The wave nature of particles causes the possible energies of an atomic-sized system to be quantized into discrete values. The system can exist only in one of these distinct energy states.* This situation—quantized energy levels—is precisely what we observed via the discrete wavelengths of light emitted from atoms. Next we will see how these quantum ideas determine the structure of real atoms.

The Quantum Mechanics of Real Atoms

What does this rather far-fetched example have to do with real atoms such as the hydrogen atom? You have already seen that the hydrogen atom energy levels are quantized (Fig. 24.2). These levels can be experimentally determined using the wavelengths of light emitted from hydrogen. The point is that, starting with the Schrödinger equation, these levels can be calcualted. That procedure is much too mathematical to do here. Instead, I want to appeal to your sense of reason: It is reasonable that if the wave nature of our particle-in-a-box leads to quantized energy levels, then the wave nature of electrons in a hydrogen atom should lead to quantized energy levels for the atom. And, in fact, a quantum mechanical calculation via the Schrödinger equation gives just those energies shown in Figure 24.2 that are observed for hydrogen.

Let's carry our analogy a bit further and see how other ideas from our hypothetical example carry over into the atomic realm and actually provide the foundation for understanding all atoms. Refer back to Figure 24.7, which shows the wave functions for a particle in a box. Notice that each function is labeled with a subscript—1, 2, 3, and so on, and that the wavelength decreases in inverse proportion to this number. Thus, the energy of the particle depends on this integer. We call this number that labels the wave functions and determines the energy of the state of the system a **quantum number**. A valid wave function exists for the particle-in-a-box only for *integer* values of the the quantum numbers. (A physicist would find these wave functions by determining the mathematical function that correctly satisfies the Schrödinger equation.)

A similar but considerably more complicated situation exists for the hydrogen atom. For it there are four quantum numbers rather than just one. A valid wave function exists corresponding only to specific integer values of these four quantum numbers. One of these quantum numbers determines the energy of the atomic level. It is symbolized by n and used to label the level in our energy-level diagram, Figure 24.2. The other three quantum numbers correspond to other physical properties of the atom. Keep in mind that a valid wave function exists only for appropriate values of these numbers. An electron in the hydrogen atom can have any set of these quantum numbers, but the ground state corresponds to the lowest set of these numbers.

If you have ever studied chemistry, you have probably learned that the electron can exist in certain "orbitals," a given orbital being that electron configuration described by one set of these four quantum numbers. The allowed values have a firm foundation in the Schrödinger equation: It's only for these values of the quantum numbers that a valid wave function exists.

But what about other atoms; the universe is not made of hydrogen! We can understand the makeup of all the atoms in the periodic table, Figure 24.8, by considering one electron at a time and assigning it appropriate values of the quantum numbers. A very important guideline must be followed—no two electrons can have the same four quantum numbers. This is a statement of the Pauli exclusion principle that you have studied if you have taken chemistry. This principle follows naturally from quantum mechanics: If any two electrons had the same set of quantum numbers, the wave function for them would be *zero everywhere*— that is, they would not exist.

So what's the big deal? It doesn't seem very impressive that quantum mechanics let's you calculate something that we already knew to start with—the quantized energy levels of the atom. Well, that is only the beginning. The Schrödinger equation serves as the foundation for understanding small-scale systems in the same way that Newton's second law of motion let's us understand the motion of ordinary-sized objects. Using Newton's law, you can find out what any object will do when subject to a set of known forces. Similarly, the Schrödinger equation lets you

Figure 24.8 The periodic table of the elements.

calculate all the properties of atoms, how they interact with each other, and how they can be combined to form useful substances.

24.4 THE QUANTIZED NATURE OF THINGS

When measured in everyday amounts, most physical properties seem to be able to exist in any arbitrary amount. However, on an atomic scale, quantization into discrete small bundles is an inherent feature of nature. Let's review the things we've discussed so far that are *quantized*—that come in discrete amounts.

1. All matter is quantized into atoms.

2. The charge on the atom's constituents is quantized into units of electron charge.

3. Light energy is quantized into photons.

4. The energy levels of atoms are quantized.

5. Angular momentum is quantized. (Even though we discussed this in connection with the Bohr model, it is true in general.)

Quantum mechanics treats this quantization in a natural way, without artificial assumptions as in the Bohr model. Some physicists have even theorized that *time* is quantized—that we can choose time intervals only in whole number multiples of some very small time interval. No one has yet given an experimental verification of this hypothesis.

In the next chapters, as we look inside the atom's nucleus, we'll see that it is quantized into components. These components are further quantized into subcomponents. The energy levels, which turn out to be much *larger* in magnitude, continue to be quantized.

· ·

Key Concepts

All matter consists of **atoms**—about 0.1 nm across—in which the entire positive charge and most of the mass is concentrated in a **nucleus** of about 1/100,000 to 1/10,000 this size. The **Bohr model** of the atom—a planetary model with **quantized** orbits—correctly describes the energy levels of the hydrogen atom but not of the more complicated atoms.

A moving particle has a **de Broglie wavelength** given by h/mv. **Quantum mechanics** uses the wave nature of particles in the *Schrödinger equation* to produce a **wave function** that gives all the information possible about the state of a particle. The square of the wave function gives the *probability distribution*, which tells the probability of finding the particle at a given location.

According to the *Heisenberg uncertainty principle*, the product of uncertainty in momentum and uncertainty in position must be greater than or equal to $h/2\pi$.

The wave function for an electron in an atom depends on a set of integer *quantum numbers*, which determine the electron's energy and other physical properties. No two electrons in an atom can have the same set of quantum numbers: The wave function for such a situation would be zero everywhere.

Questions

The Rutherford-Bohr Model of the Atom

1. Explain why a "plum pudding" model of the atom would not result in alpha particles being scattered directly backward from an atom, whereas a nuclear model would.

2. How would the light emitted from hydrogen be different if the energy levels of hydrogen atoms were not quantized?

3. What is the relation between the colors you see at night from a lighted Coca Cola sign and the energy levels of the atoms of the gases in the lights?

4. A hydrogen atom de-excites from level $n = 4$ to level $n = 3$ and then to $n = 2$. How does the energy of the photons in these two transitions compare with that if there were a direct transition from level 4 to level 2? Does that mean that the wavelength in the second case equals the sum of the wavelengths in the first case?

The Quantum Theory of the Atom

5. What similarities do you see between the requirement that the Bohr orbit be a whole number of de Broglie wavelengths and the condition for standing waves on a string?

6. As the speed of an electron increases, does its de Broglie wavelength increase or decrease?

7. To observe the smallest possible object using an electron microscope, would you want the speed of the electrons to be high or low? Why?

8. If you used a proton microscope instead of an electron microscope, but kept all other conditions such as focusing ability and particle speed the same, how would this effect the size of the object that you could "see"? (The proton mass is about 1800 times that of the electron.)

9. Compare the de Broglie wavelengths of electrons, protons, and neutrons all moving at the same speed ($m_p \simeq m_n \simeq 1800\ m_e$).

10. Compare the probability of finding the electron of a hydrogen atom in a small volume near the nucleus if it is in its ground state versus in the excited state shown in Figure 24.6(b).

11. What is wrong with the following statement? "In the Bohr model the electron moves in a smooth circular path, while in reality the electron wiggles up and down in a wavy pattern as it goes around; therefore, since it's moving so fast, it looks as if it's smeared out in a cloud-like distribution."

12. When you're playing pool, does the Heisenberg uncertainty principle influence your ability to get the eight ball into the side pocket? Why so or why not?

13. True or false: The reason that we cannot simultaneously know an electron's momentum and position precisely is that no instruments have been invented that can accurately measure them at the same time.

Quantum Mechanics: The Foundation of Chemistry

14. What would happen to the particle-in-a-box if it experienced friction? Since the energies of the hydrogen atom stay fixed over time, what does that tell you about whether or not the hydrogen atom electron experiences any friction in its motion?

15. Figure 24.9 is a graph of the *square* of the first three wave functions for the particle-in-a-box of Figure 24.7. What are the points at which the particle is most likely to be found? Are there

points, other than at the ends, where you would never find the particle. Where are they?

16. Draw an imaginary vertical line through the center of Figure 24.6(a) and 24.6(b). Based on the density of the "electron cloud" shown, sketch an approximate graph for each of these states of how the square of the wave function varies with distance along the vertical lines you have drawn.

17. What would happen to the wave function for a carbon atom if two electrons had the same four quantum numbers? Can a carbon atom exist with two electrons having the same set of four quantum numbers?

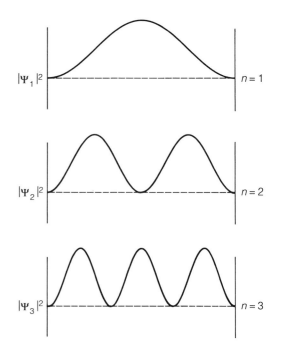

Figure 24.9 The square of the wave functions of Figure 24.7 for a particle-in-a-box.

Problems

Note: The mass of the electron (needed in several of these probelms) is 9.11×10^{-31} kg.

1. If the nucleus of a sodium atom is about 1/50,000 the size of the overall atom, and you made a scale model of the atom using a golf ball (diameter = 4.0 cm) as a nucleus, what would the diameter of the atom be?

   Answer: 2000 m ($1\frac{1}{4}$ mi)

2. The yellow light you see from a mercury vapor lamp when you view it through a prism has a wavelength of 577 nm. What is the difference in energy of a mercury atom before and after a photon of such light is emitted?

   Answer: 2.15 eV

3. Calculate the energies and wavelengths in Question 4 and quantitatively verify the answers to that question.

4. Use the electron orbit radii given in Figure 24.3 and the fact that the angular momentum, mvr, can only have values given by $nh/2\pi$ to calculate the velocity an electron would have in the $n = 2$ level for the Bohr model.

   Answer: 1.1×10^6 m/s

5. Calculate the de Broglie wavelength of an electron moving at 1/100 the speed of light—3.0 Mm/s.

   Answer: 0.24 nm

6. If the magnitude of a particle's wave function inside a small cube is twice what it is inside an identical second cube, what is the relative probability of finding the particle inside the cubes?

   Answer: 4 times as great for the first cube

7. A 0.17-kg baseball is moving at a speed of 40.2 m/s (90.0 mi/h). If its speed is known within an accuracy of 0.5 m/s, to what precision does the Heisenberg uncertainty principle limit your knowledge of its position?

   Answer: 1.2×10^{-33} m

8. A 0.17×10^{-26}-kg neutron is moving at a speed of 40.2×10^6 m/s. If its speed is known within an accuracy of 0.5×10^6 m/s, to what precision does the Heisenberg uncertainty principle limit your knowledge of its position?

   Answer: 1.2×10^{-13} m

9. If an electron is to be scattered from an atom, it needs to have a de Broglie wavelength com-

parable to the size of the atom, or smaller. What minimum speed must an electron have to give it a wavelength of 0.10 nm? What would its kinetic energy be? Compare this energy with what we calculated in Section 24.2 for a *photon* of that wavelength. Would such an electron be likely to disturb the state of the atom from which it is scattered?

Answer: 7.3 Mm/s; 150 eV; yes

10. Suppose the one-dimensional box of Figure 24.7 is 0.30-nm long (a typical atomic diameter). What are the wavelength, speed, and kinetic energy of an electron in its ground ($n = 1$) state in this box? What is the energy difference between the $n = 1$ and the $n = 2$ states?

Answer: 0.60 nm; 1.2 Mm/s; 4.2 eV; 13 eV

Home Experiment

1. Use yourself as one of the electrons in a scale model of the carbon atom. Place a golf ball on a tee in a large, open field. Walk away from it a distance of about 10,000 times its radius—about 220 paces. Imagine five of your friends to be distributed as electrons at comparable distances around the golf-ball nucleus. Look back at the nucleus. Contemplate the openness of the atom, which is typical of the atoms of your body. Are you as "solid" as you have believed yourself to be?

References for Further Reading

Hoffmann, B., *The Strange Story of the Quantum* (Dover Publications, Inc., New York, 1959). An interesting account of quantum mechanics written for the layperson.

Keller, O. L., Jr., D. C. Hoffmann, R. A. Penneman, and G. R. Choppin, "Accomplishments and Promise of Transplutonium Research," *Physics Today*, March 1984, p. 35. A review of new elements and their positions in the periodic table of elements.

O'Dwyer, J. J., *College Physics*, Second Ed. (Wadsworth Publishing Co., Belmont, California, 1984). One of many good textbooks that use algebra and trigonometry to discuss the quantum mechanics of the atom.

Roller, D. H. D., "Greek Atomic Theory," *American Journal of Physics*, March 1981, p. 206. A review of the first known atomic theories.

Ruby, L., "Modern Architects of the Periodic System," *American Journal of Physics*, Jan. 1984, p. 14. A history of the periodic classification of the elements.

THE NUCLEUS OF THE ATOM

The **nucleus** of the atom—that tiny speck at its center—is far more influential than its small size might lead you to believe. The potential for release of vast amounts of energy from atomic nuclei became evident to the whole world when "the bomb" was dropped on Hiroshima and Nagasaki. The social controversy that has long centered around "the bomb" has, in recent years, extended to the use of nuclear energy for generation of electricity. Aside from these two areas that have the spotlight, there are many noncontroversial nuclear processes that are extremely useful and important to our way of life.

Nuclear applications include medical diagnosis and treatment of cancer and various other diseases, establishing the date at which particular ancient cultures existed, detection of smoke by household smoke alarms, identification of small quantities of pollutants in our environment, and lighting of some LCD watches for nighttime viewing. Industries apply nuclear processes in hundreds of useful ways. We'll employ some of these applications to illustrate the physics of the nucleus.

In this chapter we focus on the atomic nucleus and the processes that a nucleus can undergo without outside influence. We start by looking at the components and structure of the nucleus, followed by a consideration of nuclear forces and energies. We will then consider radioactivity—the emission of radiation from the nucleus. We will also investigate the effect on biological systems of exposure to this radiation. The final section will summarize the search for the most elementary particles that compose our universe.

25.1 THE STRUCTURE OF THE NUCLEUS

The nucleus consists of two types of particles: **protons** and **neutrons**. We often use the word **nucleon** to mean *either* a neutron *or* a proton—one of the nuclear particles. Protons have a positive electrical charge equal to the negative charge of the electron, 1.60×10^{-19} C. In a neutral atom, there are as many electrons surrounding the nucleus as there are protons inside, making the *net* charge of the overall atom *zero*. The mass of the proton is 1.67×10^{-27} kg, about 1850 times that of the electron.

The other nuclear constituent, the neutron, has no electrical charge, and its mass is just slightly greater than that of the proton. Most stable nuclei have at least as many neutrons as protons.

We now see why most of the mass of the atom is in the nucleus. Nucleons have a mass of about 1850 times that of the electron, and there are usually at least twice as many nucleons as electrons. Result: About 99.97% of the mass of the atom is in the nucleus.

Neutrons and protons are actually made up of component particles called **quarks**. It takes three quarks to make up either a neutron or a proton. According to our present understanding, quarks cannot exist as free particles, but they are always bound together in combinations that form some composite particle. We can describe most of the known characteristics of nuclei by considering neutrons and protons to act as individual entities. (A lot can be known about a brick house without considering that it is made of individual bricks.) In most of this chapter we will consider nuclei as made up of neutrons and protons, and leave a discussion of quarks for Section 25.5.

We describe particular nuclei by symbols such as $^{60}_{27}$Co. *Co* is the *chemical symbol* for the element cobalt, which means the nuclei belong to cobalt atoms. The *27* is the **atomic number**: the number of protons in the nucleus. This number gives redundant information and is often left off, since cobalt is *the* element with 27 protons in its nuclei, and 27 electrons in its neutral atoms. In general, the atomic number identifies the element. Known nuclei range in their number of protons from hydrogen, with only one proton, up to element 109, which hasn't been named yet.

The *60* in $^{60}_{27}$Co is the **mass number**. This gives the number of protons *plus* neutrons in the nucleus. It's called "mass" number because it tells you the approximate value of the mass in a particular unit, the atomic mass unit (abbreviated amu). The amu, which is equivalent to 1.66×10^{-27} kg, is defined so that the mass of a nucleon is *approximately* 1 amu, and that of any nucleus is *approximately* equal in value to its mass number.* Actually, the neutron mass is 1.008665 amu, and that of ^{60}Co is 59.933813 amu. We get the number of neutrons by subtracting the

---

*The amu is defined as $\frac{1}{12}$ the mass of the carbon-12 atom. Thus ^{12}C is the only atom with mass in amu exactly equal to its mass number.

atomic number from the mass number. Thus, there are 60 − 27, or 33 neutrons in ^{60}Co.

A particular nuclear species has the name **nuclide**. Thus, $^{235}_{92}$U is a nuclide with 92 protons and 143 neutrons (U = uranium). Nuclides of the same element can have different numbers of neutrons. For example, $^{88}_{38}$Sr has 50 neutrons, $^{90}_{38}$Sr has 52 neutrons (Sr = strontium). Nuclides of the same element having different numbers of neutrons are called **iso-topes**. Hydrogen, the simplest element has three isotopes: $^{1}_{1}$H; $^{2}_{1}$H; and $^{3}_{1}$H. The ^{1}H nucleus is just a single proton. The ^{2}H and ^{3}H nuclei have the name deuteron and triton, and bulk quantities of these materials are called deuterium and tritium. Isotopes of other elements do not have special names.

· ·

QUESTION   How many protons and how many neutrons are in each of the nuclides $^{2}_{1}$H, $^{3}_{1}$H, $^{4}_{2}$He, $^{35}_{17}$Cl, $^{37}_{17}$Cl, $^{137}_{55}$Cs, $^{238}_{92}$U, and $^{239}_{94}$Pu? (H = hydrogen, He = helium, Cl = chlorine, Cs = cesium, U = Uranium, Pu = plutonium)

25.2 **THE NUCLEAR FORCE AND NUCLEAR ENERGIES**

The nucleons in a nucleus form a cluster somewhat like eggs in a basket (Fig. 25.1). That is, they don't overlap, but each nucleon takes up roughly the same amount of space. This means that a ^{234}Th nucleus occupies about 78 times the volume of a ^{3}He nucleus.

Nucleons are typically about 1 to 2 fm apart in a nucleus. You can interpret "fm" to mean either femtometer (10^{-15} m) or *fermi*, the name often used for this unit in honor of Enrico Fermi, a renowned nulcear physicist about whom we'll have more to say later. Because it's about the right size, a fermi is a handy unit for talking about nuclear dimensions.

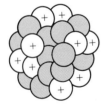

Figure 25.1 In an approximation that is about as valid as the Bohr model of the atom, you can think of the nucleus as a cluster of protons (labeled "+") and neutrons.

What do you suppose holds this cluster together? Remember that the electrons are held in the *atom* by the electrical force between the negative charge of the electron and the positive charge of the nucleus. But the *nuclear* particles are either uncharged or positively charged. The electrical force is either zero (between neutron and neutron or between neutron and proton) or repulsive (between proton and proton). As a matter of fact, if you use Coulomb's law (Chapter 14) and calculate the *electrical* force between two protons 1.5 fm apart, you get 100 N (22 lb). That doesn't seem like much force, until you think about the mass of the particles on which the force is acting. The following exercise may put this force into perspective. You can see that there is a tremendous electrical repulsion in a heavy nucleus like uranium, with 92 protons.

EXERCISE   Use Newton's second law of motion ($F = ma$) to calculate the acceleration of a proton free to move, and acted on by a net force of 100 N.

  Answer: 7.0×10^{28} m/s² (7,000,000,000,000,000,000,000,000,000,000 times your acceleration if you fall off your rocker).

What about the other ordinary force—the gravitational force? Could it hold the nucleus together? It would act to attract the nucleons to each other. But if you calculate its strength between our protons 1.5 fm apart, you get a force of only 8×10^{-35} N. That would do about as much good in overcoming the electrical repulsion between protons as your trying to hold back a tidal wave with a surfboard.

Obviously, there must be *some* very strong force that acts between nucleons, attracting them to each other enough to overcome the electrical force. This force, called the **nuclear force** or sometimes just the **strong force**, exists only between nuclear particles. Experiments show that this force is the same between any two kinds of nucleons, and that it doesn't act if the nucleons are separated by more than just a few fermis. For example, if a proton, or a group of nucleons including at least one proton, gets separated from the nucleus by a small amount, it will be repelled by the electrical force and move off at an extremely high speed.

This nuclear force is one of nature's fundamental kinds of force, and it is distinct in nature from the other two we've discussed in some detail: the electomagnetic and the gravitational forces.

One of the functions of the neutrons is to act as a sort of nuclear "glue." Since the nuclear force doesn't care what kind of nucleon it acts on, the neutrons can provide some added nuclear force to overcome the electrical repulsion. Figure 25.2 is a plot of the number of neutrons versus the number of protons for the nuclei that exist stably in nature. As the graph shows, for light nuclei up to about the calcium region (atomic number 20), the number of neutrons and protons is nearly the same. For heavier nuclei, the percentage of neutrons steadily increases to compensate for the higher electrical repulsion.

Since nuclear forces are at least strong enough to overcome the electrical repulsion between protons, we can get a *general feel* for the minimum strength of the nuclear force. From our earlier calculation, it must be at least about 100 N between nucleons. Let's compare this with the force holding an electron in orbit around the nucleus. A Coulomb's-law calculation for an electron in the first Bohr orbit of hydrogen gives the force to be about 10^{-7} N, one-billionth that for the protons in the nucleus. That gives us a hint at why nuclear energies are so much larger than chemical energies. The burning of coal—a chemical reaction—involves the rearrangement of orbital electrons as the atoms form different combinations. The nuclear reactions that occur in the reactor of a nuclear

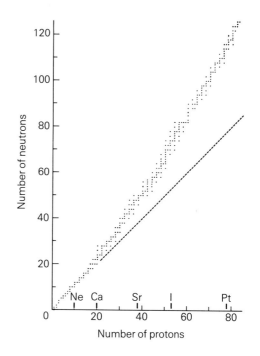

Figure 25.2 Number of neutrons versus number of protons in stable nuclei. The approximate location of some of the elements is shown (in order, neon, calcium, strontium, iodine, platinum). The dashed line shows where the number of neutrons equals the number of protons.

power plant involve the rearrangement of nuclear particles. Considering the vastly stronger forces holding the nuclear particles in place, you might expect vastly more energy to be involved in their rearrangement. In fact, single nuclear reactions typically involve about a million times more energy than single chemical reactions.

There's another way we can get an indication of the amount of energy associated with nuclear particles. Recall our particle-in-a-box illustration from Section 24.3. Let's calculate the energy such a particle can have if we let the box be about the size of a medium-sized nucleus and let the particle have the mass of the neutron (or proton). We'll use a box size of 12 fm, about the diameter of a nucleus of silver. Then the de Broglie wavelength for the ground state is 24 fm. From $\lambda = h/mv$, we have

$$v = \frac{h}{m\lambda} = \frac{(6.62 \times 10^{-34} \text{ J·s})}{(1.67 \times 10^{-27} \text{ kg})(24 \times 10^{-15} \text{ m})} = 1.65 \times 10^7 \text{ m/s}$$

Then the kinetic energy of the nucleon is

$$KE = \tfrac{1}{2}mv^2 = \tfrac{1}{2}(1.67 \times 10^{-27} \text{ kg})(1.65 \times 10^7 \text{ m/s})^2$$

$$= 2.27 \times 10^{-13} \text{ J}$$

$$= (2.27 \times 10^{-13} \text{ J})/1.60 \times 10^{-19} \text{ J/eV}$$

$$= 1.4 \times 10^6 \text{ eV}$$

If you compare this with the result of Problem 10 of Chapter 22, which used typical *atomic* dimensions and the electron mass, you see that the energy for the nuclear-type quantities is about a *million times* as much!

Certainly a neutron or proton in a *nucleus* is not the same as in a hypothetical one-dimensional box. Nevertheless, using this very simple situation and the wave nature of the particles involved, we get a measure of the relative energies that is the right order of magnitude. In the next chapter we'll see how this much higher nuclear energy can be harnessed.

25.3 RADIOACTIVITY

It's possible *in principle* to combine *any* number of neutrons with *any* number of protons. However, that does *not* mean that *any* combination of nucleons will form a stable, or near-stable, nucleus.

Figure 25.2 shows the combinations that we know to be stable. If by some means, you managed to create a nucleus with relative neutron and proton numbers far from this "stability line," it would simply fly apart. But what if we have a nucleus with only slightly too many or too few neutrons for its number of protons: a nucleus that is *almost* stable? Then, that nucleus merely *decays* to a more stable arrangement by the emission of some particle. We call this emission of particles **radioactivity**.

You'll notice that uranium, and the other heavy elements with atomic numbers higher than 83, are not on the graph of Figure 25.2. That's because there are *no* known *stable* nuclides with atomic numbers higher than 83. All heavier elements are radioactive.

The term *radioactivity* may seem bothersome because its use to describe particle emission doesn't seem to fit with the word *radio*. The dictionary, though, defines the prefix *radio-* to indicate the emission of radiation, and defines *radiation* as the emission and propagation of waves *or* particles. Radiation, then, can include both the radio waves broadcast from your favorite FM station *and* the particles emitted from uranium nuclei. (You already know that electromagnetic waves sometimes act as particles, and particles sometimes act as waves.) Radioactivity *is* just what its name implies.

Types of Radioactivity

Radioactivity was first discovered in uranium by Henri Becquerel in 1896. Within a couple of years, Marie and Pierre Curie discovered the radioactive elements thorium, polonium, and radium.

By making the radiation pass through a magnetic field, early investigators showed the radiation to be of three types. They labeled them α (*alpha*), β (*beta*), and γ (*gamma*) rays. The direction of the bending of the

Before decay   After decay

$^{226}_{88}$Ra $^{222}_{86}$Rn $^{4}_{2}\alpha$

paths of α and β rays as they passed through a magnetic field indicated that they were charged positively and negatively, respectively. Since the γ rays passed through undeflected, they had to be uncharged. More detailed studies showed that the α rays were actually $^{4}_{2}$He (helium) nuclei: clusters of two neutrons and two protons very tightly bound together. The β rays were *electrons*, and γ rays were *photons* of electromagnetic radiation. We'll look at each type decay in more detail.

α **decay.** If alpha decay involves the emission of two neutrons and two protons, the nucleus doing the emitting is obviously *transmuted* into a nucleus of a different nuclide. Consider the α decay of radium-226, symbolized in Figure 25.3. A ^{226}Ra nucleus spontaneously breaks up into an α particle and a nucleus, with two less neutrons and two less protons: radon-222. We write this in a reaction equation, somewhat as chemists write chemical reactions:

$$^{226}_{88}\text{Ra} \rightarrow ^{222}_{86}\text{Rn} + ^{4}_{2}\alpha$$

Sometimes people write the $^{4}_{2}\alpha$ as $^{4}_{2}$He; the two symbols mean the same thing. After the decay, the α moves off at high speed, being repelled by the electrical force between it and the nucleus. But the principle of momentum conservation requires the ^{222}Rn to recoil in the opposite direction just as a rifle recoils—kicks—when you fire it. The relative speeds are inversely proportional to the particle masses.

Radioactive decay reactions have to pay their own way; even there you cannot get something for nothing. That is, the principles of conservation of energy and momentum apply. In addition, there are two more conservation laws that apply in any nuclear interaction:

1. *The total number of nucleons remains constant.*

2. *The total charge remains constant.*

The reaction equation we wrote includes these laws. The first law makes the total number of nucleons on the right, 222 + 4, equal the total on the left, 226. The conservation of charge is taken care of in the protons: the total number on the right, 86 + 2, equals the total number on the left, 88. From these two laws, you can find out the third member for any reaction in which you know two of the nuclides involved. For example, if you know that some nuclide α-decays to $^{209}_{82}$Pb (lead), the initial mass number has to be 209 + 4, and the initial proton number (atomic number) has to be 82 + 2. The element with atomic number 84 is polonium. Thus, the reaction is

$$^{213}_{84}\text{Po} \rightarrow {}^{209}_{82}\text{Pb} + {}^{4}_{2}\text{He}$$

. . . .

QUESTION   Uranium-234, $^{234}_{92}\text{U}$, α-decays to Th, which α-decays to Ra, which α-decays to Rn, which α-decays to Po, which α-decays to $^{214}_{82}\text{Pb}$. Write the reaction equation for each reaction in this decay "chain."

We often call the initial nuclide before decay the **parent**, the one left after decay, the **daughter**. This terminology applies to all types of radioactivity, not just α decay.

β **decay.**   Did you notice anything strange about the definition of beta rays—*electrons* emitted from the nucleus? There are *no* electrons in the nucleus! That sounds like picking potatoes from an orange tree.

What effectively happens is that a neutron in the nucleus converts to a proton, kicking out an electron in the process. In reaction terms,

$$^{1}_{0}\text{n} \rightarrow {}^{1}_{1}\text{p} + {}^{0}_{-1}\text{e}$$

Notice that the number of nucleons is the same—1—on both sides of the equation, and the total charge is 0 on both sides. Since the electron has a *negative* charge, we write its "atomic" number, or *charge number*, as -1. Since it is not a nucleon, its mass number is 0.

This last equation, as it stands, is *not* correct. It violates both the principles of conservation of energy and conservation of momentum. Nuclear physicists observed in the late 1920s that if they measured the energy and momentum of the particles before and after beta decay, some of each seemed to be lost. Not wanting to abandon these conservation principles, Wolfgang Pauli postulated in 1930 that another particle was emitted during beta decay that carried off the missing energy and momentum. After a period of skepticism, physicists accepted the existence of this particle, called the *neutrino*, for nearly 26 years before it was confirmed experimentally. Thus, our neutron decay equation including the neutrino, $\bar{\nu}$, becomes

$$^{1}_{0}n \rightarrow {}^{1}_{1}p + {}^{0}_{-1}e + \bar{\nu}$$

For our purposes now and for most current applications of nuclear physics, the neutrino emission is immaterial. Therefore, we will not write it in our β-decay equations. In Section 25.5 we'll see how the neutrino fits into the overall scheme of "fundamental" particles, and we'll discover that the one occurring in ordinary beta decay is actually an *antineutrino*—

that's the reason for the bar over the Greek letter *nu* that symbolizes the neutrino.

One naturally occurring β emitter is lead-212, $^{212}_{82}\text{Pb}$. Its reaction equation is

$$^{212}_{82}\text{Pb} \rightarrow {}^{212}_{83}\text{Bi} + {}^{0}_{-1}\text{e}$$

Notice that, since a *negative* charge is given off, the charge of the nucleus *increases* by 1 unit: the nucleus becomes a bismuth-212 nucleus with the same *total* number of nucleons. But the daughter has one less neutron and one more proton than the parent.

You might wonder what could cause a neutron to convert to a proton with the accompanying *creation* of an electron. To account for this conversion, there is a fourth *general* type of force that is distinct from the other three we've discussed: the gravitational, the electromagnetic, and the strong (nuclear) interactions. This force, that goes by the name **weak interaction**, is responsible for the β-decay process. So far as we know, these four general types of forces account for every specific force that can exist in the universe.

Physicist proposes strange new particle

TÜBINGEN, GERMANY (December 7, 1930)—A group of physicists gathered here for a conference heard today a proposal for the existence of a strange, new particle, as yet unobserved. Wolfgang Pauli, a noted Austrian physicist, made the proposal in a letter from Zürich, Switzerland, to the conferees.

Pauli suggests the particle because a certain kind of radioactivity, called beta decay, seems to violate the principle of conservation of energy and certain other accepted conservation principles. When an atom beta-decays, it emits a particle called an electron, and in the process becomes an entirely different kind of atom. The problem is this: When experimenters measure the total energy of the electron and new atom, it adds up to less than the energy of the original atom. Rather than abandon the hallowed principle of energy conservation, Pauli suggested that another particle is emitted along with the electron,

and that it carries away the rest of the energy. He admits that the particle would need to have some strange properties: very small mass, probably less than that of the electron; no electrical charge, since it otherwise would have been detected already; and very high speeds. The name "neutrino" has been suggested for this proposed particle.

Most of the physicists at the conference remain highly skeptical of Pauli's proposal. Although they do not like the idea that energy might not be conserved, they find Pauli's suggestion outlandish and highly questionable.

Pauli himself was unable to attend the conference because of social commitments. He said in his letter, "Unfortunately, I cannot personally appear in Tübingen, since I am indispensable here on account of a ball taking place in Zürich on the night from 6 to 7 of December."

Scientists confirm existence of neutrino

AIKEN, SOUTH CAROLINA (July 20, 1956)—American physicists reported today in the journal *Science* that they have confirmed the existence of the neutrino, the particle proposed by Wolfgang Pauli nearly 26 years ago. The scientists, Clyde Cowan and Frederick Reines and their coworkers from the University of California's Los Alamos Scientific Laboratory in New Mexico, detected a small fraction of the billions and billions of neutrinos given off during the beta-decay processes inside a nuclear reactor. They carried out the experiment at the U.S. Atomic Energy Commission's Savanah River Laboratory in Aiken, South Carolina.

Pauli, who won the Nobel Prize in 1945 for his discovery of what is now called the Pauli exclusion principle, proposed in 1930 that the neutrino must exist. Otherwise, beta-decay processes would violate the principle of the conservation of energy, as well as other conservation laws of physics. Even though this elusive particle interacts so weakly with matter that it had always escaped detection, physicists have accepted its existence for some time. Nevertheless, directly observing it has reaffirmed their faith in the energy conservation principle.

γ decay. We said that gamma rays are photons of electromagnetic radiation. They have no electric charge and contain no nucleons. A typical reaction is

$$^{209}_{82}\text{Pb}^* \rightarrow \,^{209}_{82}\text{Pb} + \,^{0}_{0}\gamma$$

The nuclide does not change in γ decay. Rather, this is the way a nucleus in an *excited* state (indicated by the symbol *) can get rid of some excess energy. The process is identical in principle to the way the orbital electrons of an atom go to a lower energy state by the emission of light or x-ray photons. The wavelength regions of γ rays and x rays overlap, meaning that you can't tell a γ ray from an x ray. Their only difference is in where they come from: an x ray from electron transitions in the atom, a γ ray from the nucleus.

When a nucleus decays by α or β emission, this process is almost always followed by γ decay because the final nucleus in α or β decay is usually left in an energy state higher than its ground state. It gets rid of this excess energy by γ decay.

Decay Rate and Half-life

Imagine that your doctor suspects that you have thyroid cancer. The function of your thyroid gland is to remove iodine from your blood and use it in the manufacture of thyroid hormones. The food you eat each

(a)

(b) (c)

Figure 25.4 (a) A thyroid scanner in operation. (b) A normal thyroid scan. The darkness is proportional to the activity from that spot. (c) An abnormal thyroid scan.

day probably contains an average of about 150 μg of iodine. Your thyroid doesn't care which isotopes of iodine it gets; it just takes whatever iodine comes down the pipe. However, cancerous regions of the thyroid usually have a lower-than-normal uptake of iodine.

Among the other tests your doctor might perform to check out his or her suspicion, there is a fair chance that you will be asked to ingest something containing a small amount of an isotope of iodine that is radioactive—maybe ^{131}I. After waiting a specific period for your thyroid to perform its function, the doctor will put you in front of a scanner (Fig. 25.4-a) that can detect the *number and direction* of γ rays from the ^{131}I. If your thyroid is normal, the map of the γ activity—the *scan*—will look something like Figure 25.4(b). The darkness of each spot is proportional to the amount of γ radiation coming from that direction. The relatively even distribution in (b) indicates no regions of excessively high or low iodine concentration, such as those in Figure 25.4(c).

Now look at the nuclear physics of the procedure we've just described. First of all, even though ^{131}I is radioactive, not all the nuclei decay immediately. Instead, they decay at some given rate; that is, some fraction of the ^{131}I nuclei decay each second. *The total rate of decay is proportional to the number of radioactive nuclei present.* The scanning technique depends heavily on this fact.

The number of nuclei of a substance decaying per unit of time is called its **activity**. The customary unit used to measure activity is the *curie* (Ci), defined as 3.7×10^{10} decays per s.* The SI unit of activity, the becquerel (Bq), defined as the number of decays per second, hasn't yet gained widespread use. The usual dose of ^{131}I used for a thyroid scan is from 50 to 100 μCi: 2 to 4 MBq ($\mu = 10^{-6}$, M $= 10^6$).

• •

EXERCISE How many γ rays are emitted from a 3.7-MBq (100-μCi) ^{131}I sample during the 20 min of a typical thyroid scan? Note: The scanner detects only a small fraction of this number because not all the ^{131}I goes to the thyroid, and the scanner itself intercepts only a tiny portion of the γs emitted equally in *all* directions   *Answer*: 4.4 billion

---

It's interesting that, even though radioactive ^{131}I is used in the scanning technique, the γ rays do not actually come from I nuclei. The I nuclei β-decay through the reaction

$$^{131}_{53}\text{I} \rightarrow \, ^{131}_{54}\text{Xe} + \, _{-1}^{0}\text{e}$$

leaving the xenon nuclei in an excited state. These ^{131}Xe nuclei give up their excess energy by emitting the observed γ rays.

Half-life. As ^{131}I converts to ^{131}Xe, the supply of ^{131}I is depleted. Therefore, the activity must decrease proportionately. Suppose the technician in charge of radioisotope storage in a medical clinic measures the activity of his bottle of ^{131}I and finds he has 20 MBq (5.4 mCi). He comes back 8 days later, and finds he has only 10 MBq (2.7 mCi). Then 8 days later 5 MBq, and in another 8 days 2.5 MBq. A graph of this activity (Fig.

The curie, named after the discoverers of radioactivity in radium, was originally defined as the activity of 1 g of ^{226}Ra.

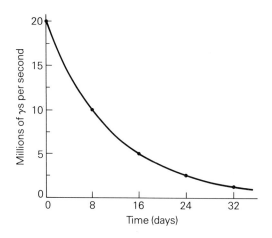

Figure 25.5 The activity of an ^{131}I sample that initially has a decay rate of 20 MBq.

25.5) shows that the decay rate after *any* 8-day period is $\frac{1}{2}$ what it was at the beginning of that period.

Suppose the technician does the same with his supply of ^{123}I, also sometimes used for thyroid studies. He would get an identical looking graph, except that the activity would cut in half every 13 hours.

This change in decay rate is typical of *all* radioactive nuclides. There is a time period for each nuclide, called its **half-life**, during which time its decay rate decreases by 50%. Remember that the decay rate is proportional to the amount of the nuclide present. Therefore, *the half-life is the time needed for half the nuclei in a given sample of a radioactive nuclide to decay*. The half-life varies from nuclide to nuclide.

• •

EXERCISE

If a patient receives a 3.7-MBq (100-μCi) dose of ^{131}I, how long will it take for the activity in her body to decrease to 0.037 MBq (1 μCi)? How long if she gets the same dose of ^{123}I? Can you think of an advantage of using ^{123}I rather than ^{131}I? Unfortunately, it's not as readily available.

Answer: 56 days; 3.8 days (to nearest half-life)

The measured half-lives of radioactive nuclides vary from extremely small fractions of a second to billions of years: 4.5 billion for ^{238}U and 14 billion for ^{232}Th. Table 25.1 gives the half-lives of a few selected nuclides, most of which have some practical application.

When radioactive nuclides, often called **radioisotopes**, are used in following the path of some element such as iodine in the body, they are called **tracers**. The name comes from the fact that you can *trace* their paths.

Table 25.1 Half-lives of selected nuclides

Nuclide	Half-life	Nuclide	Half-life
neutron	11 min	strontium-90	28.9 yr
tritium (^{3}H)	12.3 yr	iodine-131	8.1 day
beryllium-8	10^{-16} s	radium-226	1620 yr
carbon-14	5730 yr	uranium-235	7.1×10^6 yr
sodium-24	15.0 h	plutonium-239	24,400 yr
cobalt-60	5.3 yr	americium-241	433 yr

CAUTION

RADIOACTIVE MATERIALS

Figure 25.6 The symbol used universally to identify the presence of a radioactive source or a general radiation area.

An interesting industrial use of tracers is done by soap manufacturers. They like to know how effective their soap is in getting dirt out. One way they measure this effectiveness is by adding a radioactive tracer to the dirt and then checking the activity of the dirty water.

Do you have a smoke detector (alarm) in your house? Take the cover off and look inside. If it's the "ionization" type, the most popular, you'll see the "radiation" symbol inside (Fig. 25.6). This symbol is used to identify either the presence of a radioactive source or an area in which you may be exposed to radiation. The smoke detector contains a weak radioactive source, usually Americium-241, located as shown schematically in Figure 25.7. The α particles from the source ionize the air; that is, they knock out electrons leaving positive ions and free electrons. The battery then causes a weak electric current through the air. When smoke enters the measurement chamber, the ions attach themselves to the smoke particles. Their very slow movement reduces the current, causing the alarm to sound. Some brands advertize *dual* ionization chambers. The second "reference" chamber lets no smoke in, but compensates for changes in temperature, air pressure, and humidity, thus reducing false alarms.

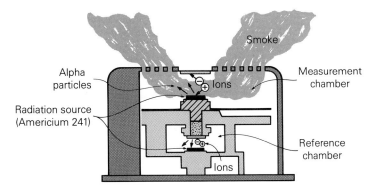

Figure 25.7 An ionization-type smoke detector uses a radioactive source.

25.4 BIOLOGICAL EFFECTS OF RADIATION

When radiation passes through matter, it knocks out electrons, thereby forming *electron-ion pairs*. This **ionization** process allows the detection of radiation in instruments such as geiger counters. But radiation also ionizes the atoms of biological cells that it traverses. This ionization can displace or chemically change the molecules of a cell, thereby doing damage to it. Radiation effects may harm the person exposed, or if the damage is to genetic cells, the offspring of the exposed person may suffer the consequences.

We are using the word *radiation* here to mean *ionizing* radiation such as x rays, γ rays, electrons, and nuclear particles. Heavy exposure to other types of radiation—infrared and ultraviolet, for example—may do surface damage, but these forms of radiation do not penetrate the body and produce ionization deep inside.

Radiation Exposure

The unit usually used for radiation **dose** is the rem, which takes into account both the energy deposited per kilogram of exposed matter and the differing effects of different types of radiation. Ordinary doses are often given in mrem (0.001 rem).

Most of the radiation exposure the average person gets is from natural unavoidable sources. These include cosmic radiation from outer space and naturally radioactive nuclides in the earth. These quantities vary drastically from place to place. For example, people in higher altitudes or who fly in airplanes get more cosmic rays; people in brick houses get more exposure than those in wooden houses because of the natural radioactivity in the sand of the bricks. Table 25.2 shows how you can compute your own radiation dose. The column on the right tells the percentage of the *average* annual dose in the United States attributable to each of the sources. If you receive medical x rays, your exposure might be twice this average annual dose of 148 mrem.

Notice that 1 mrem/yr is equivalent to moving to an elevation 100 ft higher; increasing your diet by 4%; or taking a 4- to 5-day vacation in the Sierra Nevada. The people living within a 10-mile radius of the Three Mile Island nuclear power plant near Harrisburg, Pa., received an average dose of about 10 mrem from the highly publicized accident there in 1979.

How much radiation exposure is too much? Federal regulations stipulate that individuals working with radiation may receive no more than $1\frac{1}{4}$ rems per calendar quarter. This limit is somewhat arbitrary, but no *measurable* ill effects are observed at this level. However, it is known that *any* amount of radiation *can* be harmful. Therefore, any *unnecessary* exposure is *too much*. In the case of voluntary exposure, such as medical x rays, the doctor weighs the potential benefits against the risks.

Table 25.2 Compute your own radiation dose[a]
Fill in the blanks with the numbers appropriate to your situation.

	Common Source of Radiation	Your Annual Inventory (mrem)	Percent[b]
WHERE YOU LIVE	Location: Cosmic radiation at sea level Elevation: Add 1 for every 100 feet of elevation Typical elevations (feet): Pittsburgh 1200; Minneapolis 815; Atlanta 1050; Las Vegas 2000; Denver 5280; St. Louis 455; Salt Lake City 4400; Dallas 435; Bangor 20; Spokane 1890; Chicago 595. (Coastal cities are assumed to be zero, or sea level.)	44 ——	29.7
	House construction (based on $\frac{3}{4}$ time indoors) Brick 45 Stone 50 Wood 35 Concrete 45	——	27.0
	Ground (U.S. average based on $\frac{1}{4}$ of the time outdoors)	15	10.1
WHAT YOU EAT, DRINK, AND BREATHE	Water Food } U.S. average Air	25	16.9
	Weapons test fallout	4	2.7
HOW YOU LIVE	X-ray diagnosis: Chest x ray ——— × 9 Gastrointestinal tract x ray ——— × 210	——	13.5
	Jet air travel: Number of 6000-mile flights ——— × 4.	——	0.1
	Television viewing: Number of hours per day ——— × 0.15.	——	0.1
HOW CLOSE YOU LIVE TO A NUCLEAR PLANT	At site boundary: Annual average number of hours per day ——— × 0.2 One mile away: Annual average number of hours per day ——— × 0.02 Five miles away: Annual average number of hours per day ——— × 0.002 Over 5 miles away: None	—— —— ——	Less Than 0.1%
	Total mrem	——	100

[a]From *Nuclear Power and the Environment—Questions and Answers* (American Nuclear Society, La Grange Park, Ill., 1976).

[b]Percentage of the average annual U.S. dose of 148 mrem.

Benefits of Radiation Exposure

There are numerous benefits to radiation "damage." For example, certain foods are exposed to high doses of radiation to kill microorgaisms and extend shelf life. Some medical supplies are sterilized by exposure to radiation.

You no doubt know of people given radiation treatment for cancer. The cancerous tissue is exposed to heavy doses of high-energy radiation in order to destroy the cancerous cells. In some cases this radiation is administered from "within," and in other cases, from outside the body.

We discussed earlier the use of ^{131}I in *detecting* thyroid disorders. In some cases of thyroid cancer, the thyroid still has the ability to concentrate iodine. The doctor might then administer large doses—up to 7 GBq (200 mCi)—of radioactive iodine. The thyroid will then concentrate the iodine there and hopefully destroy the cancerous tissue by radiation exposure.

The more common radiation treatment uses a source outside the body, often referred to as "cobalt" treatment because most hospitals use the γ rays from ^{60}Co for this purpose. The source is kept in a lead box with a small opening so that the γ rays emerge in a narrow beam. The radiologist aims the beam at the tumor. Unfortunately the γ rays damage healthy cells as well as diseased ones. For this reason the beams are often brought in from different directions so that they intersect at the position of the tumor (Fig. 25.8). This procedure maximizes the energy deposited there and minimizes that delivered to surrounding healthy cells.

Figure 25.8 In cancer therapy, the γ-ray beam is directed from different angles to concentrate the radiation at the point of the tumor.

25.5 ELEMENTARY(?) PARTICLES

A major thrust of natural science and philosophy over the centuries has been the search for the fundamental building blocks of matter. The extensive amount of time, energy, and money devoted to that search during the past century has uncovered some rather interesting components in the structure of our universe. Let's look at some of the highlights in that journey.

Three-fourths of the way through the nineteenth century, the physical world seemed pretty well understood. The existence of the atom was accepted, and the various types of atoms were categorized into a periodic table that formed the basis for an understanding of chemistry. The atom was assumed to be indivisible and the ultimate constituent of all matter. However, during a period of three years in the last decade of that century, the discoveries of x rays, radioactivity, and the electron dropped the bottom out of that neat package, and no one has yet claimed to put it back on.

Rutherford's experiments around 1911 (Section 24.1), showing the atom to consist of a tiny heavy nucleus surrounded by electrons, paved the way for the beginning of nuclear physics—the study of the nucleus.

After the discovery of the neutron in 1932, our understanding of the nucleus as a collection of neutrons and protons evolved over the next few decades. This understanding spawned many practical applications of nuclear physics, some of which we have mentioned in this chapter.

While nuclear physicists were concentrating on the nucleus, another field called "particle" or "high-energy" physics sprang up. In 1935 Hideki Yakawa predicted the existence of a particle that, according to his theory, should have a major role in conveying the nuclear force between nucleons. He predicted its mass to be about 270 times that of the electron. A year later a new particle, now called the *muon*, was discovered. It had almost the right amount of mass, but didn't respond to the nuclear force as Yukawa had predicted. It merely behaved like a heavy electron. Yukawa's particle, now called the *pion*, was discovered 12 years after its prediction. These were but the first two of a long list of particles discovered by high-energy physicists either in cosmic rays or by bombarding matter with high-energy protons or electrons—now up to hundreds of *billion-electron-volts*—and observing the particles produced out of this energy. The list of strange-sounding names now totals more than 200.

And there's another side to this record. For every known particle, there is a corresponding *antiparticle*. The observations of antimatter began with the discovery in 1932 of the *positron*, which is just like the electron, except that it has positive charge. For example, some nuclei beta-decay by positron emission. Neutrons, protons, muons, pions, and every other known particle have corresponding antiparticles with the same mass but opposite electric charge and magnetic properties. These particles have the rather unusual characteristic that, whenever an antiparticle meets up with one of its mate particles, they both annihilate, converting their entire mass to energy. Antiparticles, once produced, live very short lifetimes because of the abundance of ordinary particles with which to annihilate.

Caging the Zoo Animals—Quarks

The vast array of particles that had appeared by the late 1950s has often been referred to as a particle "zoo." It was clear that many of these particles play a role in the nuclear force, but the underlying function of so many particles was most mysterious. A scheme called "the eightfold way" (proposed by Murray Gell-Mann in 1961) successfully categorized the known particles in a way similar to the categorizing of atoms by the periodic table. This scheme led to the prediction and later detection of previously unobserved particles. But that classification of particles did not explain their existence.

The orderliness of the eightfold way gave a hint that there may be another "layer" of matter—that there might be more fundamental particles of which this array of particles is made. Perhaps by combining these fundamental particles in different ways, each of the known particles could be formed—in analogy to the way different atoms are formed by different combinations of electrons, protons, and neutrons. In 1964 Gell-Mann observed that all the particles known to interact strongly with nuclei could be built from just *three* types of particles that he called **quarks**.

The term used to distinguish between quark types was *flavor*, with the three original flavors being **up** quarks, **down** quarks, and **strange** quarks. The electric charge of the up quark needed to be $+\frac{2}{3}e$—$\frac{2}{3}$ the charge of the electron and positive in sign. The charges of the down and strange quarks were both proposed to be $-\frac{1}{3}e$. Neutrons and protons are formed using *only* up and down quarks, as illustrated in Figure 25.9.

(a) *Quark Flavors:* (U)– up quark (charge $+\frac{2}{3}e$)

(D)– down quark (charge $-\frac{1}{3}e$)

(S)– strange quark (charge $-\frac{1}{3}e$)

(b) Proton = (U)+(U)+(D)

(charge $= \frac{2}{3}e + \frac{2}{3}e - \frac{1}{3}e = e$)

(c) Neutron = (U)+(D)+(D)

(charge $= \frac{2}{3}e - \frac{1}{3}e - \frac{1}{3}e = 0$)

Figure 25.9 (a) Three quark flavors, called *up, down,* and *strange* could explain the particles known in 1964. (b) Protons are formed from two up quarks and a down quark, while (c) neutrons come about from an up quark and two down quarks.

If up and down quarks are all that are needed to make neutrons and protons, what good are *strange* quarks? We've mentioned that over 200 other types of particles are known to exist. We don't need to get involved in a description of these particles. However, to compose all of them, it is necessary to have the strange quark, as well as the *antiquark* of each of the three flavors.

Unfortunately, three flavors of quarks are not enough to tell the whole story. Soon after the introduction of the idea of quarks, it became clear that all the particles could not be formed from these three quarks without violating some cherished laws of physics—among them the Pauli exclusion principle. It was necessary to distinguish further between the three flavors. Each flavor is characterized by another property that has been given the name *color*. For each flavor there are *three* colors—usually designated red, blue, and green, although different physicists use different sets of colors. This color has nothing to do with the usual use of the word. The name is used here merely to distinguish between the three characteristics of quarks.

Furthermore, three flavors of quarks are not enough to explain all particles now known. Instead, we seem to need six. The other three flavors have been called *charm* quarks, *top* quarks, and *bottom* quarks. (There is not yet general agreement on what to call the last two. Some people use the names "truth" and "beauty" for these flavors of quarks.) It seems that the number of quarks has proliferated until we no longer have the simple idea of a few building blocks, which was embodied in the original proposal. We'll summarize all this and put it into perspective after we've had a chance to consider another breed of particles called leptons.

Before considering leptons, however, let's ask a question that has probably been on your mind: Do quarks *really* exist, or is this just a scheme physicists have dreamed up without any valid basis? Has anybody ever *seen* a quark?

Quarks do not seem to exist as free particles outside the larger particles that they compose. Despite extensive attempts to observe them, *no particles of electric charge less than e have been observed by themselves.* However, in experiments similar in principle to the Rutherford experiments that showed the existence of the atomic nucleus, high-energy electrons have been scattered from protons. The scattering pattern is consistent with the proton being made of tiny particles whose properties agree with those proposed for quarks. A great deal of experimental evidence points to the existence of quarks bound inside other particles as well. But try to pull them apart into individual quarks and the force holding them together becomes infinite.

Even though we can't have quite the same kind of confidence in quarks that we do in particles that can be isolated and whose individual properties can be studied separately, physicists accept quarks as the next layer below protons and neutrons in the building blocks of matter.

Leptons—the Weak Cousins of Quarks

The particles that interact strongly with the nucleus are the ones made up of quarks. That includes *most* of the known particles. But there are a few important particles that do not respond to the nuclear force—the force that holds the nucleus together. We've already discussed, or at least mentioned, several of these.

One such particle is the ordinary electron, as well as its antiparticle, the positron. Another is the *neutrino* that we introduced in connection with the form of radioactivity we call beta (β) decay. Let's take another look at β decay.

Recall that in β decay the nucleus emits an electron, thereby converting one of its neutrons into a proton (Section 25.3). A neutrino (actually an antineutrino) is also emitted in this process. The decay equation for this neutron is

$$\,^1_0n \rightarrow \,^1_1p + \,^0_{-1}e + \bar{\nu}$$

We can now describe this process in terms of the quarks that make up the neutron and proton. A neutron consists of two down quarks and an up quark, while a proton consists of one down quark and two up quarks. In other words, in β decay one down quark converts to an up quark with the emission of an electron and neutrino. We can symbolize this decay by

$$_{-1/3}d \rightarrow {}_{+2/3}u + {}_{-1}e + \bar{\nu}$$

Notice that the amount of electric charge before and after the decay remains the same, $-\frac{1}{3}e$.

Experiments show that the forces that act between quarks and the electron or the neutrino are much weaker than the strong nuclear force that acts between quarks. This force, referred to as the *weak* force, is responsible for β decay. Particles that interact with quarks, and thus with neutrons and protons, by way of this weak force are referred to as **leptons**. Leptons, like quarks, have no measurable size that can be detected in present-day experiments. Leptons and quarks seem to share roughly equal status as fundamental building blocks for matter.

Aside from the electron and its associated neutrino, there are two other pairs of leptons. One is the *muon* mentioned earlier, which seems to behave as a heavy electron. Reactions that produce muons are accompanied by the emission or absorption of a different neutrino—called the "muon neutrino" to distinguish it from the electron neutrino that accompanies the electron in β decay. More recently another pair of leptons—the *tauon* and its accompanying neutrino—have been observed. The tauon has a mass of about 3500 times that of the electron, and thus only appears in very high-energy collisions between particles.

Quarks and Leptons—Three Generations

Let's now put all these particles into perspective by summarizing our present understanding of the "fundamental" particles that make up our world. The universe seems to be built of two categories of particles that we call *quarks* and *leptons*. The particles in each of these categories can be grouped into six *flavors*. Furthermore, the flavors can be broken down into pairs that, to some extent, tell us their degree of "ordinariness." These three *generations*, as they are often called, of quarks and leptons are summarized in Table 25.3. The electric charge of each particle is given in parentheses.

The first generation contains the components of the ordinary particles. Neutrons and protons are composed of up quarks and down quarks. The other ordinary atomic component, the electron, is in the first generation of leptons. And the electron neutrino is the one that appears in ordinary β decay.

The second generation contains the additional quarks needed to make up the hundred or so particles known at the time quarks were proposed. Interactions involving the muon and its neutrino were also known at that time. The third generation contains the components of the more exotic high-mass particles discovered in more recent experiments.

Unfortunately, life is not quite as simple as implied by Table 25.3.

Table 25.3 Three generations of quarks and leptons

	Quark Flavors	Lepton Flavors
First generation	up $(+\frac{2}{3})$	electon (-1)
	down $(-\frac{1}{3})$	electron neutrino (0)
Second generation	charm $(+\frac{2}{3})$	muon (-1)
	strange $(-\frac{1}{3})$	muon neutrino (0)
Third generation	top $(+\frac{2}{3})$	tauon (-1)
	bottom $(-\frac{1}{3})$	tauon neutrino (0)

Recall that each of the quarks and leptons has a corresponding antiparticle. Furthermore, each flavor of quark comes in three *colors*. Altogether this makes 36 quarks and 12 leptons. We would certainly have liked it better if there were fewer fundamental particles. For example, at the next level up it takes only three kinds of particles—protons, neutrons, and electrons—to make up all the known atoms. But nature seems not to be that simple.

Do not get the impression that physicists claim now to have discovered the ultimate fundamental particles that make up matter. What we claim is to have uncovered another layer in the "cosmic onion," as one writer has labeled the various levels of the makeup of matter. *At this time* quarks and leptons *seem* to be "fundamental" point-like particles that make up bulk quantities of matter. But who knows how many more layers this "onion" may have? In this section we've only scratched the surface of particle physics. Some of the end-of-chapter references do an excellent job of giving a more complete description of this fascinating search.

Practical Uses of Particle Physics?

This book focuses on the physics *around you*—those areas of physics that have the most influence on our everyday lives. What place do quarks, leptons, and the search for the ultimate building blocks of matter have in such a book? Do such ideas affect the everyday lives of you and me? The answer is that we don't yet know the answer. But if history is any guide, the chances are that these discoveries will have a major impact on societies of the future. For example, when Maxwell was studying electromagnetic waves in the nineteenth century, did anyone have an idea of the role such waves would play in twentieth-century communications? When Rutherford and his co-workers discovered the atomic nucleus, did anyone foresee the effect a knowledge of the atom would have on the chemistry needed to make thousands of products we enjoy today? And did the discoverers of the neutron have any idea of the practical applications of nuclear physics that would follow?

We can only wait and see what the ingenuity of humankind will bring about concerning the practical uses of today's particle physics.

Key Concepts

The atomic *nucleus* consists of **neutrons** and **protons**, that is, **nucleons** held together by the **strong**, or **nuclear**, force. A particular nuclear species is called a **nuclide**, and it is symbolized by $_Z^A X$, where X is the chemical symbol for the element, Z is the *atomic number* (number of protons), and A is the *mass number* (number of protons plus neutrons). **Isotopes** are nuclides of the same element having different numbers of neutrons.

For light, stable nuclei, the number of neutrons is approximately equal to the number of protons; the ratio of neutrons to protons increases with total number of nucleons. Energies associated with the nucleus are typically about a *million* times those associated with the electrons in the atom.

Unstable nuclei decay to more stable forms by **radioactivity**: the emission of particles from the nucleus. In naturally occurring radioactive nuclides, these particles are α (^{4}He nuclei), β (electrons), and γ (photons) rays. In radioactive decay, *total charge and total number of nucleons remain constant*. The **half-life** is the time for half of the nuclei in a radioactive sample to decay.

An important application of radioactivity is in the use of *tracers* to follow the path of a particular element.

When radiation passes through matter, it *ionizes* the atoms. This ionization allows the detection of radiation, and it can destroy living cells. The largest radiation exposure for most people is from natural sources.

Neutrons and protons are composed of *up quarks* and *down quarks*. Four other *flavors* of quarks compose the other particles that interact strongly with the nucleus.

Questions

Nuclear Structure

1. If the proton and electron had equal mass, how would these two particles behave in a Bohr model picture of the hydrogen atom?

2. If you're comparing ^{60}Co and ^{131}I, why is it inappropriate to use the term *isotopes*, whereas if you're comparing ^{123}I and ^{131}I, it is all right to use that term? What term should be used in the first case?

3. How many neutrons are in the nuclei of ^{40}Ca (calcium) and ^{197}Au (gold)? You may refer to the periodic table of elements, Figure 24.8, for the atomic numbers.

The Nuclear Force and Nuclear Energies

4. The nuclear force is a force of attraction because it holds the nucleus together. Does it also have the ability to cause nucleons to repel each other? Hint: Based on Figure 25.1, can nucleons get in each other's space?

5. Nuclear physicists use the symbol N for the number of neutrons and Z for the number of protons in a nucleus. There are 160 stable nuclides in which both N and Z are even numbers, 53 with N odd and Z even, 49 with N even and Z odd, and only 4 with both N and Z odd. From these data, what would you suggest about the strength of the nuclear force when nucleons can pair off two-by-two of *like* particles?

6. Iodine has only one *stable* isotope: ^{127}I. From its atomic number, 53, estimate the number of neutrons it has using Figure 25.2. Compare your result with the number it must have based on the actual mass number.

7. Very roughly, how much more energy would you expect to be released via 10^{23} nuclear reactions in a nuclear reactor than via 10^{23} chemical reactions while burning coal in a boiler.

8. If you could pull a nucleus apart into its individual nucleons, work would be needed to pull each nucleon out. Any time you do work, you

produce some form of energy; in this case, it goes to mass. Would you expect the mass of all the separated nucleons to be more *or* less than the mass of the nucleus? This difference in mass energy is called the *binding energy* of the nucleus.

Radioactivity

9. Why do alpha particles and beta particles bend in opposite directions when they pass perpendicularly through a magnetic field? Why do gamma rays not bend at all?

10. The wavelength of most observed gamma rays is less than 1/10,000 that of the shortest-wavelength radiation given off by the hydrogen atom. Why does this fact indicate that nuclear energy changes are much, much greater than energy changes in the outer atom?

11. X rays and γ rays are the same stuff; they just come from different places. Explain how a ^{60}Co γ-ray source could be used to "x ray" inaccessible equipment—such as the pipes of a boiler—to check for cracks.

12. Complete the following radioactive decay equations (use an X for any unknown chemical symbols):
    a. $^{235}_{92}U \rightarrow {}^{231}_{90}Th +$
    b. $\rightarrow {}^{237}_{93}Np + {}^{4}_{2}He$
    c. $^{60}_{27}Co \rightarrow \qquad + {}^{0}_{-1}e$
    d. $^{60}_{28}Ni \rightarrow \qquad + {}^{0}_{0}\gamma$
    e. $^{14}_{6}C \rightarrow {}^{14}_{7}N +$

13. Suppose nuclear bomb tests leave a certain area contaminated with strontium-90, which has a half-life of 29 years. How long would potential residents of the area need to wait for *all* of the ^{90}Sr to decay?

14. If your smoke detector quits working, should you suspect that its ^{241}Am source has all decayed away? See Table 25.1.

Applications of Radioactivity

15. Is there any difference in what ^{131}I does for a doctor in searching for thyroid cancer and what a cowbell does for a farmer?

16. A patient complains to her doctor that her left foot is cold and numb. The doctor suspects that a blockage in an artery is impeding circulation. To pinpoint it, he injects into her blood a small amount of salt solution (sodium-chloride) in which a minute fraction of the sodium is radioactive ^{24}Na. What would his next step in the search be?

17. A fractured or diseased area of a bone takes in calcium more rapidly than normal bone does. Strontium behaves chemically in almost exactly the same way as calcium. How could physicians use radioactive ^{85}Sr (half-life 65 days) or ^{87}Sr (half-life 2.8 h) to locate bone lesions before they are large enough to show up on x-ray pictures? What is wrong with using ^{90}Sr (half-life 29 yr)?

18. When a new batch of oil is started down the Alaskan pipeline right behind an old one, how can radioactive tracers be used to let people at the other end know when the new batch gets there?

19. Some brands of LCD digital watches have a tritium lighting system that continually gives off enough green light that you can read the display at night. A small amount of tritium (^{3}H that β-decays with a half-life of 12 yr) is tightly sealed in a glass capsule coated on the inside with a phospher that glows when hit by electrons. What form of energy produces this light? If you bought such a watch today, how would the light output 12 years from now compare with that now?

20. A *nuclear battery* could be constructed as in Figure 25.10. Electrons coming off the β source on the inner cylinder travel out and hit the outer cylinder, charging it negatively. A voltage of 10,000 to 100,000 V can develop and can pro-

Metal cylinder

Cylinder coated with β emitter

Figure 25.10 A nuclear battery.

duce a few microamperes of current in an external circuit. If strontium-90 is used as a β source, how would you expect the output current of this battery to change with time?

21. When a γ-ray beam passes through a layer of matter, the fraction of the γ rays absorbed by the matter increases with the thickness of the layer. Draw a diagram showing a scheme that uses a γ-ray source and a radiation detector to continuously monitor the thickness of plastic sheet coming off a production line.

Biological Effects of Radiation

22. The radiation dose to the exposed area during a chest x ray should not exceed about 30 mrem. Since this is considerably less radiation than you get from natural sources every year, is there any reason for not routinely having a chest x ray each year?

23. The flying distance from Atlanta to Denver is roughly 1500 miles. How many times per year could Sharon, who lives in Atlanta, fly to see Charlie, who lives in Denver, and not receive any more additional radiation exposure than Charlie gets just by living in Denver? See Table 25.2.

24. Baggage at airports is inspected by x-ray beams that give an exposure of no more than 1 mrem per examination. It takes about 5 mrem to noticeably fog your photographic film. What does this say about the number of times you should let a given film go through airport inspection?

Problems

1. Use Coulomb's law (Section 14.6) to verify that the electrostatic force between two protons 1.5 fm apart is 100 N.

2. Suppose you have 4.0 mg of sodium-24. A small radiation detector nearby counts 3200 gamma rays per second. If 30 hours later the same detector in the same location registers 800 counts per second, how much ^{24}Na do you have?

Answer: 1.0 mg

25. In xeromammography, a process developed by Xerox Corporation, the film used in breast x-ray examinations is replaced by a sheet of selenium backed by photographic film. X rays hitting the selenium screen cause it to fluoresce, giving off visible light that exposes the film. With this method, about $\frac{1}{20}$ the x-ray dose gives the same film exposure as when it is exposed directly by x rays. Because of the two-step process, do you expect some loss in sharpness of the image?

26. Certain foods are pasteurized by a radiation dose of 200,000 to 500,000 rem and sterilized by a dose of 2 to 5 million rem. Since radiation is known to cause genetic damage, is there any way you could genetically damage your offspring by eating irradiated food?

Elementary Particles

27. Positively charged pions are composed of two quarks—an up and an antidown. Show that these quarks give the correct charge for this pion: $+1e$. Show that the antiparticles of these quarks give the correct charge for the antiparticle of this pion—the negative pion.

28. A particle called the J/ψ, discovered in 1974, was the first example of a particle containing charm quarks—specifically, a charm plus an anticharm. The leaders of the research teams that made the discovery received the Nobel Prize two years later. What is the eletric charge of the J/ψ?

3. A detergent manufacturer adds a new ingredient to the detergent and washes a load of greasy clothes in which the grease contains a known amount of radioactive material. If the wash water emits four times as much radiation after the wash when using the new ingredient, how much more effective is the modified detergent in removing grease?

Answer: removes four times as much grease

4. A person standing near a 5.0-MBq gamma-ray source intercepts 5% of the radiation emitted from the source. How many gamma rays strike the person's body in 1.0 hour?

Answer: 9.0×10^8

5. A clinical technician measures the activity of a ^{24}Na sample to be 200 MBq. What is the activity of the sample a week later? (The half-life of ^{24}Na is 15 h.)

Answer: 100 kBq

6. Suppose you take a sample of ^{130}Te (Te = tellurium) to your friendly neighborhood reactor operator and convince her to make you some $^{131}_{53}$I. What she will make is $^{131}_{52}$Te, which has a half-life of 25 min. How long will it take for 99% of the ^{131}Te to decay to ^{131}I?

Answer: about 2 hr, 55 min

7. Suppose your uncle has two sets of x rays performed on his gastrointestinal tract. During the first he receives a radiation dose of 180 mrem, but during the second, more extensive test he receives 320 mrem. Assuming that he would otherwise have received the average annual dose of 150 mrem, by how much is his radiation exposure increased during that year?

Answer: 4.3 times as much

8. How many round trips could an airline pilot make between New York and San Francisco without exceeding the 5 rem per year limit set for individuals working with radiation?

Answer: 1200

References for Further Reading

Allman, W. F., "Irradiated Food," *Science 81*, Oct. 1981, p. 14. The use of radiation in food preservation.

Bushong, S. C., "Radiation Exposure in Our Daily Lives," *The Physics Teacher*, March 1977, p. 135. Discusses the sources of radiation exposure to United States citizens.

Close, F., *The Cosmic Onion* (American Institute of Physics, New York, 1986). An interesting, readable account of the search for the ultimate structure of matter.

Fisher, A., "Cosmic Order," *Popular Science*, May 1983, p. 72. How the discovery of the W particle brings us closer to a grand unification theory—a synthesis of the four basic forces of nature into one mathematical structure.

Lindenfeld, P., *Radioactive Radiations and Their Biological Effects* (American Association of Physics Teachers, College Park, Md., 1984). A short paperback that effectively summarizes many aspects of the problems associated with radiation.

Maynard, C. D., *Clinical Nuclear Medicine* (Lea and Febiger, Philadelphia, 1969). Describes clinical uses of nuclear physics.

Taubes, G., "Detecting Next to Nothing," *Science 85*, May 1985, p. 59. Some of the devices used to detect particles crucial to modern particle theories.

Trefil, J. S., *From Atoms to Quarks: An Introduction to the Strange World of Particle Physics*. (Scribner's, New York, 1980).

N U C L E A R R E A C T I O N S A N D N U C L E A R E N E R G Y

Nuclear energy has the potential for changing the life of every person on this planet. How does this tiny chunk of matter at the center of each atom have the ability to release such awesome amounts of energy to be used for good or evil? That is the question pursued in this chapter.

We start our study of this topic by considering the fundamentals of nuclear reactions. We'll then find the general condition necessary for the release of nuclear energy in a reaction. We can extend that condition to discover the categories of reactions that can be put to use in the practical, large-scale release of nuclear energy: namely *fission* and *fusion*.

After considering the basic mechanisms in fission and fusion reactions, we'll consider the devices that harness nuclear energy. In addition, we'll consider some of the risks and societal implications of using nuclear energy.

26.1 NUCLEAR REACTIONS

Let's get a little more "mileage" from our ^{131}I examples discussed in the last chapter to introduce radioactivity. Where do you suppose the ^{131}I comes from? We didn't mention that it does not exist naturally. Evidently, there is some way of artificially converting one element to another. You may remember that converting one element to another was the goal of the ancient alchemists; they wanted to convert base metals to gold. We haven't quite perfected that art yet, but we can transmute a fraction of the nuclei in a sample to a nuclide with a nearby mass and atomic number.

Causing Nuclear Reactions

We get this change of nuclide by bombarding appropriate nuclei with nuclear particles. For example, we can produce ^{131}I by bombarding the naturally available nuclide tellurium-130 with protons. Figure 26.1 pictures what happens to one Te nucleus. The **reaction** *equation* is

$$^{130}_{52}\text{Te} + ^{1}_{1}\text{H} \rightarrow ^{131}_{53}\text{I} + ^{0}_{0}\gamma$$

The resulting ^{131}I nuclei are left in an excited state after the proton *capture*, and quickly de-excite by γ (gamma) emission, as the equation shows.

Reaction equations are similar to those for radioactive decay. The difference is the presence of the bombarding particle—in this case the proton, which is sometimes written $^{1}_{1}\text{p}$. Notice that, as for radioactive decay, both charge and mass numbers are *conserved*: Their sums on both sides of the equation are equal.

We can also get ^{131}I in a two-step process by bombarding ^{130}Te with neutrons:

$$^{130}_{52}\text{Te} + ^{1}_{0}\text{n} \rightarrow ^{131}_{52}\text{Te} + ^{0}_{0}\gamma$$

$$^{131}_{52}\text{Te} \rightarrow ^{131}_{53}\text{I} + ^{0}_{-1}\text{e}$$

The ^{131}Te β-decays to ^{131}I with a half-life of 25 min. After several hours, most of the tellurium has decayed, leaving the wanted iodine isotope.

Aside from the use of nuclear reactions as a way of producing radioisotopes for everyday use, such reactions provide nuclear physicists with their main tool for learning about the nucleus. We said in describing Rutherford's experiments that showed the existence of the nucleus, that the nucleus is so small that we can't get "close enough" to observe it. So we throw "rocks" at it—nuclear projectiles—and use radiation detectors to find out what comes out. We can infer what some of the nuclear features are by what comes out and by the direction and kinetic energy of the outcoming particle relative to what went in.

There's another analogy we can use to explain how we study the

Figure 26.1 The nuclear reaction that transmutes a ^{130}Te nucleus to an ^{131}I nucleus.

nucleus by nuclear reactions. If you want to see the skunk stealing your cookies from the picnic table outside your camping tent, you shine a flashlight on him. The light scattered from the skunk into your eyes lets you identify the thief. The "light" you shine on a *nucleus* to "see" it is a beam of protons, deuterons, α (alpha) particles, or some other particles. You see the effect on the scattering by radiation detectors rather than your eyes. You could not use visible light because the wavelength is much longer (100 million times longer) than the object you're trying to see. High-speed nuclear particles have a wavelength (remember de Broglie) about the right size to scatter from nuclei. Researchers use machines called *particle acclerators* to produce beams of high-speed charged particles that initiate nuclear reactions. For bombardment by neutrons, they immerse the sample in the high neutron population of a nuclear *reactor* (Section 26.3).

· ·

APPLICATION   ## CARBON-14 DATING

Have you ever wondered how we can know the age of a 5000-year-old Egyptian mummy? One method of dating that can tell its age within 100 years makes use of both nuclear reactions and radioactive decay.

The story starts in the upper atmosphere where neutrons from cosmic-ray interactions produce carbon-14 in the reaction

$$^{14}_{7}N + ^{1}_{0}n \rightarrow ^{14}_{6}C + ^{1}_{1}p$$

(N = nitrogen). This radioactive ^{14}C, with a half-life of 5700 years, combines with oxygen to form carbon dioxide. When the carbon dioxide migrates to earth, it constitutes about 1 out of every 10^{12} carbon dioxide molecules taken in by plants. These plants then get eaten by people and animals.

When the plant, animal, or person dies, it quits taking in ^{14}C, and the relative percentage of its carbon that is ^{14}C decreases by radioactive decay. By measuring the activity and thus the fraction of ^{14}C in the remains, we can measure the time since the organism's death. Inherent in this measurement is the assumption that the percentage of ^{14}C in natural carbon remains constant over the period to be dated. The method is not very accurate for ages beyond several 5700-year half-lives of ^{14}C. But it works well for things that are from a thousand or so (the Dead Sea scrolls, for example) to 15 or 20 thousand years old.

· ·

APPLICATION   ## NUCLEAR FINGERPRINTS

Some people now suspect that Napoleon was slowly poisoned to death with arsenic, rather than dying of cancer, as some of us learned in History 1111. Some English scientists not long ago found unusual amounts of arsenic in a sample of Napoleon's hair. (For some reason arsenic taken into the body concentrates in the hair.) They irradiated the sample in a reactor, and found γ rays characteristic of ^{76}As, which you get by irradiating natural arsenic, ^{75}As, with neutrons.

Neutron activation analysis, the name for this technique, is a sensitive way of finding extremely small amounts of almost any substance in a sample. Each different nuclide, after neutron irradiation, emits γ rays with energies and half-lives characteristic of that nuclide only. The procedure does not destroy the sample; the small fraction of nuclei made radioactive merely decay back to stable nuclides.

This method has been used in thousands of broad-ranging applications—matching small bits of blood found on a murder suspect with that of the victim; identifying small flakes of paint in hit-and-run automobile accidents; identifying elements in samples of moon rocks; looking for lead concentrations in the hair from deer that live near highways; protecting against counterfeiting of products by deliberately including traces of known elements that can be easily activated for identification— just to name a few.

26.2 ENERGY FROM NUCLEAR REACTIONS

The world has never been the same since large-scale exploitation of nuclear energy came onto the scene during World War II. That energy has been put to both constructive and destructive uses. Since even the constructive uses often have undesirable side effects, both uses have spawned strong feelings and concerns. At this point, let's not take sides on either issue, but instead let's look at the conditions under which we can get energy from a nuclear reaction. That will, we hope, put us in a more informed position for choosing sides.

By now, you have a good understanding of the principle of conservation of energy: Energy can neither be created nor destroyed. If energy comes from a nuclear reaction, it is not created, but must result from the loss of *some* form of energy by the nuclei involved.

We'll examine a particular reaction that can liberate energy,

$$^2_1H + {}^2_1H \rightarrow {}^3_2He + {}^1_0n$$

Figure 26.2 The mass of two deuterons (^{2}H) is larger than that of a neutron plus a ^{3}He. The diagram is purely symbolic: You can't actually *weigh* nuclei with a balance. You find their mass indirectly by methods such as measuring the curvature of their paths in a known magnetic field.

³He + n + n 4.0247 amu

²H ²H + n + n 4.0282 amu

and see both where this energy comes from and in what form it shows up. This reaction, which incidentally might supply part of the energy to light your house and cook your food in twenty or thirty years, combines two deuterons (hydrogen = 3) to form a neutron and a helium-3 nucleus.

The masses of the ^{2}H, ^{3}He, and n are 2.0141 amu, 3.0160 amu, and 1.0087 amu, respectively. As illustrated *symbolically* in Figure 26.2, this means that the total mass *before* the reaction exceeds that *after* the reaction by 0.0035 amu. So you're not impressed by a trivially small difference in the masses of things that have an unbelievably small mass to start with!

In Einstein's theory of relativity, which we'll consider in more detail in the next chapter, he showed that mass can be converted to energy, and vice versa. The relationship between mass and energy is given by the popularized equation

$$E = mc^2$$

where m is the mass in question and c the speed of light, 3×10^8 m/s. On this basis, if you could convert the $\frac{1}{2}$ kg of food you probably had for dinner yesterday entirely to energy, you'd get 4.5×10^{16} J, or about 12 billion kWh. This quantity is about $\frac{1}{5}$ of the total energy used per day by the *entire* United States for *all* purposes! This example makes it clear that small mass changes can provide significant energy changes.

Now go back to our nuclear reaction. The energy equivalent of our 0.0035 amu mass difference is 5.2×10^{-13} J, or 3.2 million electron volts (MeV). We can get this by converting the 0.0035 amu to kilograms, multiplying by c^2 to get energy in J, and converting this to MeV. If we lump all these conversion factors into one number, we find the *energy equivalent of 1 amu is 931 MeV*. In other words, if we have a mass in amu, we can get its energy equivalent in MeV by multiplying by 931:

$$E = 0.0035 \text{ amu} \times 931 \text{ MeV/amu}$$

$$= 3.2 \text{ MeV}$$

This result tells us that 3.2 MeV of energy is released in each such reaction. *This energy shows up as kinetic energy of the reaction products.* In other words, the total *kinetic energy* of the ^{3}He and n exceeds that of the two deuterons by 3.2 MeV. This is an *extremely* small amount of energy by everyday standards: it would take 4.4×10^{18} such reactions to give enough energy to boil away a liter of water. But as we'll see, there are ways of getting many times that number of reactions.

Any nuclear reaction for which the mass of the final products is less than that of the initial parts releases energy in the reaction process. To use that energy for practical purposes, there must be a way of causing large numbers of the reaction to occur without using more energy to produce the reactions than is gained *from* those reactions.

To get an idea of the kinds of reactions that might satisfy these conditions, we need to look at the way actual nuclear *masses* relate to the *mass number* of the nuclei. Figure 26.3 is a graph of total *mass per nucleon* versus mass number for all the stable nuclides. Since the number of nucleons does not change in a nuclear reaction, a reaction that leads to a lower mass *per* nucleon leads to a lower total mass. (An analogy might be useful here: If you're buying cookies to feed 100 people, the lowest overall price results from buying packages with the lowest price *per* cookie, whether they're packaged in bags of one-dozen or two-dozen.)

Figure 26.3 A graph of mass per nucleon versus mass number for stable nuclides. The dots show the values for selected nuclides.

Thus, we want to look for kinds of reactions that lead to lower overall mass by looking for packages of nucleons—nuclei—that give a lower mass per nucleon.

Figure 26.3 shows that the mass per nucleon drops sharply with increasing mass for the light nuclides, reaches a minimum in about the iron (Fe) region, and then gradually increases on up to the uranium (U) region.

Look at what happens if you start with very light nuclei—^{2}H and ^{3}H, for example—and join, or *fuse*, them to form nuclei with roughly twice their mass number. The mass per nucleon, and thus the overall mass, is much less. The reaction we discussed a little earlier is just such a *fusion* reaction.

Now consider the other end of the graph. Suppose you take a heavy nucleus in the uranium region and split it—*fission* it—into two parts of roughly equal size. This process results in nuclei in the strontium (Sr) to xenon (Xe) region, where the mass is much lower.

Since both of these types of reactions, **fission** of heavy nuclei and **fusion** of light nuclei, result in a lower total mass, energy is released as they occur. We'll look next at these two processes and at the devices to harness the energy released as they occur.

26.3 NUCLEAR FISSION

Heavy nuclei may once in a while fission on their own, but to get any significant fraction of the nuclei of a sample to fission, we need to add energy by hitting the nucleus with some kind of particle, for example, a neutron.

Figure 26.4 When a ^{235}U nucleus captures a neutron, the energy it gets causes it to oscillate and sometimes fission into two parts with a few neutrons left over.

Fission fragments

If a neutron hits a ^{235}U nucleus, it gives energy to the nucleus that can cause it to vibrate and oscillate much as you might expect a water drop to do if you hit it with a smaller droplet. As Figure 26.4 depicts, these oscillations often cause the nucleus to take an elongated "barbell" shape, and then split into two parts. Usually a few extra neutrons splatter out in the process. The particular **fission fragments** and the number of neutrons emitted vary from fission to fission. One such reaction is

$$\,^{1}_{0}n + \,^{235}_{92}U \rightarrow \,^{236}_{92}U \rightarrow \,^{94}_{38}Sr + \,^{139}_{54}Xe + 3\,^{1}_{0}n$$

In the intermediate stage between neutron absorption and fission, the nucleus is ^{236}U. Even though the fission fragments sometimes have nearly equal mass, usually they are somewhat unequal: typically in the 100 and 140 mass-number region. Each fission usually releases 2 or 3 neutrons.

The mass change during fission of ^{235}U has an average energy equivalence of about 200 MeV. Each fission, then, releases an average of about 200 MeV of kinetic energy to the fission fragments and neutrons. To get enough energy from fission for practical use, we need billions of such reactions.

This goal can be reached by using the neutrons released in each fission to interact with other heavy nuclei to cause other fissions: a **chain reaction**. Figure 26.5 symbolizes this effect. We show two neutrons from each fission going on to cause other fissions. If this situation were left uncontrolled, it would build up *very rapidly* with a very quick release of energy. We would have a bomb! This effect was just the one used in the "atomic" bombs dropped on Hiroshima and Nagasaki. As long as the average number of fission-released neutrons that go on to produce fissions is larger that 1 per fission, the reaction rate will increase indefinitely.

Under what conditions would such a reaction sustain itself? It might seem that one stray neutron could set off a chain reaction, making a bomb out of *any* quantity of uranium that happened to be around. Not so. First of all, only the odd isotopes of uranium have a significant probability of fission as a result of neutron capture. And only 0.7% of natural uranium is an odd isotope—^{235}U. Most of it is ^{238}U. Other odd-mass-number heavy nuclides, such as plutonium-239, will also work. Only if you have a certain minimum—**critical**—mass of highly purefied fissionable fuel can you produce a nuclear explosion. Otherwise, enough neutrons leak out that less than one neutron per fission causes another fission, and the reaction dies out.

So how do you "light the fuse" on such a bomb; that is, how is it

Figure 26.5 A fission chain reaction. An incoming neutron causes the fission of a uranium nucleus, which releases three neutrons. In each fission two of the neutrons (black dots) released go on to cause other fissions.

possible to maintain a critical mass of fissionable material and have it explode only at the desired time? The method used in the bomb that devastated Hiroshima kept the fissionable material separated into small enough segments that no one segment had a critical mass. TNT surrounding the uranium was then exploded to bring the segments of uranium together to form the critical mass.

The Nuclear Reactor

To use fission–released energy for everyday purposes, we need to control the fission chain reaction and force it to proceed at a manageable rate. The device that does this is a **nuclear reactor**. The reactor provides the ability to keep the reaction rate at a fixed level: on the average *one* neutron from each fission going on to cause another fission. The neutrons either escape or are absorbed without producing a fission.

Figure 26.6 is a schematic diagram of the main components of a nuclear reactor. Neutrons released in the fission of uranium or plutonium in the fuel elements are slowed down by a *moderator* (usually water or graphite), so that they have a higher probability of interacting with another fuel nucleus.

The control rods are made of a material such as cadmium that easily absorbs neutrons, without fissioning. The reactor operator controls the reaction rate by inserting or withdrawing control rods to slow down or speed up the chain reaction. The safety rods are good neutron absorbers

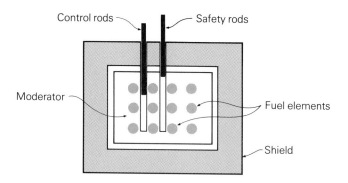

Figure 26.6 Schematic diagram showing the main components of a nuclear reactor.

that can be inserted quickly to shut down—**scram**—the reactor in an emergency.

When a reactor is operated at a fixed rate, with exactly one neutron per fission causing another fission, we say it is **critical**. When more than one neutron per fission causes another fission, the chain reaction is building up, and we say the reactor is *supercritical*. Below the critical rate, we say it's *subcritical*.

As a source of useful energy, a reactor is just a nuclear furnace. The kinetic energy of the fission fragments and neutrons heats the materials in the reactor core. A liquid *coolant* pumped through the core absorbs and extracts the heat in the same way water does in a coal-fired boiler. In most present power reactors, this coolant is water, which also serves as the moderator. The schematic diagram of a nuclear power plant would be similar to Figure 20.1, except that a reactor with plutonium or uranium fuel elements would replace the boiler and fossil fuel. There's one additional step: Rather than steam being produced directly inside the reactor, the coolant from the reactor usually goes to a steam generator, where it gives up heat to water to produce the steam.

Football and the "Atomic" Age

Even though football brings out more emotion on many campuses than any other activity, its most ardent enthusiasts (excluding the coaches and players) think of it as a diversion rather than the stuff that shapes their lives. But the events that took place in 1942 *under* the football stadium at the University of Chicago have changed the course of world affairs. It was in that unsuspected location, beneath the stands of Stagg Field (Fig. 26.7), that Enrico Fermi and his co-workers secretly produced the world's first self-sustaining chain reaction. They accomplished this feat in a reactor containing uranium fuel embedded in graphite moderator.

While the processes in that "atomic pile" were more akin to those in a reactor for peaceful uses of nuclear energy, the project demonstrated the feasibility of sustaining a nuclear chain reaction. It thereby paved the way for the Manhattan Project, which developed the first nuclear bombs.

Figure 26.7 Enrico Fermi and his co-workers produced the first self-sustaining chain reaction in a reactor under the stands of Stagg Field at the University of Chicago.

ON DECEMBER 2, 1942
MAN ACHIEVED HERE
THE FIRST SELF-SUSTAINING CHAIN REACTION
AND THEREBY INITIATED THE
CONTROLLED RELEASE OF NUCLEAR ENERGY

If you ever have a chance to hear that fascinating story recounted by a member of the team involved, don't miss it! The story has been recorded by many people, including Fermi's wife, Laura.*

Nuclear Energy from Fission: Benefits and Risks

The awesome destructiveness of nuclear weapons is evident to anyone even slightly attuned to the major concerns of our society today. The nuclear bombs of World War II were mere sling-shots compared with modern "hydrogen" bombs that use the fusion process. (More on that later.) We as citizens of this world urgently need to do everything within our grasp to reduce the dangers posed by the many thousandfold overkill capacity of the nuclear arsenals.

However, the use of reactors to channel nuclear energy into generating electricity for everyday purposes clearly has a different objective. Unfor-

*Fermi, Laura, Atoms in the Family: My Life with Enrico Fermi (University of Chicago Press, Chicago, 1954).

tunately, the two categories of nuclear energy use—nuclear weapons and nuclear power—are often not distinguished by those concerned with nuclear risks. Even though there are interconnections between these two areas that we'll point out, nuclear power carries its own set of risks as well as clear benefits. In this section we will examine both the benefits and risks of nuclear power generation. First, let's look at the benefits.

The most obvious advantage of nuclear energy over other sources lies in the use of nuclear fuels rather than fossil fuels. The energy available is many times the world's total supply from fossil fuels, and its use will not deplete the finite supply of fossil fuels. As the technology improves and fossil fuels become more scarce, nuclear power will undoubtedly be *relatively* less expensive.

The "burning" of nuclear fuels does not release the vast amount of smoke and chemical carcinogens into the air that fossil-fuel plants release. Neither does it release the extensive amounts of carbon dioxide that are a possible source of world climate change. And the devastation of land per year for uranium mining is only a small fraction of that stripped in coal mining.

Now for the problems: A question that probably occurs to everyone is, Can a runaway power reactor become an atomic bomb? We can answer that question with an emphatic **no**. Fuel used in power reactors contains only 3% to 5% fissionable nuclides such as ^{235}U. This percentage is much lower than that needed for a bomb.

However, in the event of an accident resulting in loss of coolant, the reactor core could become hot enough to melt everything inside. Reactors in the United States and Western Europe are always built with numerous redundant independent safety systems to shut down—"scram"—the reactor and prevent such a *meltdown*. Nevertheless, every reactor core is surrounded by two containment vessels strong enough to contain the radioactive hodgepodge that would result in the unlikely event of a meltdown. Even though the probability is very low, there remains the possibility that the vessels could rupture and release radioactive material into the environment.

Pollution from nuclear power plants is of two forms: radiation and thermal. Small amounts of radioactive gases, mostly krypton-85, are released from some reactors. Even though the total radiation produced is small compared with natural sources, any increase in radioactivity in the environment is undesirable. Thermal pollution comes from the heat deposited in the atmosphere or nearby body of water in condensing the steam from the turbines. For a nuclear plant, this thermal pollution is about 1.5 times that for a fossil-fuel plant.

The problem of radioactive waste disposal is a nagging one whose final solution has not yet been found. Let's see why we have this problem.

Remember that the more massive the nucleus, the higher the ratio of neutrons to protons. Therefore, when a heavy nucleus fissions, the fission fragments have too many neutrons to be stable. They get rid of these excess neutrons by β-decaying until they reach stability. For example, the ^{139}Xe in the fission reaction we wrote earlier β-decays to lanthanum-139 through the reactions

$$^{139}_{54}\text{Xe} \xrightarrow[40 \text{ s}]{\beta} {}^{139}_{55}\text{Cs} \xrightarrow[9.3 \text{ min}]{\beta} {}^{139}_{56}\text{Ba} \xrightarrow[83 \text{ min}]{\beta} {}^{139}_{57}\text{La}$$

In this more compact notation, the half-life for each decay is shown below the arrow. Notice in these reactions something that is typically, but not always, true: The closer the nucleus is to stability, the longer its half-life.

Most of the fission fragments and their decay products have fairly short half-lives and decay to stability within days. A few hang around for quite a long time. A couple of the more notorious ones are strontium-90 and cesium-137, which have half-lives of 29 and 30 years, respectively. Some fragments have half-lives of thousands of years, but their relative abundance is small.

After spent fuel elements are reprocessed, and all useful materials such as unused fuels are removed, there remains some radioactive garbage that will be dangerously radioactive for *hundreds* of years. This material has to be maintained and monitored to be sure it doesn't contaminate the environment. Various agencies, both private and governmental, are studying alternatives for long-term handling of this material. It's now being kept in large, heavily shielded underground tanks.

Finally, there's the *plutonium* problem. Because the uranium in most reactors is at least 95% nonfissionable ^{238}U, a significant amount of ^{239}Pu is produced by neutron absorption in the uranium. The ^{239}U formed by neutron absorption β-decays to ^{239}Np, which again β-decays to ^{239}Pu. This plutonium can be recovered by chemical separation when the fuel is reprocessed, and later it can be used for nuclear fuel.

Plutonium is a concern because it is highly toxic. It's chemically poisonous, but its main toxicity is associated with the radiation. The most likely source of radiation exposure in working with plutonium is from breathing plutonium-contaminated air, some of which attaches itself to lung tissue where it may later cause lung cancer. The plutonium that gets into the blood usually deposits in the skeleton where it can produce bone disease, including cancer. For these reasons, workers need to be extremely careful in handling plutonium. Currently Environmental Protection Agency regulations state that the maximum fraction of plutonium handled by a plant that may escape as airborne dust is about one part in a billion.

There is one more concern related to plutonium, and this is a way in which nuclear power could have a bearing on the nuclear weapons situation. The ^{239}Pu that is recoverable from spent reactor fuel can be used as fuel for nuclear weapons. A relatively poor country, unable to obtain bomb-quality fissionable material in other ways, may use a nuclear reactor to build up a supply of material that *could* be used to make a nuclear bomb. Such a situation could escalate the danger of a nuclear confrontation that could spread to more powerful nations. However, because of the large size, extreme complexity, enormous cost, and high visibility of a reactor system for generating electric power, it is impractical to use a power reactor for such a purpose. A country intending to use a reactor to produce plutonium would undoubtedly use a small, relatively inexpensive reactor that could be kept secret and dedicated to that purpose.

The Big Question

The decision of whether or not our country should commit itself to additional large-scale use of nuclear energy from fission depends on two questions:

1. Is the potential benefit of nuclear power worth the risk involved in using it?

2. Are the alternatives any less risky?

Taking risks is part of living. Most of us regularly ride in cars, participate in athletic events, work in or live near factories, and commit ourselves to being involved with other people, all of which require some measure of risk. We do these things because we feel the potential benefit is worth the risk.

Several of the end-of-chapter references address the problem of the risks of using nuclear energy. Some compare probability of injury from nuclear power plants with the probability of injury in other risks that we regularly accept. Other references address our second question more directly; they compare the risk of nuclear versus other energy sources. I urge you to read these references.

26.4 NUCLEAR FUSION

We pointed out in Section 26.2 that when light nuclei are fused to form heavier ones, the final mass is usually less than the initial. Some reactions that meet this condition are:

$$^2_1H + {}^2_1H \rightarrow {}^3_2He + {}^1_0n \text{ (3.2 MeV)}$$

$$^2_1H + {}^3_2He \rightarrow {}^4_2He + {}^1_1H \text{ (18.3 MeV)}$$

The energy released per reaction by loss of mass is listed in parentheses.

These reactions can take place *only* if the initial particles come together with *enough kinetic energy to overcome their mutual electrical repulsion*. The only known way to do this is to raise the gas to an intensely high temperature, at least 40 million °C. At these temperatures, thermal energy provides the needed kinetic energy, and collisions knock off all electrons from the atoms. We call this completely ionized gas a *plasma*. This high-temperature plasma presents a "slight" problem: What kind of pot do you put it in? Any material medium would immediately vaporize as soon as the plasma touched it.

Reactions in which thermal energy provides the needed kinetic energy for the nuclear reactions to occur are called **thermonuclear** reactions. A significant amount of energy is needed to raise the plasma temperature high enough for the reactions to occur in appreciable numbers. The object of thermonuclear programs is to produce a controlled thermonuclear reaction that gives off more energy than it takes to get it going. Efforts to do so center on three main objectives: (1) achieving high enough plasma temperatures, (2) compressing and confining the plasma long enough for a net energy output, and (3) recovering the energy in a useful form.

There are two general approaches being used to solve these problems: **magnetic confinement** of the plasma, and **laser bombardment** of the fuel pellets. In magnetic confinement, the plasma carries a large electric current and is confined by magnetic fields away from any material objects. Figure 26.8 shows *one* way of achieving this confinement. The current in the plasma produces a magnetic field around it as shown by the curved arrow. The charged plasma particles moving through this field experience a force toward the center. The plasma thus gets "pinched" into a small region, which both confines it and heats it by compression. This process sounds easy but is quite complicated in practice, and the machines for carrying it out are much more complex than the simple schematic of Figure 26.8.

In the *laser fusion* scheme, pellets about 1 mm in diameter of frozen deuterium-tritium are bombarded with high-intensity laser beams from many directions simultaneously. The intense pulse of light energy heats the pellet to thermonuclear temperatures before it has time to expand and disperse. The approach is sometimes called **inertial confinement** because the inertia of the pellet particles themselves confines them there long enough to be heated. The idea is that pellets are continually dropped and hit with laser pulses, thereby releasing energy in pulses.

A **fusion reactor** for producing useful energy from fusion would need some means of getting the energy out. A likely approach is to surround the reactor core with a blanket of liquid coolant. Neutrons released in the fusion process would be slowed in the coolant, giving up their kinetic energy to heat the coolant. The coolant could be circulated to a heat exchanger to produce steam and turn a turbine to generate electricity in the usual way.

As of this time neither magnetic confinement nor laser fusion have succeeded in getting more energy out than is put into the machines. Nevertheless, steady progress is being made in both areas. Fusion researchers have a fairly detailed timetable, according to which larger and larger machines are to be built and operated, each one being based

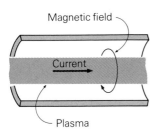

Figure 26.8 The "pinch effect" used in confinement of a plasma.

on the knowledge gained from its predecessors. We *expect* to have working fusion reactors by the early twenty-first century.

One big plus for using fusion energy is that its fuel supply—deuterium—is virtually limitless. About one in every 6000 hydrogen atoms in sea water is deuterium. That doesn't sound like much, but the potential fusion energy in each liter of water is about 100 times the combustion energy in a liter of gasoline. A second plus: Fusion reactions won't leave the long-half-life radioactive fission fragments left by fission reactors.

Some drawbacks of fusion include the following: (1) The extreme neutron density in the reactor will irradiate all components of the reactor structure, making them radioactive and probably weakening them considerably. This problem is an engineering one that presumably *can* be solved. (2) Large supplies of tritium will be generated in a fusion reactor. Tritium is the radioactive isotope of hydrogen of mass number 3, with a half-life of 12 years. Tritium behaves chemically like ordinary hydrogen, and can form water and other compounds easily taken into the human body. It will be absolutely necessary to design fusion reactors to minimize the leakage of tritium. (3) Feasibility of fusion reactors has not yet been proven, even though researchers in the field are quite optimistic about success.

The use of thermonuclear reactions in bomb production is not subject to the limitation of needing confinement. In that situation, once the fuel for fusion has been raised to the temperature needed for thermonuclear processes to occur, the reaction proceeds with an enormous release of energy. These temperatures are achieved by the use of fission bombs. The thermonuclear fuel is surrounded by small fission bombs which, when exploded, compress the deuterium and tritium fusion fuel and raise it to the temperature needed for fusion to start. The energy released in such a "hydrogen" bomb has reached as high as 58 megatons (equivalent to 58 million tons of TNT), compared with about 20 *kilo*tons for the fission bomb dropped on Hiroshima. I cannot stress enough our need for the elimination of such weapons of mass destruction!

• •

Key Concepts

A **nuclear reaction** results from bombardment of a nucleus with a nuclear particle, causing the formation of a new nucleus and some additional outgoing particle. Nuclear reactions provide our major tool for understanding the nucleus.

Nuclear reactions result in the *release* of energy if the total mass after the reaction is less than the total mass before the reaction. Reactions useful in the large-scale practical release of nuclear energy include **fission** of heavy nuclei and **fusion** of light nuclei.

Nuclear **reactors**, which use fission, depend on chain reactions in odd-mass-number heavy nuclides such as ^{235}U and ^{239}Pu. Practical release of energy from nuclear fusion is expected to depend on **thermonuclear** reactions, which use the thermal energy associated with exceedingly high temperatures to bring the light nuclei together to react. Major research in thermonuclear reactions centers on *magnetic confinement* and *inertial confinement*.

Questions

Nuclear Reactions and Applications

1. Complete the following nuclear reaction equations, using an X for any chemical symbols you don't know.

    a. $^{59}_{27}Co + ^{1}_{0}n \rightarrow \quad + ^{0}_{0}\gamma$

    (used in producing a nuclide for therapy)

    b. $^{6}_{3}Li + ^{1}_{1}H \rightarrow ^{3}_{2}He +$

    (a potential fusion reaction for fusion reactors)

    c. $^{235}_{92}U + ^{1}_{0}n \rightarrow ^{236}_{92}U \rightarrow \quad + \quad 2^{1}_{0}n$

    (a reaction in a fission reactor)

    d. $^{7}_{3}Li + ^{1}_{1}H \rightarrow ^{4}_{2}He +$

    (the first reaction ever produced using a particle accelerator)

2. Tracer methods are very useful in wear studies. For example, gasoline engine manufacturers like to know how fast their piston rings wear. They can use radioactive piston rings and then after a certain period of running the engine, look for radioactivity in the lubricating oil. How can they make the piston rings radioactive?

3. Natural sodium is all ^{23}Na. How could you use a nuclear reactor to produce the ^{24}Na used in medical diagnoses?

4. The sheriff has uncovered an illegal gunpowder manufacturing gang. He wants to know if *their* gunpowder was involved in the recent murder of one of his officers. Having the murder weapon and knowing the exact composition of their powder, how could he use neutron activation analysis to get his answer?

5. An oil slick is found floating in the Gulf of Mexico. Investigators are not sure where it came from. How can they use neutron activation analysis to find out whose oil it is?

6. An alleged Rembrandt painting is suspected of being a twentieth-century forgery. In order to use neutron activation analysis for evidence one way or the other, what would you need to know about the paint used in Rembrandt's time?

Nuclear Energy and Reactors

7. If you calculate the energy equivalent of the mass difference before and after fission of ^{235}U, you get about 170 MeV for a typical fission. (See, for example, Problem 5.) The *average total* energy released per fission is about 200 MeV. The rest is released gradually after fission. Where does it come from? (Hint: Remember what happens to the fission fragments.)

8. When a reactor is critical, is it necessary that exactly one neutron released in *every* fission cause another fission?

9. A loss of coolant in a reactor can cause a reactor core *meltdown*. Even if the fission chain reaction stops immediately, there is still enough heat being produced to cause a meltdown. Why? (Hint: What happens to the fission fragments? That's why reactors have an emergency cooling system to inject coolant if the normal coolant system ruptures.)

10. Here is a potentially realistic chain of events: Lightning knocks out a power line from a nuclear power plant; the line no longer carries current; the generator no longer produces current and therefore uses little steam; the steam generator doesn't have to produce steam and therefore doesn't cool the reactor coolant; the reactor coolant doesn't cool the reactor. Explain this sequence from a *conservation of energy* viewpoint. In designing a nuclear power plant, nuclear engineers have to provide a system for preventing meltdown from such an event.

11. We sometimes use the word *hot* to describe the fact that a substance is radioactive. Radioactive wastes from a nuclear reactor also stay very hot *thermally*. Why?

12. Discuss the feasibility of providing each home in the country with a well-encased spent reactor fuel element to put in the middle of the family room in winter and provide heat like a potbellied stove. (See Question 11.)

13. Carbon is often used as a moderator in nuclear reactors because its nuclei are light relative to those of the fuel. Water, whose nuclei are hydrogen and oxygen, is even more effective in slowing down the neutrons. Explain why materials with light nuclei make better moderators. To do so, draw an analogy with the kinetic energy a golf ball would lose if it collided with another golf ball or a billiard ball compared with what it would lose if it collided with a bowling ball.

14. People often use the word *atomic* in phrases such as "atomic energy" and "atomic bomb." Explain why atomic energy is really the kind you get from burning coal; why an atomic bomb is the kind made from TNT; and why the word *nuclear* is the one that should be used instead of the usual *atomic*.

Problems

1. A hard-working physics graduate student, while skin-diving at Wakula Springs, Florida, discovers an old mastodon bone. If he takes it back to his campus and finds that the ratio of ^{14}C to stable carbon is $\frac{1}{4}$ what it is in the backbone of his goldfish that died that morning, how long ago did the mastodon drink from Wakula Springs?

   Answer: 11,500 years

2. In the radioactive α decay

   $$^{226}_{88}Ra \rightarrow {}^{222}_{86}Rn + {}^{4}_{2}He$$

   what is the total kinetic energy of the Rn and α particle after decay? Since the α is much lighter, it carries away most of the kinetic energy. (The masses are 226.02536 amu, 222.01753 amu, and 4.00260 amu, respectively.)

   Answer: 4.87 MeV

3. Find the energy released in the fusion reaction

   $$^{2}_{1}H + {}^{3}_{1}H \rightarrow {}^{4}_{2}He + {}^{1}_{0}n$$

   (The masses in order of appearance are 2.0141 amu, 3.0160 amu, 4.0026 amu, and 1.0087 amu.)

   Answer: 17.5 MeV

4. In order to make the first-observed nuclear reaction

   $$^{14}_{7}N + {}^{4}_{2}He \rightarrow {}^{17}_{8}O + {}^{1}_{1}H$$

   go, how much mass energy must be made up by the kinetic energy of the incoming α particle? (The masses in order of appearance are 14.00307 amu, 4.00260 amu, 16.99913 amu, and 1.00782 amu.)

   Answer: 1.19 MeV

5. The following fission reaction releases five neutrons:

   $$^{1}_{0}n + {}^{235}_{92}U \rightarrow {}^{136}_{53}I + {}^{95}_{39}Y + 5({}^{1}_{0}n)$$

   If the kinetic energy of the incoming neutron is negligible, what is the total kinetic energy of the fission fragment and neutrons immediatley after fission? (The masses in order of appearance are 1.00866 amu, 235.04392 amu, 135.91474 amu, and 94.91254 amu.)

   Answer: 169 MeV

References for Further Reading

Cohen, B. L., *Before It's Too Late: A Scientist's Case for Nuclear Power* (Plenum Press, New York, 1983).

Feiveson, H., and F. von Hippel (*pro*) and H. W. Lewis (*con*), "Freeze on Nuclear-Weapons Deployment and Development: Pro and Con," *Physics Today*, Jan. 1983, p. 36. Scientists debate the issue.

Gough, W. C., and B. J. Eastlund, "The Prospects of Fusion Power," *Scientific American*, Feb. 1971, p. 50. The experimental programs and techniques for thermonuclear fusion reactors. Concentrates on magnetic confinement techniques.

Hollander, J., editor, *Nuclear Energy, Nuclear Weapons*

Proliferation, and the Arms Race (American Association of Physics Teachers, College Park, Md., 1982). Four articles on the arms race, including two that debate whether or not nuclear power has a significant bearing on nuclear weapons.

Lovins, A. B., "Nuclear Weapons and Power-reactor Plutonium," *Nature*, Feb. 1980, p. 817. Can power reactors produce weapons-grade plutonium?

Marion, J. B., *Energy in Perspective* (Academic Press, New York, 1974). Chapters 5 and 6 deal specifically with nuclear power and radiation.

McBride, J. P., R. E. Moore, J. P. Witherspoon, and R. E. Blanco, "Radiological Impact of Airborne Effluents of Coal and Nuclear Plants," *Science*, Dec. 1978, p. 1045. Shows that the radiation dose from radioactive nuclides in the emissions from coal-fired power plants may be greater than those from nuclear plants.

Morland, H., "The H-Bomb Secret," *The Progressive*, Nov. 1979, p. 14. How a fission bomb can trigger a hydrogen bomb.

Nuckols, J. H., "The Feasibility of Inertial-Confinement Fusion," *Physics Today*, Sept. 1982, p. 24.

Nuclear Power and the Environment—Questions and Answers (American Nuclear Society, La Grange Park, Ill., 1976). Answers many common questions relative to nuclear power and radiation.

Ruby, D., "Magnetic Fusion: How Soon? How Safe?" *Popular Science*, May 1982, p. 80.

Seaborg, G. T., and J. L. Bloom, "Fast Breeder Reactors," *Scientific American*, Nov. 1970, p. 13. Describes the principles of breeder reactors, which produce more nuclear fuel than they use.

Tsipis, K., *Arsenal: Understanding Weapons in the Nuclear Age* (Simon and Schuster, New York, 1983). A sourcebook for understanding nuclear weapons.

Weinberg, A. M., "The Future of Nuclear Energy," *Physics Today*, March 1981, p. 48. An authority on nuclear energy assesses the future of nuclear energy in the U.S.

THE THEORY OF RELATIVITY

There are some things in this world that we've always been able to count on to stay stable, no matter what state of turmoil our lives may happen to be in at the time. Life may get somewhat topsy-turvy, but the distance to home always stays the same. A mile is a mile, no matter who measures it. And good old *time*: It may *seem* that time really flies when you're having fun; but we all know that's only a perception and that, in reality, an hour is an hour is an hour.

Well, if that's your idea about time and distance, I have *bad* news for you. I'm going to do my best in the next few pages to change your mind about such matters.

We've all had enough experience with times and distances to be confident of the ideas expressed in the first paragraph *within our realm of experience*. But what happens outside this realm of experience? For example, what happens when you travel at speeds near the speed of light? Unless you happen to be from another planet and have advanced considerably beyond us earthlings in transportation capabilities, you have no personal experience in such matters. Your intuition may tell you that things *should* be the same at any speed. But guess what: They are not!

We are going to find in this chapter that there's something special about this entity that we call the speed of light. First of all, no object can move *faster* than the speed of light. And when one object does move relative to another at a speed close to that, funny things happen: Time is no longer the same for both; the distance between two points depends on who measures it; and the mass of the object increases; just to name a few. These observations are all predictions of the *special theory of relativity,* put forth by Albert Einstein in 1905. Most of this chapter will be devoted to aspects of that theory and its experimental verification. The last section will briefly review the *general theory of relativity,* which is actually a detailed theory of gravitation.

27.1 THE POSTULATES OF SPECIAL RELATIVITY

For several centuries, questions about the nature and behavior of *light* have intrigued some of the most brilliant scientists of the time. Isaac Newton, Thomas Young, James Clerk Maxwell, and Albert Einstein are among those having a major impact on the outcomes of extensive scientific investigations into these questions. The answers that have been found form the foundation of the relativity theory that has drastically changed the way we view our universe.

After Thomas Young showed in the early 1800s that light behaves as a wave, it was natural to assume that light, like all other known waves, needed a medium through which to travel. As a result, physicists postulated the existence of a medium called "luminiferous ether," that was presumed to permeate all space and to provide the medium for light's travel. Maxwell's evidence in the mid-1800s that light was an electromagnetic wave did not, in itself, contradict that notion. But, despite extensive experimental investigation, no one could observe the ether.

Most prominent among these experiments is the Michelson–Morley experiment, performed by A. A. Michelson and E. W. Morley in 1887. Their experiment made use of the speed of the earth in its orbit around the sun, that speed being about 0.0001 of the already known speed of light in vacuum, 3.00×10^8 m/s. The vacuum was presumed to be permeated by stationary ether, and the earth, therefore, moved through this ether. Since light would have to travel through the ether, earthbound observers should measure a different speed for light moving in the same direction as the earth than for light moving in the opposite direction. Their experiment used interference of light waves to compare these two speeds. At the time, they concluded that there is no *stationary* ether. Moreover, from our perspective, their result gave strong evidence that the ether *does not exist.*

The lack of evidence for either the ether or different speeds of light for different observers was very puzzling to physicists of around 1900. But Albert Einstein had the genius to separate the essential observations from prejudices and irrelevant assumptions. Based on the behavior of electric and magnetic fields, and the fact that electromagnetic radiation such as light can travel through vacuum, he developed a theory in which arguments about ether and relative light speed were meaningless. He based his theory on two fundamental **postulates** that can be stated as follows:

1. *The laws of physics are the same in all inertial reference frames.*

2. *The speed of light in empty space is the same for all observers and is independent of the motion of either the source or the observer.*

We'll elaborate on the meaning of these two postulates. First, however, we must explain the term *reference frame* in general and *inertial* reference frame in particular.

Frames of Reference

Any measurement must be made with respect to some frame of reference. For example, if you explain to a campus visitor that the nearest fast-food restaurant is 300 m from the physics building, your frame of reference for that measurement is the physics building. Or suppose that you explain to the police that, in getting away from the potential mugger, you ran at a speed of 9 m/s. The usual assumption would be that you meant *relative to the earth* as a frame of reference. But, as Figure 27.1 shows, if you were on a moving train running toward the front at that speed, your speed relative to the ground would be entirely different.

In Figure 27.1 we have two reference frames, the train and the earth, one moving at constant speed with respect to the other. Both of these reference frames happen to be *inertial* frames. Inertial reference frames are those *in which Newton's first law is obeyed.* An object on the moving train needs a force to accelerate it relative to the train, just as it does on the ground. But what if the train slammed on the brakes? Passengers on the train would seem to feel a force pushing them forward when none actually existed. Or if the train rounded a curve, they would feel a force

Figure 27.1 The runner's speed relative to the train is 9 m/s. However, relative to the *ground*, the runner's speed is 25 m/s + 9 m/s, or 34 m/s.

seeming to push them to the outside of the curve. An accelerating reference frame is *not* an inertial frame: In such a frame objects accelerate even when no force exists.

You'll now notice that I lied a little about the earth being an inertial reference frame. Since the earth rotates about an axis, any point on the earth's surface is always accelerating. But this acceleration is small enough to be negligible for most purposes.

Postulate 1

Postulate 1 tells us that the laws of physics are the same for all *inertial* reference frames. For example, Newton's laws of motion would apply to anything happening on the train moving at constant speed, just as they would on the ground. That does not imply that values of position, velocity, acceleration, and force relative to a point on the train are *all* the same as they are relative to a point on the ground. But it means that the relationships between these values are governed by the same laws in both places.

• •

QUESTION Does postulate 1 agree with your experience? For example, if you throw a candy bar to a friend while you're both riding in a bus moving at constant velocity, is anything different than if the bus were parked in the parking lot? What if the driver slammed on the brakes while the candy was in mid-air?

The first postulate implies something very important to the theory of relativity: There is no meaning to absolute motion. *Relative* motion is all that can be detected. If one object is moving relative to another, you cannot say that one is absolutely at rest and the other moving. One is at rest in its own reference frame while the other moves relative to that frame. But the second is at rest in *its* reference frame, and the first moves *relative to it*. Since the laws of physics are the same in all frames, you cannot single one out as being special—as being at rest.

Postulate 2

The second postulate is probably much harder on your common sense. It says that the speed of light in vacuum, which has the value 3.00×10^8 m/s and goes by the symbol c, is the exact same no matter who measures it or what happens to be the state of motion of either the light source *or* the measurer. Even though the astronaut in Figure 27.2 is moving at a

Figure 27.2 Both people, the one standing on the ground and the one in the spaceship moving at speed $(\frac{2}{3})c$, measure a light speed of c relative to themselves.

speed of $\frac{2}{3}c$ relative to the man holding the flashlight, he measures the speed of the light travel to have the exact same value as the man holding the light source! If the flashlight holder happened to be on a skateboard moving at $0.9c$ to the left (relative to earth), both men would still measure the same light speed. So would a kid sitting on the ground and watching.

It is not possible to *prove* the correctness of these two postulates of special relativity. They stand or fall on whether or not their predictions hold up to experimental tests. It is rather astounding that, even though special relativity has been put to some of the most rigorous tests ever conceived by the ingenuity of humankind, it has stood the test in every case thus far. That gives us a great deal of confidence in its predictions, even those not yet directly verified. We'll now look at some of the predictions that follow directly from the two postulates set forth by Einstein.

To understand some of the implications of the postulates of special relativity, we'll follow Einstein's example and examine how the postulates apply in certain *thought experiments*. A thought experiment is one in which you imagine a certain situation and determine what a postulate or law would predict for that situation. These differ from actual experiments referred to in the previous paragraph that have verified many of the predictions of the theory.

27.2 c: THE ULTIMATE SPEED

One consequence of the second postulate of special relativity is that *no* material object can travel at a speed greater than the speed of light. We'll use the thought experiment illustrated in Figure 27.3 to see how the second postulate leads to this conclusion.

Imagine that Amy is driving from Omaha to Des Moines. For the sake of argument, we'll assume she is driving at a speed of $2c$—twice the speed of light. Doug is waiting for her in Des Moines.

Just outside Omaha, it starts to get dark, and Amy turns on her headlights (Fig. 27.3-a). Assuming that the road is perfectly straight and flat, let's compare the front of the light beam as viewed by Amy and by Doug just as Amy gets to Des Moines. First, Amy: As part (b) of the figure shows, Amy sees nothing unusual. The light beam has moved away from her at speed c and the front of it is long gone when she gets to Des Moines.

From Doug's frame of reference, as Amy passes him, the situation is entirely different (Fig. 27.3-c). Amy and her car are beside him, but the

Figure 27.3 If Amy were moving at speed $2c$, a contradiction would appear as she got to Des Moines. As viewed from the same point, Amy and Doug would not agree on the relative position of the car and the front of the light beam.

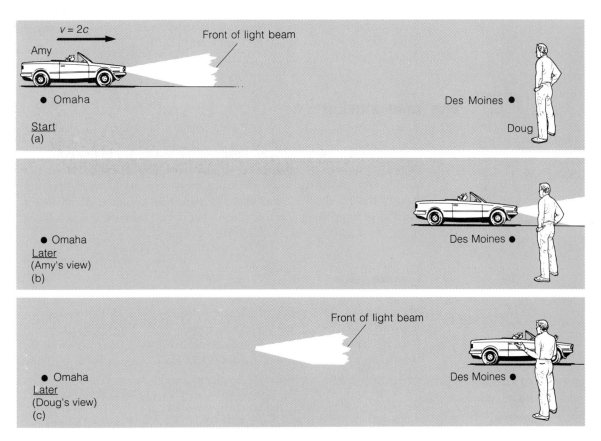

front of the light beam is only halfway from Omaha. (Remember, we're assuming Amy is traveling twice as fast as the light.) Since both people are observing from the same point, they both have to agree on the relative position of the car and light beam. Conclusion: Our assumption of speed $2c$ for Amy is not possible. Neither she, nor any other mass, can move faster than the speed of light.

So let's slow Amy down to a leisurely pace of $0.9c$. Now, what if Amy passes Brenda along the way traveling at $0.8c$ in the opposite direction (Fig. 27.4)? Surely Amy must be moving at $1.7c$ *relative to Brenda*. Putting Brenda in the role of Doug in the earlier thought experiment, we see that Amy cannot exceed speed c relative to Brenda either. When the speeds involved are a significant fraction of the speed of light, we cannot just add the velocities as we did in connection with the low speeds used in Figure 27.1. For such high speeds the method of addition, which is mathematically somewhat more complex than we need to consider here, *always gives a relative speed that is less than c.*

• •

QUESTION If the person running through the train in Figure 27.1 runs at a speed of $0.3c$ and the train moves at speed $0.8c$, what can you say about the speed of the person relative to the railroad track?

27.3 TIME IS RELATIVE

When Einstein introduced the theory of relativity, it produced a great simplification in our description of the behavior of light. But it has necessitated giving up some cherished ideas—among them the notion that time is an absolute that moves forward independent of whatever else may be happening. This, too, follows directly from the independence of the speed of light for all observers.

Simultaneity

One feature of the time at which events occur relates to the concept of *simultaneity*. If two events are simultaneous, they happen at the same time. It's natural to assume, based on our experience, that if events are simultaneous for one person, they must be simultaneous for any other. Let's examine that assumption.

Our thought experiment to analyze this situation finds Don inside the empty cargo bay of a space shuttle moving at a speed of $0.7c$ relative to his friend Ward. Standing exactly in the center of the open area, Don strikes a match and observes the light that heads toward the two ends of the bay (Fig. 27.5-a). As part (b) of the figure shows, Don sees the light pulse strike the two ends of the cargo bay *simultaneously*. However, from Ward's viewpoint (Fig. 27.5-c and d), the left end of the shuttle is moving toward one light pulse, and the right end is moving away from the other. In his frame of reference, the light gets to the left end before it gets to the right. Remember, the light speed is the same for both observers. The events are not simultaneous from the second frame of reference. In general, *events that are simultaneous in one reference frame are not necessarily simultaneous in a reference frame moving at constant velocity with respect to the first.*

Time Dilation

If simultaneity has fallen by the wayside, then perhaps time *intervals* are different in different reference frames that move with respect to each other. Again, the two postulates show this situation to be true.

Let's go back to our thought experiment with Don in the space shuttle traveling at a high speed relative to Ward. There is a mirror overhead to reflect that part of the light from Don's match that travels straight up. Don measures the time it takes the light pulse to travel to the mirror and back to the starting point (Fig. 27.6-a). Call that time interval T_0. From Ward's frame of reference, if the light is to strike the mirror and return to the center of the bay, it must follow the much longer path shown in Figure 27.6(b). Since both observers measure the same speed for the light, the time interval T measured by Ward must be much longer than that measured by Don. Some algebra shows these two time intervals to be related by

$$T = \frac{T_0}{\sqrt{1 - \dfrac{v^2}{c^2}}}$$

If we think of the apparatus we have been describing as a sort of *clock*, and the time between ticks of the clock as the time from departure to return of the light pulse, we can say something about the relative speeds of clocks in the two reference frames. In Don's reference frame, the clock is at rest. We call the time interval in the rest frame of the clock the *proper time*, labeled T_0. From Ward's frame of reference, the clock is *moving* relative to him. Since the time interval between ticks of the clock is longer for the moving clock than for one at rest relative to him, he concludes that *moving clocks run slower.* That conclusion is a general one, whether the clock is the kind we have described, a quartz wristwatch, or a human heartbeat.

Proof of time dilation. As unlikely as it may seem, time dilation is one of the predictions of the theory of relativity that has been experimentally verified to a very high precision. Even though humans have not acquired

Figure 27.4 It would seem that Amy's speed *relative to Brenda* should be 1.7c. The second postulate of special relativity requires that the relative speed be *less than c.*

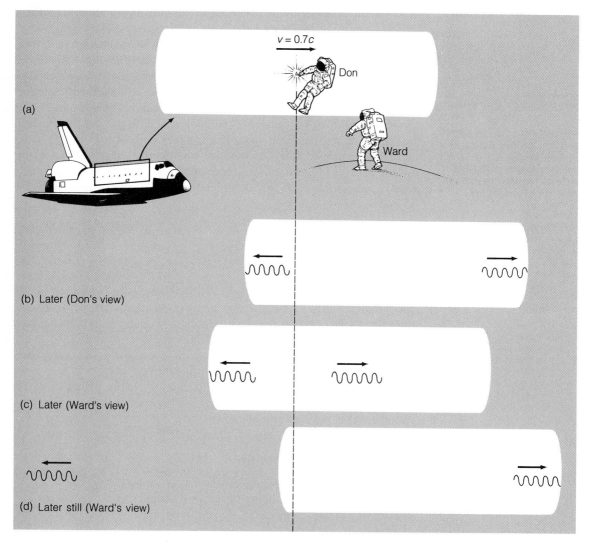

Figure 27.5 Don's shuttle moves past Ward at a speed of 0.7c. (a) Don strikes a match at the center of the cargo bay as Ward watches. (This is a hypothetical match that burns briefly without oxygen.) (b) Later Don sees the light pulses strike both ends of the bay at the same time. (c) Slightly earlier than in (b), Ward sees the light strike the left end of the bay. (d) Somewhat later, Ward sees the light strike the right end. Conclusion: Events that are simultaneous in one reference frame are not necessarily simultaneous in another.

the skill to accelerate themselves to speeds approaching that of light, we routinely accelerate atomic and subatomic particles to speeds very close to c in machines called particle accelerators. When these particles collide with matter, other particles are produced that have very short lifetimes—10^{-6} s or less. Because these particles are moving at such high speeds, time dilation extends their lifetimes in the laboratory so that they travel much farther than they could otherwise do. These travel times and distances are easy to measure with the right instruments.

However, we don't need to look at human-made particles to find that *moving clocks run slower*. Nature provides the "clocks" in the form of *cosmic rays*. High-speed protons coming in from outer space toward the earth collide with atomic nuclei high in the atmosphere, producing "'showers" of particles that travel toward the earth at speeds near the speed of light. Among these particles are *muons* that have a half-life of

$v = 0.7c$

Don

Ward

(a)

(b) Later (Don's view)

(c) Later (Ward's view)

(d) Later still (Ward's view)

Chapter 27   The Theory of Relativity

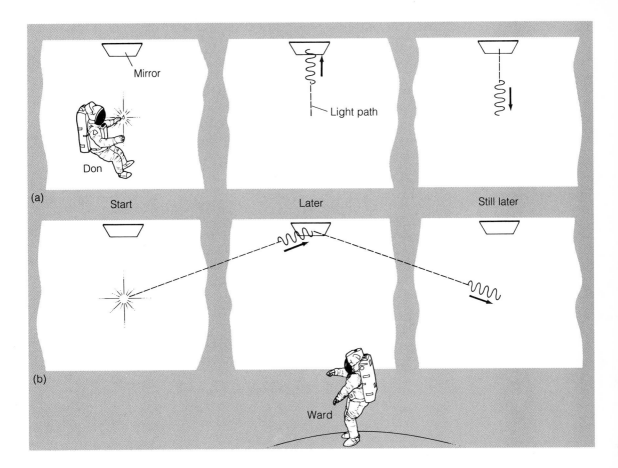

(a) Start — Later — Still later

(b) Ward

Figure 27.6 Light travels from the match at the center of the cargo bay to the mirror and back to the starting point as seen (a) by Don and (b) by Ward. Notice that the horizontal distance the ship travels in this interval is greatly exaggerated to clarify the figure. The horizontal distance between each successive position cannot be greater than the distance from match to mirror. *Why?*

1.5×10^{-6} s. Thus, after 1.5 μs, half of any given number of muons should have decayed. The number of these muons passing per hour through a region of the atmosphere can be measured.

Suppose a particular detection system measures 1000 muons per hour at an elevation of 1800 m. At a speed *near* 3×10^8 m/s, muons would travel about 450 m in one half-life—1.5 μs, or the 1800 m in about 4 half-lives. Without time dilation, only about 63 muons should survive to sea level. However, measurement shows that about 740 out of each 1000 reach sea level. That means that only 0.67 μs have passed as far as the muons are concerned. That's just the time difference predicted by our time–dilation equation if the muon speed is 0.994c. *Moving clocks do run slower!*

27.4 LENGTH CONTRACTION

If you thought time dilation was too much to swallow, another blow to your common sense is on the way. Not only do time intervals depend on the relative speed, but so do distances. To understand why, we'll go back to Amy driving the perfectly straight and flat road from Omaha to Des Moines, while Doug looks on from Des Moines. Figure 27.7 pictures the situation. This time Amy will stay within nature's speed limit and travel at speed $v = 0.8c$.

We need to compare the distance D between the cities as measured in the moving frame—Amy's frame—with the distance D_0 between them as measured in the rest frame of the cities—Doug's frame. In Doug's frame, the distance Amy travels is related to her speed and the time T it takes by

$$D_0 = vT$$

In Amy's frame of reference, the *road* is passing underneath her in the *opposite* direction but with the *same* speed v. The distance between the cities as measured by her is related to the time T_0 for them to pass by

$$D = vT_0$$

However, since Amy's clock is moving relative to Doug's, her time is related to his by

$$T_0 = T\sqrt{1 - v^2/c^2}$$

Therefore, the distance between the points as measured by Amy is

$$D = vT_0 = vT\sqrt{1 - v^2/c^2}$$

or

$$D = D_0\sqrt{1 - v^2/c^2}$$

In summary, *an object moving relative to a given reference frame* (such as the road moving past Amy) *is contracted, or shortened, by the factor* $\sqrt{1 - v^2/c^2}$ *relative to its length in its own rest frame.* The situation is completely symmetrical. In our example, the car moving past Doug is contracted, as seen from his frame of reference, by the same factor. The length of an object in its own rest frame is called its *proper length*.

It may seem confusing to try to keep straight whose lengths are contracted and whose are not; whose times are dilated and whose are not. Remember, the proper length, the length in the rest frame of the object, is completely normal. But the length of the *moving* object is measured to be shorter. Similarly, the proper time is that measured in the rest frame of the clock, and it is normal to observers in that frame. But clocks *moving* relative to that frame run slower. The equations just mentioned designate the proper time and proper length with subscripts "0." However, it's usually easier to remember these guidelines and fit the equations to them than to try to memorize which T or D goes where in the equations.

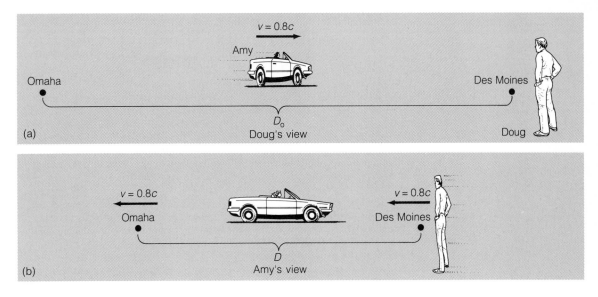

Omaha

Des Moines

$v = 0.8c$

Amy

D_0
Doug's view

Doug

(a)

$v = 0.8c$
Omaha

$v = 0.8c$
Des Moines

D
Amy's view

(b)

Figure 27.7 (a) In Doug's frame of reference, the distance between cities is D_0, and Amy moves with speed v. (b) In Amy's reference frame the road, the cities, and Doug are moving past her at speed v. The distance D is less than D_0.

Do not get the idea that what we're talking about is some sort of optical illusion. I've used the term *see* several times to describe some effect in the reference frame of an observer. Using that term has described what *actually exists* in the particular reference frame. The contracted length *is* the length of the road in Amy's frame; the contracted length *is* the length of the car in Doug's frame. The same idea holds for time intervals. The time interval for the trip in Amy's frame of reference is less than it is in Doug's by the factor $\sqrt{1 - v^2/c^2}$.

Length contraction applies only in the direction parallel to the relative motion. In the examples we've considered, Doug sees the car shorter than Amy sees it, but its width and height are the same for both observers. Amy sees the distance between the two cities as shortened, but the road is the same width, and the people on the side of the road are the same height.

• •

QUESTION Do the people on the side of the road seem as fat to Amy as they do to Doug? Are they skinnier in *both* their horizontal dimensions?

You may be wondering why you have never before observed any of these strange effects—time dilation and length contraction. Keep in mind that measureable time dilation and length contraction both depend on the factor $\sqrt{1 - v^2/c^2}$. Unless v is an appreciable fraction of c, this factor is essentially 1 and has no effect. For all ordinary speeds, even of the

Einstein right: Traveling twin now younger

WASHINGTON D.C., (October 7, 1971)—One of two identical twins returned from a trip today to find its twin who stayed home considerably older. To the scientists accompanying the traveler or waiting with the twin that stayed home, this was a happy confirmation of exactly what Einstein's theory of relativity predicted early in this century.

The theory of relativity tells us that if one of two identical twins travels around at high speeds while the other stays home, the returning traveler will be younger than the twin who stayed home. This effect is more pronounced, the closer the traveler's speed is to the speed of light.

Scientists J. C. Hafele and R. E. Keating tested this theory by an experiment. They sent one of two identical twins on commercial jet flights around the world while the other stayed home. After the flight, the home-bound twin was found to be older than the other by 0.000000059 second, just the amount Einstein's theory predicted. Hafele and Keating had to use a rather special breed of twins in order to detect such a small age difference: They were twin atomic clocks capable of measuring time accurately to within a few billionths of a second.

Scientists consider this result to be strong evidence that, *for example,* if: (1) the twins were named Ralph and Maxine; (2) Maxine went to Alpha Centauri, four light-years away, and returned; (3) she traveled at a speed of 8/10 the speed of light; and (4) the twins were 20 years old when she left, then Ralph would be 30 on her return, but Maxine would be only 26.

fastest rocket available today, v/c is a very small number. Thus, all these new "formulas" reduce to the old familiar ones when the speeds are low compared with c.

EXERCISE

If the straight, flat road in Figure 27.7 is 130 miles long and Amy travels at speed $0.8c$, how much will her odometer advance in going from Omaha to Des Moines?

Answer: 78 miles

APPLICATION

THE TWIN PARADOX

Imagine your surprise if, on returning from vacation, you discovered that all your friends were old, wrinkled, and grey-haired; your campus had tripled in size, with almost all new buildings and new students dressed rather strangely; and everybody was zipping around in strange vehicles that looked as if they had come from another planet. And what if you had to go to a museum to find anything that looked remotely familiar! Well, that situation is not as far-fetched as it may seem. In principle it appears entirely possible. All we need to do is develop the technology to move ourselves at speeds near the speed of light relative to earth—something that is *no small feat*.

Figure 27.8 pictures the situation. Annie leaves on a space cruise and travels at a speed of 0.9995*c* in a straight line for a year. Realizing she may be missed, she quickly turns around and heads home in a straight line at the same speed. In the meantime, Scott is waiting for her at home. Both Annie and Scott were 21 years old when she left. Scott waits faithfully for her during the *63 years* it takes her to return. He is delighted to see that Annie is as beautiful at age 23 as the day she left. Annie, however, isn't thrilled to be greeted by an 84-year-old claiming to be her boyfriend.

Let's look at the situation from Scott's point of view. Relative to his frame of reference, Annie is moving and therefore her "clocks run slower." If T_0 is the 2 years in Annie's frame of reference—her clocks are at rest in her frame—then the time T in Scott's frame is

$$T = \frac{T_0}{\sqrt{1 - v^2/c^2}} = \frac{2 \text{ yrs}}{\sqrt{1 - (0.9995)^2}} = 63 \text{ yrs}$$

Thus, Scott has aged 63 years while Annie has aged only 2.

But wait! Relative to Annie, Scott is moving and *his* clocks should be running *slower* relative to her. Isn't relativity completely symmetric? That's why this situation is referred to as a paradox. But it really isn't a paradox. It is true that, while Annie is moving at a constant speed, Scott sees her clocks running slower, and she sees Scott's clocks running slower. However, their circumstances are not identical. The laws of special relativity apply to *inertial* reference frames only. And Annie does not remain in an inertial frame; she accelerates at the beginning, in turning around, and at the end of the trip. Since Scott stays in an inertial frame, we can view the situation from his frame. Assuming the acceleration, turn-around, and deceleration times are short compared with the rest of the trip, we can neglect them and calculate the time according to Scott's clocks.

We can also see that Annie's time must be much shorter than Scott's from the standpoint of length contraction. On each leg of the trip, Scott sees Annie travel for 31.5 years at a speed of almost *c*. She, therefore, goes a distance of 31.5 *light years*. But Annie sees that distance contracted by the factor of

$$\sqrt{1 - (0.9995)^2}$$

so that the distance is only 1 light year. At a speed near *c*, she travels that far in *one* year.

Figure 27.8 (a) Annie leaves the earth and travels at a speed of 0.9995c in a straight line for a year (her time). (b) She turns around and returns at the same speed in a straight line. (c) She arrives home 2 years after leaving to discover that Scott is 63 years older than when she left.

27.5 TIME: THE FOURTH DIMENSION

We have discovered using thought experiments that we can no longer consider time to be an invariant quantity that marches on independently of what else may be happening. Time intervals and distances measured by two observers differ according to their state of relative motion.

Suppose an astronaut traveling at high speed relative to us decides to take a 20-minute nap. From the earth, we would observe the nap to last longer—about 30 minutes for an astronaut traveling at speed $(3/4)c$. But if the astronaut were 6 feet tall and lying parallel to her direction of motion, we would observe her to be only 4 feet tall. From *our* frame of reference, her time intervals are longer than she observes, and her space intervals—her lengths—are shorter. In her frame, she observes them to be normal.

These two phenomena, time dilation and length contraction, are not separate effects but result from the *four-dimensional spacetime* in which we live. Let's illustrate with a fable. The Hacketts and the Clacketts are planning a new town in which all the streets are to be planned and laid out in advance. The center of town is to be the intersection of Hackett Drive and Clackett Road. A controversy has arisen because the Hacketts want the avenues to be parallel to Hackett Drive, while the Clacketts feel certain that Clackett Road should form the main street. Figure 27.9 shows the situation.

Meanwhile, there are two crows named Jill and Heather who are already permanent residents of oak trees that will be inside the town limits. The crows want the controversy to be settled so that they can inform their friends of their addresses. Heather, for example, will live at Fourth Avenue S and Fourth Street W if the Hacketts win, but her address will be First Avenue N and Sixth Street W if the Clacketts have their way.

There is only one quantity that will remain the same, no matter how the streets are laid out: the distance Heather will have to fly to visit Jill. If you remember the Pythagorean theorem, you know that distance in blocks is

$$\sqrt{(\text{no. streets})^2 + (\text{no. avenues})^2}$$

where the number of streets or avenues is the number of blocks you would need to travel to get there along the sides of a right triangle following Heather's avenue to Jill's street.

· ·

EXERCISE　　Using the Pythagorean theorem, verify that the straight line distance from Heather's nest to Jill's is 11.4 blocks using both street layouts.

Answer: $\sqrt{9^2 + 7^2} = \sqrt{11^2 + 3^2}$

Figure 27.9 A map of the town streets with (a) the avenues parallel to Hackett Drive and (b) with the avenues parallel to Clackett Road. Jill and Heather have different addresses in the two plans.

In our fable, the two layouts for the town are two different reference frames in which the two crows may measure their addresses. Their addresses vary between reference frames, but the distance between their addresses is *invariant* in the two frames—it is independnet of the choice of reference frame and would be the same in *any* reference frame.

A similar situation exists in regard to distance and time. Einstein discovered the idea of an *interval* defined by

$$\text{interval} = \sqrt{(ct)^2 - x^2}$$

where t is the time and x is the position of an event. Even though observers in different reference frames would measure different values of time and distance—that is, different values for t and x—the *interval remains*

Chapter 27   The Theory of Relativity

the same for any two inertial reference frames. Notice the similarity between the form of the interval and the distance between two points on our two-dimensional map. The quantity ct is a distance—the distance light travels in time t.

The situation puts space and time on an equal footing and indicates that they are both part of a single entity called *spacetime*. Another way of saying the same thing is that time, considered as a *fourth dimension*, and the three dimensions of space form a four-dimensional spacetime.

This idea of four dimensions is not very easy to visualize. Another analogy might help some with this problem: You may ordinarily be an active person, accustomed to running, jumping, climbing stairs, and generally moving around in any direction in your three-dimensional world. But if you break your leg and are confined to a wheelchair, your world becomes two-dimensional. You can only move on flat surfaces. (Elevators are not allowed in this story!) It's only when you recover and have the necessary physical ability that you can reenter the three-dimensional world.

It's only when we humans gain the physical ability to accelerate ourselves to speeds near the speed of light that we'll be able to move around in a four-dimensional world in which space and time are intertwined. That isn't likely to happen in the foreseeable future. However, it is already happening for subatomic particles routinely given high speeds in particle accelerators.

27.6 MASS AND ENERGY

If you were to use our previous definitions of momentum and kinetic energy—mv and $\frac{1}{2}mv^2$, respectively—to solve a problem in which speeds were near the speed of light, you would find that the principles of conservation of energy and conservation of momentum are no longer valid. Neither of these principles is one we would like to give up. Apparently, we need a new definition of kinetic energy and momentum for "relativistic" speeds.

Our expression for momentum remains valid if we define the mass of a particle to be

$$m = \frac{m_0}{\sqrt{1 - v^2/c^2}}$$

where m_0 is the *rest mass* of the particle, or the mass the particle has in its own rest frame. In other words, the mass of a body increases as its velocity increases, approaching infinity as v approaches c.

This expression for mass gives us another way of seeing that no material object with mass can ever reach the speed of light. Accelerating a body to speed c would mean that its mass would become infinite. Thus, it would take an infinite amount of energy to accomplish this acceleration.

Einstein also showed that the postulates of special relativity imply that mass and energy are equivalent and are related by the famous equation

$$E = mc^2$$

We have already used this equation in Chapter 26 to calculate the energy released as the mass decreases during a nuclear reaction.

This last equation tells us that there is an energy associated with the rest mass of a body, given by

$$E = m_0 c^2$$

Recall that *kinetic energy* is the energy associated with the *motion* of a particle. If mc^2 is the total energy of a body, and $m_0 c^2$ is the energy of the body at rest, then evidently the kinetic energy of a body is the difference of these two quantities:

$$KE = mc^2 - m_0 c^2$$

This expression for kinetic energy is consistent with the principle of conservation of energy.

27.7 THE GENERAL THEORY OF RELATIVITY

The *special* theory of relativity applies only to the behavior of objects in *inertial* reference frames—reference frames in which there is no relative acceleration. The general theory, published by Einstein in 1916, has no such limitations. Instead, it shows that the laws of physics can be formulated so that they are valid for all observers, no matter what their state of motion.

The foundation of the general theory of relativity is a postulate called the *principle of equivalence*. It states that *the effect of a gravitational force and the effect of an acceleration are indistinguishable.*

Figure 27.10 illustrates this point. If an astronaut in the space shuttle drops a ball while the shuttle is resting on its launch pad *on earth*, the ball falls with an acceleration *g*. However, if the shuttle is far out in space where the gravitational force is negligible, the astronaut observes the exact same effect if the shuttle is *accelerating* with acceleration *g*. In fact, according to the principle of equivalence, no experiment can be done inside a closed compartment that can distinguish the effect of a gravitational field from an acceleration of the compartment.

The legendary experiment in which Galileo is supposed to have dropped rocks from the leaning tower of Pisa is actually a confirmation of this principle within the experiment's limit of accuracy. The experiment demonstrated that heavy objects and light objects fall with the same acceleration if air resistance can be neglected. We showed in Chapter 2 that this effect occurs because (1) the gravitational force on an object is proportional to its mass and (2) the force needed to give an object a

Figure 27.10 From inside the shuttle, the astronaut cannot distinguish between the effect of (a) dropping the ball in the gravitational field of the earth and (b) dropping the ball in a gravity-free environment but with the shuttle accelerating forward with acceleration g. In the astronaut's frame of reference, the ball accelerates with acceleration g in both cases.

certain acceleration is proportional to its mass. These two effects combine to give the same acceleration to any object in free fall. This result illustrates that the mass used in the law of gravitation, the *gravitational mass*, and the mass used in Newton's second law ($F = ma$), the *inertial mass*, are equivalent.

If an acceleration and a gravitational field produce the same effect, then it should be possible to cancel the effect of a gravitational field by means of a suitable acceleration. In the earth's gravitational field, that acceleration would be the acceleration due to gravity g. An object accelerating downward with acceleration g should cancel the effect of gravity and produce *weightlessness*. In fact, that is exactly the situation in an orbiting spacecraft. The spacecraft and its passengers are kept in orbit by the gravitational field and, at the same time, are freely falling toward the earth (Chapter 3). The acceleration cancels the effect of gravity, and the astronauts are in a gravity-free, or *weightless*, environment as observed from the spacecraft reference frame. However, from the earth's frame gravity exists and maintains the spacecraft and its occupants in orbit.

Tests of the General Theory of Relativity

The general theory does not lend itself to verification as readily as the special theory does. However, the theory has been verified experimentally in several ways, and we will briefly describe some of those pieces of evidence for its validity.

The orbit of the planet Mercury. Einstein was able to explain, using the general theory of relativity, an observation from astronomy that had previously been very puzzling. The lowest point in the orbit of the planet Mercury is shifted slightly in position on each revolution. The amount of this shift could not be explained on the basis of Newtonian mechanics. Einstein's calculations of the amount of this shift and experimental observations of it were in excellent agreement, providing one of the outstanding successes of the general theory.

The weight of light. Recall that the special theory of relativity shows that energy and mass are equivalent and related to each other through $E = mc^2$. Since light carries energy, it should therefore have an associated mass. You might, then, expect light to be attracted by the gravitational field of a massive object. The general theory of relativity does indeed predict a precise value for this attraction that can be tested experimentally.

One consequence of such an attraction would be that light from a distant star would have its direction bent if it passed near the sun. We can look for this bending only during an eclipse of the sun, when the sun's light does not overpower the light from the star. An observation of this bending during an eclipse in 1919 provided an early impressive experimental verification of a prediction of the general theory of relativity.

The effect of a gravitational field on a light beam shows up in another way. Light emitted from a massive light source, such as the sun, is gravitationally attracted to its source as it moves away. It, therefore, loses energy just as a particle would in moving away from a mass. However, since light cannot decrease in speed, it loses energy by a decrease in *frequency*. A decrease in frequency means an *increase* in wavelength: The light shifts toward the *red* end of the visible spectrum. This *red shift* can be observed by comparing certain radiations from the sun with those from the same atoms on earth. Even though observations on sunlight provided the first verification of the red shift, more accurate measurements have made use of modern technology to measure, for example, the shift in wavelength of radiation directed toward earth from a source carried on a rocket.

Black holes.   Our sun and similar stars use nuclear fusion reactions as the source of their energy. Pressure due to the outward radiation from these reactions balances that from the inward gravitational forces attracting the nuclei to each other. When a star exhausts its supply of nuclear fuel, the outward pressure diminishes, and the star contracts into itself. If the mass is greater than about three times that of our sun, the collapse can continue until the field becomes too strong for anything to get out. The red shift is so great that the frequency shifts to *zero*: Not even light can get out! Such a star is known as a *black hole*. Since *nothing* can get out, there can be no communication with the rest of the universe from within a black hole. Anything that approaches within a certain radius of such a system will be sucked in and never be heard from again.

Curved Spacetime

In special relativity, we consider time and the three spatial dimensions to form the four-dimensional spacetime in which we live. As long as we stick to the conditions for special relativity—inertial reference frames—the usual laws of geometry apply—in their four-dimensional form, of course. Such a four-dimensional spacetime is said to be *flat*. But the spacetime in the vicinity of a mass producing a gravitational field is considered to be curved. Thus, general relativity replaces the concept of a gravitational field with that of *curved spacetime*.

Visualizing this curved spacetime is virtually impossible, but an analogy may help to explain the concept. In our travels around town, we're accustomed to thinking in terms of plane—flat—geometry. For example, to walk around the block, it takes *four* right turns to get back to where you started. But suppose you start at the north pole and walk south to the equator. (Never mind that you may not feel like that long a hike today!) If you then turn right, walk a quarter of the way around the equator, turn right again, and walk straight, you will end up where you started after only *three* right turns. The rules of geometry are different because the space is curved. The relationshp between the flat spacetime of special relativity and the curved spacetime in a gravitational field is similar to the relationship between the flat space of a small area of earth's surface and the curved space of a spherical earth.

. . . .

Key Concepts

The **special theory of relativity** is based on two **postulates**:

1. The laws of physics are the same in all inertial reference frames.

2. The speed of light in empty space is the same for all observers and it is independent of the motion of either the source or the observer.

An inertial reference frame is one in which Newton's first law applies. The special theory of relativity applies only to inertial reference frames.

Some consequences of the postulates of special relativity are:

1. No material object can move faster than the speed of light.

2. Events that are simultaneous in one reference frame are not necessarily simultaneous in another.

3. The time interval T between two events in a moving reference frame is related to the *proper time* interval T_0 between these events as measured in the rest frame of the events by:

$$T = \frac{T_0}{\sqrt{1 - v^2/c^2}}$$

4. The length D of a moving object is related to its *proper length* D_0 measured in the rest frame of the object by:

$$D = D_0 \sqrt{1 - v^2/c^2}$$

Time and the three dimensions of space make up the *four-dimensional* **spacetime** in which we exist.

Energy and mass are equivalent and related by the equation:

$$E = mc^2$$

The **general theory of relativity** is actually a theory of gravity and is based on a postulate called the *principle of equivalence: The effect of a gravitational force and the effect of an acceleration are indistinguishable.* Some of the consequences are the following: Inertial mass and gravitational mass are equivalent. Light is attracted by a gravitational field. The effect of a gravitational field can be described by considering the spacetime to be *curved*.

Questions

The Postulates of Special Relativity

1. If you're in a train on a straight track at night and another train passes you in the opposite direction, can you tell who is moving and who isn't? Does this question have any meaning?

2. Taking the earth's surface to be an inertial reference frame, which of the following are also inertial reference frames: a car moving at constant speed on a smooth, level road? a car moving at constant speed around a curve? a car moving at constant speed on a straight but hilly road? a car moving up a long, straight hill at constant speed?

3. An earth-bound javelin thrower exerts a force of 100 N on a javelin, giving it acceleration a relative to the ground. An astronaut in a spaceship traveling by at constant velocity $0.5c$ throws an identical javelin, giving it the same acceleration a relative to the spaceship. What force does the astronaut exert on her javelin?

4. A spaceship heading toward you with speed $0.9c$ turns on its headlights. The passengers measure the speed of this light headed toward you to be c relative to themselves. If you measured the speed of this light, what would you get?

5. Suppose you are traveling in your spaceship toward a large mirror at speed $0.7c$. If you aim your flashlight at the mirror and turn it on, with what speed relative to your spaceship does the light reflected from the mirror approach the ship?

The Ultimate Speed

6. A polo player rides at speed $0.4c$ and hits the ball forward at speed $0.7c$ relative to himself. Is the speed of the ball relative to the goal less than, equal to, or greater than c?

7. How would the game of baseball be different if the speed of light were 90 miles/hour? (Major league pitchers typically pitch fastballs in excess of 90 miles/hour.)

Time Is Relative

8. A long high-speed train moves at a speed near the speed of light. A passenger near the center of the train observes lightning bolts to hit both the front and back of the train at precisely the same time. Would an observer on the ground observe the lightning flashes to be simultaneous?

9. Most stars are at least hundreds of light years away—meaning it would take light at least hundreds of years to go between such a star and here. Does that fact prevent an earthling from ever visiting a planet near such a star and returning in his lifetime?

10. If the answer to Question 9 is *no*, does that mean an earthling could visit such a planet and return to describe his trip to his friends *who saw him off*?

11. Pions, particles that can be produced during collisions between high-energy protons in a particle accelerator, have a half-life at rest of 1.8×10^{-8} s. In this amount of time light can travel 5.4 m. Since particles cannot move faster than light, must half of these particles decay by the time they move 5.4 m across the laboratory?

12. If a friend the same age as you left on a trip and traveled at a speed near that of light, she would be younger than you when she returned. Since you would be moving relative to her, why wouldn't you be younger than she on her return?

Length Contraction

13. In a year light travels about 9×10^{12} km. If you visited a planet 36×10^{12} km away, would you necessarily have to travel that far to do so? Would it necessarily take you 4 years or more to get there?

14. If you could throw a Frisbee horizontally at a speed of $0.8c$, what would it look like from above?

15. If you were traveling in a spaceship and passed an identical spaceship going the other way at a speed near that of light, the other spaceship would look shorter than yours. Does that mean the passengers in the other ship would think your ship was *longer* than theirs?

16. How would our lives be different if the speed of light were 25 miles per hour?.

Mass and Energy

17. How much energy would it take to accelerate a 0.15-kg baseball to a speed *equal* to the speed of light?

18. If a particle is moving at a speed such that its mass is twice its rest mass, how does its kinetic energy compare with the energy associated with its rest mass?

The General Theory of Relativity

19. If you were in an elevator in a tall building and the cable broke, why would you feel weightless if there were no friction as the elevator fell freely?

20. A laser in a satellite far above the earth aims its beam at the earth. Would the wavelength of the light be longer or shorter at the earth's surface than when it left the laser?

21. Would it make any difference to you if the gravitational force on an object were proportional to the *square* of its mass, but Newton's laws of motion remained the same as we know them?

Problems

1. While you're taking a 50-minute exam, your professor takes a ride in her new spaceship. She accelerates quickly to speed $0.75c$, rides for 25 minutes at constant speed, turns around quickly, rides for 25 minutes at the same constant speed, and stops at the classroom. Assuming her acceleration, turn-around, and deceleration times are negligible, how much time do you actually have to take the exam?

   Answer: 75 min

2. Suppose you're on a rocketship traveling at speed $0.6c$ away from the earth. If you're watching a soccer match on TV that is being played on earth, how long do you have to watch to see a 90-minute match?

   Answer: 112.5 min

3. In the situation of Problem 2, how long do you observe the soccer field to be if the players measure it to be 120 yards?

   Answer: 96 yards

4. If you were to make the trip described in Question 13 traveling at speed $0.8c$, how far would you travel one way? How long would the one-way trip take?

   Answer: 21.6×10^{12} km; 2.85 years

5. After a 9000-kg (weight ≈ 10-tons) rocket is accelerated to speed $0.8c$, what is its mass in the rest frame of its launch pad?

   Answer: 15,000 kg

6. What is the kinetic energy of the rocket of Problem 5? Compare this with the *total* annual *world-wide* energy consumption of about 3×10^{20} J, and the 75×10^{12} J released in the "atomic" bomb dropped on Hiroshima. If this rocket were fueled by gasoline, how many gallons would be needed to get it to that speed? (See Table 8.2; gasoline density = 2.8 kg/gal)

   Answer: 5.4×10^{20} J (1.29×10^{17} kcal); 7 *million* Hiroshima-type "atomic" bombs; 4,000,000,000,000 gal

References for Further Reading

Bartusiak, M., "Sensing the Ripples in Space-Time," *Science 85,* April 1985, p. 58. A search for one of the predictions of general relativity—gravity waves.

French, A. P., *Special Relativity* (W. W. Norton & Co., New York, 1968).

Hall, D. E., "The Hazards of Encountering a Black Hole," *The Physics Teacher,* Dec. 1985, p. 540. "Part II: A Variety of Ways to Die" contains an interesting discussion of what might happen to you if you approached a black hole.

Taylor, E. F., and J. A. Wheeler, *Spacetime Physics* (W. H. Freeman and Co., San Francisco, 1966).

REVIEW
OF ALGEBRA

Aside from arithmetic, the mathematics that you need in order to understand this book and work the problems in it includes only simple algebra.
In case you are somewhat rusty in this area, we will review here the
procedures that you'll need.

The algebra required can be summarized in three rules, listed here in
the order of most frequent use.

Rule 1: *You may multiply or divide both sides of any equation by the same
number.*

Rule 2: *You may add the same number to or subtract the same number from
both sides of any equation.*

Rule 3: *You may square or take the square root of both sides of any equation.*

We will now illustrate the application of these rules using some of the
equations and symbols that appear in parts of this book. But remember
that the rules apply *whatever* the symbol used as well as when some of
the symbols are actual numbers.

Multiplication and Division

We use Rule 1 when we have an equation such as

$$F = ma$$

and we want to find one of the quantities on the right side of the equation—let's say we want a. If we divide both sides of the equation by m, we get

$$\frac{F}{m} = \frac{ma}{m} = \frac{m}{m}a = 1a = a$$

The $1a$ comes about because any quantity divided by itself, such as $\frac{m}{m}$, is 1. Reversing the two sides of the equation, we get

$$a = \frac{F}{m}$$

Thus, we have solved the original equation for a.

As another example, suppose we have the equation

$$a = \frac{v}{t}$$

and we want to solve it for t. Multiply both sides of the equation by t to get

$$at = \frac{v}{t}t = v\frac{t}{t} = v$$

or

$$at = v$$

Now divide both sides by a:

$$\frac{at}{a} = \frac{v}{a}$$

and thus,

$$t = \frac{v}{a}$$

As these two examples illustrate, the overall effect of Rule 1 is the following: *A factor may be moved from the numerator of one side of an equation to the denominator of the other, or from the denominator of one side to the numerator of the other.* After you've had some practice with these procedures, it's okay to go ahead and use this simplified version of Rule 1. However, do *not* do so until it is so well-embedded in your mind that you are, in effect, following the full-blown Rule 1. When in doubt, go back to the basic rule. Furthermore, if one or both sides of the equation contain a sum or difference of terms, you must be careful to multiply or divide the *entire* side, as we'll illustrate in the next section.

Addition and Subtraction

Consider an equation of the type

$$K + P = E$$

that you would like to solve for P. Using Rule 2, subtract K from both sides of the equation:

$$K + P - K = E - K$$

The left-hand side is just $K - K + P$, or just P, since $K - K$ is zero. Therefore, our equation solved for P is

$$P = E - K$$

Suppose we have an equation like this one, but the K and P are themselves made up of several factors. For example,

$$\tfrac{1}{2}mv^2 + mgh = E$$

and we want to know h. We can subtract $\tfrac{1}{2}mv^2$ from both sides of the equation, as we did above, to get

$$mgh = E - \tfrac{1}{2}mv^2$$

We can now divide both sides of this equation (Rule 1) by the product gh and get

$$\frac{mgh}{gh} = \frac{E - \tfrac{1}{2}mv^2}{gh}$$

or

$$m = \frac{E - \tfrac{1}{2}mv^2}{gh}$$
$$= \frac{E}{gh} - \frac{mv^2}{2gh}$$

A word of caution that we pointed to at the end of the last section: When you divide (or multiply) both sides of an equation by some quantity, you *must* divide (or multiply) the entire side—all terms—not just one term such as the E on the right above.

Finally, we can summarize Rule 2 in a simpler form: *You may move an entire term from one side of an equation to another if you change the sign of the term.* When the $\frac{1}{2}mv^2$ moved from the left to the right of the above equation, its sign changed from plus to minus. But you *must* move the *entire* additive or subtractive term—not just the *m* in our illustration, for example. As we said for the simplified version of Rule 1, do not use this procedure until you are convinced that it is equivalent to Rule 2 itself.

Squares and Square Roots

Rules 1 and 2 will handle most of your algebra needs in this book. However, there may be times when you need to solve for some factor that is squared in an equation. One example would be the equation

$$d = \frac{1}{2}at^2$$

in which you want to find *t*. First use Rule 1 to multiply both sides by 2 and divide both sides by *a*. The result is

$$\frac{2d}{a} = t^2$$

Now take the square root of *both* sides to get

$$\sqrt{\frac{2d}{a}} = \sqrt{t^2} = t$$

We can reverse the sides of the equation to finally have

$$t = \sqrt{\frac{2d}{a}}$$

If one side of an equation has more than one term, and you want the square root of both sides, you must take the square root of the *entire* side as a unit, *not* the square root of the individual terms. The answer to the following exercise illustrates this point.

• •

EXERCISE   Solve the equation, $\frac{1}{2}mv^2 + mgh = E$, for *v*.

$$Answer: v = \sqrt{\frac{2\,(E - mgh)}{m}}$$

In taking the square (or any other power) of the two sides of an equation, the same rules apply.

HANDLING NUMBERS

If we follow certain general guidelines when we use numbers in calculations, the calculations will be easier and the answers will have more meaning. We'll discuss guidelines in two areas: properly rounding off numbers, and using exponential notation.

A. ROUNDING OFF NUMBERS

In the numerical calculations we do in this book, we round off almost all answers to two "significant digits." (Zeros that only locate the decimal point—for example those in 0.0013—are not significant digits.) To illustrate why this rounding off is appropriate, consider the distance-of-travel problem at the beginning of Section 1.3. Imagine that you rode for 2 hours traveling at 80 kilometers/hour (km/h).

Suppose you had measured the time with an accurate digital stopwatch that showed you had traveled for 7200.15 seconds, but you had measured the speed by watching the speedometer. When you converted seconds to hours and figured out the distance on your calculator you got 160.00333 kilometers. However, your speed measurement was not very accurate. It might, for example, have been anywhere between 78 and 82 km/h. That would mean the distance was somewhere between 156.00325 and 164.00342 km. In other words, there is no real significance to those last six digits in the distance. The best we can do is say the distance is 160 km.

This example illustrates two points. First, the number of significant digits tells you roughly how accurate the number is: The six digits in the time measurement tell you that you know the time to one or two hundredths of a second; the two digits in the speed measurement (assuming the zero here is significant) tell you that you know the speed to one or two km/h.

Second, this example shows that *when you multiply two numbers together, there is no significance to leaving more digits in the answer than you have in the multiplier with the fewest number of significant digits.* Since our examples usually have the input information given to two digits, it's wrong (or at least misleading) not to round off answers obtained by multiplication or division to two digits. For intermediate answers that are used in subsequent calculations, we usually carry along another digit to avoid ambiguities later.

B. EXPONENTIAL NOTATION

The sizes of physical quantities that are important in our lives vary over a *tremendous* range. For example, your height is about 300 trillion times the diameter of the nucleus of one of the carbon atoms in your left little toe. Rather than carry along a bunch of zeros in large numbers such as this one, or small numbers such as the nuclear diameter in meters, we usually use *exponential* notation, in which the above number would by 3×10^{14}. We won't go into the mathematical meaning of the notation but will merely summarize how to use it.

If the *power*, or *exponent*, of ten is a positive number, you get the usual notation by moving the decimal point to the right the number of places given by the exponent. For example,

$$3.2 \times 10^8 = 320{,}000{,}000$$

$$94 \times 10^2 = 9400$$

$$6.8 \times 10^0 = 6.8$$

If the power of ten is negative, you move the decimal point to the left that many places. For example,

$$3.2 \times 10^{-8} = 0.000\,000\,032$$

$$84 \times 10^{-3} = 0.084$$

$$2.8 \times 10^{-1} = 0.28$$

Should you need to multiply numbers with exponents, remember that to find the product of two powers of 10, you *add* the exponents. For example,

$$(3.2 \times 10^8) \times (9.7 \times 10^3) = (3.2 \times 9.7) \times (10^8 \times 10^3)$$
$$= 31 \times 10^{(8 + 3)}$$
$$= 3.1 \times 10^{12}$$

To add or subtract numbers in exponential notation, be sure they are written to the same power of ten before performing the operation. For example,

$$(8.41 \times 10^3) - (2.3 \times 10^2) = (8.41 \times 10^3) - (0.23 \times 10^3)$$
$$= 8.18 \times 10^3$$

Notice that the exponent stays a *3*, not *0*.

Both of these two operations are handled *automatically* by a calculator that handles exponential numbers. You do not need to do them unless you're carrying out the calculation by hand.

UNITS AND CONVERSION FACTORS

A. METRIC UNITS

Metric units have a big advantage over the traditional "British" units because, in the metric system, there is only one basic unit for each quantity being measured. The basic metric units for length, mass, and time are the meter (m), gram (g), and second (s), respectively. We specify larger or smaller units of each quantity by attaching a prefix to the basic unit that merely moves the decimal point a prescribed number of places. These prefixes and their abbreviations are given in Table III.1.

Table III.1 Metric prefixes

Prefix	Abbre-viation	Multiplier	Prefix	Abbre-viation	Multiplier
exa	E	10^{18}	deci	d	10^{-1}
peta	P	10^{15}	centi	c	10^{-2}
tera	T	10^{12}	milli	m	10^{-3}
giga	G	10^{9}	micro	μ	10^{-6}
mega	M	10^{6}	nano	n	10^{-9}
kilo	k	10^{3}	pico	p	10^{-12}
hecto	h	10^{2}	femto	f	10^{-15}
deca	da	10	atto	a	10^{-18}

Table III.2 The international system of units

Quantity Measured	Unit	Symbol	Equivalent Units
length	meter	m	—
mass	kilogram	kg	—
time	second	s	—
force	newton	N	$kg \cdot m/s^2$
work, energy	joule	J	$N \cdot m$
power	watt	W	J/s
temperature	kelvin	K	—
electric charge	coulomb	C	—
electric current	ampere	A	C/s
electric potential difference	volt	V	J/C
electric resistance	ohm	Ω	V/A
magnetic field	tesla	T	$N/A \cdot m$
frequency	hertz	Hz	cycle/s
light output	lumen	lm	—
radioactivity	becquerel	Bq	decay/s

We use these prefixes to express a quantity in units of the appropriate size. Some examples: The radius of a carbon nucleus is 2.7 fm (2.7×10^{-15} m); the average distance of the earth from the sun is 149.6 Gm (1.496×10^{11} m); the mass of a typical flea is 450 μg (4.5×10^{-4} g); and it takes about 12 ms (0.012 s) for a fastball to pass over home plate.

B. SI UNITS*

When doing a calculation, each quantity in the calculation must usually be expressed in the same system of units. The subgroup of metric units used internationally for scientific purposes is the SI system. Table III.2 summarizes the SI units used in this book.

C. CONVERSION FACTORS

You can convert non-SI metric units to SI units merely by moving the decimal point according to the appropriate prefix. Notice that the SI unit of mass—the kilogram—is not a basic metric unit, but is 1000 of the base units, the gram. In converting other metric units of mass to SI units, you need to allow for this factor of 1000.

The International System of Units.

Table III.3 Conversion factors between units

time:	1 second (s)	= 1/60 minute (min)
		= 1/3600 hour (h)
length:	1 meter (m)	= 39.37 inches (in)
		= 3.281 feet (ft)
	1 inch (in)	= 2.54 cm
	1 kilometer (km)	= 0.621 mile (m)
	1 mile (mi)	= 1609 meters (m)
		= 1.609 kilometer (km)
volume:	1 cubic meter (m^3)	= 1000 liters (l)
	1 liter	= 0.001 m^3
		= 1.057 quart
mass:	1 atomic mass unit (amu)	= 1.66×10^{-27} kilogram (kg)
	1 kilogram (kg)	*weighs* 2.20 pounds (lb)
force:	1 newton (N)	= 0.2248 pound (lb)
		= 1×10^5 dyne (dyn)
	1 pound (lb)	= 4.448 newton (N)
pressure:	1 pascal (Pa) = 1 N/m^2	= 1.45×10^{-4} lb/in^2
	1 lb/in^2	= 6895 N/m^2
	1 atmosphere (atm)	= 101.3 kPa
		= 76.0 cm of mercury
energy:	1 joule (J)	= 0.239 calorie (cal)
		= 2.39×10^{-4} kcal
		= 2.778×10^{-7} kilowatt-hour (kWh)
		= 6.24×10^{18} electron volts (eV)
		= 9.48×10^{-4} British thermal units (Btu)
		= 0.737 foot-pound (ft·lb)
	1 kilocalorie (kcal)	= 4186 joules (J)
		= 1.16 watt-hour (Wh)
	1 electron volt (eV)	= 1.60×10^{-19} joule (J)

Table III.3 gives conversion factors from other often-used units to SI units.

D. CONVERTING UNITS

When we do a calculation to convert from one unit to another, or to work out a problem, the units themselves can give us a strong hint about whether or not we have done the right thing. We can handle units as if they were algebraic symbols. If we come out with the right units after this manipulation, we can be confident we've gone about the calculation correctly. Some examples will illustrate this point.

· ·

EXAMPLE III.1   Convert 65 in to centimeters. Noting that there are 2.54 cm/in, we get

$$65 \text{ in} = 65 \text{ in} \times 2.54 \text{ cm/in}$$
$$= 165 \text{ in·cm/in}$$

Notice that just as $x \cdot y / x = (x/x) \cdot y = 1 \cdot y = y$, then in·cm/in = cm. Thus,

$$65 \text{ in} = 165 \text{ cm}$$

Since we wanted the distance in cm, we know we have done the conversion correctly. If we mistakenly divide by the conversion factor instead of multiplying, we get units of in/(cm/in) = in²/cm. Since this is not the unit we wanted, we know we made a mistake.

· ·

EXAMPLE III.2   Convert a speed of 55 mi/h to meters/second.

Since 1 mi = 1609 m and 1 h = 3600 s,

$$55 \text{ mi/h} = \frac{55 \text{ mi/h} \times 1609 \text{ m/mi}}{3600 \text{ s/h}}$$

$$= 24 \frac{\text{mi/h} \times \text{m/mi}}{\text{s/h}} = 24 \text{ m/s}$$

Notice that the miles and hours cancel, leaving meters/second.

· ·

EXERCISE   Show that if you either divide by the distance conversion factor or multiply by the time conversion factor in Example III.2, you will not get the right units, meters/second.

You can use the same technique to find out if you're solving a problem correctly. We'll illustrate.

EXAMPLE III.3   How long does it take you to drive 320 km at a speed of 80 km/h? From your experience you know that

$$time = \frac{distance}{speed}$$

$$= \frac{320 \text{ km}}{80 \text{ km/h}} = 4.0 \text{ km}/(\text{km/h})$$

$$= 4.0 \ (\cancel{km}/\cancel{km})h = 4.0 \text{ h}$$

If we had mistakenly divided speed by distance, the resulting units—1/hour rather than hour—would have told us we did not solve the problem correctly.

The examples we have worked illustrate how to manipulate units algebraically in any problem, no matter how simple or how complicated.

SUGGESTIONS FOR SOLVING PROBLEMS

The ability to solve problems is not a skill we are born with. It is a skill we acquire. The way to acquire that skill is to develop a systematic, logical approach to every problem, from the simplest to the most difficult. Such a procedure should break the problem down into manageable chunks that can be handled easily.

I'm going to outline a procedure that many good problem-solvers find useful. I suggest that you follow this outline, or something close to it, for *every* problem. It may seem unnecessary to follow all the steps for an easy problem. But using the same procedures for easy problems as for hard ones gives you the practice and reasoning ability needed for the difficult ones. Clearly, some problems will require adapting this procedure to fit the given situation; however, try to be consistent with these *general* guidelines.

1. Read the problem and form a mental picture of the situation. Let us take the following problem (which uses principles from Chapter 2) for an example: Jack is on roller skates. Harry pushes him forward with a force of 30 N and Alex pushes him backwards with a force of 50 N. If Jack's mass is 40 kg, what is his acceleration?

2. Draw a simple diagram of the situation, indicating symbolically what is happening. You don't need to draw a likeness of the objects in the problem: squares or circles to represent the objects will do:

3. Using appropriate symbols, make a list of the quantities that are given in the problem. For our example that would be

$F_H = 30$ N (forward)

$F_A = 50$ N (backward)

$m_J = 40$ kg

4. Using symbols, list the quantities you are asked to find. In our example, we are asked to find a_J. (Having performed steps 1 through 4, you have determined in your mind what the problem is all about. The solution should now be straightforward.)

5. If the problem includes several objects, draw a simplified diagram showing only the object whose behavior you are focusing on in this step. If forces are involved, then show them acting on the object.

6. Write down the relevant equations that relate the given quantities to the one you want to find—in this case,

$F = ma$

7. Use algebra to rearrange the appropriate equation to solve for the particular unknown you want:

$$\frac{F}{m} = \frac{m a}{m}$$

or

$$a = \frac{F}{m}$$

This procedure usually simplifies the calculation, reducing the chance of error. In addition, sometimes certain symbols will cancel, and you won't need them in the calculation. These symbols may be some you thought you needed for the solution but didn't have values for.

8. Solve for one unknown at a time. For our example, the force is the *net* force of 50 N in one direction and 30 N in the other, or

$$F = 50 \text{ N} - 30 \text{ N} = 20 \text{ N (backward)}$$

Next, we substitute the other numerical values *and units* (see Appendix III.D):

$$a = \frac{F}{m} = \frac{20 \text{ kg·m/s}^2}{40 \text{ kg}} = 0.50 \text{ m/s}^2 \text{ (backward)}$$

If you have more than one unknown, break the problem into parts and find other relationships between the knowns and unknowns. (For example, one unknown could equal twice the other, as in $y = 2x$. Then you could substitute $2x$ for y in the problem and thus reduce the number of unknowns you have.) This procedure will rarely be needed for problems in this book.

9. State the answer, *with units*. Use the correct number of significant digits (see Appendix II).

10. Think about your result. If it isn't reasonable, recheck your calculation. How would the result be different if the values of the input quantities had been different? What insight does the result provide about the physical situation related to the problem?

The following example, along with its solution, illustrates the use of the guidelines given here. It uses principles from Chapter 4.

• •

EXAMPLE IV.1 The driver of a 1000-kg car traveling at a speed of 55 miles/hour (25 m/s) slams on brakes for a distance of 20 m. A frictional force of 6000 N opposes the motion during the braking. (a) How fast is the car moving afterwards? (b) How much heat was generated by friction as the car slowed down? (See the solution on the next page.)

Given:

$v_1 = 25 \text{ m/s}$ $v_2 = ?$

$d = 20 \text{ m}$

$m = 1000 \text{ kg}$

$F_f = 6000 \text{ N}$

Find: (a) v_2

   (b) $E = $ heat generated

Solution:

(a) Work done $=$ change in KE

$$F \times d = \tfrac{1}{2} m v_1^2 - \tfrac{1}{2} m v_2^2$$

$$Fd + \tfrac{1}{2} m v_2^2 = \tfrac{1}{2} m v_1^2$$

$$\tfrac{1}{2} m v_2^2 = \tfrac{1}{2} m v_1^2 - Fd$$

$$m v_2^2 = m v_1^2 - 2Fd$$

$$v_2^2 = \tfrac{1}{m}(m v_1^2 - 2Fd) = v_1^2 - \tfrac{2Fd}{m}$$

$$v_2 = \sqrt{v_1^2 - \tfrac{2Fd}{m}}$$

$$\approx \sqrt{(25 \text{ m/s})^2 - \frac{2(6000 \text{ kg·m/s}^2)(20 \text{ m})}{1000 \text{ kg}}}$$

$$= \sqrt{625 \text{ m}^2/\text{s}^2 - 240 \text{ m}^2/\text{s}^2}$$

$$\underline{v_2 = 20 \text{ m/s}}$$

(b) $E = $ heat generated $= $ work done $= Fd$

$E = (6000 \text{ N})(20 \text{ m}) = 120{,}000 \text{ N·m}$

$\underline{E = 120 \text{ kJ}}$

. . . .

EXERCISE   Show that if we had made an algebra mistake in the solution of Example IV.1 and gotten

$$v_2 = \sqrt{v_1^2 - 2mFd}$$

the units would not have worked out correctly.

INDEX